Dynamic Electromagnetics

Paul Diament

Professor of Electrical Engineering
Columbia University

Prentice Hall
Upper Saddle River, NJ 07458

Library of Congress Cataloging-in-Publication Data

Diament, Paul.

    Dynamic electromagnetics / Paul Diament.

      p. cm.

    Includes bibliographical references and index.

    ISBN 0–02–328760–8

    1. Electromagnetic theory.   I. Title.

QC760.D47 2000

530.14'1—dc21

99-055286

CIP

Editor-in-chief: Marcia Horton
Publisher: Tom Robbins
Associate editor: Alice Dworkin
Executive managing editor: Vince O'Brien
Managing editor: David George
Editorial/production supervision: Rose Kernan
Art director: Jayne Conte
Manufacturing buyer: Pat Brown

 © 2000 by Prentice-Hall, Inc.,
Upper Saddle River, New Jersey 07458

The author and publisher of this book have used their best efforts in preparing this book. These efforts include the development, research, and testing of the theories and programs to determine their effectiveness. The author and publisher make no warranty of any kind, expressed or implied, with regard to these programs or the documentation contained in this book. The author and publisher shall not be liable in any event for incidental or consequential damages in connection with, or arising out of, the furnishing, performance, or use of these programs.

Printed in the United States of America

10 9 8 7 6 5 4 3 2

ISBN 0-02-328760-8

Prentice-Hall International (UK) Limited, *London*
Prentice-Hall of Australia Pty. Limited, *Sydney*
Prentice-Hall Canada Inc., *Toronto*
Prentice-Hall Hispanoamericana, S.A., *Mexico*
Prentice-Hall of India Private Limited, *New Delhi*
Prentice-Hall of Japan, Inc., *Tokyo*
Pearson Education Asia Pte. Limited, *Singapore*
Editora Prentice-Hall do Brazil, Ltda., *Rio de Janeiro*

*In honor of
my mother
and in memory of
my brother*

Contents

7   Impedance Matching Techniques and Oblique Waves   306

8   Poynting Theorems and Lossy Transmission Lines   363

Preface

It is a commonly held view, and one often vehemently expressed by students—particularly recent survivors of a course on the subject—that electromagnetics is a difficult, complicated, mysterious discipline. It requires mastery of abstruse mathematical techniques, they say; it entails juggling a bewildering variety of equations and laws and rules, they decide. Even intense study has left them with only a superficial grasp of the concepts. Few see the beauty of electromagnetics; not many appreciate the simplicity and economy of its fundamental laws. A minority of its practitioners realize its wide-ranging utility, the breadth and scope of its applications. Only a minority master it enough to be able to use its principles to understand or predict the capabilities and limitations of the engineering systems they need to analyze or design.

Instructors of the subject are in a better position to assess the grandeur and importance of electromagnetics but are plagued by other demons as they plan the presentation of the subject. There is never enough time to cover all the aspects of electromagnetics that they deem essential, indispensable, and obligatory. They are forced to make coldhearted choices of their topics and suffer the agony of having to discard this or that favorite theme. Some find that they must turn their course into lectures on applied mathematics; some concentrate on mastery of electrostatics and magnetostatics, leaving little time for the radically different phenomena of dynamics. Agreement on what topics must be included and which may be discarded, what areas are to be emphasized and which to be glossed over, is indeed rare.

The present work will not lay these matters to rest but offers an approach and a pedagogic philosophy aimed at making a respectable contribution to the debate over an appropriate syllabus for the subject. It is intended as a textbook for a one-semester first course in electromagnetics for engineers and physicists. It is an outgrowth of decades of experience in teaching the subject to juniors in electrical engineering at the School of Engineering and Applied Science of Columbia University. The students come to the course with preparation in calculus, including a smattering of vector analysis, and with some exposure to notions of electricity and magnetism in a physics course, usually with emphasis on electrostatics and magnetostatics. Several elective courses follow this one, but at Columbia this is the only required course in electromagnetics in a crowded undergraduate curriculum in electrical engineering.

What approach underlies this work and how is it different? The pedagogic preferences expressed in this text include the following.

1. *Time-varying fields.* Time variation is central and paramount virtually from the start. The emphasis is on distributed systems, featuring action at a distance and after a delay, so that time development and dynamics are deeply involved and inescapable. A modern curriculum should avoid the classic textbook's concen-

tration on statics. Few practicing engineers hang pith balls and rub combs to transfer charges. Even the engineers or designers who appear to deal with purely static systems usually intend to have them process time-varying signals. Those practitioners who design devices and systems based on only electrostatics and magnetostatics principles, or on software that deals only with statics, remain uncertain, if not actually ignorant, of the dynamic response or bandwidth of their systems. They are likely to be left behind, wondering why their designs misbehave or fail at today's required speeds or data rates.

2. *Simple mathematics.* The mathematics is kept simple, at the level of integral calculus. The text avoids approaching the subject from the standpoint of solving partial differential equations and boundary value problems that invoke esoteric mathematical functions and apply to only a few types of geometry. Useful as they undoubtedly are, Bessel and Hankel functions, Legendre polynomials, elliptic integrals, and the like are likely to overwhelm students at the level of this text, keep them from appreciating the underlying phenomena, and leave them helpless when confronted with some unfamiliar geometry. The special functions of mathematical analysis should be taught, but at a later stage. There are enough important concepts to learn without adding these special functions into the mix, at this elementary level.

   Instead of handling differential equations, the present text makes extensive use of Maxwell's equations in integral form. Note that this does not mean we are dealing with integral equations, any more than asking to find the derivative of a given function is the same as solving a differential equation. We just need integrals of familiar functions; the integrals need not involve functions more complicated than powers or logarithms of the variable. The use of equations in integral form allows us to deal with any geometry, at least approximately. As an added bonus, the integral form applies, unchanged, to moving media as well as to stationary ones.

3. *Quasistatics for high frequencies.* The text approaches high-frequency phenomena by way of quasistatic analysis. This furnishes a more user-friendly, gentle transition from the more intuitive low-frequency circuit theory results to those of faster circuitry. Quasistatic analysis imparts insights difficult to gain from formal solutions to differential equations that feature advanced, special functions of mathematical analysis. Does one readily extract insights into the skin effect in a cylindrical wire by expressing the fields in terms of Kelvin functions or Bessel functions with complex arguments? Quasistatics emphasizes deviations from the results of circuit theory that are due to interactions between the electric and the magnetic fields in the configuration. It can be applied to geometries that defy analytic solutions. It is particularly helpful in readily furnishing approximate answers in complicated cases. Most important, it provides the engineer with estimates of the frequency range of applicability of possible designs.

4. *Transmission lines paramount.* The main focus of the work is on wave propagation, exemplified particularly by the behavior and response of transmission

lines, both transient and steady state. Such structures are realistic and practical signal-transmission systems. Transmission lines illustrate time delay, reflections, standing waves, matching procedures, the effects of a mismatch, measurement techniques, and power transfer. The simplicity and far-ranging applicability of harmonic plane waves are also stressed, however. The two are carefully integrated to support and complement their concepts and applications.

5. *Electric and magnetic fields on a par.*   Electric and magnetic fields are treated as twins, with virtually equal importance. Too many textbooks and courses stress the electric field and introduce the magnetic field as an afterthought or oddity. While this can be sensible in the static case, the two types of fields are inextricably entwined in the time-varying case. We seek to treat them on an equal footing and emphasize how each affects and actually creates and maintains the other.

     In this connection, it is interesting that modern electronic circuit design has emphasized capacitive elements in integrated circuits; these are based on the behavior of electric fields. More recently, the design of inductive elements, based on magnetic field behavior, has regained importance, driven particularly by the exigencies of wireless communication circuits and their tuning elements.

6. *Avoid potentials!*   We studiously and deliberately avoid introducing potentials and the notion that the electric field is the gradient of a scalar potential. This radical attitude and apparent heresy calls for some explanation.

     We contend that the student who hasn't learned or has forgotten the notions that electric fields are conservative and that conductors are equipotentials is more ready to tackle time-varying situations than is the rival who must unlearn and abandon these incorrect assertions when dealing with dynamic fields. Avoiding the potentials is, in fact, a consistent element of the approach that emphasizes dynamics over statics.

     Potentials are enormously helpful when dealing with purely electrostatic or magnetostatic effects, but they become a major obstacle to making the transition to the time-varying case, when we need to unlearn almost all about the scalar potential. For example, students who have been taught that all points in a circuit that are connected by conductors must be at the same voltage typically have a difficult time with the fact that the time-varying voltages at various points of a circuit or structure may be, and usually are, different even when they are connected by perfect conductors.

     That we avoid introducing potentials is not to say that we don't deal with voltage; we do of course, but as an electromotive force (emf), not as a potential. The distinction is vital: Static potentials are conservative (zero closed-loop integral of the field) but emf (voltage) is not. The nonzero voltage around a closed curve is crucial to time-varying fields and to circuits with energy sources in them; it is incompatible with fields derived from a scalar potential. Our experience has been that it is easier to impart an understanding of dynamic field effects when the scalar potential has not been inculcated prior to the need to modify or abandon it and the limitations it imposes. If we insist on deriving the electric field from a scalar potential, we find that it is only part of the field and we need to add the time derivative of the vector potential. The new potential fields are further removed from the physical interaction, and this complicates the situation to an extent that may make the potentials more of a pedagogic liability than an asset. We choose to postpone discussion of potentials to the end, when we look back on the entire development and review the overall principles of electromagnetics.

7. *Interpretations stressed.* As a matter of habit, the examples in the text do not end when the answer has been found. Those answers are extensively discussed and interpreted, a practice that should be made routine. Students should be encouraged not merely to extract a number or formula from the equations but to look at the answer, examine it from the standpoint of physical plausibility and logic, and especially to interpret the result. This book features careful derivations and emphasizes interpretations. It is also often helpful to draw analogies among different results; we have deliberately emphasized such analogies and similarities of seemingly disparate aspects of the subject, such as the electric and the magnetic versions of certain effects.

8. *Realistic figures.* We have felt strongly that figures should never be misleading, and we have kept them realistic whenever it may matter. Curves have been properly calculated and plotted, not merely sketched. In some cases, perspective views are the only renderings that can avoid confusion, and these have been carefully computed and presented as realistically as possible.

9. *Answers to all problems.* Answers to all problems are provided at the back of the book. Obviously, judicious use of the answers can be of great help to the student; indiscriminate abuse of the answer key is just as obviously harmful to the learning process. Answers are most often sought in formulaic or symbolic form; such forms are usually much more informative than a numerical answer that applies to only a specific instance. In the guise of a formula, the answer provides information on how the result varies with the parameters of the problem. Nevertheless, we do ask for a numerical value when appreciation of typical magnitudes is important.

   For the instructor, a Solutions Manual has been prepared; it provides complete solutions to every problem.

10. *Some topics treated in problems.* We have sought to make the problems instructive rather than mere drill. A variety of problems has been provided, some easy, some more challenging, some illustrative, some to fill in gaps in the text, some for practice, some particularly instructive, some to pique the student's interest.

   In a few cases where we have omitted formal demonstrations of certain assertions, guidance as to how to prove them has been relegated to the problems. A few topics that might seem to be missing from the text may be found among the sets of problems. Instructors may prefer to lecture on some of these topics, rather than leave them as homework assignments. The Solutions Manual can help guide such lectures.

11. *So many equations, so little time?*   Although the level of the mathematics used has been kept at that of elementary integral calculus, there are occasions when the onslaught of equations may seem like that of a blizzard. There are several reasons for it when the mathematical development becomes intense.

Show all steps.   The primary reason is that we prefer to err on the side of showing all the steps in developing a result; our experience is that most students appreciate seeing the intermediate steps between the starting point and the final equation.

Derive parameters.   Another cause is that we prefer, whenever practical, to show where a result comes from, rather than merely assert its truth. One example for which this leads to an intensive development is the discussion of the parallel-wire transmission line. This development leads to an important engineering design equation for the characteristic impedance of transmission lines, specifically the parallel-wire line, in terms of the geometry and electrical constitution of the structure. This is something an engineer should know how to obtain and use, for whatever configuration may be under design. Most textbooks don't even attempt to calculate this; they may simply furnish a ready-made formula for the characteristic impedance, descendent from heaven or from the manufacturer of the transmission line—but then who will be hired by the manufacturer to design that line, if engineers are not taught how such formulas are developed?

Oblique waves.   Yet another instance is represented by the development of the results for oblique incidence of plane waves. In many textbooks, the derivation is confined to Snell's laws and the polarizations are merely quoted. In this text, we use the concept of impedance to unify, simplify, and develop the full set of equations. As a result, many high-frequency or optical system design methods reduce to special cases of impedance matching or transforming techniques.

Types of boundary conditions.   Another section that may seem overly elaborate is that on boundary conditions. Our experience has been that this subject often causes confusion. For most applications, the boundary conditions are comfortably simple; for other cases, they can become baffling. We have divided the presentation of how to select and apply boundary conditions into "ordinary" and "extraordinary" types to clarify this circumstance.

Corrections to circuit theory.   The material on quasistatic analysis is extensive, though not intensive, and may be controversial for that reason alone. The aim is to present a systematic, tabular approach to the adjustments and corrections that must be made when a design based on statics is subjected to time-varying signals. Pushing electromagnetic designs to higher and higher frequencies is currently an important aim of engineers, particularly for communications applications. If an instructor feels the material on quasistatics is superfluous or too extensive, it can be skipped or truncated and still leave a coherent treatment of electromagnetics. However, this topic has much that is of great conceptual value, especially to unveil, correct, and extend certain circuit-theory results. It also serves as a transition to wave phenomena, an approach that avoids handing students the wave solution from on high and merely asking or allowing them to verify that it satisfies the equations.

A few words on some technical aspects of the text are in order. We have used the MKSA subset of the SI (Système Internationale) system of units; we do not want confusion over units to cloud the meaning of the concepts. Examples have been set off from the rest of the text by a vertical line; within the example, the end of the question has been marked with a symbol ◈, to separate it from the start of the solution. Each chapter ends with a summary or review of the material covered.

Special efforts by some and special patience on the part of others have been indispensable to the production of this textbook. Particular recognition and thanks are due to Dr. Perry Malouf, who graciously undertook to attempt and to criticize the problem sets. The book was years in the making, but there were peaks of sometimes frantic activity that deprived family members of the attention they deserved. I take this opportunity to recognize and applaud their patience, forbearance, and assistance.

Paul Diament

Columbia University

CHAPTER 1

Gauss's Law, Surface Integrals, and Electric Fields

We wish to describe and study phenomena that vary from point to point in space and from instant to instant in time. We may already be familiar with observations of signals at discrete points in space, such as time variations of voltages probed at just a few isolated points in an electronic circuit, but the need to recognize that signals may differ significantly at *all* the points of some region of space may be less obvious. The point-to-point variation of various effects is a crucial feature of numerous physical systems, notably those of concern in electromagnetics—including transmission lines, waveguides, microwaves, antennas, lightwave systems—as well as in continuum mechanics, acoustics, fluid dynamics, heat transfer, quantum mechanics, plasma physics, optics, and solid-state physics. Important phenomena in such "distributed" systems are wave propagation, radiation, and diffusion processes. The discipline that deals with effects that are distributed in space and time is *field theory*, and we will focus specifically on the interactions of *electromagnetic* fields.

MOTIVATION

We aim at the ability to answer questions such as the following one. Take a pair of wires and apply a voltage source to the pair of terminals at one end, leaving the terminals at the other end open circuited, as in Figure 1-1. What happens?

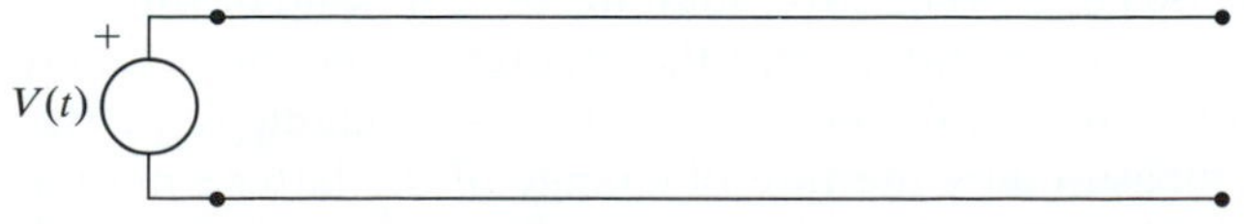

$V(t)$

Figure 1-1 Voltage source applied to a pair of wires.

On the basis of elementary circuit theory, one might answer "Nothing, because the circuit is not complete." But the wires know better than to obey mere circuit theory; something does indeed happen.

A more advanced answer, based on elementary physics considerations, might be that the wires get "charged up" and that a brief current flows, at least until the charging process ends. That is a better answer, but we may be a little hazy about the details of the charging process and how long it may last.

A truly sophisticated answer would entail descriptions of exceedingly involved phenomena, including propagation effects, energy storage, energy loss, radiation into space, transmission of energy and information along the wires, reflection from the open end, oscillation of the electric charge, loading of the source, mechanical motion of the wires, and perhaps more effects (such as random noise, heating, magnetic effects, possible sparking between the wires) that might also be of interest from an engineering point of view.

We do not intend to go into all these phenomena herein, but we do seek no less than practical engineering answers to such questions as the following.

- What current flows, for how long?
- How much power must the source supply?
- What may be observed at the other end?
- How can use be made of the effect at the other end?
- For such use, how should the wires be spaced? How thin should they be? How long should they be? Is it important that they be straight and parallel?
- Will the wires deform?
- What happens if some electrical load is attached to the other end?
- What if the wires were short-circuited at their end, instead?
- What if the wires were to touch, somewhere?

We intend to become adept at answering many of these questions, although advanced study would be required to describe all that does happen in this example. Much background and introductory material will have to be reviewed, however, before we can cope with even the simple features of the matters we have outlined.

FIELD CONCEPT

We begin with the concept of a *field*, which is needed to deal with interactions that operate *at a distance* and *after a delay*. An important purpose of this concept is to simplify the description of actions that occur over intervals of space and of time.

It is a basic observation that matter influences other matter; in particular, it may alter the state of motion of other matter. We say then that the first clump of matter *exerts a force* on the second, measured by the rate of change of the latter's motion (technically, of its *momentum*). The simplest version is that of a force exerted *on con-*

tact, at the point and at the instant of contact, as when a bat hits a baseball. Forces can also be exerted *at a distance*, as in the case of gravitational forces, such as the force that keeps a planet in orbit around the sun. Forces can become enormous under certain circumstances, notably when matter acquires an electric charge.

What is observed then is that electrically charged matter exerts a force on other such matter, at a distance and after a delay. Cause and effect are thereby separated in both space and time. For example, we may jiggle a charged particle here and now and observe that another charge, elsewhere and later, may jump in response, as suggested in Figure 1-2. An application of this delayed effect at a distance is made when mission control sends a command to a far-away space probe, perhaps initiating some critically timed maneuver that must take into account the transmission delay.

It is too complicated to keep track of delayed and removed effects between charges in motion in any direct way. Instead, we use an indirect description. We invent an intermediary agency between the cause and the effect; that agency is the *force field*. We think of the one charged particle, here and now, as generating a force field; we consider the field, elsewhere and later, to be exerting the force on the other charge. It is the field, then, that exists within the space and time intervals between the cause and the effect.

The utility of the field concept goes beyond the fact that this approach simplifies the description of a complicated phenomenon by breaking it down into easy stages. It turns out that fields are not merely helpful figments, created by and for our feeble minds just to help us deal with complex events. Beyond that role, fields have their own properties and laws of behavior; they constitute legitimate physical entities on their own, without reference to the charges whose interaction they are intended to mediate. That this must be so follows from the fact that particles that exert forces on each other can do work and therefore expend energy. If the effect is far removed, in space and in time, from its cause, then there is a need for some intermediary agent to be the repository of the energy in transition between the excitation and the response; that agent is the field.

Energetic fields can even dominate the behavior of physical systems; their own behavior can be more important than that of the charges that generate them or are affected by them. The fields can carry and convey not only energy but also other mechanically significant attributes, particularly momentum and angular momentum, and can communicate information as well. We wish, therefore, to develop an understanding of the laws governing the dynamic behavior of electromagnetic fields.

Here, now   Elsewhere, later

Figure 1-2   Excitation of source charge and response of test charge.

DEFINITION OF FIELD

The concept of a field is intended to generalize that of a signal. While a signal is a physical quantity that varies in time, a field is its spatial counterpart: *A field is a physical quantity that varies in space.* A field therefore associates a value of the quantity with each point in space, or at least within some region of space. An example is the temperature in a room; this may be higher near the radiator than it is at an open window. We can measure the temperature at any one point in the room by placing a thermometer there; to specify the temperature field within the room, however, we would have to furnish the totality of all the temperature values, at each point of the room, describing particularly their variation from point to point.

Physically, then, a field is a quantity that varies from point to point. Mathematically, therefore, a field is represented by a function of position, such as $f(x, y, z)$ if position is identified by the rectangular coordinates x, y, z of a point. The functional dependence of the quantity f on its arguments x, y, z specifies the field.

> **Example 1.1**
>
> An example of a mathematically specified field may be the distance from some particular reference point. If that reference point is the origin (or if we choose to place the origin of our coordinate system at the location of the specified reference point), then the field is given by
>
> $$f(x, y, z) = \left(x^2 + y^2 + z^2\right)^{1/2}. \tag{1.1}$$
>
> If the reference point were at (x_0, y_0, z_0) instead of at the origin, then what we may consider physically (or geometrically) to be the *same* field would be given by
>
> $$g(x, y, z) = \left[(x - x_0)^2 + (y - y_0)^2 + (z - z_0)^2\right]^{1/2}. \tag{1.2}$$
>
> We usually choose our coordinate system to simplify the representation of the field, not only by locating the origin judiciously but also by selecting a set of coordinates that fits the field as well as possible. For this example, the spherical coordinate system (r, θ, φ) is ideal, since the radial coordinate r represents precisely the distance from the origin, so that the same field is
>
> $$f(x, y, z) = h(r, \theta, \varphi) = r. \tag{1.3}$$
>
> Better yet, we often wish to avoid reference to any particular coordinate system at all, by using the position vector $\mathbf{r}$, which points from the origin to the point in question and whose rectangular components are therefore just $x, y,$ and z. The same field we have been describing is then, if the reference point is either at the origin or at the point $\mathbf{r}_0$,
>
> $$f = f(\mathbf{r}) = (\mathbf{r} \cdot \mathbf{r})^{1/2} \quad \text{or} \quad g = g(\mathbf{r}) = \left[(\mathbf{r} - \mathbf{r}_0) \cdot (\mathbf{r} - \mathbf{r}_0)\right]^{1/2}. \tag{1.4}$$

Mathematically, therefore, *a field is any function of the position vector*, $\mathbf{r}$. Calculations we perform with fields are those that are appropriate to functions of a vector argument.

The field quantity—that is, the function of the position vector $\mathbf{r}$—may itself be a vector, but it need not be; it will often be a scalar, or sometimes even a higher-order quantity, a tensor. If the function is a vector, $\mathbf{F} = \mathbf{F}(\mathbf{r})$, we will need to specify both its magnitude and its direction (or else its components) at every point of space. If it is a scalar, we give its algebraic value (i.e., with a sign) at each point. It is important to note that whether a quantity is itself a scalar or a vector is not relevant to whether it may be a field; only its dependence on the position vector **r** makes it a field. An example of a vector field is the air velocity in a room, perhaps a room with air flowing in at an open window and out at an open door. The air velocity may well vary from point to point, being strong along the direct line from window to door but less so in a far corner, where the air may be stagnant.

> **Example 1.2**
>
> As a specific mathematical example, consider the vector field $\mathbf{v}(\mathbf{r}) = \mathbf{a} \times \mathbf{r}$, where **a** is a fixed, given vector. We wish to visualize both the magnitude and direction of this vector at many different points in space. Figure 1-3 is an attempt to show the construction of several typical vectors $\mathbf{a} \times \mathbf{r}$ at some representative points **r** and also to suggest the circulating nature of the field **v**. If the vector field represents air flow, we might be justified in referring to the space as containing a whirlwind. The significance of the fixed vector **a** in relation to such air motion is that its direction is that of the axis of the whirlwind.

We will often wish to consider several fields in the same space. One field may influence another; for example, the air velocity field in a room may well affect the temperature field within that room, as time elapses. Because field interaction entails time variation, we merge the concept of a signal in time with that of a field in space. The general field we will deal with will be a function of both space and time, $f(\mathbf{r}, t)$,

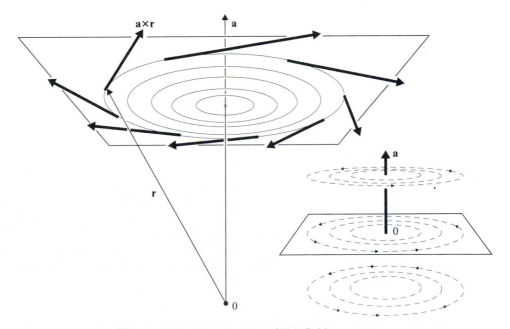

Figure 1-3   Visualization of the field $\mathbf{v} = \mathbf{a} \times \mathbf{r}$.

which assigns some physical attribute to each point in space and each instant of time. We may also think of $f(\mathbf{r}, t)$ as the time variation of a field, or else as the spatial distribution of a signal. Most generally, however, it is any quantity that varies in both space and time.

ELECTRIC FIELD

Electrified bits of matter interact; we say that one charge generates a field and that the field exerts the force on the other charge. We can manifest the existence of the field, therefore, by placing a *test charge* at the point in question and measuring the force exerted on it (exclusive of gravity or other disturbing forces). The distinction between a *source charge* and a *test charge* lies only in our attitude toward their roles; the charge that we use to make the existence of the field evident is the test charge, while the one that causes the field we are measuring to exist is termed the source charge.

We find that the force exerted on the test charge is, first, proportional to the amount of test charge used and, second, oriented in one of two opposite directions. We conclude from these two observations, first, that the relevant attribute of the point in space where the action takes place is the ratio of the force on the test charge to the amount of that test charge and, second, that there are two kinds of (test) charge, one that is pushed and another that is pulled when subjected to the same field. We therefore assign an algebraic scalar quantity, with a magnitude and a sign, to the measure of the test charge itself, with the sign of the quantity indicating which of the two forms of electric charge is in use. Note that the signs used are a matter of convention; the type of charge left on glass after rubbing with silk is termed positive, and the type acquired by rubber from fur is called negative. The natural unit of electric charge is that of an electron (which has a negative charge), but that unit is inconveniently small. The practical (MKSA) unit of charge is the coulomb; there are approximately $10^{18.8}$ electron charges in one coulomb.

The ratio $\mathbf{E} = \mathbf{F}/q$ of the force vector $\mathbf{F}$ to the algebraic amount of the test charge q is the *electric field* $\mathbf{E}$ at the point where the test charge has been placed. This field is attributed to the existence of the "source" charges, somewhere else; the test charge is not the cause but merely what manifests the existence of the electric field. The field referred to is the preexisting field, which is considered to be present even before we make its existence evident by use of the test charge at the point. As in all cases of physical measurement, the process of measurement disturbs the quantity to be measured. To mitigate this, the intent of the definition is made clearer by specifying that the (preexisting) electric field $\mathbf{E}$ at the point is the ratio $\mathbf{F}/q$, provided that the presence of the test charge q does not alter the field to any significant extent. To avoid the possibility of disturbing what we are measuring, we declare the definition of the electric field to be

$$\mathbf{E} = \lim(\mathbf{F}/q), \qquad \text{as } q \to 0, \tag{1.5}$$

where we mean $q \to 0$ both in amount of charge, so as not to disturb the sources of the field, and in the physical size of the test charge, so that we may clearly identify the point in space at which the measurement has been made. The units of an electric field are those of the ratio that defines them: newton/coulomb (N/C), although another unit is more commonly used, the volt/meter, to be discussed later.

To this point, the use we can make of knowledge of an electric field is to obtain the force exerted on any charge that may be placed within the field space, provided that the presence of that charge does not disturb the field. Applying Newton's laws of motion, with that force, can then yield the motion of the charge in the field region.

Example 1.3

Suppose that a uniform (in space) and constant (in time), downward electric field $\mathbf{E}(\mathbf{r}, t) = \mathbf{E}_0$ *has been set up within a region of space and that an electron enters that space at position* $\mathbf{r}_0$, *with horizontal velocity* $\mathbf{v}_0$, *at time* $t = 0$. *Where will the electron be at time* t? *As a practical example, the varying location of the electron when it reaches the screen is the essence of operation of the cathode ray tube of an oscilloscope.* ◈

Wherever the electron may find itself at time t, the force on it there and then will be $\mathbf{F} = -e\mathbf{E}_0$, so that the equation of motion is simply: mass × acceleration = force, or

$$m \, d^2\mathbf{r}/dt^2 = -e\mathbf{E}_0. \tag{1.6}$$

Since the right side of this differential equation is a known function of t (actually, it is merely a constant), we need only integrate, twice, to arrive at the trajectory of the electron. The first integration yields the velocity at any time,

$$d\mathbf{r}/dt = \mathbf{v}_0 - (e/m)\mathbf{E}_0 t, \tag{1.7}$$

where the given initial velocity $\mathbf{v}(0) = \mathbf{v}_0$ has been used as the integration constant. The second integration then gives the desired orbit of the electron,

$$\mathbf{r}(t) = \mathbf{r}_0 + \mathbf{v}_0 t - \tfrac{1}{2}(e/m)\mathbf{E}_0 t^2, \tag{1.8}$$

where the new integration constant has been evaluated as the given initial position $\mathbf{r}(0) = \mathbf{r}_0$.

To interpret this result, note that if there were no electric field, the electron would have drifted at constant velocity in a straight line, given by

$$\mathbf{r}(t) = \mathbf{r}_0 + \mathbf{v}_0 t, \tag{1.9}$$

and would, in particular, cover the horizontal distance s in a time s/v_0. Instead, in the presence of the downward field, the electron has been deflected upward from the unaccelerated, horizontal, straight-line path into a parabolic orbit, as indicated in Figure 1-4.

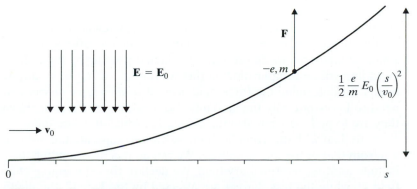

Figure 1-4   Parabolic deflection of electron in uniform electric field.

Note that in the time of horizontal drift over a distance s, which is still the time s/v_0, there is an upward deflection of amount $\frac{1}{2}(e/m)E_0(s/v_0)^2$; the vertical displacement varies as the square of the horizontal one, s, yielding a parabolic trajectory. Note also that if s is a fixed distance to a screen (at which the arrival of the electron is made visible), then the deflection is proportional to the strength of the electric field $\mathbf{E}_0$ and can therefore serve to measure that field. More realistically, the electron would be allowed to drift, along a (slanted) straight-line path in a field-free region, after exiting the field space, before striking the screen. The ultimate displacement on the screen is then still a measure of the deflecting field strength.

It is important to note that the effect of the electric field on the charge is determined by the value of the field at the location of the charge, at the instant when it is there. In this respect, the force exerted by the field on the charge is like a contact force. There is no need to consider the sources, elsewhere and earlier, of the field that is present and acting here and now. The result is that if charge q, of mass m, is in the presence of electric field $\mathbf{E}(r, t)$ and is at location $\mathbf{r}$ at time t, then its location $\mathbf{r}$ will change in time (as the field exerts its force on the particle) to become a trajectory $\mathbf{r}(t)$ that obeys the equation of motion

$$\frac{d^2\mathbf{r}(t)}{dt^2} = \frac{q}{m}\,\mathbf{E}\big(\mathbf{r}(t), t\big). \tag{1.10}$$

The time t is the same everywhere in the equation; also, the position $\mathbf{r} = \mathbf{r}(t)$ is the same on both sides of the equation. This means that the field $\mathbf{E}$ is to be evaluated at time t at the instantaneous location $\mathbf{r} = \mathbf{r}(t)$ of the charge, not elsewhere. There is no action at a distance, and there is no delay, in the interaction of the field with the charge. The actual spatial and temporal separation between the cause (the source charges) and the effect (on the test charge) is dealt with by the separate process of generation of the field in the space around the source, at a distance from the source charge, not by the field's effect on the test charge, at the location of the test charge.

EFFECT OF THE MEDIUM

We now wish to address the question of how the electric field is generated by the source charges. Suppose we are given the amounts, locations, and motions of various source charges; can we determine the field that is set up by these charges? There arises an immediate complication, in that we find the result to be different in different media. To determine the force field around some source charges, it seems that we need to know not only the amounts and locations of the sources, where and when they are to be found, but also the medium in which they are immersed.

To deal with this complication, we seek to separate the influence of the medium from that of the source charges. We can do this by extending the field concept to include another intermediary field. We declare that the source charges generate a field $\mathbf{D}$, while the test charges are affected by the $\mathbf{E}$ field; the medium determines how the fields **E** and **D** are related to each other. The **D** field is called *electric flux density*, or also *displacement flux density*, for reasons to be explained later.

Depending on the nature of the material medium, the relation between **D** and **E** may be quite simple or exceedingly complicated. In what we will term simple media, on an average basis that ignores very fine detail on the scale of atomic dimensions and provided the time variation of the fields is sufficiently slow, the medium relates **D** to **E** locally and instantaneously. This means that, to determine the **D** field at some point and some instant, it suffices to know the field **E** at only that point and instant, not anywhere else or at other times. In such media, the **D** field is just some function of the **E** field, $\mathbf{D} = \mathbf{D}(\mathbf{E})$. In the most simple types of medium, this function is merely a proportionality, with a factor ε called the permittivity of the medium, as

$$\mathbf{D} = \varepsilon \mathbf{E}. \tag{1.11}$$

On a macroscopic basis, this electrical description of a medium by a permittivity is often quite adequate, and we will not hesitate to use that simplification. We may on occasion have to allow the permittivity ε to be different at different points, $\varepsilon = \varepsilon(\mathbf{r})$, if the medium is not some unique, homogeneous material throughout the space of interest. We may even need it to be a function of time, $\varepsilon = \varepsilon(\mathbf{r}, t)$, if this variation is very slow, as when the material is changing (melting, perhaps) and the fields' time variation is also sufficiently slow. We may also have to allow for the direction of the **D** vector to differ from that of the **E** field, while still retaining the simple, linear proportionality, as happens in crystals; the permittivity then becomes a tensor $\boldsymbol{\varepsilon}$ instead of merely a scalar ε (a tensor applied to a vector can change both its magnitude and its direction, while a scalar can alter only the magnitude of the vector it multiplies, or also the sense along the direction of the vector if the scalar is negative). In all these cases, the proportionality between **D** and **E** is still local and instantaneous, expressible as

$$\mathbf{D}(\mathbf{r}, t) = \boldsymbol{\varepsilon}(\mathbf{r}, t)\, \mathbf{E}(\mathbf{r}, t). \tag{1.12}$$

This involves the fields and the permittivity at only the point **r** and the time t at which **D** is to be found. The more realistic description of a real medium, on a more microscopic basis, is nonlocal and noninstantaneous (dynamic), in that the **D** field here and now depends on the **E** field elsewhere and at other times as well as here and now. For example, a dynamic relation for a linear medium may be expressed as a *convolution* of the electric field with the *impulse response* of the medium. In a convolution, the result is not merely a multiplication but also an integration over time τ, ranging over the past history of the electric field $\mathbf{E}(\mathbf{r}, \tau)$ up to the present time $\tau = t$ at which the $\mathbf{D}(\mathbf{r}, t)$ field is to be evaluated. Refer to the Appendix in Chapter 10 for more explicit details about convolutions.

An even more general linear relation would entail a spatial convolution as well as a temporal one, involving an integration over all space as well as over time. The expedient way to deal with the dynamics embodied in a convolution is to express the relation between **D** and **E** in the frequency domain; that is, in the sinusoidal steady state, one frequency at a time. The relation thereby reverts to a simple multiplicative factor ε, rather than a convolution integral in the time domain, although the factor $\varepsilon = \varepsilon(\omega)$ then depends on the frequency ω.

For sufficiently strong fields, the relation between the two electric fields **D** and **E** can become nonlinear, whereupon the deviation from a simple proportionality has to be described by some more general function

$$\mathbf{D} = \mathbf{D}(\mathbf{E}). \tag{1.13}$$

In some exotic materials, called ferroelectrics, this nonlinear relation is not even single-valued, exhibiting hysteresis; in that case, the value of **D** depends on how **E** got to have its present value.

We will allow for the dynamic linear relationship of **D** to **E**, without losing the simple proportionality factor ε, when we consider the sinusoidal steady state, later in the text. For the most part, we will restrict ourselves to media simple enough to be described electrically by just a permittivity ε. Depending on the context, this will be considered to apply in the time domain, or else at one specified frequency in the steady state.

In a vacuum (that is, in the absence of any material medium), there is no need to distinguish between the **E** and **D** fields, yet it is inconvenient to make an exception of that particular medium, so we retain both fields even in that case. For a vacuum, the distinction between **D** and **E** becomes only that the units in which we measure the two types of field are different, so that a units conversion factor is all that is needed to convert from one to the other. This conversion factor is designated ε_0 and is called the *permittivity of free space*, as if empty space exhibited the physical property of permittivity that material media do.

$$\mathbf{D} = \varepsilon_0 \mathbf{E} \qquad (\text{vacuum}). \tag{1.14}$$

As illustrated in Figure 1-5, the scheme that was to express the interaction of one charge with another through the intermediary of the electric field **E** has been extended to incorporate an additional intermediary field, the **D** field, as well as the influence of the medium. To complete the chain from cause to effect, there remains to determine the laws governing the generation of the **D** field by the source charges.

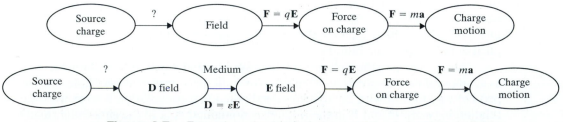

Figure 1-5   Extension of schematic interaction of charges via fields.

FLUX OF VECTOR FIELD

The law governing the generation of the **D** field by source charges is found to involve a certain property of that field, called its flux. It is the flux that is directly related to the source charges; the actual field distribution throughout space is determined, indirectly, from the flux. What is the flux of a vector field?

While a field is specified on a pointwise basis, by associating field values with *individual* points in space, there are significant quantities that pertain to a field that can be associated with certain *collections* of points in space. Those collections are not, however, some random sprinkling of points. Rather, they form a curve, or a surface, or a volume in space.

An important relationship between a vector field, such as **D**, and a collection of points that forms a surface S in space is expressed by the *surface integral of* **D** *over the surface S*, also referred to as the *flux of* **D** *through S*. The flux is a scalar quantity ψ, associated with both the field and the surface; it is denoted

$$\psi = \int_S \mathbf{D} \cdot d\mathbf{S}. \qquad (1.15)$$

To emphasize the dependence of the flux on the surface, we will sometimes append a description or designation of the surface, in brackets, to the symbol for the flux, as in $\psi\{S\}$. We must now define and illustrate the significance and methods of evaluation of this quantity.

We will proceed to review the formal definition of the surface integral and both a formal and an informal evaluation procedure for it, with an example that we can solve in several ways. To specify or calculate the flux of a vector field through a surface, we must furnish both a vector field and an oriented surface. The field must be given, as a vector (by its magnitude and direction, or else by its components), everywhere, or at least at every point of the surface. The surface must be identified, by specifying all its points (including its edge or boundary), and an orientation for the surface must be selected, so that one of its two sides is identified as the positive side. (Yes, one-sided surfaces like the Moebius strip do exist, but we can readily ensure that a positive side be selected at each of its points and avoid ambiguity by not counting any points more than once.) As with any integral, we have a formal definition for the flux as a limit of approximating sums; that definition allows us to apply it to physical situations of interest but it is usually ill suited to actual evaluation. We will present the formal definition and the evaluation procedures separately. Refer to Figure 1-6.

Definition of flux: The definition involves several steps.

1. Break up the surface S into a large number N of small segments. Any small area has both a magnitude and a direction; the direction is the one that the area faces, with the sense along that direction taken so as to agree with the specified orientation of the surface. Consequently, the mth segment of the surface is given by an infinitesimal vector $d\mathbf{S}_m$ that defines its size and orientation.

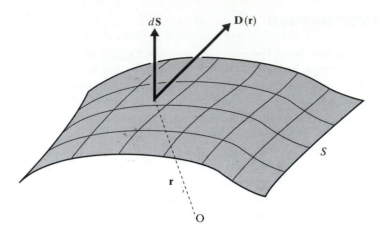

Figure 1-6   Surface S partitioned into many segments.

2. At a typical point $\mathbf{r}_m$ of the mth segment, evaluate the field $\mathbf{D}(\mathbf{r}_m)$.

3. Evaluate the (infinitesimal) scalar $\mathbf{D}(\mathbf{r}_m) \cdot d\mathbf{S}_m$. Since $d\mathbf{S}_m$ is normal to the surface, this scalar quantity involves only the perpendicular component of $\mathbf{D}$ at the point $\mathbf{r}_m$ and measures both the strength of the field and how well it "penetrates" or "crosses" the surface S at the point.

4. Evaluate this dot product at each of the N segments and form the sum

$$\sum_{m=1}^{N} \mathbf{D}(\mathbf{r}_m) \cdot d\mathbf{S}_m.$$

This sum is a number, one of many that could be obtained by this process, by different decompositions of the surface and different choices of the typical point in each segment.

5. The flux ψ is defined as the limit of such sums, as $N \to \infty$, but subject to the condition that all the segments be infinitesimal, in the sense that

$$\max|d\mathbf{S}_m| \to 0.$$

That the limit of the sums exists, for well-behaved fields and nonpathological surfaces, is demonstrated in courses on mathematical analysis; here we will be content with the usual intuitive notions of integration that justify the limiting process. We do want to note, however, that discontinuities in either the field or the surface do not necessarily vitiate the definition, since we can have the partitions follow the lines of discontinuity, so that there is no ambiguity at any segment of the partition. We can therefore calculate the flux of a discontinuous field as well as of a continuous one, and the surface may have corners.

We should realize from the process that defines it that, physically, the flux measures how strongly the field "cuts across" the surface. The more perpendicular the field is to the surface, the greater the numerical value obtained for the flux; conversely, if the field tends instead to lie parallel to the surface, the dot product at each such point becomes small and the flux is then slight. Besides this effect of the direction of the field in relation to the surface, the strength of the field and the size of the surface both affect the overall value of the flux. If the integrand were a flow field, the integral would measure the transport (of whatever is flowing) across the surface, from its negative to its positive side; the word *flux* originally means "flow."

The preceding formal definition of the flux tells us what we are after and guides us toward applications but it is not well suited to actual evaluation. We proceed therefore to a formal evaluation procedure for the flux.

Formal evaluation: Change the surface integral to an ordinary double integral, with respect to two variables that parametrize the surface, as follows.

1. Parametrize surface S as $\mathbf{r} = \mathbf{r}(u, v)$.

   This means that $\mathbf{r}$, which is a typical point of the surface, is expressed as a function of two parameters, u and v, that serve to distinguish one point of the surface from another. That two parameters suffice to locate a point on the surface corresponds to the fact that a surface is a two-dimensional object. The parameters u and v act as coordinates, like latitude and longitude on the surface of the earth, to locate points on the surface.

2. On the surface S, the integrand $\mathbf{D}(\mathbf{r})$ becomes $\mathbf{D}(\mathbf{r}(u, v))$, which is a (vector) function of the two parameters, u and v.

3. Express the differential element of area $d\mathbf{S}$ in terms of u and v, with $du\,dv$ as the second-order differential:

$$d\mathbf{S} = \frac{\partial \mathbf{r}}{\partial u} \times \frac{\partial \mathbf{r}}{\partial v}\, du\, dv. \tag{1.16}$$

   This is the cross product of two infinitesimal vectors $d\mathbf{r}_1 = (\partial \mathbf{r}/\partial u)\, du$ and $d\mathbf{r}_2 = (\partial \mathbf{r}/\partial v)\, dv$. These tiny vectors represent small changes of position on the surface, corresponding to infinitesimal changes du and dv imposed on the coordinates u and v. Both of the vectors $d\mathbf{r}_1$ and $d\mathbf{r}_2$ are tangential to the surface at the point $\mathbf{r}(u, v)$ and their cross product is perpendicular to the surface at that point. The sense along the perpendicular direction should conform to the selected orientation of the surface; if it does not, the order of u and v should simply be reversed, so that the cross product in the indicated order will point properly in the direction of the positive side of the surface.

4. Form the required dot product of $\mathbf{D}$ with $d\mathbf{S}$, as

$$\mathbf{D} \cdot d\mathbf{S} = \mathbf{D}(\mathbf{r}(u, v)) \cdot \frac{\partial \mathbf{r}}{\partial u} \times \frac{\partial \mathbf{r}}{\partial v}\, du\, dv \equiv f(u, v)\, du\, dv. \tag{1.17}$$

   The integrand $f(u, v) = \mathbf{D} \cdot (\partial \mathbf{r}/\partial u) \times (\partial \mathbf{r}/\partial v)$ in front of the infinitesimal $du\,dv$ has by now been expressed entirely as a function of u and v.

5. The flux ψ is now evaluated as the double integral of $f(u, v)$:

$$\psi = \int_S \mathbf{D} \cdot d\mathbf{S} = \iint \mathbf{D}(\mathbf{r}(u, v)) \cdot \frac{\partial \mathbf{r}}{\partial u} \times \frac{\partial \mathbf{r}}{\partial v} \, du \, dv \equiv \iint f(u, v) \, du \, dv, \quad (1.18)$$

with limits on the u and v integrations suitably set to cover the full surface S.

In practice, we perform a combination of the informal and the formal evaluation procedures: We break up the surface S into large, not infinitesimal, segments over which the normal component of $\mathbf{D}$ is constant, or at least easily integrated, and we add the contributions from each such segment. We can use different parametrizations for different portions of the entire surface.

Before proceeding to an example, we note that an especially significant version of a flux integral is the *closed surface integral*, which is denoted $\oint \mathbf{D} \cdot d\mathbf{S}$. The only distinction here is that the surface S is closed, has no edge. For this special case, the convention is that the orientation of the closed surface is, invariably, to be chosen outward.

Example 1.4

In order to illustrate the calculation of a surface integral, we need to give both a vector field as the integrand and a surface as the domain of integration. We ask for the closed surface integral of the field $\mathbf{F}(\mathbf{r}) = \mathbf{r}$ over the surface of a right circular cylinder of radius a and height h, placed with its axis along the z-axis and the center of its lower circular face at the origin. That is, the field is the vector displacement from the center of the bottom face to any point on the surface and the closed surface is that defined by the cylinder $\rho = a$, $0 \leq z \leq h$; here, (ρ, φ, z) are cylindrical coordinates. Figure 1-7 shows the cylinder and one position vector to a typical point on the curved part of the surface. We want to calculate $\psi\{S\} = \oint \mathbf{r} \cdot d\mathbf{S}$ over this surface S. «»

Before proceeding, let's be certain that we all agree that the surface is closed, meaning that it has no edge. If it did have an edge, the surface would end at that edge.

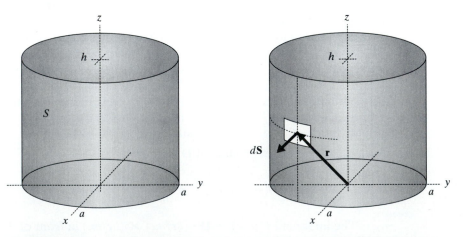

Figure 1-7 Closed cylindrical surface; position vector to typical point.

For example, if the upper circular face (at $z = h$) were excluded, then the surface would consist of only the bottom face and the curved side; it would have the circle of radius a at height h as its edge. Instead, we have specified that the surface is closed, so that it does include the upper circular disk. It therefore has no edge or end. It does have corners (at which the direction faced by the surface suddenly changes, from horizontal to vertical) but it has no edge, no boundary.

The infinitesimal vector $d\mathbf{S}$ has a magnitude denoted dA and its direction is given by the unit vector $\hat{\mathbf{n}}$. This unit vector is different at various points of the surface, but it is always perpendicular to the surface and, by the convention for closed surfaces, always points outward. Hence, $d\mathbf{S} = \hat{\mathbf{n}}\, dA$ and $\mathbf{r} \cdot d\mathbf{S} = \mathbf{r} \cdot \hat{\mathbf{n}}\, dA$. Therefore, we need to evaluate $\mathbf{r} \cdot \hat{\mathbf{n}}$ all over the surface, multiply this by the area element dA, and sum all the contributions.

Using first the "practical" approach, we divide the closed surface into its three major portions, the top and bottom end faces and the cylindrical side. Over each of these portions, the expression $\mathbf{r} \cdot \hat{\mathbf{n}}$ is simple and easy to obtain, as in Figure 1-8. On the bottom face, which is centered at the origin, the unit vector $\hat{\mathbf{n}}$ points uniformly vertically downward (vertically because the surface is horizontal and downward because it must point outward from the closed cylindrical surface). However, the position vector $\mathbf{r}$ at any point of the bottom face is horizontal, since its tail is at the center of the disk (the origin) and its head is at the point of the bottom face. As a result, the position vector and the unit normal vector anywhere on the bottom face are perpendicular to each other and the dot product $\mathbf{r} \cdot \hat{\mathbf{n}}$ is zero at every point of the bottom portion. There is therefore no contribution to the surface integral from the bottom face alone. This means only that the field $\mathbf{r}$ does not cross the bottom face of the cylinder.

On the curved side surface of the cylinder, both $\hat{\mathbf{n}}$ and $\mathbf{r}$ vary from point to point but, fortunately for us, the dot product $\mathbf{r} \cdot \hat{\mathbf{n}}$ does not: Since $\hat{\mathbf{n}}$ is horizontal (and directed outward) at every point of the side surface, what the dot product evaluates is the horizontal component of the position vector at each point of the side surface. But, on the cylindrical side surface, the *horizontal* component of the position vector from the center of the bottom face to any point on the side must be the radius of the cylinder, a, no

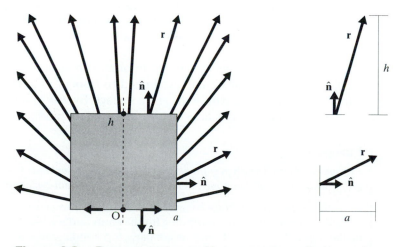

Figure 1-8   Representative position vectors on cylinder surface.

matter how high the vertical part of the position vector may be. Consequently, the dot product $\mathbf{r} \cdot \hat{\mathbf{n}}$ is a constant, a, for every point of the side of the cylinder. It follows that the contribution to the complete surface integral from the side of the cylinder is the integral of the constant, a, over the area of the cylinder's side:

$$\int \mathbf{r} \cdot d\mathbf{S} = \int \mathbf{r} \cdot \hat{\mathbf{n}} \, dA = \int a \, dA = a \int dA = aA = a(2\pi ah) = 2\pi a^2 h. \quad (1.19)$$

There remains to obtain the contribution to the full surface integral from the top face of the cylinder. On the top portion of the surface, the unit vector $\hat{\mathbf{n}}$ points uniformly vertically upward (vertically because the surface is horizontal and upward because it must point outward from the closed cylindrical surface). The unit vector $\hat{\mathbf{n}}$ is a constant on the top face, but the position vector $\mathbf{r}$ at points on the top is a variable. Nevertheless, the dot product $\mathbf{r} \cdot \hat{\mathbf{n}}$ is constant at every point of the top portion: Since $\hat{\mathbf{n}}$ is vertical (and directed outward) at every point of the top surface, what the dot product evaluates is the vertical component of the position vector at each point of the top surface. But, on the top face, the *vertical* component of the position vector from the center of the bottom face to any point on the top must be the height of the cylinder, h, no matter what the horizontal part of the position vector may be. Consequently, the dot product $\mathbf{r} \cdot \hat{\mathbf{n}}$ is a constant, h, for every point of the top of the cylinder. It follows that the contribution to the complete surface integral from the top of the cylinder is the integral of the constant, h, over the area of the top disk:

$$\int \mathbf{r} \cdot d\mathbf{S} = \int \mathbf{r} \cdot \hat{\mathbf{n}} \, dA = \int h \, dA = h \int dA = hA = h(\pi a^2) = \pi a^2 h. \quad (1.20)$$

Now that we have the contributions to the closed surface integral from each of the three segments into which we broke up the surface, we need only add the three values obtained, for the bottom, side, and top, respectively:

$$\psi = \oint \mathbf{r} \cdot d\mathbf{S} = 0 + 2\pi a^2 h + \pi a^2 h = 3\pi a^2 h. \quad (1.21)$$

This is the value of the flux of the position vector over the closed surface of the cylinder, evaluated by the "practical" approach of breaking up the surface into large portions over each of which the integral is easily evaluated.

We will not always be lucky enough to deal with easily integrated functions over simple surfaces, so that we may well need to resort to the formal integration procedure over at least some part of the given surface. By way of illustration, we evaluate the same flux integral, this time by formal parametrization.

Example 1.5

Evaluate $\psi\{S\} = \oint \mathbf{r} \cdot d\mathbf{S}$ over the surface S defined by the cylinder $\rho = a, 0 \leq z \leq h$, together with the top and bottom disks $z = h, z = 0$, by parametrization. ◈

Since the closed surface comprises a curved and two planar surfaces, it would be difficult to use a single parametrization for the entire surface. We therefore again break up the closed surface into its three major parts, top, bottom, and side. On the bottom, we recognize immediately that the position vector $\mathbf{r}$ and the normal to the surface $\hat{\mathbf{n}}$ are orthogonal vectors, so that $\mathbf{r} \cdot d\mathbf{S} = 0$, there is no contribution from the bottom, and parametrization is superfluous. We proceed to the curved, side surface.

To parametrize any surface, we need two variables that distinguish readily between one point of the surface and another. To locate ourselves at a typical point of the side of the cylinder, we need to tell how high up from the bottom we are, which is measured by the z-coordinate, and also where around the circular periphery of the cylinder we are situated, which is conveniently measured by the azimuthal angle φ. These two coordinates furnish a suitable parametrization of the cylinder's side and we need only express the position vector $\mathbf{r}$ from the origin to a *typical* point of the curved surface as some function of the two parameters: $\mathbf{r} = \mathbf{r}(\varphi, z)$.

We can give the position vector $\mathbf{r}$ by specifying its three rectangular components x, y, and z in $\mathbf{r} = x\hat{\mathbf{x}} + y\hat{\mathbf{y}} + z\hat{\mathbf{z}}$, but all expressed in terms of the two parameters φ and z. As indicated in Figure 1-9, x and y on the circle $\rho = a$ are given in terms of the azimuth φ as $x = a \cos \varphi$ and $y = a \sin \varphi$, so that the parametrization of the side of the cylinder is given by

$$\mathbf{r} = \mathbf{r}(\varphi, z) = a \cos \varphi \hat{\mathbf{x}} + a \sin \varphi \hat{\mathbf{y}} + z\hat{\mathbf{z}}. \tag{1.22}$$

This is to be accompanied by a specification of the range of values of the two parameters that will span the entire side of the cylinder. Here, we need

$$0 \le \varphi < 2\pi \quad \text{and} \quad 0 \le z \le h. \tag{1.23}$$

Having parametrized the surface, we can proceed with the calculation of the infinitesimal surface element $d\mathbf{S}$. For this we need the two partial derivatives of the $\mathbf{r}(\varphi, z) = a \cos \varphi \hat{\mathbf{x}} + a \sin \varphi \hat{\mathbf{y}} + z\hat{\mathbf{z}}$ function:

$$\partial \mathbf{r}/\partial \varphi = -a \sin \varphi \hat{\mathbf{x}} + a \cos \varphi \hat{\mathbf{y}} \quad \text{and} \quad \partial \mathbf{r}/\partial z = \hat{\mathbf{z}}. \tag{1.24}$$

The surface element is now given by

$$d\mathbf{S} = \frac{\partial \mathbf{r}}{\partial \varphi} \times \frac{\partial \mathbf{r}}{\partial z} \, d\varphi \, dz = (a \sin \varphi \hat{\mathbf{y}} + a \cos \varphi \hat{\mathbf{x}}) \, d\varphi \, dz, \tag{1.25}$$

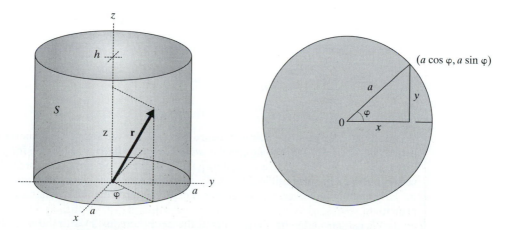

Figure 1-9 Parametrization of a point on the side of the cylinder.

obtained by evaluating the cross product of the two vectors in Eq. (1.24). We hasten to check that this vector does point outward, so that we did choose the right order of the two parameters, φ and z; for example, at $\varphi = 0$ the vector points along $\hat{\mathbf{x}}$, which is outward at that azimuth.

Now that both the field, here $\mathbf{r}$, and the surface element $d\mathbf{S}$ have been expressed entirely in terms of the two parameters φ and z (and their differentials), there remains to take the dot product of the two, which is

$$\mathbf{r} \cdot d\mathbf{S} = (a \cos\varphi\hat{\mathbf{x}} + a \sin\varphi\hat{\mathbf{y}} + z\hat{\mathbf{z}}) \cdot (a \sin\varphi\hat{\mathbf{y}} + a \cos\varphi\hat{\mathbf{x}}) \, d\varphi \, dz$$

$$= (a^2 \cos^2\varphi + a^2 \sin^2\varphi) \, d\varphi \, dz = a^2 \, d\varphi \, dz, \tag{1.26}$$

and to integrate the result over the range of values of the parameters that spans the surface, as already noted in Eq. (1.23). We obtain

$$\int_{\text{side}} \mathbf{r} \cdot d\mathbf{S} = \int_0^h \int_0^{2\pi} a^2 \, d\varphi \, dz = 2\pi a^2 h, \tag{1.27}$$

as found in the previous example by the less elaborate method.

We are now ready for the top surface, a flat disk of radius a in the $z = h$ plane. To locate ourselves at any typical point on that disk, starting at the origin, we need to go up a distance h along z and then tell how to get from the center of the upper disk to the point. The polar coordinates ρ and φ are convenient to distinguish one point from another on the circular area, and they can therefore serve as our two parameters for the top surface. If we express the position vector $\mathbf{r}$ by giving its rectangular components, we find

$$\mathbf{r} = \mathbf{r}(\rho, \varphi) = h\hat{\mathbf{z}} + \rho \cos\varphi\hat{\mathbf{x}} + \rho \sin\varphi\hat{\mathbf{y}}, \tag{1.28}$$

with the range of parameters

$$0 \leq \rho \leq a \quad \text{and} \quad 0 \leq \varphi < 2\pi. \tag{1.29}$$

From this we get the partial derivative vectors

$$\partial\mathbf{r}/\partial\rho = \cos\varphi\hat{\mathbf{x}} + \sin\varphi\hat{\mathbf{y}}, \tag{1.30}$$

$$\partial\mathbf{r}/\partial\varphi = -\rho \sin\varphi\hat{\mathbf{x}} + \rho \cos\varphi\hat{\mathbf{y}}, \tag{1.31}$$

whose cross product gives the area element $d\mathbf{S}$ as

$$d\mathbf{S} = \frac{\partial\mathbf{r}}{\partial\rho} \times \frac{\partial\mathbf{r}}{\partial\varphi} \, d\rho \, d\varphi = (\rho \cos^2\varphi\hat{\mathbf{z}} + \rho \sin^2\varphi\hat{\mathbf{z}}) \, d\rho \, d\varphi,$$

$$= \hat{\mathbf{z}}\rho \, d\rho \, d\varphi. \tag{1.32}$$

We verify at once that this points outward (upward) at the top surface, thereby confirming our choice of ρ preceding φ in the two parameters (had we needed $d\mathbf{S}$ for the bottom face, we would have had φ preceding ρ there, to get $d\mathbf{S}$ to point downward). We also recognize in $d\mathbf{S}$ the magnitude $dA = \rho \, d\rho \, d\varphi$ of the area element in polar coordinates; we have merely rederived that well-known expression for the area element from first principles.

We can now take the dot product of the field (here just $\mathbf{r}$) with the area element $d\mathbf{S}$ and integrate over the top face:

$$\mathbf{r} \cdot d\mathbf{S} = (h\hat{\mathbf{z}} + \rho \cos \varphi \hat{\mathbf{x}} + \rho \sin \varphi \hat{\mathbf{y}}) \cdot \hat{\mathbf{z}} \rho \, d\rho d\varphi = h\rho \, d\rho \, d\varphi; \qquad (1.33)$$

$$\int_{top} \mathbf{r} \cdot d\mathbf{S} = \int_0^{2\pi} \int_0^a h\rho \, d\rho \, d\varphi = 2\pi h(\tfrac{1}{2} a^2) = \pi a^2 h, \qquad (1.34)$$

as found in the previous example by the more direct method. Once again, we get for the total surface integral

$$\psi = \oint \mathbf{r} \cdot d\mathbf{S} = 0 + 2\pi a^2 h + \pi a^2 h = 3\pi a^2 h. \qquad (1.35)$$

This is the value of the flux of the position vector over the closed surface of the cylinder, evaluated by the formal approach of parametrizing large portions of the surface.

As a final assault on this same problem, we illustrate its solution by an advanced method that takes advantage of the *divergence theorem*, which allows conversion of any closed surface integral of a field into a volume integral of the divergence of that field, over the volume enclosed by the surface.

Example 1.6

Evaluate $\psi\{S\} = \oint \mathbf{r} \cdot d\mathbf{S}$ over the surface S defined by the cylinder $\rho = a, 0 \le z \le h$, together with the top and bottom disks $z = h, z = 0$, by use of the divergence theorem. ◈

We need the divergence of the vector field; here we need $\nabla \cdot \mathbf{r}$. In rectangular coordinates, the divergence operator is the sum of the x, y, and z derivatives of the x, y, and z components of the vector field. Here this means

$$\nabla \cdot \mathbf{r} = \frac{\partial}{\partial x} x + \frac{\partial}{\partial y} y + \frac{\partial}{\partial z} z = 1 + 1 + 1 = 3. \qquad (1.36)$$

Consequently, application of the divergence theorem requires merely the following conversion to a volume integral, in this case of a constant:

$$\oint_S \mathbf{r} \cdot d\mathbf{S} = \int_V \nabla \cdot \mathbf{r} \, dV = \int_V 3 \, dV = 3V = 3(\pi a^2 h) = 3\pi a^2 h, \qquad (1.37)$$

where V is the volume of the cylinder; this confirms the result found by the other two methods.

Of course, this approach works only for a closed surface integral and is useful when the divergence of the field is a sufficiently simple expression to integrate over the enclosed volume. We sometimes may use this approach also for an open surface integral, provided we can close the open surface with a conveniently simple additional auxiliary surface, readily evaluate the volume integral of the divergence, and then subtract from this the excess contribution from the auxiliary surface.

We close this discussion by noting that the flux $\Psi = \int_S \mathbf{D} \cdot d\mathbf{S}$ of the vector field $\mathbf{D}$ is itself no longer a vector field. It is a scalar and it is not a field. It would be a field if it were associated with individual points in space; instead, it is associated with an entire surface S, a particular collection of points in space. It therefore represents a higher-order creature than does a field.

GAUSS'S LAW

Now that we are adept at evaluating surface integrals, we can describe the law governing the generation of the **D** field by source charges. This is an experimentally derived result, expressed by *Gauss's law*:

For *any closed* surface S, the total electric flux $\psi = \oint_S \mathbf{D} \cdot d\mathbf{S}$ equals the total charge Q enclosed by the surface:

$$\psi\{S\} = Q_{\text{encl}} \qquad \text{or} \qquad \oint_S \mathbf{D} \cdot d\mathbf{S} = Q. \qquad (1.38)$$

The flux is a scalar quantity associated with both the **D** field and the surface S; the total charge enclosed by S is the algebraic sum of all the enclosed charges, or the net charge, with the negative charges subtracted from the positive ones. The area element on the surface, as always for a closed surface, points outward.

Interpretation. There are two aspects to the interpretation of Gauss's law. One applies when the field **D** is known over a closed surface, whereupon the law allows us to calculate the net amount of charge enclosed by the surface. The other aspect reigns when it is the charge enclosed by any surface that is known, while the **D** field is unknown; the law then becomes an integral equation to be solved for the integrand, given the integral. Both aspects concern us, but the second one is the more challenging and is the one that justifies thinking of Gauss's law as governing the generation of the **D** field by the source charges. We see that this process is an indirect one: According to the law, the source charge determines the electric flux ψ for *any* closed surface surrounding it. This *indirectly* determines what the detailed point-by-point **D** field distribution in space must be. Gauss's law represents a severe restriction on the way the field $\mathbf{D}(\mathbf{r})$ can vary in space, because it holds for *any* closed surface.

Because the surface integral of the **D** field is referred to as the electric flux ψ (or, historically, the displacement flux ψ), the integrand $\mathbf{D}(\mathbf{r})$ is called the *electric flux density*, or also the *displacement flux density*.

As a direct application of Gauss's law in its first aspect, consider the following not particularly practical example.

Example 1.7

Suppose we knew that the **D** field were $\mathbf{D}(\mathbf{r}) = k\mathbf{r}$ (with k a constant, with appropriate units) and we wanted to know how much charge is contained in a cylinder of radius a and height h with its bottom face centered at the origin. Gauss's law immediately provides the answer as

$$Q_{\text{encl}} = \oint \mathbf{D} \cdot d\mathbf{S} = \oint k\mathbf{r} \cdot d\mathbf{S} = k3\pi a^2 h, \qquad (1.39)$$

based on the calculation in the preceding examples.

This seems more than a little whimsical but, because the geometrical parameters a and h could be considered variables and yet the result is valid for *any* radius and height of the cylinder in the space in which $\mathbf{D} = k\mathbf{r}$ and because $\pi a^2 h$ is the volume of the cylinder for any values of a and h, the result indicates that the charge in any (cylindrical) volume V must total $3kV$, so that we feel justified in concluding that there is charge throughout the space, uniformly distributed with density $\rho = 3k$ C/m^3. We have thereby learned about the distribution of charge in space by using Gauss's law directly, for a given $\mathbf{D}$ field, by integrating that field over a closed surface of variable dimensions.

There is a subtlety worth pursuing briefly at this point. While it is correct that if the displacement density field is $\mathbf{D} = k\mathbf{r}$ throughout some portion of space, there must be a uniform net charge of density $\rho = 3k$ within that region, we cannot quite conclude that, if we wanted to set up this field pattern, it would suffice to spread charge uniformly with this density throughout space. What is lacking is some provision for making the origin a special point in space, the point from which the field seems to emanate and the only point where the field is zero. Clearly, a completely uniform charge distribution throughout space makes no point more special than any other.

What would be needed to set up this particular field pattern is not only the uniform charge distribution throughout the space in which the field is desired, but also some means for imposing the symmetry that makes the origin a special point. One possibility is to construct a spherically symmetric uniform charge distribution of density $3k$, or to delimit the region of uniform charge with a spherical boundary, in each case setting the center of the sphere at the origin. Imposing suitable symmetry is one way of supplementing Gauss's law with enough information to determine the desired field.

Gauss's law is of the sort that can give us information about the electrical contents of a region of space from measurements made only at its boundary. Consider the following example, which, again, is illustrative but not particularly realistic.

Example 1.8

A spherical "core" region of radius R is surrounded by a thin "crust" extending from radius R to radius $(1.01)R$. Suppose we have discovered an electric field within the crust and have mapped it out to be adequately represented by the electric flux density field $\mathbf{D}(\mathbf{r}) = \rho_0\big[x\hat{\mathbf{x}} + (z^3/R^2)\hat{\mathbf{z}}\big]$, where ρ_0 is a constant. We have no access to the core and whatever field it may contain. Nevertheless, determine in terms of ρ_0 and R

  (a) *the total charge within the core;*

  (b) *the ratio of the total charge within the thin crust to the charge in the core.* ◈

  (a) By Gauss's law, the charge in the core equals the flux across the spherical inner surface of the crust, at radius R. This is $Q = \oint \mathbf{D} \cdot d\mathbf{S}$, with $d\mathbf{S} = \hat{\mathbf{r}}\, dA$ and $dA = R^2 \sin\theta\, d\theta\, d\varphi$. We need $\mathbf{D} \cdot \hat{\mathbf{r}} = \rho_0\big[x\hat{\mathbf{x}} \cdot \hat{\mathbf{r}} + (z^3/R^2)\hat{\mathbf{z}} \cdot \hat{\mathbf{r}}\big]$ or

$$\mathbf{D} \cdot \hat{\mathbf{r}} = \rho_0\big[R\sin\theta\,\cos\varphi\,\sin\theta\,\cos\varphi + R\cos^3\theta\,\cos\theta\big]$$

$$= \rho_0 R\big[\sin^2\theta\,\cos^2\varphi + \cos^4\theta\big]. \tag{1.40}$$

From this,

$$Q = \rho_0 R^3 \int_0^{2\pi}\int_0^{\pi}\big[\sin^2\theta\,\cos^2\varphi + \cos^4\theta\big]\sin\theta\, d\theta\, d\varphi$$

$$= (32/15)\pi\rho_0 R^3. \tag{1.41}$$

(b) The charge q in the crust is the difference between the flux at radius $(1.01)R$ and the flux at R. At the outer surface, $x = (1.01)R \sin\theta \cos\varphi$ and $z^3/R^2 = (1.01)^3 R \cos^3\theta$; the area element becomes $dA = (1.01)^2 R^2 \sin\theta\, d\theta\, d\varphi$. Integrating over the outer surface gives us the sum of the core charge and the crust charge:

$$Q + q = \rho_0 R^3 \int_0^{2\pi} \int_0^{\pi} \left[(1.01)^3 \sin^2\theta \cos^2\varphi + (1.01)^5 \cos^4\theta \right] \sin\theta\, d\theta\, d\varphi$$

$$= \pi\rho_0 R^3 \left[(1.01)^3 4/3 + (1.01)^5 4/5 \right]$$

$$= 4\pi\rho_0 R^3 \left[(1.01)^3/3 + (1.01)^5/5 \right]. \tag{1.42}$$

We could, of course, just evaluate this and subtract $Q = 4\pi\rho_0 R^3(8/15)$ to get q, but let's take advantage of the fact that the crust is thin to get an approximate answer with less arithmetic. Since Q is what we get using 1 instead of 1.01 and since $[(1 + x)^n - 1]/n \approx x$ for small x, we have $q \approx 4\pi\rho_0 R^3(0.02)$ and the required ratio is $q/Q \approx (0.02)/(8/15) = 0.038$. The crust charge is nearly 4% of the core charge, although the crust volume is only 3% of that of the core.

Now let's look at an example of the use of Gauss's law to calculate a time-varying electric flux.

Example 1.9

The closed surface S is a right-circular cylinder, defined by the curved portion $\rho = a$, $-h \leq z \leq h$ and the two disks at the ends, $z = \pm h$, $0 \leq \rho \leq a$. A point charge q follows the orbit $\mathbf{r}(t) = be^{-\sigma t}\hat{\mathbf{x}} + ut\hat{\mathbf{z}}$, where b, σ, and u are given constants. Find the electric flux $\Psi = \Psi\{S\}$ as a function of time t,

(a) *if $b/a = 0.2$ and $\sigma h/u = 1.5$;*
(b) *if $b/a = 2.0$ and $\sigma h/u = 0.8$;*
(c) *if $b/a = 3.0$ and $\sigma h/u = 1.0$.*

Sketch plots of $\Psi(t)$ versus σt for each case.

Warning: Even if you recall Coulomb's law (to be discussed later), don't apply it here. Disciples of Coulomb and of Gauss can both find the result, but the latter will spend only minutes on it, while the former will need hours! ◇

Although we may be capable of calculating the flux as a surface integral at each instant of time, using the field of the charge q (the familiar field of a point charge is to be discussed in full detail later), that approach would be exceedingly tedious. Gauss's law comes to the rescue, alerting us to the fact that, since the electric flux over the closed surface always equals the enclosed charge, the only question to answer is whether the charge is inside the cylinder or outside it.

The electric flux $\Psi\{S\}$ is q during the time the charge is inside the cylinder S; it is zero while the charge is outside. The charge reaches the plane $z = -h$ at time $t = -h/u$, at radius $\rho = be^{\sigma h/u}$; it reaches the plane $z = h$ at $t = h/u$, at radius $\rho = be^{-\sigma h/u}$; it attains radius a at time $t = (1/\sigma)\ln(b/a)$, at $z = (u/\sigma)\ln(b/a)$. Figure 1-10 shows the orbit of the charge, in relation to the xz-plane cross section of the cylinder, for each of the three cases.

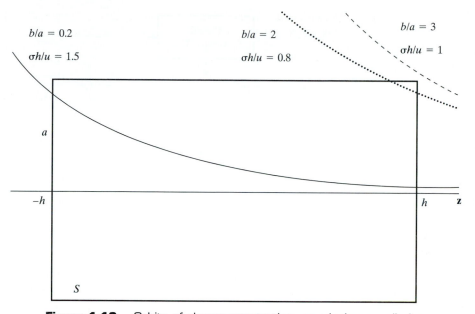

Figure 1-10 Orbits of charge penetrating, or missing, a cylinder.

(a) At $z = -h$, $\rho/a = (b/a)e^{\sigma h/u} = 0.2e^{1.5} = 0.90 < 1$, so that the charge enters and exits at the two ends. We get a flat-top pulse of height q, lasting from time $t_1 = -h/u = -1.5/\sigma$ to $t_2 = h/u = 1.5/\sigma$.

(b) At $z = -h$, $\rho/a = (b/a)e^{\sigma h/u} = 2.0e^{0.8} = 4.5 > 1$, so that the charge does not cross that disk, but at $z = h$, $\rho/a = 2.0e^{-0.8} = 0.9 < 1$, so that the charge does exit at the end disk. The charge crosses the side of the cylinder at $z/h = \ln(b/a)/(\sigma h/u) = \ln(2.0)/0.8 = 0.866 < 1$. We get a flat-top pulse of height q, lasting from time $t_1 = \ln(b/a)/\sigma = 0.693/\sigma$ to $t_2 = h/u = 0.8/\sigma$.

(c) At $z = h$, $\rho/a = 3.0e^{-1.0} = 1.10 > 1$, so that the charge misses the end disk and never enters the cylinder. We get $\Psi(t) = 0$ for all t.

Figure 1-11 presents plots of the electric flux as a function of normalized time, for the two cases in which it is not forever zero.

Figure 1-11 Pulses of electric flux over a cylinder penetrated by a charge.

DETERMINATION OF FIELD FROM FLUX

Knowing Q in the relation $\psi = \oint \mathbf{D} \cdot d\mathbf{S} = Q$, can the integrand $\mathbf{D} = \mathbf{D}(\mathbf{r})$ be extracted? Generally not, in the absence of further information about the field, as indicated in one of the examples. We are vitally interested in developing the missing link

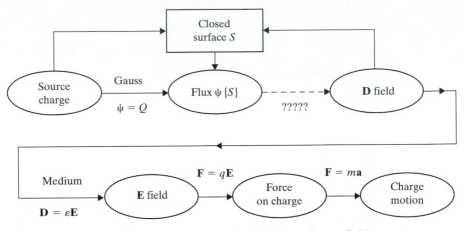

Figure 1-12   The missing link from flux to field.

from the flux ψ to the field **D**, as indicated in Figure 1-12. In general, we need additional restrictions on the field, besides its flux across any closed surface; this can be a symmetry condition or, as we shall see, it can be a superposition principle or, most generally, information about another type of field integral, the line integral.

There are situations in which *almost* everything is known in advance about the field and Gauss's law furnishes the missing item of information. This is the case particularly when we know the pattern of a field but not its strength. A prime example is the case of the electric flux density field generated by a point charge.

POINT CHARGE

Consider charge q, concentrated within a region small enough to be considered a point. There is nothing else in the universe and the point charge has no internal structure. This point source generates flux ψ, in accordance with Gauss's law, which demands that there be a flux, amounting to exactly q, over any surface surrounding the point charge. How can the field **D**(**r**) that integrates to this flux be distributed in space?

The **D** field is a vector at any point in space; we need to determine both its magnitude and its direction everywhere. Its direction is readily deduced from symmetry considerations. Since the source, having no internal structure, looks the same from all directions, the *direction* of the generated **D** field at any point away from the source point *must* be radial. Any deviation from the radial direction would imply that there is something in the universe that prefers that particular deviation over another. We have stipulated, however, that there is nothing in the universe, either internal to the charge or external to it, that determines any preferred direction in space. At any observation point (other than the source point itself), the only direction that can possibly matter is the one defined by the line that joins the observation point and the source point.

Figure 1-13 illustrates the futility of contemplating any direction other than the radial one as a possible field direction at the observation point. If any deviation from

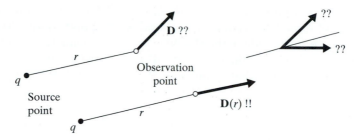

Figure 1-13   The field of a point charge can only be radial.

the radial direction were suggested, there would arise the question, Why not another deviation instead? We conclude from this symmetry argument that, upon adopting a spherical coordinate system centered at the source point, the generated **D** field will have to be of the form $\mathbf{D}(\mathbf{r}) = D(r, \theta, \varphi)\hat{\mathbf{r}}$, where (r, θ, φ) are the spherical coordinates of the observation point. We now need more information about the magnitude of the field, $D(r, \theta, \varphi)$. How can this vary with the coordinates?

For the same symmetry reason, the magnitude of the **D** field cannot be different when observed in different directions, for what would make the field stronger in one direction than in another? The only way one observation point can be distinguished from another is in terms of its distance from the source point. Consequently, the field strength D can differ at a far observation point from the strength at a near one, but there can be no difference in field strength between two observation points equidistant from the source. This symmetry argument imposes the condition that $D = |\mathbf{D}(\mathbf{r})| = D(r, \theta, \varphi)$ cannot depend on the angular coordinates θ and φ, which distinguish directions in space, but only on the radial distance coordinate $r: D = D(r)$ only. The functional dependence of the field strength D on r, say $f(r)$, is as yet unknown, but we know from symmetry alone that $\mathbf{D}(\mathbf{r}) = \mathbf{D}(r, \theta, \varphi) = \hat{\mathbf{r}}f(r)$. Note that the symmetry argument does not demand that the *vector* **D** be independent of direction and we make no such claim; in fact, we claim the vector does depend on the angular coordinates θ and φ, since we have convinced ourselves that it must be radial everywhere and the radial unit vector does change with elevation θ and azimuth φ. We claim only that the vector's *magnitude*, $|\mathbf{D}|$ can depend only on the radial distance from the point charge, as some unknown function $f(r)$. Gauss's law provides the means for determining the missing information about the functional dependence $f(r)$.

The reason this miracle can come about is that Gauss's law holds for *any* closed surface. This means *we* can choose the closed surface over which to integrate the **D** field in order to apply the law. Since we will be called upon to integrate a field $f(r)$ that we do not know, it will be wise to choose a surface on which this unknown function will at least not have an unknown variation; that is, let us try to keep this function constant, even though that constant will have an unknown value. The function $f(r)$ will remain constant if its argument r is kept constant; this occurs on the surface of any sphere centered on the point charge. By choosing such a sphere as our closed surface, we maintain the symmetry of the configuration and reduce the unknown function to an unknown constant.

We therefore choose the closed surface S to be used in Gauss's law to be a sphere of some fixed radius r_0, centered on the point source. At every point of this surface, the magnitude of the $\mathbf{D}$ field is the same—namely, the unknown but constant value $f(r_0)$. Note that the field vector $\mathbf{D}$ does vary from point to point on the sphere, being everywhere radial; it varies as $\mathbf{D} = f(r_0)\hat{\mathbf{r}}$, but its magnitude $f(r_0)$ does not. Furthermore, the area element on the sphere we have chosen is also radial at each point of the sphere, being perpendicular to the surface, and it points outward, as required for any closed surface: $d\mathbf{S} = \hat{\mathbf{r}}\, dA$. We can therefore evaluate the surface integral in terms of the still unknown but constant value $f(r_0)$, as follows:

$$\psi\{S\} = \oint_S \mathbf{D} \cdot d\mathbf{S} = \int f(r_0)\hat{\mathbf{r}} \cdot \hat{\mathbf{r}}\, dA = \int f(r_0)\, dA$$

$$= f(r_0) \int dA = f(r_0)A = f(r_0)4\pi r_0^2, \tag{1.43}$$

where $A = 4\pi r_0^2$ is the surface area of the sphere of radius r_0. We have been able to pull the unknown $f(r_0)$ out of the integral only because we arranged for it to be a consant; constants are the only quantities that can be pulled out of integrals.

By Gauss's law, this flux $\psi\{S\} = f(r_0)4\pi r_0^2$ [which is still unknown because $f(r_0)$ is still not known] must be equal to the total charge enclosed by the surface S of the sphere. Because the sphere is centered on the point charge, that point charge q is certainly enclosed by the surface; all of it is enclosed and there is no other charge within the sphere, so we conclude that

$$\psi\{S\} = q \qquad \text{or} \qquad f(r_0)4\pi r_0^2 = q, \tag{1.44}$$

which yields the unknown constant $f(r_0)$ for this particular sphere as

$$f(r_0) = q/4\pi r_0^2. \tag{1.45}$$

But this is the result for *any* radius $r_0 > 0$. Hence the function $f(r)$ must be

$$f(r) = q/4\pi r^2 \tag{1.46}$$

and it follows that the $\mathbf{D} = \hat{\mathbf{r}}f(r)$ field generated by the point charge q is

$$\mathbf{D}(\mathbf{r}) = \hat{\mathbf{r}}\,\frac{q}{4\pi r^2} = \frac{q\mathbf{r}}{4\pi r^3}, \tag{1.47}$$

where r is the distance from the source point at the origin to the field point at $\mathbf{r} = r\hat{\mathbf{r}}$. Figure 1-14 attempts to illustrate the "pincushion" aspect of the direction of the field at an arbitrary distance from the point charge. At half this arbitrary distance, the field is four times stronger; at three times that distance, it is nine times weaker.

COULOMB'S LAW

We can now determine the force between two point charges, say charge Q at the origin and charge q at position $\mathbf{r}$. The source charge Q generates the $\mathbf{D}$ field $\mathbf{D} = \hat{\mathbf{r}}Q/4\pi r^2$ at $\mathbf{r}$; in a vacuum, the electric field $\mathbf{E}$ is related to this $\mathbf{D}$ field by

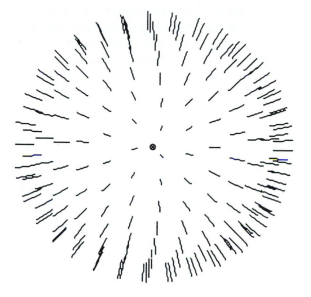

Figure 1-14 Electric flux density field around a point charge.

$\mathbf{E} = \mathbf{D}/\varepsilon_0 = \hat{\mathbf{r}}Q/4\pi\varepsilon_0 r^2$ at that point; with the test charge q located there, the force on that charge is $\mathbf{F} = q\mathbf{E}$, or

$$\mathbf{F}(\mathbf{r}) = \hat{\mathbf{r}}\frac{Qq}{4\pi\varepsilon_0 r^2} = \frac{Qq\mathbf{r}}{4\pi\varepsilon_0 r^3}, \tag{1.48}$$

where r is the distance from the source charge Q at the origin to the test charge q at the field point $\mathbf{r}$. This is the force exerted on the charge q at $\mathbf{r}$ by the charge Q at the origin. This result expresses Coulomb's law; it is a prime example of an *inverse square law* (that is, the magnitude of the force varies with separation distance inversely as the square of that distance). We are not fooled by the cubic dependence on distance in the denominator of the second form of the equation, because there is then a distance factor r in the position vector $\mathbf{r}$ in the numerator as well.

We should note the units of the various quantities we have encountered; we use the MKSA system of units. Force is measured in newtons (N) and charge in coulombs (C). It follows from $\mathbf{F} = q\mathbf{E}$ that the electric field $\mathbf{E}$ is to be measured in newtons/coulomb (N/C). From Gauss's law $\psi = \oint \mathbf{D} \cdot d\mathbf{S} = q$, we see that the electric flux density $\mathbf{D}$ must be measured in coulombs/m^2 (C/m^2), since the flux ψ must have the same units as charge. Hence, from the relation $\mathbf{D} = \varepsilon_0 \mathbf{E}$, we determine the units of ε_0 to be C^2/Nm^2. This is correct but awkward; we will see later that this unit is the same as a farad/m (F/m).

The value of the units conversion factor ε_0 is, to five places,

$$\varepsilon_0 = (8.8542)10^{-12}F/m = 8.8542 \text{ pF/m} \tag{1.49}$$

and we note that the factor $1/4\pi\varepsilon_0$ in Coulomb's law is roughly $9 \cdot 10^9 \text{ Nm}^2/C^2$.

Coulomb's law gives the force exerted by one charge, the source charge, on another, the test charge. Of course, it is only our attitude toward the two charges that designates one or the other to be the source charge. The force vector **F** found previously was the force exerted by Q on q. If we ask for the force exerted by q on Q, we need only designate q as the source and Q as the test charge and the same law gives us the force **F′** on Q. Since the product Qq of the charges is unchanged and since their separation r is the same, the only change in the expression of Coulomb's law is that the position vector now points from q to Q instead of from Q to q. Hence the new position vector **r′** is merely −**r** and we find that **F′** = −**F**, meaning that the force exerted on Q by q is equal in strength and opposite in direction to the force exerted by Q on q.

Finally, we note that Coulomb's law makes sense only for nonzero separation between the charges. If we were to allow the separation to become zero, the expression for the force would become infinite in magnitude and undefined in direction as the two charges coalesced. We dispose of this singularity by noting the physical fact that a point charge cannot exert a force on itself. Equivalently, we can say that the field of a point charge at its own location is zero.

As an example of the application of Coulomb's law, consider the following rudimentary electrometer, an (obsolete) instrument that detects charge.

Example 1.10

Two identical small metal spheres (radius a, mass m) are suspended from a common pivot point by insulating threads (each of length l, with l ≫ a). When a charge Q is transferred to the spheres, it divides equally between them and they fly apart. Their new equilibrium positions make the threads form an angle φ that can be read on a graduated scale; see Figure 1-15. This allows measurement of the charge Q.

(a) *What is the charge Q, in terms of the given parameters and the angle φ?*

(b) *What is the smallest charge Q_{min} that could be measured with this device, if a = 1 mm, m = 33 mg, l = 3 cm?* ◈

(a) Upon charging, the distance between the spheres becomes $2l \sin(\varphi/2)$ and the electrical force of repulsion between them is $F = (Q/2)^2 / [4\pi\varepsilon_0 4l^2 \sin^2(\varphi/2)]$, by

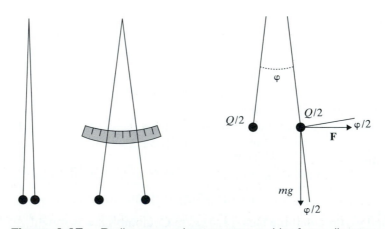

Figure 1-15   Rudimentary electrometer and its force diagram.

Coulomb's law. The component of this force perpendicular to the thread is $F \cos(\varphi/2)$, and this is balanced by the weight's component $mg \sin(\varphi/2)$ perpendicular to the thread. Equating these gives $Q^2 \cos(\varphi/2)/[64\pi\varepsilon_0 l^2 \sin^2(\varphi/2)] = mg \sin(\varphi/2)$ or

$$Q = \sqrt{64\pi\varepsilon_0 l^2 mg \sin^3(\varphi/2)/\cos(\varphi/2)}. \tag{1.50}$$

(b) To be discernible, the angle must exceed the one for which the spheres touch: $\sin(\varphi/2) = a/l$. With $a \ll l$, $\cos(\varphi/2) \approx 1$ and

$$Q_{min} = \sqrt{64\pi\varepsilon_0 mg\, a^3/l} = 0.14 \text{ nC}. \tag{1.51}$$

SUPERPOSITION OF FIELDS

Gauss's law speaks of the electric flux over a closed surface as being equal to the total (net) charge enclosed by the surface. If there are several charges within the enclosed space, not necessarily all of the same sign, the total charge is the algebraic sum of the individual ones. The flux, too, is an additive quantity, in that the flux of a sum of individual **D** fields is the sum of the individual fluxes of each one, by the rules of integration. For example, if $\mathbf{D}_0 = \mathbf{D}_1 + \mathbf{D}_2$, then the flux ψ_0 of $\mathbf{D}_0$ over some surface S is indeed the sum:

$$\psi_0 = \int \mathbf{D}_0 \cdot d\mathbf{S} = \int (\mathbf{D}_1 + \mathbf{D}_2) \cdot d\mathbf{S} = \int \mathbf{D}_1 \cdot d\mathbf{S} + \int \mathbf{D}_2 \cdot d\mathbf{S} = \psi_1 + \psi_2, \tag{1.52}$$

where ψ_1 and ψ_2 are the fluxes of the individual vector fields $\mathbf{D}_1$ and $\mathbf{D}_2$ over the same surface S.

If now the surface S is closed and encloses a total charge q_0 comprised of two portions q_1 and q_2 and if $\psi_1\{S\} = q_1$ (as if q_2 were not present) and $\psi_2\{S\} = q_2$ (as if q_1 were absent), then $\psi_0 = \psi_1 + \psi_2 = q_1 + q_2 = q_0$. This corresponds to the additivity of fluxes: The flux of a sum of charges is the sum of the fluxes of the individual charges, each treated as if the others were not present. The electric flux density field **D** also satisfies this principle of superposition: The **D** field generated by a collection of charges is obtainable by superimposing the fields generated by each charge individually (each considered in the absence of the others). The individual contributions are added vectorially.

Before citing examples and consequences of this *principle of superposition*, we should clarify briefly how we are using it. That the source of a sum of fluxes is the sum of the sources of each individual flux is an obvious result of the linearity of the flux integration process. But we want to assert that the principle works the other way as well: that the flux of a sum of sources is the sum of the individual fluxes from each source. This follows from the linearity of Gauss's law only if we can count on the uniqueness of the flux from any source (that is, if there is only one possible flux for a given set of source charges). If so, then the fact that the superposition of fluxes or fields certainly does satisfy the equation makes it the one and only solution and the principle holds. But do we have uniqueness, or could another field also satisfy the equation?

In fact, we can count on uniqueness only when certain restrictions, beyond that of Gauss's law, apply. In the case of the point charge (with no internal structure and with nothing else in the universe), we could invoke symmetry to guarantee ourselves that the Coulomb field is the only solution. In the absence of symmetry, some other constraint would have to be imposed to make the solution to Gauss's law (for given source charges) unique. Without a sufficiently stringent constraint, it is not at all difficult to come up with alternate solutions that also satisfy Gauss's law. All we need is to add to the Coulomb field $\mathbf{D}$ any field $\mathbf{D}'$ that has zero flux over any closed surface; the sum of the two fields clearly has the same source charge as does $\mathbf{D}$ alone. One example of an additional field $\mathbf{D}'$ would be any constant vector; another, less simple, example would be the "whirlwind" field $\mathbf{D}' = \mathbf{a} \times \mathbf{r}$ (for constant $\mathbf{a}$) that we considered earlier. These additional fields integrate to zero over any closed surface and can therefore be superimposed with the Coulomb field of a given source without affecting Gauss's law.

We hasten to note that these additional fields must be rejected for the solution to the problem of the point charge, because they destroy the symmetry that we must have in that case. They require the existence of a preferred direction in space (that of the constant vector itself, or that of the vector $\mathbf{a}$ for the nonconstant "whirlwind"). Furthermore, unlike the inverse-square Coulomb field, these additional fields persist or even grow stronger at ever larger distances from the given source charges. Mathematically, these fields are said to have sources at infinity; practically, they must terminate at some finite distance and there will have to be some source charges for them wherever they do end.

If we exclude source charges at infinity (and if we also exclude a supremely important magnetic effect that can also contribute to the flux, as will be discussed later), we can rely on the superposition of Coulomb fields to be the unique solution for a given set of source charges, and the problem of electric field generation can be considered thereby to be solved.

Example 1.11

How does the electric flux density field vary on the midplane between two equal and opposite point charges? ◇

Since we know the $\mathbf{D}$ field of a single point charge, we can get the field of a pair of point charges by a vector superposition of their individual fields. Let the positive charge q be located at $\mathbf{r}_1$ and the negative one, $-q$, be at $\mathbf{r}_2$. If the observation point is at $\mathbf{r}$, then the position vector of the observation point relative to the location of the first charge is $\mathbf{r} - \mathbf{r}_1$ and the field of the positive charge alone is

$$\mathbf{D}_1(\mathbf{r}) = \frac{q(\mathbf{r} - \mathbf{r}_1)}{4\pi |\mathbf{r} - \mathbf{r}_1|^3}, \tag{1.53}$$

while that of the second charge alone is

$$\mathbf{D}_2(\mathbf{r}) = \frac{-q(\mathbf{r} - \mathbf{r}_2)}{4\pi |\mathbf{r} - \mathbf{r}_2|^3}, \tag{1.54}$$

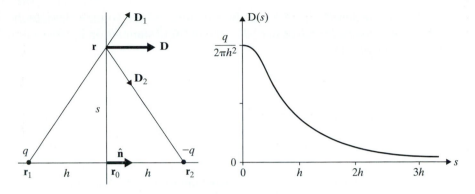

Figure 1-16   Field variation on midplane of a pair of charges.

so that the total field at any observation point $\mathbf{r}$ in space (except at $\mathbf{r}_1$ and $\mathbf{r}_2$) is

$$\mathbf{D}(\mathbf{r}) = \frac{q(\mathbf{r} - \mathbf{r}_1)}{4\pi|\mathbf{r} - \mathbf{r}_1|^3} - \frac{q(\mathbf{r} - \mathbf{r}_2)}{4\pi|\mathbf{r} - \mathbf{r}_2|^3}. \tag{1.55}$$

The midplane passes through the midpoint $\mathbf{r}_0 = \frac{1}{2}(\mathbf{r}_1 + \mathbf{r}_2)$ of the line that joins the two charges, and any observation point on the midplane is equidistant from both source charges. If the two source charges are at distance h from the midpoint and an arbitrary observation point on the midplane is at distance s from the midpoint, then $|\mathbf{r} - \mathbf{r}_1| = |\mathbf{r} - \mathbf{r}_2| = (h^2 + s^2)^{1/2}$ and the variation of the field from point to point on the midplane is given by

$$\mathbf{D}(s) = \frac{q(\mathbf{r} - \mathbf{r}_1)}{4\pi(h^2 + s^2)^{3/2}} - \frac{q(\mathbf{r} - \mathbf{r}_2)}{4\pi(h^2 + s^2)^{3/2}} = \frac{q(\mathbf{r}_2 - \mathbf{r}_1)}{4\pi(h^2 + s^2)^{3/2}}$$

$$= \frac{q2h\hat{\mathbf{n}}}{4\pi(h^2 + s^2)^{3/2}} = \frac{q\hat{\mathbf{n}}}{2\pi h^2[1 + (s^2/h^2)]^{3/2}}, \tag{1.56}$$

where $\hat{\mathbf{n}}$ is the unit vector perpendicular to the midplane, in the direction of the vector $(\mathbf{r}_2 - \mathbf{r}_1)$, which is the displacement from the positive to the negative charge. Figure 1-16 shows the configuration and a plot of the field-strength variation $D(s)$ versus s, the distance of a point on the midplane from the midpoint.

It is worth noting, for later purposes, that the electric field on the midplane of two equal and opposite point charges is perpendicular to that midplane at each point of the midplane.

For any number, say N, of individual point charges, of amount q_m at location $\mathbf{r}_m$, the superposition of the individual fields of each source gives for the field at some observation point $\mathbf{r}$:

$$\mathbf{D}(\mathbf{r}) = \sum_m \frac{(\mathbf{r} - \mathbf{r}_m)q_m}{4\pi|\mathbf{r} - \mathbf{r}_m|^3}, \tag{1.57}$$

where the discrete summation is, of course, a vector sum over each source charge. For the case of a continuous distribution of charge, smeared over some region of space with a density $\rho(\mathbf{r}) \, C/m^3$, so that an infinitesimal volume element dV' at $\mathbf{r}'$ contains an amount of charge $\rho(\mathbf{r}') \, dV'$, the discrete summation is replaced by a continuous sum, or integral:

$$\mathbf{D}(\mathbf{r}) = \int_V \frac{(\mathbf{r} - \mathbf{r}')\rho(\mathbf{r}') \, dV'}{4\pi |\mathbf{r} - \mathbf{r}'|^3}; \qquad (1.58)$$

the integral is over whatever region of space has a nonzero charge density. Note that in this integral, the observation point $\mathbf{r}$ is fixed, while the source point $\mathbf{r}'$ ranges over the volume of integration.

If the charges are distributed over a region of lesser dimension than a volume, say on a surface or on a line, then the summation becomes a surface or line integral instead of a volume integral. If some charges are concentrated and some smeared, we perform both a summation and an integration.

We now have the means for obtaining the electric field for any distribution of charges, either from Gauss's law or by superposition of Coulomb fields. When do we use Gauss's law and when an integration or summation? Gauss's law is much simpler to use but requires additional constraints, such as may arise from symmetry considerations. We use Gauss's law directly when there is sufficient symmetry or when we know (or can guess) the field pattern but not its strength. Otherwise, we resort to the superposition integral or sum.

Example 1.12

Charge is uniformly distributed along an infinitely long straight line, with a density of $Q_l \, C/m$. Obtain the electric field around this line charge. ◈

We have here a highly symmetrical charge distribution, because one observation point in space is distinguishable from another not by the azimuthal angle around the line charge and not by the axial distance along its infinite length, but only by how far the point is from the line. Since all angular directions are equivalent, the **D** field can only be directed radially, perpendicular to the line charge; no deviation from a radial direction can occur in the absence of a preferred azimuth (Figure 1-17). Also, the field strength can only depend on the radial distance from the line charge, not on azimuthal angle and not on axial location. We conclude that the **D** field pattern at an observation point $\mathbf{r}$ must be

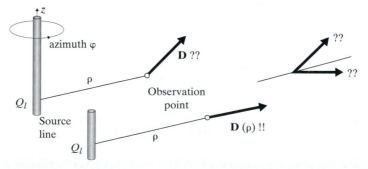

Figure 1-17   The field of a uniform line charge can only be radial.

of the form $\mathbf{D}(\mathbf{r}) = \hat{\boldsymbol{\rho}} D(\rho)$, where ρ is the radial coordinate of the cylindrical (ρ, φ, z) coordinate system that has its axis along the line charge and $\hat{\boldsymbol{\rho}}$ is the unit vector along the radial direction at the observation point. We know the field pattern but not the strength $D(\rho)$, and Gauss's law is ideal for extraction of this missing item of information.

To use Gauss's law, we choose a closed surface that encloses a known amount of charge and that conforms to the known field pattern without destroying the existing cylindrical symmetry. Since we will be called upon to integrate a field $D(\rho)$ that we do not know, we choose a surface on which this unknown function reduces to an unknown constant. The function $D(\rho)$ will remain constant if its argument ρ is kept constant; this occurs on the surface of any circular cylinder with axis along the line charge.

We therefore choose the closed surface S to be used in Gauss's law to be a cylinder of some fixed radius ρ_0, with its axis along the line source and of some length l, closed at each end by a transverse circular disk of radius ρ_0. The two end disks do not contribute to the flux over the closed surface, however, since their normals are directed axially, while the field is radial, so that $\mathbf{D} \cdot d\mathbf{S} = 0$ on the two ends. At every point on the curved part of the cylinder, the magnitude of the $\mathbf{D}$ field is the same—namely, the unknown but constant value $D(\rho_0)$. The area element on the cylinder we have chosen is also radial at each point of the curved side, being perpendicular to the surface, and it points outward, as required for any closed surface: $d\mathbf{S} = \hat{\boldsymbol{\rho}} \, dA$. We can therefore evaluate the surface integral in terms of the still unknown but constant value $D(\rho_0)$, as follows:

$$\psi\{S\} = \oint_S \mathbf{D} \cdot d\mathbf{S} = \int D(\rho_0)\hat{\boldsymbol{\rho}} \cdot \hat{\boldsymbol{\rho}} \, dA = \int D(\rho_0) \, dA$$

$$= D(\rho_0) \int dA = D(\rho_0)A = D(\rho_0)2\pi\rho_0 l, \qquad (1.59)$$

where $A = 2\pi\rho_0 l$ is the surface area of the curved part of the cylinder of radius ρ_0 and length l. We have been able to pull the unknown $D(\rho_0)$ out of the integral only because we arranged for it to be a constant.

By Gauss's law, this flux $\psi\{S\}$ must be equal to the total charge enclosed by the surface S of the cylinder. Because the cylinder's axis is coincident with the line charge, an amount of charge $q = Q_l l$ is enclosed by the surface, for any radius $\rho_0 > 0$. We conclude from $\psi\{S\} = q$ that

$$D(\rho_0)2\pi\rho_0 l = Q_l l, \qquad (1.60)$$

which yields the unknown constant $D(\rho_0)$ for this particular cylinder as

$$D(\rho_0) = Q_l/2\pi\rho_0. \qquad (1.61)$$

But this is the result for *any* radius $\rho_0 > 0$. Hence the function $D(\rho)$ must be

$$D(\rho) = Q_l/2\pi\rho \qquad (1.62)$$

and it follows that the $\mathbf{D} = \hat{\boldsymbol{\rho}} D(\rho)$ field generated by the line charge q is

$$\mathbf{D}(\mathbf{r}) = \hat{\boldsymbol{\rho}} \, \frac{Q_l}{2\pi\rho}, \qquad (1.63)$$

where ρ is the perpendicular distance from the line charge to the field point. Figure 1-18 attempts to illustrate the "pincushion" aspect of the direction of the field at an arbitrary

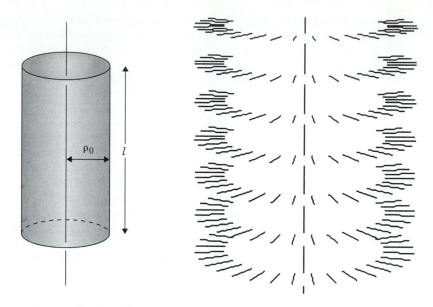

Figure 1-18   Gaussian surface and field pattern for line charge.

distance from the line charge. At half this arbitrary distance, the field is twice as strong; at three times that distance, it is three times weaker.

  The **D** field we have found is valid for any uniform medium. The electric field is just $\mathbf{E} = \mathbf{D}/\varepsilon$ if the medium has permittivity ε, or $\mathbf{E} = \mathbf{D}/\varepsilon_0$ for a vacuum.

Although there are easier methods of superposition and although we will soon find that electric field generation by charge distributions is of less importance than generation by magnetic means, it is worthwhile to illustrate a vector superposition for a case with symmetry insufficient to reveal the field pattern in advance.

Example 1.13

Charge Q is uniformly distributed along a straight line of finite length l. Find the electric field around this charge distribution. ⬫

  Unlike the case of the infinitely long line charge, this one does not let us argue that the field pattern must be radial. We do still have azimuthal symmetry: No azimuth is more special than any other. We can therefore predict that the magnitude of the field cannot depend on azimuthal angle φ and that the direction of the **D** field cannot have an azimuthal component. The form of the field must be $\mathbf{D}(\mathbf{r}) = D_\rho(\rho, z)\hat{\boldsymbol{\rho}} + D_z(\rho, z)\hat{\mathbf{z}}$, but it would require more insight into the solution than we now have to come up with a surface over which both D_ρ and D_z would necessarily be constants, to make the flux easy to calculate. We therefore resort to the superposition of Coulomb fields from infinitesimal bits of the line charge, each acting as a point source.

  For a line charge extending along the z-axis from z_1 to z_2 (with $z_2 - z_1 = l$), an element of length dz' at z' comprises charge $dq = (Q/l)\,dz'$ and generates a field at an observation point at (ρ, φ, z) in the direction of the vector $\mathbf{r} - \mathbf{r}' = \mathbf{r} - z'\hat{\mathbf{z}} =$

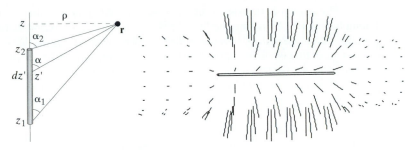

Figure 1-19   Field of finite line charge, seen on a cylinder.

$\rho\hat{\boldsymbol{\rho}} + (z - z')\hat{\mathbf{z}}$, as indicated in Figure 1-19. (Note that we use the primed letter z' for the distance along the line charge, rather than simply the unprimed letter z, because the unprimed one is already in use to mean the axial distance z from the xy-plane to the observation point.) The squared distance between source and observation point is $\rho^2 + (z - z')^2$, so that the superposition of Coulomb fields becomes

$$\mathbf{D}(\mathbf{r}) = \int_{z_1}^{z_2} \frac{[\rho\hat{\boldsymbol{\rho}} + (z - z')\hat{\mathbf{z}}](Q/l)\, dz'}{4\pi[\rho^2 + (z - z')^2]^{3/2}}. \tag{1.64}$$

This is not so formidable as it looks, because the unit vectors $\hat{\boldsymbol{\rho}}$ and $\hat{\mathbf{z}}$ are constants with respect to the integration variable z' and because the sum of squares suggests a trigonometric substitution like $z - z' = \rho \cot \alpha$. With foresight, we could have recognized in advance that α would have been a more convenient choice of parameter for the superposition integral than z'; lacking the foresight, however, we have used z' and we only now invoke the simplifying substitution of integration variable. The substitution implies

$$dz' = \rho \csc^2 \alpha \, d\alpha \tag{1.65}$$

and we are left with

$$\mathbf{D}(\mathbf{r}) = \int_{\alpha_1}^{\alpha_2} \frac{[\rho\hat{\boldsymbol{\rho}} + \rho \cot \alpha\hat{\mathbf{z}}](Q/l)\rho \csc^2 \alpha \, d\alpha}{4\pi[\rho^2 + \rho^2 \cot^2 \alpha]^{3/2}}, \tag{1.66}$$

which, since $1 + \cot^2 \alpha = \csc^2 \alpha$, reduces to

$$\mathbf{D}(\mathbf{r}) = \frac{Q}{4\pi l \rho} \int_{\alpha_1}^{\alpha_2} [\sin \alpha \hat{\boldsymbol{\rho}} + \cos \alpha \hat{\mathbf{z}}] \, d\alpha$$

$$= [Q/4\pi l \rho][(\cos \alpha_1 - \cos \alpha_2)\hat{\boldsymbol{\rho}} + (\sin \alpha_2 - \sin \alpha_1)\hat{\mathbf{z}}]. \tag{1.67}$$

Note that both $\hat{\boldsymbol{\rho}}$ and $\hat{\mathbf{z}}$ are constant vectors, independent of the integration variable α, and so each can come out of the integral. As indicated in Figure 1-19, α_1 and α_2 are the angles between the line charge and the observation point, as seen at the bottom and top ends of the line; these angles depend on the location of the observation point in relation to the source distribution.

We can understand why, as suggested in the figure, the field is relatively weak at points beyond the ends of the line charge, compared to the field at points such that $z_1 < z < z_2$. In the latter region, α_2 is obtuse and $\cos \alpha_2$ is negative, so that the difference of cosines actually adds strength to the radial component; beyond the ends, the cosines really do subtract from each other, leaving a weak field.

We can verify that our result does, as it should, reduce to the one for an infinitely long line charge by taking the limit of $z_1 \to -\infty$ and $z_2 \to \infty$, while retaining a finite charge per unit length Q/l even as both Q and l become infinite. In this limit, $\alpha_1 \to 0$ and $\alpha_2 \to \pi$, so that the field approaches $\mathbf{D}(\mathbf{r}) = [Q/4\pi l\rho][2\hat{\boldsymbol{\rho}} + 0\hat{\mathbf{z}}] = \hat{\boldsymbol{\rho}}(Q/l)/2\pi\rho$, which is the result we found for the infinitely long line charge.

We should also confirm that the opposite limit, $l \to 0$, for which the total charge Q is concentrated at one point, does revert to the case of the field of a point charge. We need the limit of $z_1 \to 0$ and $z_2 \to 0$, for which α_1 becomes nearly equal to α_2, so that the differences $\Delta \cos \alpha$ and $\Delta \sin \alpha$ in the result for $\mathbf{D}(\mathbf{r})$ become differentials, proportional to $d\alpha = \alpha_2 - \alpha_1$. Since $z_2 - z_1 = l$ even as $l \to 0$, we can relate $d\alpha$ to $dz' = l$ by differentiating the relation that defines α:

$$\tan \alpha = \rho/(z - z') \qquad \text{yields} \qquad \sec^2 \alpha \, d\alpha = [\rho/(z - z')^2] \, dz' = \rho l/z^2, \qquad (1.68)$$

at $z' = 0$ in the limit. Since $z \sec \alpha = r$, the distance from source point to field point, we get $d\alpha = \rho l/r^2$ in the limit and the field approaches

$$
\begin{aligned}
\mathbf{D}(\mathbf{r}) &= [Q/4\pi l\rho][(-\Delta \cos \alpha)\hat{\boldsymbol{\rho}} + (\Delta \sin \alpha)\hat{\mathbf{z}}] \\
&= [Q/4\pi l\rho][\sin \alpha \, d\alpha \, \hat{\boldsymbol{\rho}} + \cos \alpha \, d\alpha \, \hat{\mathbf{z}}] \\
&= [Q/4\pi r^2][\sin \alpha \, \hat{\boldsymbol{\rho}} + \cos \alpha \, \hat{\mathbf{z}}] = Q\hat{\mathbf{r}}/4\pi r^2. \qquad (1.69)
\end{aligned}
$$

This is, of course, the field of a point charge at the origin.

Let us illustrate superposition of charge elements for the calculation of fields yet again, for a charged-ring harmonic oscillator. By a leap of faith, this problem can be made applicable to a crude atomic model that can tell us something about the frequencies of atomic oscillations.

Example 1.14

A thin wire is bent into a large horizontal circular ring of radius a and uniformly charged to total charge $-q$. A point charge of mass m and charge $+q$ is placed at the exact center of the ring. If this point charge is displaced precisely vertically, but only slightly ($z \ll a$), from the center and then released, it will oscillate about the center point.

   (a) *Verify that the ensuing motion is that of a simple harmonic oscillator and find the frequency ω of the oscillation.*

   (b) *If the charge q and mass m were those of an electron and the radius a of the ring were the size of an atom, what frequency $f = \omega/2\pi$ could be expected of this oscillator? What part of the electromagnetic spectrum has such frequencies?*

Note: Relevant universal constants: Electron charge $e = (1.60)10^{-19}$ C; electron mass $m = (9.11)10^{-31}$ kg; Bohr radius $a_0 = (5.29)10^{-11}$ m. ◇

   (a) An infinitesimal element of the ring at azimuth φ and of length $a \, d\varphi$ has charge $dq = -q(d\varphi/2\pi)$ and the observation point on the axis at $z\hat{\mathbf{z}}$ is displaced from this source point by $\mathbf{R} = z\hat{\mathbf{z}} - a\hat{\boldsymbol{\rho}}$. It therefore contributes the infinitesimal field $d\mathbf{E} = [-q/4\pi\varepsilon_0][\mathbf{R}/R^3] \, d\varphi/2\pi$. For slight vertical displacements $z \ll a$, R^3 reduces to a^3. Integrating $\mathbf{R}$ around the ring, the term $a\hat{\boldsymbol{\rho}}$ contributes nothing, because the radial unit vector points in all directions as all azimuths are traversed. That is, by symmetry, the total field has only the vertical component. The other term, $z\hat{\mathbf{z}}$, is independent of azimuth, so that integrating around the ring gives the vertical field as $\mathbf{E}(z) = [-q/4\pi\varepsilon_0 a^3]z\hat{\mathbf{z}}$.

Thus, the point charge experiences a restoring force $\mathbf{F} = -[q^2/4\pi\varepsilon_0 a^3]z\hat{\mathbf{z}}$, proportional to the vertical displacement z. This results in a simple harmonic oscillator, because $\mathbf{F} = m\mathbf{a} = m[d^2z/dt^2]\hat{\mathbf{z}}$ gives

$$d^2z(t)/dt^2 = -\omega^2 z(t) \quad \text{with} \quad \omega^2 = q^2/4\pi\varepsilon_0 ma^3, \tag{1.70}$$

and this yields a sinusoidal oscillation at frequency ω.

(b) Substituting e for q and a_0 for a gives the oscillation frequency $f = \omega/2\pi = (e/2\pi)/\sqrt{4\pi\varepsilon_0 ma_0^3} = (6.6)10^{15}$ Hz, which is in the ultraviolet range. We note that atoms do emit radiation in that vicinity of the electromagnetic spectrum.

SUMMARY

We have pursued the concept of a field, which is a quantity associated with points in space, as an intermediary agent between cause and effect in the interaction of electric charges. We found it convenient to define two types of electrical field, the electric field $\mathbf{E}$, which acts on a (test) charge, and the electric flux density field $\mathbf{D}$, which is generated by a (source) charge. This division into two fields allows us to relegate the influence of the medium in which the interaction takes place to a separate phenomenon. We even agree to close our eyes to the more subtle microscopic effects of a material medium and reduce our description of the medium to merely a permittivity factor ε.

The action of the $\mathbf{E}$ field on a charge occurs at the point and instant at which they coexist, as would a contact force: If charge q is at position $\mathbf{r}$ at time t in the presence of an electric field, the force exerted on the charge there at that time is $\mathbf{F} = q\mathbf{E}(\mathbf{r}, t)$. Whatever values the $\mathbf{E}$ field may have elsewhere and at other times are irrelevant to the force it exerts here and now on the charge.

The process of generation of the $\mathbf{D}$ field by a charge, however, involves creation of an electric flux on any surrounding surface, no matter how large. We therefore have action at a distance, because the source charge generates flux over a distant surrounding surface, in accordance with Gauss's law, and the existence of flux requires the existence of a $\mathbf{D}$ field over at least some part of the distant surface. This action at a distance is reflected in Coulomb's law, which involves an inverse-square distance dependence.

Although we already have action at a distance, we do not yet see a time delay in charge interaction. Gauss's law does not entail any difference in time of evaluation of the flux over a closed surface and the enclosed charge: In

$$\psi(t) = \oint_S \mathbf{D}(\mathbf{r}, t) \cdot d\mathbf{S} = Q(t), \tag{1.71}$$

the time t is the same on both sides of the equation. Even if the point on the surface S nearest to the most far-flung portion of the enclosed charge Q is light-years away from the source, the flux at time t equals the enclosed charge at the same time t. Nevertheless, there is no inconsistency between Gauss's law and the requirements of relativity. To resolve this issue, as well as to discover how time delay comes about, we need to account for time variation of charges and fields. The simplest version of this is motion of charges, which we address next.

We will also find that the direct process of generation of electric fields by source charges is overshadowed by the far more important and ubiquitous one of generation by magnetic effects.

We have deliberately avoided introducing the notion of an electric potential, which would simplify the superposition of fields of combinations of source charges. Although the potential is very useful in the static case, it becomes more of a liability than an asset as soon as we allow for time variation, which is our primary focus. We have enough to study without having to learn and then unlearn the use of potentials!

PROBLEMS

Geometrically defined field

1.1 Write expressions for the mathematical field $f(\mathbf{r})$ that represents the sum of the distances from two fixed points in space separated by distance s,

(a) in rectangular coordinates [i.e., giving $f(x, y, z)$ if the two fixed points are at $(0, 0, s/2)$ and $(0, 0, -s/2)$];

(b) in spherical coordinates [i.e., giving $f(r, \theta, \varphi)$ for the same fixed points];

(c) in terms of position vector $\mathbf{r}$ and the vector separation $\mathbf{s}$, with the origin midway between the two fixed points.

Electron orbit in nonuniform electric field

1.2 Suppose we can set up an electric field pattern in space, described by $(e/m)\mathbf{E}(\mathbf{r}) = -\Omega^2(x\hat{\mathbf{z}} + z\hat{\mathbf{x}})$, where Ω is a constant. An electron is at the origin at time $t = 0$ and is moving at speed u_0 along the x-axis; see Figure P1-1. Find the orbit $\mathbf{r}(t)$ of the electron.

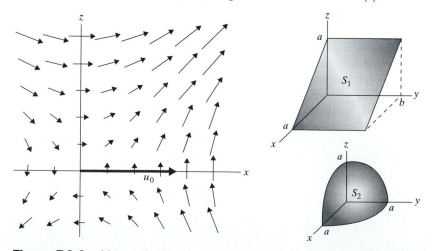

Figure P1-1   Vector field $(x\hat{\mathbf{z}} + z\hat{\mathbf{x}})$; two surfaces for flux calculations.

Flux over planar surface

1.3 The electric flux density $\mathbf{D}(\mathbf{r})$ is given by $\mathbf{D}(x, y, z) = \rho_0(x\hat{\mathbf{z}} + z\hat{\mathbf{x}})$, where ρ_0 is a constant. The surface S_1 illustrated in Figure P1-1 is a slanted planar rectangle (of width $a\sqrt{2}$ and height b). Find the electric flux $\Psi\{S_1\}$ over this surface, which is oriented away from the origin.

Flux over spherical surface

1.4 The electric flux density $\mathbf{D}(\mathbf{r})$ is given by $\mathbf{D}(x, y, z) = \rho_0(x\hat{\mathbf{z}} + z\hat{\mathbf{x}})$, where ρ_0 is a constant. The surface S_2 illustrated in Figure P1-1 is the portion of the sphere $r = a$ within the first octant. Find the electric flux $\Psi\{S_2\}$ over this surface, which is oriented away from the origin.

Flux of "whirlwind" field

1.5 Find the flux of the "whirlwind" field $\mathbf{a} \times \mathbf{r}$ over the rectangular surface whose base extends from the origin to a distance b along the x-axis and whose height extends from the origin to h along the z-axis, if $\mathbf{a}$ is along the z-axis.

Closed surface integral

1.6 Given the field $\mathbf{F}(\mathbf{r}) = y\hat{\mathbf{x}} + x\hat{\mathbf{y}} + z\hat{\mathbf{z}}$ and that the closed surface S is the hemisphere defined by $r = a, z > 0$ and closed by the equatorial plane $z = 0, \rho < a$, find the flux of $\mathbf{F}$ over the closed surface S, $\Psi\{S\} = \oint_S \mathbf{F} \cdot d\mathbf{S}$.

Application of Gauss's law

1.7 To simulate an exponentially decaying field, a sphere of radius R is built up of N regions, comprising $(N - 1)$ concentric spherical layers around a spherical core. The core has radius R/N; the nth layer ($n = 1, 2, \cdots, N - 1$) extends from radius nR/N to radius $(n + 1)R/N$; see Figure P1-2. By implanting charges in each layer appropriately, we contrive to make the electric flux density at radius nR/N be $\mathbf{D}_n = \hat{\mathbf{r}}D_0 e^{-n}$, for $n = 1, 2, \cdots, N$. Assume the charges are immobile. Let Q_0 be the charge within the core.
   (a) Express D_0 in terms of Q_0, R, N.
   (b) What is the total charge Q_n embedded within the nth layer, in terms of Q_0?
   (c) What is the total charge Q on the entire sphere, in terms of Q_0?

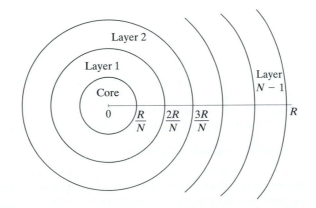

Figure P1-2 Sphere composed of concentric charged core and layers.

Time-varying electric flux

1.8   The closed surface S is an imaginary sphere of radius $R = 36.8$ cm, centered at the origin. A small pellet of charge $q = 0.2$ C and mass $m = 0.4$ kg is released from a point at height $h = 100.5$ cm above the origin and $s = 29.4$ cm horizontally away from the vertical axis through the origin, at time $t = 0$. The acceleration of gravity is $g = 9.8$ m/sec^2. Find the electric flux $\Psi = \Psi\{S\}$ as a function of time t and sketch a plot of $\Psi(t)$ versus t.

Electric and gravitational forces compared

1.9   Suppose there existed elementary charged particles whose mutual gravitational attraction exactly balanced their electrical repulsion.
   (a) If their charge were that of an electron, how much heavier than an electron would they have to be?
   (b) How much stronger is the electrical force between electrons than the gravitational one?
   Note: Relevant universal constants: Gravitational constant $G = (6.67)10^{-11}$ Nm^2/kg^2; electron charge-to-mass ratio $e/m = (1.76)10^{11}$ C/kg.

Charges in equilibrium

1.10   Charges Q_1 and Q_2, both positive, are immobilized, a distance l apart. A third charge q is mobile, but only along the straight line joining the other two charges. At what fractional distance x/l from Q_1 can q be placed between the two charges so that it remains in equilibrium?

Field of uniform spherical charge

1.11   A spherical region of radius a contains a total charge Q, distributed uniformly with density $Q/[(4/3)\pi a^3]$.
   (a) What symmetry requirements must the electric flux density field pattern $\mathbf{D}(\mathbf{r})$ satisfy, both outside and inside the sphere?
   (b) Apply Gauss's law to determine the $\mathbf{D}(\mathbf{r})$ field, both outside and inside the sphere.

Field of uniform planar charge

1.12   **(a)** An infinite planar sheet is charged to a uniform density of Q_s C/m^2. Find the electric field in the vacuum on both sides of the sheet.
   (b) A second sheet, parallel to the first, is given an equal but opposite uniform charge. Explain why the electric field now exists only between the planes and find its strength.

Field of nonuniform spherical charge

1.13   A spherical region of radius a contains a total charge Q, distributed nonuniformly with density $[(\lambda + 3)/(4\pi a^3)]Q(r/a)^\lambda$, where λ is some constant (not necessarily an integer, subject to $\lambda > -3$).
   (a) Verify that the total charge is Q.
   (b) Determine the $\mathbf{D}(\mathbf{r})$ field, both outside and inside the sphere.

Field of nonuniform cylindrical charge

1.14   An infinitely long cylindrical region of radius a contains a total charge Q_l per unit length, distributed nonuniformly with density $[(\lambda + 2)/(2\pi a^2)]Q_l(\rho/a)^\lambda$, where λ is some constant (not necessarily an integer, subject to $\lambda > -2$).

(a) Verify that the total charge is Q_l per unit length.
(b) Determine the $\mathbf{D}(\mathbf{r})$ field, both outside and inside the cylinder.

Field of nonuniform layered charge

1.15 An infinitely long and wide layered region of thickness a along the z-axis (perpendicular to the layering) contains a total charge Q_s per unit area, distributed nonuniformly with density $[(\lambda + 1)/a]Q_s(z/a)^\lambda$, where λ is some constant (not necessarily an integer, subject to $\lambda > -1$).
(a) Verify that the total charge is Q_s per unit area.
(b) Determine the $\mathbf{D}(\mathbf{r})$ field, both outside and inside the layered slab. The field should be zero on the plane that has half the total charge in front of it and half behind it.

Cylindrical multipole fields

1.16 The electric flux density field $\mathbf{D}(\mathbf{r})$ is given in cylindrical coordinates by

$$\mathbf{D}(\rho, \varphi, z) = D_0(\rho/a)^m[\hat{\boldsymbol{\rho}} \cos l\varphi + \hat{\boldsymbol{\varphi}} \sin l\varphi],$$

where D_0 and a are constants. How must the integers l and m be related if there is no charge within a sector-shaped region defined by any radii ρ_1 and ρ_2 and any azimuths φ_1 and φ_2 such that $0 < \rho_1 < \rho_2$ and $\varphi_1 < \varphi_2$? (*Comment:* The patterns $\hat{\boldsymbol{\rho}} \cos l\varphi + \hat{\boldsymbol{\varphi}} \sin l\varphi$ are called cylindrical multipole fields; $l = 0$ describes a monopole, $l = 1$ a dipole, $l = 2$ a quadrupole, etc.)

Field lines of cylindrical multipoles

1.17 For visualization of a given vector field $\mathbf{F}(\mathbf{r})$, we often plot a set of "field lines." These are curves in space whose tangent at each point is directed along the field $\mathbf{F}$ at that point. Mathematically, field lines are defined by curves that satisfy the differential equation $d\mathbf{r} \times \mathbf{F}(\mathbf{r}) = 0$, because $d\mathbf{r}$ is tangential to the curve. For example, in cylindrical coordinates, $\mathbf{r} = \mathbf{r}(\rho, \varphi, z) = \rho\hat{\boldsymbol{\rho}}(\varphi) + z\hat{\mathbf{z}}$ and $d\mathbf{r} = d\rho\,\hat{\boldsymbol{\rho}} + \hat{\boldsymbol{\varphi}}\rho\,d\varphi + dz\,\hat{\mathbf{z}}$, using the fact that $d\hat{\boldsymbol{\rho}}/d\varphi = \hat{\boldsymbol{\varphi}}$.

For the *cylindrical dipole field* $\mathbf{F}(\mathbf{r}) = \hat{\boldsymbol{\rho}} \cos\varphi + \hat{\boldsymbol{\varphi}} \sin\varphi$, the differential equation becomes $d\rho \sin\varphi - \cos\varphi\,\rho\,d\varphi = 0$ and $dz = 0$. The first of these two is the same as $d\rho/\rho = \cos\varphi\,d\varphi/\sin\varphi$, or

$$d(\ln\rho)/d\varphi = d(\ln|\sin\varphi|)/d\varphi \quad \text{or} \quad \ln[\rho/|\sin\varphi|] = \text{constant}.$$

The field lines therefore satisfy $\rho = b|\sin\varphi|$; each choice of the constant b yields another field line. A typical dipole field line is shown in Figure P1-3.

Plot a typical field line for the quadrupole field $\mathbf{F}(\mathbf{r}) = \hat{\boldsymbol{\rho}} \cos 2\varphi + \hat{\boldsymbol{\varphi}} \sin 2\varphi$.

Electric field of falling charge

1.18 A heavy particle with charge q is released from rest from an initial height h_0 and falls past an observer at fixed height h and fixed horizontal distance b from the vertical path of the charge.
(a) Find the electric field $\mathbf{E}(t)$ measured by the observer, as a function of time t since the release of the charge. For convenience, express the field in terms of the field strength $E_0 = q/4\pi\varepsilon_0 b^2$ and the times t_0 and t_1 such that $\frac{1}{2}gt_0^2 = h_0 - h$ and $\frac{1}{2}gt_1^2 = b$.
(b) Plot the vertical and horizontal components E_z and E_x versus t, for the case that $b = h_0 - h$.

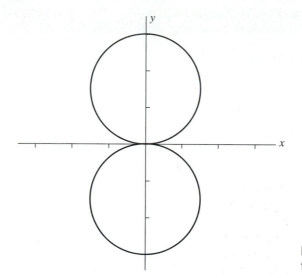

Figure P1-3   Cylindrical dipole
field line.

Electric field of a charged ring

1.19 A thin wire is bent into a large circular ring of radius a and uniformly charged to total charge q. By direct vector superposition, find the electric field at all points along the axis of the ring (the central straight line perpendicular to the plane of the ring).

Charge configuration as harmonic oscillator

1.20 Charges Q_1 and Q_2, both positive, are immobilized, a distance l apart. A third charge q, also positive, is mobile, but only along the straight line joining the other two charges. As found in Problem 1.10, there exists an equilibrium position x_0 between the two charges for this third charge. If the charge q, of mass m, is displaced slightly from the equilibrium location and then released, it will oscillate about the equilibrium point. Verify that the ensuing motion is that of a simple harmonic oscillator and find the frequency ω of the oscillation.

Maximally flat electric field

1.21 Each of two thin wires is bent into a large circular ring of radius a; one is placed parallel to the xy-plane with its center at $z = -h$ and uniformly charged to a total charge of $+q$; the other is placed parallel to the xy-plane with its center at $z = +h$ and uniformly charged to a total charge of $-q$. The resultant electric field $E(z)$ along the z-axis varies with z, but we want to make it as constant as possible in the vicinity of the origin. How should the radius a and the height h of each ring be related in order to make $E(z)$ as flat as possible at $z = 0$?

CHAPTER 2

Ampère's Law, Line Integrals, and Magnetic Fields

As soon as we allow for motion of interacting charges, we face a difficulty: The force law $\mathbf{F} = q\mathbf{E}$ is no longer valid. The force observations that led to the concept of the electric field are valid only when the test charge is stationary. Under conditions termed *magnetization*, test charges in motion fail to respond in accordance with the force exerted by the electric field (which was measured with a stationary test charge). We therefore need to extend the field concept to allow for interactions of moving charges.

MAGNETIC FIELD

Electrified moving bits of matter interact; we say that one moving charge generates, besides the electric field, an additional *magnetic* field and that both the electric and the magnetic field exert forces on the other moving charge. We can separate the two fields by comparing the forces on a test charge at the point in question when the charge is moving and when it is stationary.

We find that the additional, magnetic force exerted on the test charge is, first, proportional to the amount of test charge used and, second, dependent on both the speed and direction of the test charge's motion. With regard to the speed of the test charge, we find a simple proportionality. As to the direction of the charge's motion, we find that, at the location of the moving test charge, there is one direction of motion for which there is no additional magnetic force, regardless of the speed. We can therefore assign a vector to represent the additional magnetic force effect at any point; its direction will be that of the special direction for which there is no magnetic force and its magnitude will account for the proportionality to both the amount and the speed of the test charge. For the magnetic force alone, we now have three vectors to deal with: one in the direction of charge motion, one in the special direction that yields no magnetic force, and one in the direction of the actual magnetic force. The direction of this last vector is found to be perpendicular to the other two and its magnitude is revealed to be proportional to only the component of the motion perpendicular to the special direction. All of these effects are summarized by a magnetic force law expressed as a cross product:

$$\mathbf{F}_m = q\mathbf{v} \times \mathbf{B}, \tag{2.1}$$

where $\mathbf{F}_m$ is the additional magnetic force exerted on the test charge, q is the amount (with sign) of the test charge whose velocity is the vector $\mathbf{v}$, and $\mathbf{B}$ is the magnetic field vector, called the magnetic flux density for reasons to be explored later, whose direction is that of the special direction in space for which no magnetic force arises and whose magnitude accounts for the overall proportionality to charge, speed, and perpendicular component of the motion. The order of $\mathbf{v}$ and $\mathbf{B}$ in the cross product fixes the sense of the vector $\mathbf{B}$ along the special direction of motion that would yield zero force. Figure 2-1 shows the relation among the three vectors.

The total electromagnetic force exerted on a test charge under conditions of both electrification and magnetization of the space is given by the *Lorentz force law*, combining both the electric and the magnetic forces, $\mathbf{F}_e$ and $\mathbf{F}_m$:

$$\mathbf{F} = q\mathbf{E} + q\mathbf{v} \times \mathbf{B}. \tag{2.2}$$

The fields $\mathbf{E}$ and $\mathbf{B}$ are the preexisting fields, which are considered to be present even before we make their existence evident by use of the test charge at the point. The units of a magnetic flux density field are those of the ratio that defines it: Ns/Cm, abbreviated as the tesla (T).

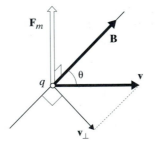

Figure 2-1   Magnetic force as cross product of velocity and field.

To this point, the use we can make of knowledge of a magnetic field is to obtain the force exerted on any charge that may be placed within the field space, provided that the presence of that charge does not disturb the field. Applying Newton's laws of motion, with that force, can then yield the motion of the charge in the field region.

Example 2.1

Suppose that a uniform (in space) and constant (in time), downward magnetic field $\mathbf{B}(\mathbf{r}, t) = \mathbf{B}_0$ *has been set up within a region of space and that an electron enters that space at time* $t = 0$ *at position* $\mathbf{r}_0$, *with horizontal velocity* $\mathbf{v}_0$. *Where will the electron be at time* t? *Assume no electric field is present to compete with the magnetic one.* ◈

Wherever the electron may find itself at time t, if its velocity there and then is $\mathbf{v}(t)$, the force on it at that point will be $\mathbf{F} = -e\mathbf{v} \times \mathbf{B}_0$, so that the equation of motion is

$$m \frac{d\mathbf{v}(t)}{dt} = -e\mathbf{v}(t) \times \mathbf{B}_0. \tag{2.3}$$

Because the right side of this differential equation involves an unknown function of t—namely, $\mathbf{v}(t)$—we cannot immediately integrate the equation to arrive at the velocity of the electron. We need to solve this vector differential equation for the unknown function $\mathbf{v}(t)$ and then we can integrate this velocity to obtain the trajectory of the electron, $\mathbf{r}(t)$.

We first note that the acceleration $d\mathbf{v}/dt$ is perpendicular to the downward $\mathbf{B}_0$ and is therefore horizontal at all times. The initial velocity is also given as horizontal, but the acceleration at $t = 0$ is at right angles to the initial velocity. The motion is therefore in the horizontal plane of the initial velocity $\mathbf{v}_0$ and the initial acceleration $\mathbf{a}_0$; the latter is a known vector and is given by

$$\mathbf{a}_0 = (e/m)\mathbf{B}_0 \times \mathbf{v}_0. \tag{2.4}$$

It is convenient to express the unknown velocity vector $\mathbf{v}(t)$ for all time as a combination of the two orthogonal vectors $\mathbf{v}_0$ and $\mathbf{a}_0$ in the horizontal plane. As illustrated in Figure 2-2, we write

$$\mathbf{v}(t) = f(t)\mathbf{v}_0 + g(t)\mathbf{a}_0, \tag{2.5}$$

where $f(t)$ and $g(t)$ are unknown scalar time functions to be found. The given initial velocity $\mathbf{v}(0) = \mathbf{v}_0$ is recovered by imposing the initial conditions

$$f(0) = 1 \quad \text{and} \quad g(0) = 0. \tag{2.6}$$

To determine the two functions $f(t)$ and $g(t)$, we substitute the trial form of $\mathbf{v}(t)$ into the differential equation, as follows:

$$\frac{d\mathbf{v}}{dt} = \frac{d[f(t)\mathbf{v}_0 + g(t)\mathbf{a}_0]}{dt} = \frac{df(t)}{dt}\mathbf{v}_0 + \frac{dg(t)}{dt}\mathbf{a}_0 \tag{2.7}$$

is to be equated to

$$\frac{e}{m}\mathbf{B}_0 \times \mathbf{v}(t) = \frac{e}{m}\mathbf{B}_0 \times [f(t)\mathbf{v}_0 + g(t)\mathbf{a}_0]$$

$$= f(t)[(e/m)\mathbf{B}_0 \times \mathbf{v}_0] + g(t)[(e/m)\mathbf{B}_0 \times \mathbf{a}_0]. \tag{2.8}$$

The first of the last two terms is just $f(t)\mathbf{a}_0$, so that we have to solve

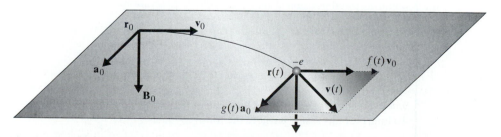

Figure 2-2   Velocity components along initial velocity and acceleration.

$$[df/dt]\mathbf{v}_0 + [dg/dt]\mathbf{a}_0 = f(t)\mathbf{a}_0 + g(t)[(e/m)\mathbf{B}_0 \times \mathbf{a}_0] \tag{2.9}$$

for the unknown functions $f(t)$ and $g(t)$.

Because the left side has been expressed as a combination of the two orthogonal vectors $\mathbf{v}_0$ and $\mathbf{a}_0$, it is convenient to separate its components by dotting the last equation with each of the two vectors. Dotting with $\mathbf{a}_0$ gives

$$[dg/dt]\mathbf{a}_0 \cdot \mathbf{a}_0 = f(t)\mathbf{a}_0 \cdot \mathbf{a}_0 \quad \text{or} \quad dg/dt = f(t), \tag{2.10}$$

since $\mathbf{a}_0$ is orthogonal to the cross product $\mathbf{B}_0 \times \mathbf{a}_0$. Dotting instead with $\mathbf{v}_0$ yields

$$[df/dt]\mathbf{v}_0 \cdot \mathbf{v}_0 = g(t)\mathbf{v}_0 \cdot [(e/m)\mathbf{B}_0 \times \mathbf{a}_0]$$

$$= g(t)[\mathbf{v}_0 \times (e/m)\mathbf{B}_0] \cdot \mathbf{a}_0, \tag{2.11}$$

since the dot and cross can be interchanged in a scalar triple product. We recognize the last term to be $g(t)[-\mathbf{a}_0] \cdot \mathbf{a}_0$, so that we get

$$[df/dt]\mathbf{v}_0 \cdot \mathbf{v}_0 = -g(t)\mathbf{a}_0 \cdot \mathbf{a}_0 \quad \text{or} \quad df/dt = -(a_0^2/v_0^2)g(t). \tag{2.12}$$

We now have two scalar equations for the two unknowns $f(t)$ and $g(t)$:

$$dg/dt = f(t) \quad \text{and} \quad df/dt = -(a_0^2/v_0^2)g(t), \tag{2.13}$$

subject to the initial conditions $f(0) = 1$ and $g(0) = 0$. The two differential equations can be combined into the familiar form

$$\frac{d^2f}{dt^2} = -\frac{a_0^2}{v_0^2}f(t), \quad \text{with } f(0) = 1 \text{ and } \frac{df(0)}{dt} = 0. \tag{2.14}$$

The solution is the trigonometric function

$$f(t) = \cos\omega_0 t, \quad \text{with} \quad \omega_0 = a_0/v_0, \tag{2.15}$$

and, from $g(t) = -(v_0^2/a_0^2)\, df/dt$,

$$g(t) = \frac{\sin\omega_0 t}{\omega_0}. \tag{2.16}$$

This gives the velocity at any time as

$$\mathbf{v}(t) = \cos\omega_0 t\, \mathbf{v}_0 + \frac{\sin\omega_0 t}{\omega_0}\mathbf{a}_0. \tag{2.17}$$

Since the velocity $\mathbf{v}(t) = d\mathbf{r}/dt$ is now known as an explicit function of time, we can obtain the orbital position at time t directly by integration:

$$\mathbf{r}(t) = \mathbf{r}_0 + \int_0^t \mathbf{v}(t')\, dt'$$

$$= \mathbf{r}_0 + \frac{\sin\omega_0 t}{\omega_0}\mathbf{v}_0 + \frac{1 - \cos\omega_0 t}{\omega_0^2}\mathbf{a}_0, \tag{2.18}$$

where the integration constant has been evaluated as the given initial position $\mathbf{r}(0) = \mathbf{r}_0$. To interpret this result, we note the following facts.

1. The motion is periodic: The electron returns to its initial location and repeats its trajectory every $2\pi/\omega_0 = 2\pi v_0/a_0$ seconds.

2. The electron's speed remains constant: since

$$\mathbf{v}(t) \cdot \mathbf{v}(t) = \cos^2 \omega_0 t \, \mathbf{v}_0 \cdot \mathbf{v}_0 + \sin^2 \omega_0 t \, \mathbf{a}_0 \cdot \mathbf{a}_0/\omega_0^2$$
$$= \left(\cos^2 \omega_0 t + \sin^2 \omega_0 t\right) v_0^2 = v_0^2, \tag{2.19}$$

the speed remains v_0 for all time.

3. If we average out the sinusoidal oscillations in time, we are left with an "average" location for the electron:

$$\langle \mathbf{r}(t) \rangle = \mathbf{r}_0 + \mathbf{a}_0/\omega_0^2. \tag{2.20}$$

The electron is never at this average location but oscillates about that point.

4. The electron's distance from this average location never changes: Its displacement from there is

$$\mathbf{r}(t) - \langle \mathbf{r} \rangle = \sin \omega_0 t \left[\mathbf{v}_0/\omega_0\right] - \cos \omega_0 t \left[\mathbf{a}_0/\omega_0^2\right], \tag{2.21}$$

so that the square of its distance from the average location is

$$\left|\mathbf{r} - \langle \mathbf{r} \rangle\right|^2 = \sin^2 \omega_0 t \left[v_0^2/\omega_0^2\right] + \cos^2 \omega_0 t \left[a_0^2/\omega_0^4\right]. \tag{2.22}$$

Because $v_0^2/\omega_0^2 = a_0^2/\omega_0^4$, this becomes

$$\left|\mathbf{r} - \langle \mathbf{r} \rangle\right|^2 = \left(\sin^2 \omega_0 t + \cos^2 \omega_0 t\right) v_0^2/\omega_0^2 = \left(v_0/\omega_0\right)^2 \tag{2.23}$$

and the distance from the average location remains forever v_0/ω_0.

We conclude from these facts that the electron's motion traces out a circle in the horizontal plane, with its center at $\mathbf{r}_0 + \mathbf{a}_0/\omega_0^2$ and with radius v_0/ω_0. It travels along the circle at constant speed, with constant angular velocity ω_0. This is

$$\omega_0 = (e/m)B_0, \tag{2.24}$$

because $\omega_0 = a_0/v_0 = \left|(e/m)\mathbf{B}_0 \times \mathbf{v}_0\right|/v_0 = (e/m)B_0$ for perpendicular $\mathbf{v}_0$ and $\mathbf{B}_0$. The stronger the externally applied magnetic field $\mathbf{B}_0$, the faster the rotation and the smaller the circle. The circular motion persists as long as the electron remains in the undisturbed, uniform applied field. Figure 2-3 depicts this type of trajectory, called *cyclotron motion*; it forms the basis for the operation of the cyclotron, a device in which electrons are maintained in circular orbits.

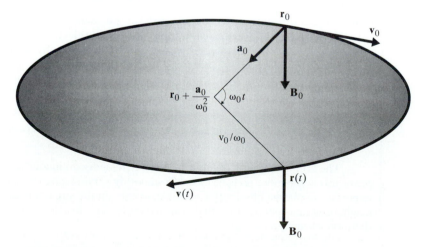

Figure 2-3 Cyclotron orbit of electron in uniform magnetic field.

This example illustrates that a magnetic field, by itself, cannot alter the speed (or energy) of a charge, as is evident from the fact that the force it exerts is perpendicular to the motion of the charge. In a subtle way, this aspect of the effect of the field is modified in the time-varying case, as an electric field must then accompany the magnetic one, as we shall see, and this can change the particle's energy.

The example also demonstrates that charges moving in a uniform magnetic field tend to be "captured," in that they revolve around some fixed location, without going off to infinity across the field. This feature is lost, however, if the magnetic field is not uniform, in that the charge can then drift away. The analysis that confirms this can be rather complicated, but the following example illustrates the drift effect in a relatively simple way.

Example 2.2

To gain some appreciation for the effects of nonuniformity of the magnetic field on charge motion, imagine that we can set up a magnetic field comprised of two uniform regions: $(e/m)\mathbf{B} = -\hat{\mathbf{z}}\omega_1$ for $x < 0$ and $(e/m)\mathbf{B} = -\hat{\mathbf{z}}\omega_2$ for $x > 0$. An electron is at the origin at time $t = 0$, with velocity $\mathbf{v}(0) = v_0\hat{\mathbf{x}}$.

  (a) *Obtain the orbit of the electron; verify that it is periodic in space.*

  (b) *Verify that the motion includes a drift in a direction perpendicular to the magnetic field.*

  (c) *How far does the electron drift within one spatial period? How long does it take to complete one period? What is the drift velocity (averaged over a period)?*

  (d) *Sketch the orbit, over a few periods, for $\omega_2/\omega_1 = 4.5$.* ◈

  (a) The motion begins as a cyclotron orbit in $x > 0$ with frequency ω_2 and speed v_0. After one semicircle, the electron enters $x < 0$ and executes a gyration with frequency ω_1, still with speed v_0. After a semicircle of this gyration, the electron reenters $x > 0$ with the same speed and repeats the process. The orbit is a periodic sequence of a semicircle of radius v_0/ω_2 followed by a semicircle of radius v_0/ω_1, in the xy-plane.

  (b) The electron enters the ω_2 region at the origin, then exits at $y = -2v_0/\omega_2$. The diameter of the next semicircle is $2v_0/\omega_1$, so that the electron reenters the ω_2 region at $y = -2v_0/\omega_2 + 2v_0/\omega_1 \neq 0$, so that there is a drift in the y-direction.

  (c) The first semicircle is traced in time π/ω_2; the next in time π/ω_1. The drift in one period is $\Delta y = 2v_0[1/\omega_1 - 1/\omega_2]$; the time is $\Delta t = \pi[1/\omega_1 + 1/\omega_2]$ for one period; the average drift speed is $\Delta y/\Delta t = (2/\pi)v_0[(\omega_2 - \omega_1)/(\omega_2 + \omega_1)]$. Note that there is no drift if $\omega_2 = \omega_1$, which would make the field uniform.

  (d) Figure 2-4 shows the trajectory, for four periods; it is formed of a succession of alternately small and large semicircles. The drift is obvious.

The drift illustrated here for a magnetic field of discontinuous spatial variation appears also in the more realistic case of a smoothly varying magnetic field. The effect is important in plasma physics (for example, to explain the streaming of electrons in the magnetized ionosphere, or in attempting to confine ions in a device intended to achieve thermonuclear fusion).

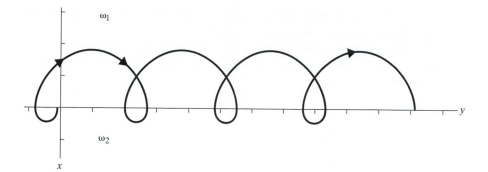

Figure 2-4   Four periods of electron in stepped **B** field, for $\omega_2/\omega_1 = 4.5$.

We note again that the effect of the magnetic field on the charge is determined by the value of the field at the location of the charge, at the instant when it is there. In this respect, the force exerted by the magnetic field on the charge is like a contact force and like the electric force. The sources of the electric and magnetic fields, elsewhere and earlier, need not be considered to find their effect here and now. If charge q, of mass m, is in the presence of electric field $\mathbf{E}(\mathbf{r}, t)$ and of magnetic field $\mathbf{B}(\mathbf{r}, t)$ and is at location $\mathbf{r}$ at time t, then its location $\mathbf{r}$ will change in time to become a trajectory $\mathbf{r}(t)$ in accordance with the equation of motion

$$m\frac{d^2\mathbf{r}(t)}{dt^2} = q\mathbf{E}(\mathbf{r}(t), t) + q\frac{d\mathbf{r}(t)}{dt} \times \mathbf{B}(\mathbf{r}(t), t). \tag{2.25}$$

The time t is the same everywhere in the equation; the position $\mathbf{r} = \mathbf{r}(t)$ is the same everywhere it appears, on both sides of the equation. There is no action at a distance and there is no delay in the interaction of the fields with the charge. Cause and effect are really separated, in both space and time, but that feature is the province of the process of generation of the fields in the space around the sources, not that of the calculation of the fields' effects on the test charge.

EFFECT OF THE MEDIUM

When we investigate how the magnetic field is generated by the source charges, we find that, just as the magnetic field exerts a force on moving charges, the source of the magnetic field is a moving charge as well. We suppose we are given the amounts, locations, and motions of various source charges and we wish to determine the field that is set up by these charges. Just as in the case of the purely electric field, we find the process to be different in different media. To determine the force field around some source charges, we need to know not only the amounts, locations, and motions of the sources but also the medium that forms their environment.

We deal with this complication, just as we did in the electric case, by extending the magnetic field concept to include another intermediary field. We assert that the moving source charges generate a field **H**, while the moving test charges are affected by the **B** field; the medium dictates how the fields **H** and **B** are related. The **H** field is called the *magnetic intensity*.

The relation between **B** and **H** is often trivially simple but can become rather complicated, depending on the constitution of the material medium. In simple media, ignoring fine detail on the scale of atomic dimensions and provided the time variation of the fields is sufficiently slow, the medium relates **B** to **H** locally and instantaneously. That is, to determine the **B** field at some point and some instant, it suffices to know the field **H** at only that point and instant, not elsewhere or at other times. In such media, the **B** field is just some function of the **H** field, $\mathbf{B} = \mathbf{B}(\mathbf{H})$. In the most simple types of medium, this function is merely a proportionality, with a factor μ called the permeability of the medium, as

$$\mathbf{B} = \mu\mathbf{H}. \tag{2.26}$$

On a macroscopic basis, describing a medium by giving its permeability may be adequate for our purposes. We may, however, need to allow the permeability μ to vary from point to point, as $\mu = \mu(\mathbf{r})$, if the medium is not homogeneous throughout space. We may even need it to be a function of time, $\mu = \mu(\mathbf{r}, t)$, if this time variation is very slow, as when the material is changing and the fields' time variation is also slow. We may also have to allow for the direction of the **B** vector to differ from that of the **H** field, while still retaining the simple, linear proportionality. This happens in crystals; the permeability then becomes a tensor $\boldsymbol{\mu}$ instead of merely a scalar μ. In all these cases, the proportionality between **B** and **H** is still local and instantaneous, as

$$\mathbf{B}(\mathbf{r}, t) = \boldsymbol{\mu}(\mathbf{r}, t)\mathbf{H}(\mathbf{r}, t), \tag{2.27}$$

which involves the fields and the permeability at only the point **r** and the time t at which **B** is to be found.

The more realistic description of a real medium, on a more microscopic basis, is nonlocal and noninstantaneous (dynamic), in that the **B** field here and now depends on the **H** field elsewhere and at other times as well as here and now. For example, a dynamic relation for a linear medium may be expressed as a convolution of the magnetic field with the magnetic impulse response of the medium. The convolution effectively sums up the effects at the present time of the magnetic field over its entire past history, up to the present time. Details about convolutions appear in the Appendix in Chapter 10.

An even more general linear relation would involve a spatial convolution as well as a temporal one. The practical way to deal with the dynamics expressed by the convolution is to rewrite the relation between **B** and **H** in the frequency domain; that is, in the sinusoidal steady state, one frequency at a time. The relation thereby reverts to a simple multiplicative factor μ, rather than the convolution integral in the time domain. The only new feature is that the factor $\mu = \mu(\omega)$ then depends on the frequency ω.

Most ordinary materials are nonmagnetic, behaving magnetically like air. For sufficiently strong fields and for ferromagnetic materials, the relation between the two magnetic fields **B** and **H** becomes nonlinear and the deviation from a simple proportionality has to be described by some more general function

$$\mathbf{B} = \mathbf{B(H)}. \tag{2.28}$$

For many materials of engineering importance, this function is not even single valued, exhibiting a phenomenon called hysteresis that makes the relation between the two fields dependent on the magnetization history of the material.

We will account for the dynamic relation between **B** and **H**, without losing the simple proportionality factor μ, when we deal with the sinusoidal steady state. We will restrict ourselves to media simple enough to be described magnetically by just a permeability μ. This will be treated as a constant if we are in the time domain, or else as a function of frequency if in the steady state.

In a vacuum, with no material medium present at all, there is no need to distinguish between the **H** and **B** fields. However, it is more convenient not to make an exception of that particular medium and to retain both fields even in that case. For a vacuum, the distinction between **B** and **H** becomes only that the units in which we measure the two types of magnetic fields are different, so that a units conversion factor is all that is needed to convert from one to the other. This units conversion factor is designated μ_0 and called the *permeability of free space*, on the pretense that empty space exhibits the physical property of permeability that material media do.

$$\mathbf{B} = \mu_0 \mathbf{H} \quad \text{(vacuum)}. \tag{2.29}$$

Air behaves nearly as does a vacuum in this respect, as do nonmagnetic materials; we will use μ_0 for all such media.

Figure 2-5 illustrates the influence of the medium and the extension of the scheme that was to express the interaction of one moving charge with another through the agency of the magnetic field, by the addition of another intermediary field, the **H** field. To complete the description of cause and effect, there remains to determine the laws governing the generation of the **H** field by the moving source charges.

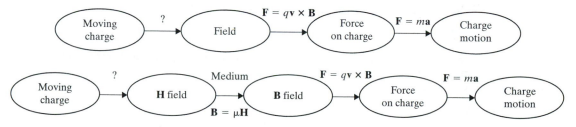

Figure 2-5　Extension of schematic magnetic interaction of charges.

MAGNETOMOTIVE FORCE

The law of generation of the **H** field by moving source charges is found to involve a particular entity pertaining to the field, called its magnetomotive force, abbreviated mmf (we should appreciate the abbreviation, as it is less misleading than the full name: The quantity in question is not a force and has no motive power!). The mmf is what is directly related to the motion of the source charges; the field distribution in space is determined, indirectly, from the mmf. What is the mmf of a magnetic field?

While a field is specified point by point, there are significant quantities that pertain to a field that can be associated with certain collections of points in space that form a curve, or a surface, or a volume. An important relationship between a vector field, such as **H**, and a collection of points that forms a curve C in space is expressed by the *line integral of* **H** *along the curve C*, also referred to as the *mmf along C*. The mmf is a scalar quantity U, associated with both the field and the curve; it is denoted

$$U = \int_C \mathbf{H} \cdot d\mathbf{l}. \tag{2.30}$$

To emphasize the dependence of the mmf on the curve, we will sometimes append a description or designation of the curve, in brackets, to the symbol for the mmf, as in $U\{C\}$. We proceed now to define and illustrate the significance and methods of evaluation of the mmf.

We will review the formal definition of the line integral and a formal and an informal evaluation procedure for it, using an example that we can solve in several ways. To specify the line integral of a vector field along a curve, we must furnish both a vector field and an oriented curve. The field must be given, as a vector (magnitude and direction, or else components), everywhere, or at least at every point of the curve. The curve must be identified, by specifying all its points, and an orientation for the curve must be selected, so that one direction along it is identified as the positive one.

As with any integral, there is a formal definition as a limit of approximating sums; that definition allows us to apply it to physical situations of interest, but it is rather ill suited to actual evaluation. We will present the formal definition and evaluation procedures separately (Figure 2-6).

Definition of line integral: The definition involves several steps.

1. Break up the curve C into a large number N of small segments. Any small line has both a magnitude and a direction; the direction is the one tangential to the line, with the sense along that direction taken so as to agree with the specified

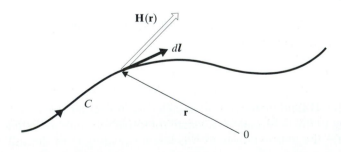

Figure 2-6   Line integration along oriented curve C.

orientation of the curve. Consequently, the mth segment of the curve is given by an infinitesimal vector $d\boldsymbol{l}_m$ that defines its size and orientation.

2. At a typical point $\mathbf{r}_m$ of the mth segment, evaluate the field $\mathbf{H}(\mathbf{r}_m)$.

3. Evaluate the (infinitesimal) scalar $\mathbf{H}(\mathbf{r}_m) \cdot d\boldsymbol{l}_m$. Since $d\boldsymbol{l}_m$ is tangential to the curve, this scalar quantity involves only the tangential component of $\mathbf{H}$ at the point $\mathbf{r}_m$ and measures both the strength of the field and how well it aligns with the curve C at the point.

4. Evaluate this dot product at each of the N segments and form the sum

$$\sum_{m=1}^{N} \mathbf{H}(\mathbf{r}_m) \cdot d\boldsymbol{l}_m.$$

This sum is a number, one of many that could be obtained by this process, by different decompositions of the curve and different choices of the typical point in each segment.

5. The line integral U is defined as the limit of such sums, as $N \to \infty$, but subject to the condition that all the segments be infinitesimal, in the sense that

$$\max |d\boldsymbol{l}_m| \to 0.$$

That the limit of the sums exists, for well-behaved fields and nonpathological curves, is demonstrated by (and for) mathematicians; here we will rely on the usual intuitive ideas of integration that justify the limiting process. We do note, however, that discontinuities in either the field or the curve need not trouble us, as we can have boundaries of the partitions at the points of discontinuity, so that there is no ambiguity at any segment of the partitioned curve. We can therefore calculate the mmf of a discontinuous field as well as of a continuous one and the curve may have corners (i.e., sudden changes of direction).

Physically, the mmf measures how strongly the field aligns itself with the curve, as should be clear from the defining process. The more tangential the field is to the curve, the greater the numerical value obtained for the mmf; if, instead, the field tends to orient itself perpendicular to the curve, the dot product at each such point becomes small and the mmf is then weak. The strength of the field and the length of the curve both also affect the value of the mmf. If the integrand were a force field, the integral would measure the work done on a particle in moving it along the curve under the action of that force.

Formal evaluation: Change the line integral to an ordinary integral, with respect to a variable that parametrizes the curve, as follows.

1. Parametrize curve C as $\mathbf{r} = \mathbf{r}(u)$.

 This means that $\mathbf{r}$, which is a typical point of the curve, is expressed as a function of a parameter, u, that serves to distinguish one point of the curve from another. That one parameter suffices to locate a point on the curve corresponds to the fact that a curve is a one-dimensional object.

2. On the curve C, the integrand $\mathbf{H}(\mathbf{r})$ becomes $\mathbf{H}(\mathbf{r}(u))$, which is a vector function of the parameter u.

3. Express the differential element of length $d\boldsymbol{l}$ in terms of u, with du as the differential:

$$dl = \frac{\partial \mathbf{r}}{\partial u}\, du. \tag{2.31}$$

This vector is tangential to the curve at the point $\mathbf{r}(u)$; the sense along the tangential direction should conform to the selected orientation of the curve, by ensuring that the parameter u increases as the curve is traversed in the positive direction.

4. Form the required dot product of $\mathbf{H}$ with $d\boldsymbol{l}$, as

$$\mathbf{H} \cdot dl = \mathbf{H}(\mathbf{r}(u)) \cdot \frac{\partial \mathbf{r}}{\partial u}\, du \equiv f(u)\, du. \tag{2.32}$$

The integrand $f(u) = \mathbf{H} \cdot (\partial \mathbf{r}/\partial u)$ in front of the infinitesimal du is entirely expressed as a function of u.

5. The line integral U is now evaluated as the ordinary integral of $f(u)$:

$$U = \int_C \mathbf{H} \cdot dl = \int \mathbf{H}(\mathbf{r}(u)) \cdot \frac{\partial \mathbf{r}}{\partial u}\, du \equiv \int f(u)\, du, \tag{2.33}$$

with limits on the integration suitably set to cover the full curve C.

In practice, we use a combination of the informal and the formal evaluation procedures: We break up the curve C into large, not infinitesimal, segments along which the tangential component of $\mathbf{H}$ is constant, or at least easily integrated, and we add the contributions from each such segment. We may use different parametrizations for different portions of the entire curve.

Before proceeding to an example, we note that an especially significant version of a line integral is the *closed line integral*, which is denoted $\oint \mathbf{H} \cdot dl$. The only distinction here is that the curve C is closed and has no end; (i.e., the end points of the oriented curve coincide). Note that the closed curve may have corners (abrupt changes in direction) without having ends.

Example 2.3

To illustrate the calculation of a line integral, we need to give both a vector field as the integrand and a curve as the domain of integration. We ask for the closed line integral of the field $\mathbf{F}(\mathbf{r}) = \mathbf{r} \cdot \hat{\mathbf{x}}\hat{\mathbf{y}}$ along the curve C formed by a semicircle and its diameter, in the xy-plane, with center at the origin. The circle has radius a and is oriented counterclockwise. ◈

The curve is shown in Figure 2-7, on the left side.

Digression on notation: The way that the field to be integrated has been expressed should not, but may, cause some confusion. We wanted to write it as an explicit function $\mathbf{F}(\mathbf{r})$ of the position vector $\mathbf{r}$; it can also be written, less explicitly so, as $\mathbf{F}(x, y, z) = x\hat{\mathbf{y}}$, since

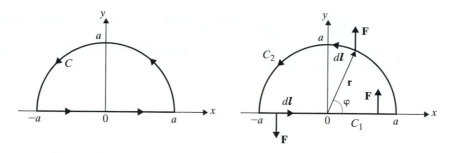

Figure 2-7   Closed semicircular curve; field at typical points.

$\mathbf{r} \cdot \hat{\mathbf{x}} = x$, and simply means that $\mathbf{F}$ at any point $\mathbf{r}$ is in the direction of $\hat{\mathbf{y}}$, with an amplitude given by the x-coordinate of the point. Figure 2-7 includes illustrations of this field at a few points.

The more mathematically inclined reader may choose to think of the expression for the row vector $\mathbf{F}(\mathbf{r})$ as $\mathbf{r} \cdot \mathbf{G}$, where $\mathbf{G}$ is a matrix, here given as $\mathbf{G} = \hat{\mathbf{x}}\hat{\mathbf{y}}$. This is neither a dot product (which would be a scalar), nor a cross product (which would be a vector), but rather an outer product, or dyadic, actually simply a matrix. It is calculated as the ordinary matrix product of the column vector $\hat{\mathbf{x}}$ with the row vector $\hat{\mathbf{y}}$, forming the square matrix $\mathbf{G}$. Upon dotting the row vector $\mathbf{r}$ with the matrix $\mathbf{G}$, we get the row vector $\mathbf{F} = \mathbf{r} \cdot \hat{\mathbf{x}}\hat{\mathbf{y}} = x\hat{\mathbf{y}}$. We therefore can interpret $\mathbf{F} = \mathbf{r} \cdot \hat{\mathbf{x}}\hat{\mathbf{y}}$ as either $(\mathbf{r} \cdot \hat{\mathbf{x}})\hat{\mathbf{y}}$ or $\mathbf{r} \cdot (\hat{\mathbf{x}}\hat{\mathbf{y}})$ and get the same result either way. Refer to the Appendix in Chapter 10 for more details about dyadics.

The less mathematically inclined reader can safely ignore all of this cultural enrichment, as long as it is understood that the field to be integrated is given by $\mathbf{F} = x\hat{\mathbf{y}}$. **End of digression.**

Following the *practical* approach, we divide the closed curve into its two major portions, the diameter and the circular arc. Along the diameter C_1, there is no contribution to the integral, because the field $\mathbf{F}$ is along $\hat{\mathbf{y}}$ and the line element is along $\hat{\mathbf{x}}$, so that the dot product vanishes. Along the semicircle C_2, the field and the line element both vary and we proceed to parametrize the curve.

An easy way to distinguish points along the circular arc is by specifying the angle φ. From the origin to a *typical* point on the arc at angle φ is a vector displacement

$$\mathbf{r} = \mathbf{r}(\varphi) = \hat{\mathbf{x}}a \cos \varphi + \hat{\mathbf{y}}a \sin \varphi, \qquad 0 < \varphi < \pi. \tag{2.34}$$

It follows that, on C_2, the line element is

$$d\mathbf{l} = [d\mathbf{r}/d\varphi]\, d\varphi = [-a \sin \varphi \hat{\mathbf{x}} + a \cos \varphi \hat{\mathbf{y}}]\, d\varphi \tag{2.35}$$

and the field becomes

$$\mathbf{F}(\mathbf{r}(\varphi)) = \mathbf{r}(\varphi) \cdot \hat{\mathbf{x}}\hat{\mathbf{y}} = a \cos \varphi \hat{\mathbf{y}}. \tag{2.36}$$

The required dot product on C_2 is therefore

$$\mathbf{F} \cdot d\mathbf{l} = a \cos \varphi \hat{\mathbf{y}} \cdot [-a \sin \varphi \hat{\mathbf{x}} + a \cos \varphi \hat{\mathbf{y}}]\, d\varphi = a^2 \cos^2 \varphi \, d\varphi \tag{2.37}$$

and the closed line integral reduces to the ordinary integral

$$\oint_C \mathbf{F} \cdot d\mathbf{l} = \int_{C_1} \mathbf{F} \cdot d\mathbf{l} + \int_{C_2} \mathbf{F} \cdot d\mathbf{l} = 0 + \int_0^\pi a^2 \cos^2 \varphi \, d\varphi = \tfrac{1}{2}\pi a^2, \quad (2.38)$$

upon recognizing that $\cos^2 \varphi$ has an average value of $1/2$, or else upon using a trigonometric identity to simplify the integrand.

For the reader who knows too much, we note that $U_1 = \int_{C_1} \mathbf{F} \cdot d\mathbf{l} = 0$ is *not* the

same as $U_2 = \int_{-C_2} \mathbf{F} \cdot d\mathbf{l} = -\tfrac{1}{2}\pi a^2$ (the notation $-C_2$ is intended to represent the path C_2 reversed in orientation), despite the fact that both line integrals begin and end at the same two points, the ends of the diameter, but merely proceed along two different paths. The line integral generally *does* depend on the path from the initial to the final point. There are important special cases, however, for which the line integral is independent of the path between two given points; this case is not one of those special ones.

As an alternative approach to this same problem, we illustrate its solution by an advanced method that takes advantage of *Stokes's theorem*, which allows conversion of a closed line integral of a field into a surface integral of the curl of that field, over a surface whose edge is the closed curve. The theorem states

$$\oint_C \mathbf{F} \cdot d\mathbf{l} = \int_S \nabla \times \mathbf{F} \cdot d\mathbf{S}, \quad (2.39)$$

where the closed curve C is the edge of the open surface S and the curve and surface are mutually oriented by the right-hand rule. That rule states that the positive orientations of the surface and of the closed curve are to be in the same relation to each other as that of the thumb and fingers, respectively, of the right hand.

Example 2.4

Evaluate $U\{C\} = \oint_C \mathbf{r} \cdot \hat{\mathbf{x}}\hat{\mathbf{y}} \cdot d\mathbf{l}$ over the curve C defined by the semicircle and its diameter, in the xy-plane, centered at the origin, of radius a and oriented counterclockwise, as in the previous example, by use of Stokes's theorem. ◈

We need the curl of the vector field; here we want $\nabla \times (x\hat{\mathbf{y}})$. We can get this by direct evaluation of the curl in rectangular coordinates, or else by conversion of this curl of the product of x and $\hat{\mathbf{y}}$ to the gradient of x crossed with $\hat{\mathbf{y}}$; either method yields

$$\nabla \times (x\hat{\mathbf{y}}) = \hat{\mathbf{z}}. \quad (2.40)$$

Consequently, application of Stokes's theorem requires merely the following conversion to a surface integral, in this case of a constant vector $\hat{\mathbf{z}}$:

$$\oint_C x\hat{\mathbf{y}} \cdot d\mathbf{l} = \int_A \nabla \times (x\hat{\mathbf{y}}) \cdot d\mathbf{S} = \int_A \hat{\mathbf{z}} \cdot d\mathbf{S} = \int dA = A = \tfrac{1}{2}\pi a^2, \quad (2.41)$$

where $d\mathbf{S} = \hat{\mathbf{z}} \, dA$ (by the right-hand rule) and A is the area of the semicircle. This confirms the result found by the other method and shows that it was no coincidence that the outcome was exactly the area of the semicircle.

This approach works only for a closed line integral and is useful when the curl of the field is a sufficiently simple expression to integrate over some surface bounded by the curve. We may use this approach also for an open line integral, provided we can close the open curve with a conveniently simple additional auxiliary curve, readily evaluate the surface integral of the curl, and then subtract from this the excess contribution from the auxiliary curve.

Note that the line integral $U = \int_C \mathbf{H} \cdot d\mathbf{l}$ of the vector field $\mathbf{H}$ is itself not a vector field. It is a scalar and it is not a field. It would be a field if it were linked with individual points in space; instead, it relates to an entire curve C, a particular collection of points in space. It therefore represents a higher-order entity than does a field.

AMPÈRE'S LAW

Now that we are expert at evaluating line integrals, we can describe the law governing the generation of the $\mathbf{H}$ field by moving source charges. We find that this involves the mmf along a closed curve and relates to the total flow of charge "through" that curve. The motion or flow of charge constitutes a current. Current is a scalar quantity associated with a surface S and with a flow of charge: It is the rate at which charge crosses the surface, from the negative to the positive side. It is measured in amperes (A); one ampere equals one C/s. The law of generation of mmf, and hence of magnetic field, is an experimentally derived result, expressed by *Ampère's law*:

For *any* closed curve C, the mmf $U = \oint_C \mathbf{H} \cdot d\mathbf{l}$ equals the total current I that crosses any surface S whose edge is the curve C and whose orientation conforms to the right-hand rule:

$$U\{C\} = I\{S\} \qquad \text{or} \qquad \oint_C \mathbf{H} \cdot d\mathbf{l} = I. \tag{2.42}$$

The curve C "surrounds" the current $I\{S\}$, in the sense that C is the edge of the surface S across which the current is I, as in Figure 2-8.

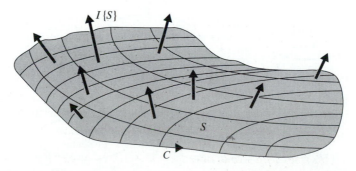

Figure 2-8 Curve C surrounds current I across surface S.

The mmf is a scalar quantity associated with both the **H** field and the curve C. The total current surrounded by C (crossing S) is the algebraic sum of all the charge flow; this yields the net current, with negative charge flow (negative charge moving in the positive direction or positive charge running in the negative direction) subtracted from the positive flow.

Interpretation. The utility and interpretation of Ampère's law comes in two guises. One applies when the field **H** is known along a closed curve, whereupon the law allows us to calculate the net current surrounded by the curve. The other reigns when it is the current across any surface that is known, while the **H** field is unknown; the law then becomes an integral equation to be solved for the integrand, given the integral. Both aspects are useful; the second one presents the greater challenge and justifies thinking of Ampère's law as governing the generation of the **H** field by the flowing source charges. This process is an indirect one: According to the law, the source current determines the mmf U for *any* closed curve surrounding it. This *indirectly* determines what the detailed point-by-point **H** field distribution in space must be. Ampère's law imposes a severe requirement on the way the field **H(r)** can vary in space, because it holds for *any* closed curve.

DETERMINATION OF FIELD FROM CURRENT

Knowing I in the relation $U = \oint \mathbf{H} \cdot d\boldsymbol{l} = I$, can the integrand $\mathbf{H} = \mathbf{H(r)}$ be extracted? Generally, knowledge of the integral is not sufficient to determine the integrand; we need further information about the field. We are concerned with developing the missing link from the mmf U to the field **H**, as indicated in Figure 2-9. In general, we need some restrictions on the field, besides its mmf along any closed curve. This can be a symmetry condition or, as we will see, it can be a superposition principle or, most generally, information about another type of field integral, the surface integral.

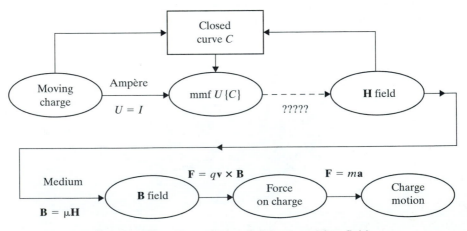

Figure 2-9   The missing link from mmf to field.

There are cases in which *almost* everything is known in advance about the field and Ampère's law furnishes the missing item of information. This is so especially when we know the pattern of a field but not its strength. One example is a uniform filament of current.

Filamentary Current. Consider current I, concentrated along a cylindrical region thin enough to be considered a filament (thread, or wire, with no thickness). There is nothing else in the universe and the filament has no internal structure. That the current is concentrated implies that any surface, no matter how small, pierced by the filament has the same current I crossing it. This line source generates mmf U, in accordance with Ampère's law, which demands that there be an mmf, exactly equal to I, along any curve surrounding the filament. How can the field $\mathbf{H(r)}$ that integrates to this mmf vary from point to point?

The $\mathbf{H}$ field is a vector at any point in space; we need both its magnitude and its direction everywhere. Its direction is restricted by symmetry considerations. The current tends to generate a magnetic field that surrounds the flowing charge. Since the source, having no internal structure, looks the same from all directions, the *direction* of the generated $\mathbf{H}$ field at any point away from the source point *must* be azimuthal, surrounding the current. Any deviation from the azimuthal direction would imply that there is something in the universe that causes that deviation to be preferred over another. We have specified, however, that there is nothing in the universe, either internal to the filament or external to it, that determines any preferred direction in space, other than the axis along the filament. At any observation point (other than on the filament itself), the only direction that can possibly matter for a field that is to surround the filament is the direction at right angles both to the axis and to the line that joins the observation point with the nearest point on the filament.

Figure 2-10 illustrates why we must reject any direction other than the azimuthal one as a possible field direction at the observation point. If any deviation from the azimuthal direction were suggested, there would arise the question, Why not another deviation instead? The symmetry precludes any answer to this question. We conclude from this symmetry argument that, upon adopting a cylindrical coordinate system with

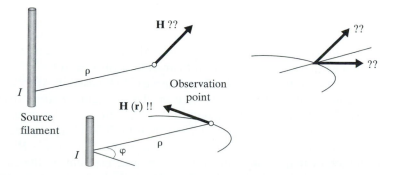

Figure 2-10   The field of a filamentary current can only be azimuthal.

the filament as its axis, the generated **H** field will have to be of the form $\mathbf{H}(\mathbf{r}) = H(\rho, \varphi, z)\hat{\boldsymbol{\varphi}}$, where (ρ, φ, z) are the cylindrical coordinates of the observation point. Having thereby settled on the direction of **H**, we now need information about the magnitude of the field, $H(\rho, \varphi, z)$. How can this magnitude vary with the coordinates?

For the same symmetry reason, the magnitude of the **H** field cannot be different when observed in different azimuthal directions, for what would make the field stronger at one azimuth than at another? The only way one observation point can be distinguished from another is in terms of its distance from the filament. The field strength H can differ at a far observation point from the strength at a near one, but there can be no difference in field strength between two observation points at different azimuths but equidistant from the filament. This symmetry argument imposes the condition that $H = |\mathbf{H}(\mathbf{r})| = H(\rho, \varphi, z)$ cannot depend on the angular coordinate φ, which distinguishes directions in space.

At the same time, the fact that our model considers an infinitely long, uniform current filament does not permit observation points to be distinguished by how far along a direction parallel to the filament they may be. The axial coordinate z of the observation point cannot matter to the field strength; only the radial distance coordinate ρ can distinguish one point from another: $H = H(\rho)$ only. The functional dependence of the field strength H on ρ, $H(\rho)$, is as yet unknown but we know from symmetry alone that $\mathbf{H}(\mathbf{r}) = \mathbf{H}(\rho, \varphi, z) = \hat{\boldsymbol{\varphi}} H(\rho)$. Note that the symmetry argument does not demand that the *vector* **H** be independent of azimuth, and we make no such claim; in fact, we claim that the vector does depend on the angular coordinate φ, since we have convinced ourselves that it must be azimuthal everywhere and the azimuthal unit vector $\hat{\boldsymbol{\varphi}}$ does change with azimuth φ. We claim only that the vector's *magnitude*, $|\mathbf{H}|$, can depend only on the radial distance from the filament, as some unknown function $H(\rho)$. Ampère's law provides the means for determining the missing information about $H(\rho)$.

That the missing information can be extracted is because Ampère's law holds for *any* closed curve. This means *we* can choose the closed curve along which to integrate the **H** field in order to apply the law. Since we will be called upon to integrate a field $H(\rho)$ that we do not know, we should choose a curve along which this unknown function will at least not have an unknown variation; that is, we try to keep this function constant, even if that constant be an unknown value.

The function $H(\rho)$ will remain constant if its argument ρ is kept constant; this occurs on a circle centered on the filament, in a plane perpendicular to it. By choosing such a circle as our closed curve, we maintain the symmetry of the configuration and reduce the unknown function to an unknown constant.

We therefore choose the closed curve C to be used in Ampère's law to be a circle of some fixed radius ρ_0, centered on the filament, at some fixed value of z. At every point of this circle, the magnitude of the **H** field is the same—namely, the unknown but constant value $H(\rho_0)$. Note that the field vector **H** does vary from point to point on the circle, being everywhere azimuthal; it varies as $\mathbf{H} = H(\rho_0)\hat{\boldsymbol{\varphi}}$, but its magnitude does not vary. Furthermore, the line element on the circle we have chosen is also azimuthal at each point of the circle, being tangential to the curve:

$d\boldsymbol{l} = \hat{\boldsymbol{\varphi}}\, dl$. We can therefore evaluate the line integral in terms of the still unknown but constant value $H(\rho_0)$, as follows:

$$U\{C\} = \oint_C \mathbf{H} \cdot d\boldsymbol{l} = \int H(\rho_0)\hat{\boldsymbol{\varphi}} \cdot \hat{\boldsymbol{\varphi}}\, dl = \int H(\rho_0)\, dl$$

$$= H(\rho_0) \int dl = H(\rho_0)l = H(\rho_0)2\pi\rho_0, \qquad (2.43)$$

where $l = 2\pi\rho_0$ is the circumference of the circle of radius ρ_0. We have been able to pull the unknown $H(\rho_0)$ out of the integral only because it is a constant, by virtue of our deliberate choice of the curve C.

By Ampère's law, this mmf $U\{C\} = H(\rho_0)2\pi\rho_0$ [which is still unknown because $H(\rho_0)$ is still being sought] must be equal to the total current across any surface S whose edge is the circle C. Because the circle is centered on the filament, the filament's current I certainly crosses any surface whose edge is the circle, such as the flat circular disk; all of the current is surrounded and there is no other current crossing the surface. We conform to the right-hand rule with our azimuthal orientation of the circle if the positive direction along the filament is taken along the positive z-axis. Assuming that the current I is along the positive z-axis, not opposite to it, we conclude that

$$U\{C\} = I \qquad \text{or} \qquad H(\rho_0)2\pi\rho_0 = I, \qquad (2.44)$$

which yields the unknown constant $H(\rho_0)$ for this particular circle as

$$H(\rho_0) = I/2\pi\rho_0. \qquad (2.45)$$

But this is the result for *any* radius $\rho_0 > 0$. Hence the function $H(\rho)$ must be

$$H(\rho) = I/2\pi\rho \qquad (2.46)$$

and it follows that the $\mathbf{H} = \hat{\boldsymbol{\varphi}}H(\rho)$ field generated by the uniform filamentary current I flowing along the entire z-axis is

$$\mathbf{H}(\mathbf{r}) = \hat{\boldsymbol{\varphi}}\,\frac{I}{2\pi\rho} = \frac{I\,\hat{\mathbf{z}} \times \mathbf{r}}{2\pi|\hat{\mathbf{z}} \times \mathbf{r}|^2}, \qquad (2.47)$$

where ρ is the radial distance from the filament to the field point at $\mathbf{r}$. Figure 2-11 attempts to illustrate the directionality of the field at a particular distance from the filament. At half this arbitrary distance, the field is twice as strong; at three times that distance, it is three times weaker.

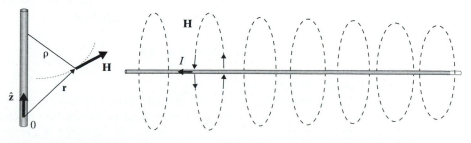

Figure 2-11 Magnetic intensity field around a current filament.

Regarding the units of the various magnetic quantities we have encountered, we note that Ampère's law $U = \oint \mathbf{H} \cdot d\mathbf{l} = I$ requires that the magnetic intensity H be measured in amperes/m (A/m), since the mmf U must have the same units as current. Hence, from the relation $B = \mu_0 H$, we determine the units of μ_0 to be Ns^2/C^2. This is correct but awkward; we will see later that this unit is the same as the henry/m (H/m).

The value of the units conversion factor μ_0 is, to five places,

$$\mu_0 = (1.2566)10^{-6} \text{ H/m} = 1.2566 \ \mu\text{H/m}. \tag{2.48}$$

This quantity is actually *defined* as $(4\pi)10^{-7}$ H/m in the MKSA system, thereby indirectly fixing the size of the ampere or coulomb in terms of mechanically defined units.

CURRENT AND CURRENT DENSITY

Qualitatively, current is the motion or flow of charge. Quantitatively, current is a scalar quantity associated with the flow rate (amount and speed) of charge and with a surface S. Given a surface S, the amount of charge crossing it (from the negative to the positive side of the oriented surface) per unit time is the current I across S. Since the charge has an algebraic sign and the motion can be along or opposite to the orientation of the surface, the current can have positive or negative contributions, in four ways: for example, negative charges flowing backward across the surface contribute positively to the current.

Current $I\{S\}$ is associated with an entire surface S; we may wish, however, to give details of the charge flow in space. A detailed, point-by-point description of the current is given by the *current density* field, $\mathbf{J}(\mathbf{r})$. This is a vector at any point, whose direction is that of the flow of charge and whose magnitude gives the flow per unit area: If we measure the current dI across infinitesimal surface $d\mathbf{S}$ at a point in space, the rate at which charge crosses the surface depends on the degree of alignment between the direction of flow and that faced by the surface element; only the component of the flow along $d\mathbf{S}$ contributes to the current dI, so that we have the dot product

$$dI = \mathbf{J} \cdot d\mathbf{S} \tag{2.49}$$

to define the current density $\mathbf{J}$ at a point.

The surface integral of the current density field over a given surface S is the (algebraic) sum of all the infinitesimal contributions to the flow across the surface and therefore defines the current $I\{S\}$ in terms of the flow field $\mathbf{J}$:

$$I\{S\} = \int_S \mathbf{J} \cdot d\mathbf{S}. \tag{2.50}$$

The units of current density are A/m^2.

On its most fundamental level, current is comprised of moving charges, so that we should be able to relate the current density field to the motion of charges that may be streaming in space. Let us immerse ourselves in the midst of a river of charge and conceptually measure the current dI across an infinitesimal area $d\mathbf{S}$ located at

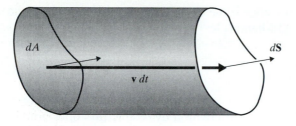

Figure 2-12 Oblique cylinder swept out by dA along **v**.

some point in the flow. Since current is the rate of charge transfer across the surface in time, we count the amount of charge that crosses dS in time dt. As indicated in Figure 2-12, the charges that cross area dA, which faces in the direction of dS, in time dt are contained in an infinitesimal volume dV formed by the oblique cylinder with base dA and height vdt, where **v** is the velocity of the moving charges. This volume would be just base dA times height vdt if it were a right cylinder but, for the oblique case, the volume swept out includes the cosine of the angle between **v** and dS as an alignment factor in $dV = \mathbf{v}\, dt \cdot d\mathbf{S}$. The total charge contained in that volume is $\rho\, dV$ if the charge has density ρ (C/m^3) and this charge dq that has crossed dA in time dt defines the current dI through

$$dq = dI\, dt = \rho\, dV = \rho\mathbf{v} \cdot d\mathbf{S}\, dt. \tag{2.51}$$

Since $dI = \mathbf{J} \cdot d\mathbf{S}$, we find

$$\mathbf{J} = \rho\mathbf{v} \tag{2.52}$$

as the expression for the current density field in terms of the charge density and charge velocity fields, ρ and **v**. Because the charge is conveyed along the flow in this river of charge, this type of current is termed *convection* current, and $\rho\mathbf{v}$ is the convection current density. Note that $\mathbf{v}(\mathbf{r}, t)$ is here the velocity of the charge species (electrons or ions or holes, perhaps) whose density is $\rho(\mathbf{r}, t)$. If there are several species flowing, we superimpose the individual products as current density vectors to form the overall vector $\mathbf{J}(\mathbf{r}, t)$; we do not merely multiply the net charge density by the overall resultant velocity vector, which would have no meaning in the context of summing all the contributions to charge transfer across a surface element.

The total electromagnetic force on a lumped charge q is $\mathbf{F} = q\mathbf{E} + q\mathbf{v} \times \mathbf{B}$ and the charge in volume element dV is $\rho\, dV$; we can therefore express the force on the charge in volume dV as $d\mathbf{F} = (\rho\mathbf{E} + \rho\mathbf{v} \times \mathbf{B})\, dV = (\rho\mathbf{E} + \mathbf{J} \times \mathbf{B})\, dV$, so that the Lorentz force density is

$$\mathbf{f} = \rho\mathbf{E} + \mathbf{J} \times \mathbf{B}, \tag{2.53}$$

in N/m^3, wherever electromagnetic fields and charge and current density fields coexist.

The Ampère law becomes, in terms of the current density field,

$$U\{C\} = I\{S\} \quad \text{or} \quad \oint_C \mathbf{H} \cdot d\mathbf{l} = \int_S \mathbf{J} \cdot d\mathbf{S}, \tag{2.54}$$

for any open surface S whose edge is the closed curve C, with the curve and surface mutually oriented by the right-hand rule.

CONSERVATION OF CHARGE

The physical law of conservation of charge states that charge cannot be destroyed or created. (Even electron–positron pair creation by cosmic rays does not violate this principle, as the two newly created particles carry equal and opposite charges.) Consequently, the only way the amount of charge in a region of space can change is by charge flow into or out of the region. The net amount of charge that crosses the boundary S of a volume V per unit time is precisely what is measured by the current across the surface, or the closed surface integral of the current density,

$$I\{S\} = \oint_S \mathbf{J} \cdot d\mathbf{S}, \tag{2.55}$$

and *conservation of charge* says that this must exactly equal the rate at which the enclosed charge is depleted:

$$I\{S\} = -\frac{dQ\{V\}}{dt} \quad \text{or} \quad \oint_S \mathbf{J} \cdot d\mathbf{S} = -\frac{d}{dt} \int_V \rho \, dV. \tag{2.56}$$

The negative sign indicates depletion, not accumulation, of charge because $d\mathbf{S}$ points outward from the enclosed volume V, by the convention that orients closed surfaces, so that the current $I\{S\}$ is the net outflow from the region, while $Q\{V\}$ is the enclosed charge. A negative outward flow across S would correspond to charge accumulation within the volume V.

A charge configuration can flow and yet be static—that is, not change in time (so that $dQ\{V\}/dt = 0$ for any volume V)—provided that the flow satisfies

$$\oint_S \mathbf{J} \cdot d\mathbf{S} = 0, \tag{2.57}$$

which means that as much current enters the enclosed region as leaves it, instant by instant. Another word that describes this unchanging state is "stationary."

For an example of the use of the law of conservation of charge, consider the charge density and electric field in the following hypothetical, mathematically perfect, hailstorm.

> ### Example 2.5
>
> *A cloud of area A and altitude h drops N hailstones per second, steadily, releasing them with negligible initial velocity. Some unspecified mechanism causes each hailstone to emerge with one electron's charge. Pretend that air resistance can be neglected, so that the stones are in free fall. As the hailstones accelerate, their numbers per unit volume, and hence also the charge density, drop from their original high values.*
>
> **(a)** *What is the hailstones' downward velocity $v(z)$ when they reach height z above the ground?*

(b) *In this steady state, what is the charge density $\rho(z)$ at height z?*

(c) *What is the electric field $E(z)$ in the space between the ground and the cloud (assuming there is no field in the cloud itself)?* ◈

The key concept here is that the current density is the product of the charge density and the charge velocity; we need only relate these two quantities to the given information.

(a) At constant acceleration g, the velocity when distance $(h - z)$ has been traversed, starting from zero speed, is $v(z) = \sqrt{2g(h - z)}$, downward. This follows directly from the equations of motion for constant acceleration, or from equating the gain in kinetic energy $\frac{1}{2}mv^2$ to the loss of potential energy $mg(h - z)$.

(b) The cloud releases current $-Ne$, so the current density is $J = -Ne/A$. Since no charge is created or destroyed in the process of falling, the charge contained between any two values of the height z remains constant and we have a stationary situation. This requires that the current at each of the two heights be the same, so that the current density J must be independent of the height z. But $J = \rho(z)v(z)$, so that $\rho(z) = J/v(z) = -[Ne/A]/\sqrt{2g(h - z)}$.

(c) The total charge between the cloud and height z is $Q(z) = \int_z^h \rho(z')A\,dz' = -Ne\sqrt{(2/g)(h - z)}$. By Gauss's law, this enclosed charge is $-\varepsilon_0 E(z)A$ if there is no field in the cloud. Hence $E(z) = (Ne/\varepsilon_0 A)\sqrt{(2/g)(h - z)}$, directed upward.

Figure 2-13 shows a random sprinkling of hailstones, dense near the cloud level and getting sparser as they accelerate downward. The density varies as the inverse square root of the distance fallen; the speed varies as the square root of that distance; the product is constant.

In contrast to the case just considered, if we encounter a nonzero net current crossing some closed surface, we know that the enclosed charge must be changing in time and that the situation cannot be static, as in the following example.

Example 2.6

Consider a spherically symmetric flow field of the same configuration as the Coulomb field of a point charge, as suggested in Figure 2-14:

$$\mathbf{J}(\mathbf{r}) = \frac{I_0\hat{\mathbf{r}}}{4\pi r^2}. \tag{2.58}$$

This outward flow has the property that its surface integral over a sphere of any radius r_0, centered at the origin, is the same (independent of r_0)—namely, I_0:

$$I\{S\} = \oint \mathbf{J} \cdot d\mathbf{S} = \int [I_0/4\pi r_0^2]\hat{\mathbf{r}} \cdot \hat{\mathbf{r}}\,dA = [I_0/4\pi r_0^2]4\pi r_0^2 = I_0. \tag{2.59}$$

But conservation of charge says that this must be the rate of diminution of the charge enclosed by the sphere of radius r_0, no matter how small r_0 may be! We conclude that there must be a point charge Q at the origin and that this amount of concentrated charge must be "dissolving" or "melting away" at precisely the rate I_0; that is, $dQ/dt = -I_0$ or $Q(t) = Q(0) - I_0 t$. The spherically symmetric outflow pattern can-

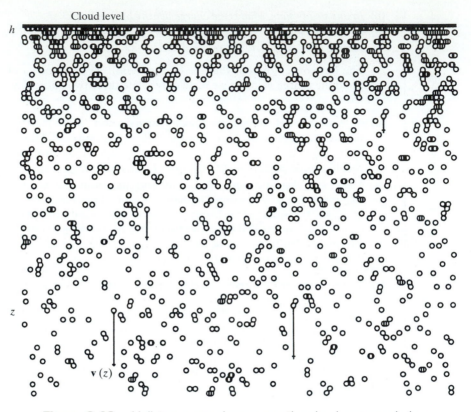

Figure 2-13   Hailstones are dense near the cloud, sparser below.

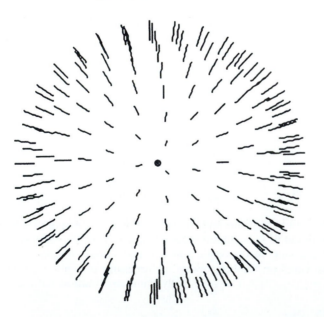

Figure 2-14   Spherically symmetric current density.

not be a stationary configuration, unless the charge at the origin is somehow replenished at just the rate of outflow.

We note, for later use, that the law of conservation of charge—namely, that $I\{S\} = -dQ\{V\}/dt$—and Gauss's law, $\psi\{S\} = Q\{V\}$, can be combined to yield, for *any* closed surface S,

$$I\{S\} + \frac{d\psi\{S\}}{dt} = 0 \quad \text{or} \quad \oint_S \mathbf{J} \cdot d\mathbf{S} + \frac{d}{dt} \oint_S \mathbf{D} \cdot d\mathbf{S} = 0. \quad (2.60)$$

It is also perhaps not too early to issue a warning about Ampère's law. That law is incomplete as stated; it is valid in the stationary case, but a more powerful and important dynamic version will supersede it as soon as we are able to appreciate the present one's inconsistencies.

We can now return to our development of the process of generating a magnetic field. We first note that the spherically symmetric current distribution in the last example has too much symmetry to generate any magnetic field that could satisfy Ampère's law. That law contemplates a magnetic field that *surrounds* a current, but the spherical symmetry precludes any particular axis or special direction around which the magnetic field could possibly form. We cannot come up with any magnetic field configuration consistent with the radial flow, absent a preferred direction in space. Only a radial magnetic field would be consistent with the symmetry, but a radial field surrounds no current.

We can already see an inconsistency rearing its ugly head here. Having convinced ourselves of the indisputable fact that the spherically symmetric flow cannot generate a magnetic field, we realize that just about any open surface would have some current across it in this flow field and should therefore have an mmf along its edge (if we could trust Ampère's law), despite the absence of a magnetic field! The only straw to grasp as we drown in inconsistency is the fact that the situation here is not static, since the charge at the origin is dwindling away at the rate I_0. This is our first warning that, at least in this dynamic or nonstationary situation, Ampère's law is faulty.

MAGNETIC FIELD OF SEMI-INFINITE FILAMENT

If we supplement the preceding spherically symmetric flow field with a filamentary current that feeds exactly I_0 amperes to the origin, we replenish the point charge at the origin at just the rate needed to keep the process stationary. This configuration is used in electrolysis and electroplating, for example.

The filament comes from infinitely far away but ends at the origin; it is straight but only semi-infinitely long. The current that was concentrated within the filament emerges at the origin and flows outward, equally in all directions. (Of course, the infinitesimal trickle in exactly the direction back up the filament is overwhelmed by the finite but concentrated flow down to the origin; we are contemplating an ideal filament of zero thickness and assume that the filament interferes negligibly with the symmetric outflow.)

Now that the situation is stationary, can we rescue Ampère's law from the inconsistency discussed previously? The law does apply to *magnetostatics*, so that we must accept the nonzero mmfs and therefore nonzero **H** fields. But nonzero fields are now acceptable, because now we do have a preferred direction in space, that of the semi-infinite filament that replenishes the disappearing charge. Therefore, the existence of a nonradial magnetic field pattern is no longer inconsistent with the spherical symmetry of the outflow alone. The spherical flow by itself still does not generate any magnetic field, but the straight semi-infinite filamentary current can and does; the field is attributable to the filament alone, although the field-free outflow is needed to bring the current back to infinity, completing the circuit. We wish now to calculate the field generated by the semi-infinite filamentary current I_0.

We still have azimuthal symmetry, as we did for the infinitely long filament, so we again expect the magnetic field to be directed azimuthally at any observation point in space (not on the filament itself). Upon adopting a spherical coordinate system with the filament as its positive z-axis, the generated **H** field is therefore necessarily of the form $\mathbf{H}(\mathbf{r}) = H(r, \theta, \varphi)\hat{\varphi}$, where (r, θ, φ) are the spherical coordinates of the observation point. (Actually, the right-hand rule suggests that the field should be directed along $-\hat{\varphi}$ since the current flows *down* the z-axis, but the negative sign can be incorporated in the coefficient H and the calculation will tell us the correct direction.) The azimuthal symmetry also demands that the magnitude of the **H** field not be different when observed in different azimuthal directions, so that H can only depend on r and θ; points can be distinguished by their distance from the origin and their angle with the axis, but not by their azimuthal angle around the axis. We therefore expect a field pattern of the form $\mathbf{H}(\mathbf{r}) = H(r, \theta)\hat{\varphi}$ and we need to find the function $H(r, \theta)$.

We can choose a closed curve through a given observation point (away from the z-axis) along which the field magnitude is sure to remain constant, thereby maintaining the symmetry of the situation and allowing the mmf to be calculated in terms of one unknown constant. For an observation point at r_0, θ_0, the circle defined by all points for which $r = r_0$ and $\theta = \theta_0$ is horizontal and centered on the z-axis; if we choose the orientation of the line element along $\hat{\varphi}$, the mmf along this circle is just the integral of the (unknown) constant $H(r_0, \theta_0)$ along the circle:

$$U\{C\} = \oint_C \mathbf{H} \cdot d\mathbf{l} = \int H(r_0, \theta_0)\hat{\varphi} \cdot \hat{\varphi}\, dl = H(r_0, \theta_0)2\pi r_0 \sin\theta_0, \quad (2.61)$$

where $2\pi r_0 \sin\theta_0$ is the circumference of the circle, as shown in Figure 2-15.

We can now apply Ampère's law to determine the value of the mmf U and thereby find the constant $H(r_0, \theta_0)$. The mmf equals the total current that crosses any open surface whose edge is the circle specified by r_0 and θ_0. Because we know the current distribution everywhere, we could evaluate this total current over *any* such surface, but we may as well make it easy for ourselves and choose a surface on which the current density field is readily integrated, or even a constant. Because of the spherical symmetry of the outflow, a spherical surface would be easiest to deal with; we therefore choose the spherical cap of radius r_0 and extending only to elevation

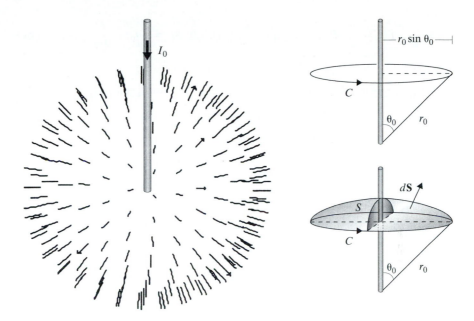

Figure 2-15 Outflow from semi-infinite filament.

angle θ_0, so that its edge is the circle C. We also orient S radially, to conform to the right-hand rule for the azimuthally directed circle, as indicated in Figure 2-15.

The total current $I\{S\}$ that crosses this surface in the radial direction comprises two contributions. There is the concentrated current I_0 in the filament, directed negatively (toward the origin), that crosses at the north pole; there is also the distributed outflow of the current density $\mathbf{J}(\mathbf{r})$. The total current is

$$I\{S\} = -I_0 + \int_S \mathbf{J}(\mathbf{r}) \cdot d\mathbf{S}. \tag{2.62}$$

Both the current density $\mathbf{J}$ and the surface element $d\mathbf{S}$ are directed radially. The area element on the spherical cap is given in our spherical coordinates by

$$d\mathbf{S} = \hat{\mathbf{r}} r_0^2 \sin\theta \, d\theta \, d\varphi, \tag{2.63}$$

with the ranges of the two parameters limited to ensure that the edge of the surface be the closed curve C:

$$0 < \theta < \theta_0, \qquad 0 < \varphi < 2\pi. \tag{2.64}$$

The current density on the surface S is $\mathbf{J} = \left[I_0/4\pi r_0^2 \right]\hat{\mathbf{r}}$, so that $\mathbf{J} \cdot d\mathbf{S}$ reduces to

$$\mathbf{J} \cdot d\mathbf{S} = \left[I_0/4\pi \right] \sin\theta \, d\theta \, d\varphi \tag{2.65}$$

and the total current is

$$I\{S\} = -I_0 + \int_0^{2\pi} \int_0^{\theta_0} \frac{I_0}{4\pi} \sin\theta \, d\theta \, d\varphi = -I_0 + \frac{I_0}{4\pi} 2\pi(1 - \cos\theta_0)$$

$$= -\tfrac{1}{2} I_0 [1 + \cos\theta_0]. \tag{2.66}$$

Equating the mmf and current, in accordance with Ampère's law, we find

$$U\{C\} = H(r_0, \theta_0) 2\pi r_0 \sin\theta_0 = I\{S\} = -\tfrac{1}{2} I_0 [1 + \cos\theta_0], \tag{2.67}$$

so that the unknown constant $H(r_0, \theta_0)$ has been found to be

$$H(r_0, \theta_0) = -\frac{I_0}{4\pi r_0} \frac{1 + \cos\theta_0}{\sin\theta_0} \tag{2.68}$$

This result is valid for *all* r_0 and θ_0 (except $\theta_0 = 0$, which is on the filament, or $r_0 = 0$, which is at its tip). We conclude that the magnetic field pattern for the semi-infinite filament is

$$\mathbf{H}(\mathbf{r}) = H(r, \theta)\hat{\varphi} = -\frac{I_0}{4\pi r} \frac{1 + \cos\theta}{\sin\theta} \hat{\varphi} = \frac{I_0}{4\pi r} \cot\frac{\theta}{2} (-\hat{\varphi}). \tag{2.69}$$

We note that this calculation confirms that the magnetic field is directed in the negative azimuthal direction. We also observe that the field becomes infinite at $\theta = 0$, which is on the filament, but approaches zero at $\theta = \pi$, which is on the z-axis but beyond the end of the filament. If we look at a point very close to the filament, for which $\theta \to 0$, the numerator $[1 + \cos\theta]$ becomes 2 while the denominator $r \sin\theta$ remains just the small radial distance ρ from the filament. We then recover the infinite-filament result $\mathbf{H} \to -\hat{\varphi} I_0/2\pi\rho$, as we would expect if our observation point is so close to the filament that we (and the field pattern) are not "aware" of the far-away tip.

Figure 2-16 hints at the variation of the field strength of the azimuthal magnetic field of a semi-infinite filament, as seen both on a cylinder whose axis includes the filament and also on a sphere centered at the tip of the filament.

CURRENT ELEMENT

To this point, we have succeeded in obtaining the magnetic field generated by an infinite and a semi-infinite current filament, both of which exhibited enough symmetry to enable us to complete the calculation with just Ampère's law. But what can we do if faced with a given current distribution that lacks this sort of symmetry? In the case of generation of electric fields by unsymmetrical charge distributions, we relied on the superposition principle and our early knowledge of the field of a point charge to decompose any distribution into point sources and sum the individual Coulomb fields. For magnetic fields generated by arbitrarily unsymmetrical current distributions, we would like again to invoke the linearity of Ampère's law and utilize superposition, but we do not yet have the magnetic equivalent of a point source and its field.

Since current has direction, even a current distribution concentrated in a single point would have a direction, so that our new point source has zero or infinitesimal

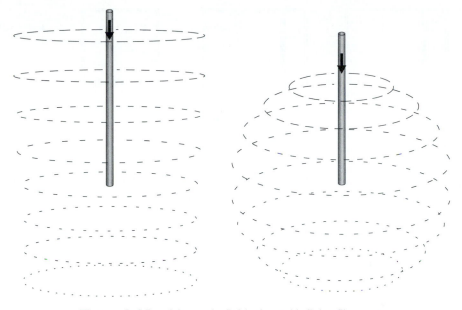

Figure 2-16   Magnetic field of semi-infinite filament.

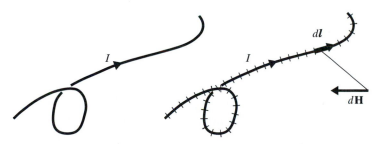

Figure 2-17   Current filament of arbitrary shape.

size but does have a built-in direction. Figure 2-17 shows a current-carrying filament that is twisted in space. To get its magnetic field, we want conceptually to break up the filament into infinitesimal pieces, each of which is to act as a point source. That tiny source retains the direction tangential to the filament at its location, even as its size approaches zero. If we knew the magnetic field generated by any one such "current element," we could obtain the field of the entire filament, or even a more general current distribution, by superposition of the individual fields.

Fortunately, we are already in a position to obtain the magnetic field of a current element: We can use superposition of two semi-infinite current filaments. As shown in Figure 2-18, the two filaments are to carry equal currents but in opposite directions, one toward the tip and the other away from it. Both filaments lie along the z-axis, but their tips are slightly displaced from each other, just above and just below

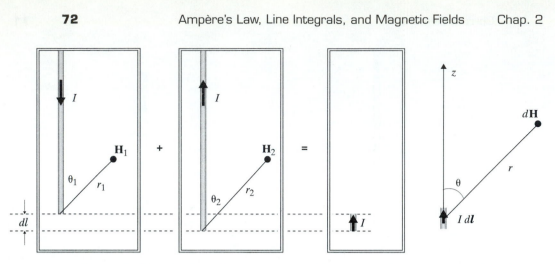

Figure 2-18   Superposition of opposed filaments to form a current element.

the origin. The result is that the two currents in the superimposed configuration *cancel* along all of the z-axis, except for an infinitesimal segment at the origin. We know the magnetic field of each semi-infinite filament, each referred to its own tip as origin for the spherical coordinate system:

$$\mathbf{H}_1(\mathbf{r}) = -\frac{I}{4\pi r_1}\frac{1 + \cos\theta_1}{\sin\theta_1}\hat{\boldsymbol{\varphi}}_1, \tag{2.70}$$

$$\mathbf{H}_2(\mathbf{r}) = +\frac{I}{4\pi r_2}\frac{1 + \cos\theta_2}{\sin\theta_2}\hat{\boldsymbol{\varphi}}_2, \tag{2.71}$$

where the reversal of the direction of the flow in the second case, away from the tip instead of toward it, is accounted for by the change of sign. The superposition of the two configurations, canceling all along the axis except where they were not allowed to overlap, leaves only the current element as source and results in the sum of the two magnetic fields at the observation point:

$$\mathbf{H}(\mathbf{r}) = \mathbf{H}_1(\mathbf{r}) + \mathbf{H}_2(\mathbf{r}) = -\frac{I}{4\pi r_1}\frac{1 + \cos\theta_1}{\sin\theta_1}\hat{\boldsymbol{\varphi}}_1 + \frac{I}{4\pi r_2}\frac{1 + \cos\theta_2}{\sin\theta_2}\hat{\boldsymbol{\varphi}}_2. \tag{2.72}$$

But we note also that, because the z-axes of the two configurations superimpose, the two azimuthal directions coincide. Furthermore, we see that the radial distances from the z-axis of each configuration to the single observation point are the same:

$$\hat{\boldsymbol{\varphi}}_1 = \hat{\boldsymbol{\varphi}}_2 = \hat{\boldsymbol{\varphi}} \quad \text{and} \quad r_1\sin\theta_1 = r_2\sin\theta_2 = r\sin\theta. \tag{2.73}$$

The resultant magnetic field therefore reduces to

$$\mathbf{H}(\mathbf{r}) = \frac{I}{4\pi}\frac{\hat{\boldsymbol{\varphi}}}{r\sin\theta}\left[\cos\theta_2 - \cos\theta_1\right]. \tag{2.74}$$

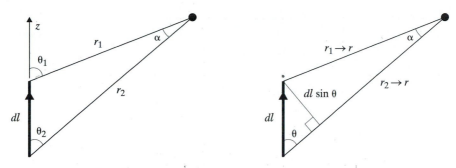

Figure 2-19   Difference between angles in terms of length element.

Now the two angles θ_1 and θ_2 approach equality, each becoming θ, as the length of the uncanceled portion of the superimposed filaments becomes infinitesimal, dl, so that the resultant of the two magnetic fields also becomes infinitesimal. We need the difference, or differential in the limit,

$$\cos\theta_2 - \cos\theta_1 = d[\cos\theta] = -\sin\theta \; d\theta = -\sin\theta[\theta_2 - \theta_1]. \qquad (2.75)$$

Figure 2-19 indicates how to express $d\theta = \theta_2 - \theta_1$ in terms of the differential length dl of the current element. From the triangle formed by the element and the observation point, with external angle θ_1 and internal angle θ_2, we find that $\theta_1 - \theta_2 = -d\theta = \alpha$, which is the angle subtended by the element, as seen from the observation point. Figure 2-19 also shows that the angle α is given by the ratio of the arc length $dl \sin\theta$ to the distance r between the source point and the observation point:

$$\alpha = [dl/r]\sin\theta. \qquad (2.76)$$

Consequently, the resultant magnetic field $\mathbf{H}(\mathbf{r})$ becomes, in the limit, the infinitesimal field $d\mathbf{H}$:

$$\begin{aligned}
d\mathbf{H}(\mathbf{r}) &= \frac{I}{4\pi} \frac{\hat{\boldsymbol{\varphi}}}{r \sin\theta}\left[-\sin\theta\left(-\frac{dl}{r}\sin\theta\right)\right] \\
&= \frac{I \, dl \sin\theta}{4\pi r^2} \hat{\boldsymbol{\varphi}}. \qquad (2.77)
\end{aligned}$$

This is the magnetic field of a current element located at the origin and directed along the z-axis, observed at the point (r, θ, φ). The current element's strength and direction is expressed by the vector *current moment*, $I d\mathbf{l}$, which comprises both the current I and the infinitesimal length dl, in the direction of the element.

We can rewrite this field explicitly in terms of the position vector $\mathbf{r}$ of the observation point with respect to the location of the element as origin, by recognizing that $\hat{\mathbf{z}} \times \mathbf{r} = r \sin\theta \, \hat{\boldsymbol{\varphi}}$ and $d\mathbf{l} = \hat{\mathbf{z}} \, dl$, so that

$$d\mathbf{H}(\mathbf{r}) = \frac{I \, d\mathbf{l} \times \mathbf{r}}{4\pi r^3}. \qquad (2.78)$$

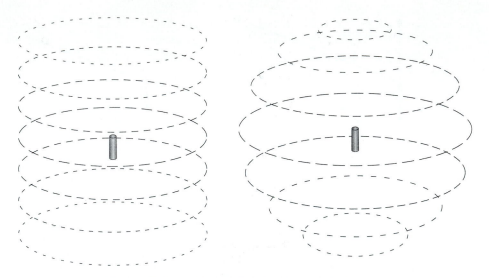

Figure 2-20   Magnetic field of current element.

This, then, is the magnetic field generated by an elementary point source of moment Idl located at the origin. Note that this is the cross product of the moment of the current element with the Coulomb field of a unit point charge. Figure 2-20 suggests how the current element's azimuthal magnetic field strength varies, as seen both on a cylinder whose axis passes through and along the element and on a sphere centered at the element. The field strength varies inversely as the square of the distance from this source but also varies sinusoidally with the angle to the element's direction, being maximum on the equatorial plane but zero along the axis.

SUPERPOSITION OF FIELDS

Ampère's law declares the mmf along a closed curve to be equal to the total (net) current surrounded by, or linked with, the curve. If there are several individual currents linked with the curve, not necessarily all in the same direction, the total current is the algebraic sum of the individual ones. The mmf, too, is an additive quantity, in that the mmf of a sum of individual **H** fields is the sum of the individual mmfs of each one, by the rules of integration. The magnetic field satisfies the *principle of superposition*, in that the **H** field generated by a collection of currents is obtainable by superimposing the fields generated by each current individually (each considered in the absence of the other). The individual contributions are added vectorially. Once again, we rely for this property on the uniqueness of the mmf from any source current.

In truth, we can count on uniqueness only when certain restrictions, beyond that of Ampère's law, apply. In the case of the current filament, we were able to invoke symmetry to guarantee ourselves that the magnetic field we found is the only solution. If we exclude a supremely important electric effect that can also contribute to the mmf, as will be discussed later, we can count on the superposition of magnetic fields of a succession of current elements to be the unique solution for a given set of source currents, and the problem of generation of magnetic fields can be considered to be solved by this approach.

For a current filament of arbitrary shape, the position of the fixed observation point **r** in space relative to the current element at a typical location **r′** on the filament is given by the displacement **r** − **r′**, and the total contribution of all such elements of the filament to the observed magnetic field is

$$\mathbf{H}(\mathbf{r}) = \int_C \frac{I\, d\mathbf{l} \times (\mathbf{r} - \mathbf{r'})}{4\pi |\mathbf{r} - \mathbf{r'}|^3}. \tag{2.79}$$

This is a line integral, along the filament C, but it differs from the line integrals we examined earlier in that it is a *vector* line integral. It is evaluated in the same manner as the scalar version, but we must ensure that the integration (or continuous summation) be a vector summation. This is usually achieved by decomposing the vector integrand into components along constant basis vectors. We must particularly remember never to move a vector (or any other factor, in fact) from inside the integral to outside, unless it is truly constant.

Example 2.7

Find the magnetic field generated by a uniform current I_0 confined to a thin filament that has been bent into a circle of radius a, for any observation point that lies on the perpendicular axis of the filament. 《》

With the z-axis of a cylindrical coordinate system coincident with the axis of the circular filament and the origin at the center of the circle, the observation point is at $\mathbf{r} = z\hat{\mathbf{z}}$ and a typical source point is at $\mathbf{r'} = a\hat{\boldsymbol{\rho}}$ for some azimuth φ. The moment of the current element is $I\,d\mathbf{l} = \hat{\boldsymbol{\varphi}} I_0 a\, d\varphi$, as indicated in Figure 2-21. The vector displacement between source and observation points is $\mathbf{r} - \mathbf{r'} = z\hat{\mathbf{z}} - a\hat{\boldsymbol{\rho}}$, with magnitude $|\mathbf{r} - \mathbf{r'}| = (z^2 + a^2)^{1/2}$, and the cross product

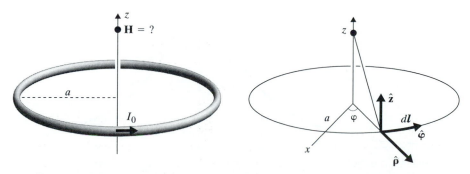

Figure 2-21   Finding the magnetic field of a circular current filament.

$$I\, d\boldsymbol{l} \times (\mathbf{r} - \mathbf{r}') = I_0 a\hat{\boldsymbol{\varphi}} \times (z\hat{\mathbf{z}} - a\hat{\boldsymbol{\rho}})d\varphi = I_0 a(z\hat{\boldsymbol{\rho}} + a\hat{\mathbf{z}})\, d\varphi \qquad (2.80)$$

gives the required field as

$$\mathbf{H}(\mathbf{r}) = \int_0^{2\pi} \frac{I_0 a(z\hat{\boldsymbol{\rho}} + a\hat{\mathbf{z}})d\varphi}{4\pi(z^2 + a^2)^{3/2}}. \qquad (2.81)$$

The entire integrand is constant, except for the vector $\hat{\boldsymbol{\rho}}$, which varies with the azimuth φ and cannot come out of the integral. The integral reduces to

$$\mathbf{H}(\mathbf{r}) = \frac{I_0 a\left\{ z \oint \hat{\boldsymbol{\rho}}\, d\varphi + a\hat{\mathbf{z}}2\pi \right\}}{4\pi(z^2 + a^2)^{3/2}}$$

$$= \frac{I_0 az \oint \hat{\boldsymbol{\rho}}\, d\varphi}{4\pi(z^2 + a^2)^{3/2}} + \frac{I_0 a^2 \hat{\mathbf{z}}}{2(z^2 + a^2)^{3/2}}. \qquad (2.82)$$

If it is not obvious that the integral of the radial vector $\hat{\boldsymbol{\rho}}$ around the complete circle must vanish, by cancellation at diametrically opposite points of the circle, then the remaining integral should be done, formally, by expressing the variable vector $\hat{\boldsymbol{\rho}}$ in terms of components along vectors that remain truly constant, such as the rectangular unit vectors $\hat{\mathbf{x}}$ and $\hat{\mathbf{y}}$:

$$\oint \hat{\boldsymbol{\rho}}\, d\varphi = \int_0^{2\pi} [\cos\varphi\hat{\mathbf{x}} + \sin\varphi\hat{\mathbf{y}}]\, d\varphi = 0\hat{\mathbf{x}} + 0\hat{\mathbf{y}} = 0. \qquad (2.83)$$

There remains only

$$\mathbf{H}(z) = \hat{\mathbf{z}}\tfrac{1}{2} I_0 a^2 (z^2 + a^2)^{-3/2} \qquad (2.84)$$

as the magnetic field at any point on the axis of the circular current filament.

This field is directed along the filament's axis, with the sense determined from the direction of the current by the right-hand rule. It is strongest at the center of the circle, at the value $H_0 = I_0/2a$; elsewhere on the axis, its strength falls off by the factor $\left[1 + (z^2/a^2)\right]^{-3/2}$, approaching $I_0\, a^2/2|z|^3$ for $|z| \gg a$. Figure 2-22 plots the variation of the axial field with axial position.

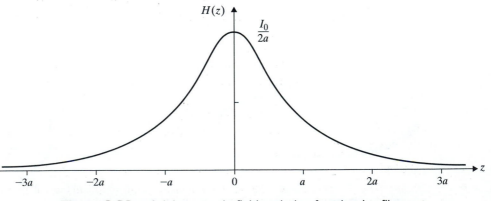

Figure 2-22   Axial magnetic field variation for circular filament.

We have avoided asking for the magnetic field at observation points off the axis of the filament because the integral involved, although readily set up, cannot be evaluated in terms of elementary functions. The result involves elliptic integrals, the type that we need when we seek the length of an ellipse of given major and minor axes. However, we have no need to pursue this now.

For the case of a continuous distribution of current, smeared over some region of space with a density $\mathbf{J}(\mathbf{r})$ A/m², an infinitesimal volume element dV' at $\mathbf{r}'$ has a current moment $\mathbf{J}(\mathbf{r}')\, dV'$, in place of the moment $I d\mathbf{l}$ for an element of a filament. This alternative expression for the moment becomes clear, if we think of $\mathbf{J}dV$ as comprising $J\, dS\, dl$, with flow along $d\mathbf{l}$ and with $J dS$ becoming I when the current is concentrated in the filament. The superposition used to obtain the magnetic field is then replaced by a volume integral:

$$\mathbf{H}(\mathbf{r}) = \int_V \frac{\mathbf{J}(\mathbf{r}') \times (\mathbf{r} - \mathbf{r}')\, dV'}{4\pi |\mathbf{r} - \mathbf{r}'|^3}; \tag{2.85}$$

the integral is over whatever region of space has a nonzero current density. In this integral, the observation point $\mathbf{r}$ is fixed, while the source point $\mathbf{r}'$ ranges over the volume of integration.

If the currents are distributed over a surface, the summation becomes a surface integral instead of a volume or line integral. In fact, the infinitesimal current moment has several guises for different degrees of concentration of the current distribution. For the most general smearing of the current, the source moment of a volume element is $\mathbf{J}\, dV$; if the currents are confined to a surface, with surface current density $\mathbf{K}$ A/m, the moment of an area element reduces to $\mathbf{K}\, dA$; upon constricting the flow to a filament, the source moment becomes $I d\mathbf{l}$; if the current is a moving charge concentrated in a point that moves with velocity $\mathbf{v}$, then the original moment $\mathbf{J}\, dV = \rho \mathbf{v}\, dV$ reduces to $\mathbf{v}\, dq$.

We now have the means for obtaining the magnetic field for any distribution of currents, either from Ampère's law or by superposition of the field of a current element. When do we use Ampère's law and when a full vector integration? Ampère's law is much simpler to use but requires additional constraints, such as may arise from symmetry considerations. We use Ampère's law directly when there is sufficient symmetry or when we know (or can guess) the field pattern but not its strength. Otherwise, we resort to the superposition integral.

The formula for the $\mathbf{H}$ field we have found is valid for any uniform medium. The magnetic flux density is just $\mathbf{B} = \mu \mathbf{H}$ if the medium has permeability μ, or it is $\mathbf{B} = \mu_0 \mathbf{H}$ for a vacuum.

Let us tackle a couple of examples of the application of Ampère's law. The first is intended to provide a crude estimate of the currents that may be circulating in the earth's core.

Example 2.8

Suppose we can ascribe the earth's magnetic field at its surface, which is about $B = (0.5)10^{-4}$ T, to a hypothetical current circulating about the earth's axis, in its core.

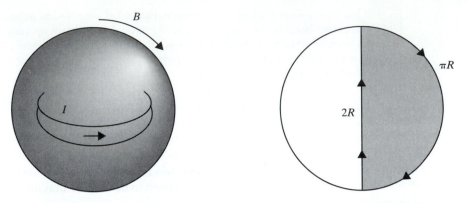

Figure 2-23   Assumed core current I generates observed field B.

Estimate this current, assuming for this purpose that the magnetic field is of the same strength on the earth's surface and along its axis. Note: *Earth's radius $R = (6.37)10^6$ m.* ◇

The mmf along the semicircle formed by a meridian on the earth's surface and closed by the earth's axis equals the current circulating inside the earth, around the earth's axis. If the magnetic intensity is B/μ_0 all along and parallel to this path (that is, ignoring any declination of the surface field or deviation from the strictly north–south direction and assuming the internal field on axis is parallel to the earth's axis), the mmf and current are the perimeter $\pi R + 2R$ times the field B/μ_0, or $(\pi + 2)RB/\mu_0 = (1.3)10^9$ A; see Figure 2-23.

The next example considers the highly practical question of how to generate a magnetic field that is as uniform as possible, using current-carrying coils. We are called upon to design the coil configuration to achieve a maximally flat magnetic field in a region of space in the vicinity of a particular point.

Example 2.9

Each of two thin wires is bent into a large circular ring of radius a; one is placed parallel to the xy-plane with its center at $z = -h$; the other is placed parallel to the xy-plane with its center at $z = +h$. Both wires carry current I_0, in the same direction (azimuthally). The resultant magnetic field $H(z)$ along the z-axis varies with z, but we want to make it as constant as possible in the vicinity of the origin. How should the radius a and the height h of each ring be related in order to make $H(z)$ as flat as possible at $z = 0$? ◇

From the calculation of the magnetic field of a current loop along its axis, as performed in an earlier example, we get immediately for the field $H(z)$ along z:

$$(2/I_0a^2)H(z) = \left[(z + h)^2 + a^2\right]^{-3/2} + \left[(z - h)^2 + a^2\right]^{-3/2}.$$

This is $2/(h^2 + a^2)^{3/2}$ at the origin, and we seek to make the curvature of this function as small as possible at $z = 0$. The function can be rewritten as

$$F(z) = f(h + z) + f(h - z), \quad \text{with} \quad f(x) = \left[a^2 + x^2\right]^{-3/2}.$$

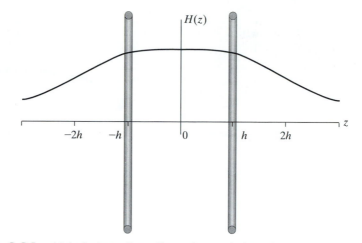

Figure 2-24   Helmholtz coil configuration and plot of magnetic field on axis.

We want to expand $f(h \pm z)$ in a Taylor series about h, so as to obtain $F(z)$ in the vicinity of $z = 0$. We note that

$$df(x)/dx = -3x(a^2 + x^2)^{-5/2}, \qquad d^2f(x)/dx^2 = (12x^2 - 3a^2)(a^2 + x^2)^{-7/2}.$$

Therefore, $F(z) = 2f(h) + 2[d^2f(h)/dx^2](z^2/2!) + 2[d^4f(h)/dx^4](z^4/4!) + \cdots$ as the coefficients of the odd powers of z cancel in the sum of the two series. This power series can be made maximally flat by causing $d^2f(h)/dx^2$ to vanish. This requires that $12h^2 - 3a^2 = 0$, or $a = 2h$ (radius equal to separation) for maximal flatness. The resultant highly practical configuration is called a pair of *Helmholtz coils*. See Figure 2-24; the $H(z)$ function has zero curvature as well as zero slope at $z = 0$.

Next, we want to compare the magnetic field generated by a continuous circular current loop with that produced by a point charge that revolves around a circle of the same size. After all, the moving charge constitutes a current, so the two cases should be quite similar. How alike are they?

Example 2.10

  (a) *Find the magnetic field $\mathbf{H}_0(z, t)$ along the z-axis for a charge q that revolves in the xy-plane around that axis, along a circular orbit of radius a, at constant angular velocity ω.*

  (b) *Find the time average of this magnetic field $\langle \mathbf{H}_0(z, t) \rangle$ (meaning the time integral of $\mathbf{H}_0$ over one period, divided by that period) and compare it with the magnetic field $\mathbf{H}_1(z)$ on the axis of a circular loop of current I in the xy-plane, of radius a, as already calculated in a previous example. See Figure 2-25.*
  For what current I is the field $\mathbf{H}_1(z)$ the same as $\langle \mathbf{H}_0(z, t) \rangle$? «

  (a) When the source is a charge moving with a given velocity $\mathbf{v}$, the element of integration $I d\mathbf{l}$ used for a current loop is replaced by $\mathbf{v} \, dq$, where dq is an element of charge, and we need to integrate $d\mathbf{H}_0 = \mathbf{v} \, dq \times \mathbf{r}/4\pi r^3$ over the source; $\mathbf{r}$ is the position of the observation point, seen from the source point. In the present instance,

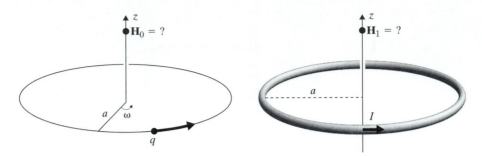

Figure 2-25   Revolving point charge compared to circular current loop.

the velocity is azimuthal, given by $\mathbf{v} = a\omega\hat{\boldsymbol{\varphi}}$. Hence, the required integration element is $\mathbf{v}\,dq = a\omega\hat{\boldsymbol{\varphi}}\,dq$, and we need its cross product with the position vector $\mathbf{r}$ from the source point at $a\hat{\boldsymbol{\rho}}$ to the observation point at $z\hat{\mathbf{z}}$, on the axis. Hence, $\mathbf{r} = z\hat{\mathbf{z}} - a\hat{\boldsymbol{\rho}}$ and we find $\mathbf{v}\,dq \times \mathbf{r} = a\omega\,dq\,\hat{\boldsymbol{\varphi}} \times (z\hat{\mathbf{z}} - a\hat{\boldsymbol{\rho}}) = a\omega\,dq\,(z\hat{\boldsymbol{\rho}} + a\hat{\mathbf{z}})$. For the denominator in the integral, we need $4\pi r^3$, where $r^2 = a^2 + z^2$. In the integration of $d\mathbf{H}_0 = \mathbf{v}\,dq \times \mathbf{r}/4\pi r^3$ over all sources (there is only the point charge), dq integrates to q itself and we get, finally, that

$$\mathbf{H}_0(z, t) = (a\omega q/4\pi)[z\hat{\boldsymbol{\rho}}(t) + a\hat{\mathbf{z}}]/(a^2 + z^2)^{3/2},$$

where $\hat{\boldsymbol{\rho}}(t)$ is the radial unit vector toward the location of the charge at time t and is the only factor that depends on time.

(b) The radial unit vector to the charge, $\hat{\boldsymbol{\rho}}(t) = \hat{\mathbf{x}}\cos\omega t + \hat{\mathbf{y}}\sin\omega t$, is a function of time that averages to zero over a period, so that $\langle\mathbf{H}_0(z, t)\rangle = (\omega q a^2/4\pi)\,\hat{\mathbf{z}}/(a^2 + z^2)^{3/2}$. From a previous example, the field $\mathbf{H}_1(z)$ of the current loop I is $(Ia^2/2)\,\hat{\mathbf{z}}/(a^2 + z^2)^{3/2}$ on the axis. The two are the same if the loop current and the charge are related by $I = \omega q/2\pi$. This is the product of the charge and the rotation frequency $\omega/2\pi$ (in revolutions per second).

BIOT-SAVART LAW

Since we now know both how a current element generates a magnetic field and how a magnetic field acts upon a current element, we can express the magnetic force between two current elements, or moving charges, as follows. Let the source element $\mathbf{J}_1\,dV_1$ be at position $\mathbf{r}_1$ and the test element $\mathbf{J}_2\,dV_2$ be at location $\mathbf{r}_2$. Then the source generates infinitesimal magnetic field

$$d\mathbf{H}_1(\mathbf{r}) = \mathbf{J}_1\,dV_1 \times (\mathbf{r} - \mathbf{r}_1)/4\pi|\mathbf{r} - \mathbf{r}_1|^3, \tag{2.86}$$

which becomes, at $\mathbf{r}_2$ in a vacuum,

$$d\mathbf{B}_1(\mathbf{r}_2) = \mu_0\mathbf{J}_1\,dV_1 \times (\mathbf{r}_2 - \mathbf{r}_1)/4\pi|\mathbf{r}_2 - \mathbf{r}_1|^3. \tag{2.87}$$

This field exerts force $d\mathbf{F}_{21}$ on the test element, given by

$$d\mathbf{F}_{21} = \mathbf{J}_2\,dV_2 \times d\mathbf{B}_1(\mathbf{r}_2)$$
$$= \mathbf{J}_2\,dV_2 \times \{\mu_0\mathbf{J}_1\,dV_1 \times (\mathbf{r}_2 - \mathbf{r}_1)/4\pi|\mathbf{r}_2 - \mathbf{r}_1|^3\}. \tag{2.88}$$

Consequently, by summing over both the source currents and the test currents in a general distribution of flows, we get the force exerted on an arbitrary volume V_2 of currents by another volume V_1 of source currents:

$$\mathbf{F}_{21} = \frac{\mu_0}{4\pi} \int_{V_1} \int_{V_2} \frac{\mathbf{J}(\mathbf{r}_2) \times \{\mathbf{J}(\mathbf{r}_1) \times (\mathbf{r}_2 - \mathbf{r}_1)\}\, dV_2\, dV_1}{|\mathbf{r}_2 - \mathbf{r}_1|^3}. \qquad (2.89)$$

This is one form of the Biot-Savart law of interaction of currents. Various other forms can be written by simplifying the triple cross product in the integrand, or by specializing to current filaments or to moving charges.

Example 2.11

Find the force exerted by one charge q moving with velocity **v** *on an identical other charge with the identical velocity, when they are in a vacuum and are separated by vector displacement* **l** *at right angles to their motion. (Figure 2-26).* ◈

Figure 2-26 Forces exerted between equal moving charges.

There are both electric and magnetic forces in this case. The electric one is the Coulomb force $\mathbf{F}_e = q^2 \mathbf{l}/4\pi\varepsilon_0 l^3$, while the magnetic one is given by the Biot-Savart law as $\mathbf{F}_m = \mu_0 q\mathbf{v} \times [q\mathbf{v} \times \mathbf{l}]/4\pi l^3$, which reduces for this case of perpendicular **v** and **l** to $\mathbf{F}_m = -\mu_0 q^2 v^2 \mathbf{l}/4\pi l^3$. Note that the electric and magnetic forces are here in opposite directions: While like charges repel (electrically), like current elements attract (magnetically). The total force is the vector sum of the two forces: $\mathbf{F} = [q^2 \mathbf{l}/4\pi\varepsilon_0 l^3][1 - \mu_0\varepsilon_0 v^2]$.

The Biot-Savart force between moving charges is correct for nonrelativistic motions; for speeds near that of light, the mechanical equations of motion must be modified to conform to relativity. The preceding result is therefore valid for charges moving at nonrelativistic speeds. In the present instance, this requires that v be small enough that $\mu_0\varepsilon_0 v^2 \ll 1$. We see that this implies that, at ordinary speeds, the magnetic force is much weaker than the electric one.

CURRENT ELEMENT AS DIPOLE FLOW SOURCE

The technique we used to obtain the magnetic field of a current element involved superposition of nearly equal and opposite configurations of current and replacing the difference between the nearly equal fields by their differential. The same approach allows us to get the flow pattern of another important point source, called a *dipole*.

Example 2.12

The dipole is a configuration of equal and opposite point sources of current, brought almost together, leaving only an infinitesimal separation between them. This comprises a

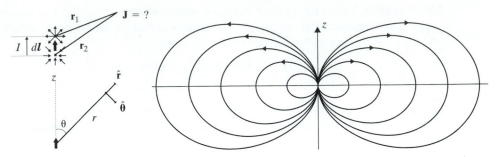

Figure 2-27   Dipole flow source in conducting medium and its flow field.

*point source of current I and a point sink of current −I; the displacement from the sink to the source is given by the infinitesimal vector d**l** (Figure 2-27).*

The combined source amounts to a current element within a conducting medium. Some unspecified agent receives the current that enters the sink and pumps it across the infinitesimal gap to the source, which emits it back into the surrounding conducting medium. Note that the dipole source ultimately has zero size but retains the direction from sink to source even in the limit, so that it has a symmetry axis. The word dipole is applicable to any superposition of equal and opposite point sources that are nearly coincident, such as infinitesimally separated equal and opposite point charges in a dielectric.

*We want the flow field **J**(**r**) at any observation point **r** in the conducting space around the current element, which is at the origin.* ◈

If the observation point is at $\mathbf{r}_1$ as seen from the source end of the dipole and at $\mathbf{r}_2$ from the sink end, then the flow field is the superposition of the two inverse-square Coulomb flows $\mathbf{J}_1 = I\mathbf{r}_1/4\pi r_1^3$ and $\mathbf{J}_2 = -I\mathbf{r}_2/4\pi r_2^3$. As $\mathbf{r}_1$ and $\mathbf{r}_2$ are nearly equal, we replace the difference of the two functions by the differential, as

$$\mathbf{J} = \mathbf{J}_1 + \mathbf{J}_2 = I\mathbf{r}_1/4\pi r_1^3 - I\mathbf{r}_2/4\pi r_2^3 = (I/4\pi)(\mathbf{r}_1/r_1^3 - \mathbf{r}_2/r_2^3)$$

$$= (I/4\pi)d(\mathbf{r}/r^3) = (I/4\pi)(d\mathbf{r}/r^3 - 3\mathbf{r}\,dr/r^4) \qquad (2.90)$$

and we need only relate the differential $d\mathbf{r}$ of the vector displacement $\mathbf{r}$ to the differential dr of the distance r. This is simple, because we know that $r^2 = \mathbf{r} \cdot \mathbf{r}$, so that, upon differentiating this definition, we obtain $2r\,dr = 2\mathbf{r} \cdot d\mathbf{r}$, or $dr = \hat{\mathbf{r}} \cdot d\mathbf{r}$. We are left with

$$\mathbf{J}(\mathbf{r}) = (I/4\pi)(d\mathbf{r}/r^3 - 3\hat{\mathbf{r}}\hat{\mathbf{r}} \cdot d\mathbf{r}/r^3). \qquad (2.91)$$

However, $d\mathbf{r} = \mathbf{r}_1 - \mathbf{r}_2 = -d\mathbf{l}$, so that the flow field of the dipole source is

$$\mathbf{J}(\mathbf{r}) = (I/4\pi r^3)(3\hat{\mathbf{r}}\hat{\mathbf{r}} \cdot d\mathbf{l} - d\mathbf{l}). \qquad (2.92)$$

Note the inverse-cube dependence on radial distance, as contrasted to the inverse-square dependence for the simple point source. Note also that the current I and the infinitesimal displacement $d\mathbf{l}$ appear only in the product of the two, called the *dipole moment*; they do not appear individually.

The result is usually expressed in spherical coordinates, with the dipole at the origin and aligned with the z-axis. Then $\mathbf{J} = J_r\hat{\mathbf{r}} + J_\theta\hat{\boldsymbol{\theta}}$ and the components are

$$J_r = \hat{\mathbf{r}} \cdot \mathbf{J} = (I/4\pi r^3)2\hat{\mathbf{r}} \cdot d\boldsymbol{l} = (Idl/4\pi r^3)2\cos\theta, \tag{2.93}$$

$$J_\theta = \hat{\boldsymbol{\theta}} \cdot \mathbf{J} = (I/4\pi r^3)(-\hat{\boldsymbol{\theta}} \cdot d\boldsymbol{l}) = (Idl/4\pi r^3)\sin\theta. \tag{2.94}$$

Note that the two components in $\mathbf{J}(\mathbf{r}) = (Idl/4\pi r^3)(\hat{\mathbf{r}}\,2\cos\theta + \hat{\boldsymbol{\theta}}\,\sin\theta)$ are inseparable; together, they form a field that satisfies $\oint_S \mathbf{J} \cdot d\mathbf{S} = 0$ for any closed surface S that either encloses or completely excludes the dipole. If the closed surface were to cut through the dipole, enclosing only the source or only the sink, then the flux would be $\oint_S \mathbf{J} \cdot d\mathbf{S} = \pm I$. The two components individually do not satisfy $\oint_S \mathbf{J} \cdot d\mathbf{S} = 0$ for any closed surface. The figure shows some flow lines of the dipole field.

SUMMARY

We have examined the concept of a magnetic field as an intermediary agent between cause and effect in the interaction of electric charges in motion. We defined two types of magnetic field, the magnetic flux density **B**, which acts on a moving (test) charge, and the magnetic intensity field **H**, which is generated by a moving (source) charge. The introduction of two types of magnetic field allows us to relegate the influence of the medium in which the interaction takes place to a separate process. We also agree to reduce the description of the medium to merely a permeability factor μ.

The action of the **B** field on a charge occurs at the point and instant at which the field and charge coexist, as would a contact force: If charge q is at position **r** with velocity **v** at time t in the presence of a magnetic field, the force exerted on the charge there at that time is $\mathbf{F} = q\mathbf{v} \times \mathbf{B}(\mathbf{r}, t)$. Whatever values the **B** field may have elsewhere and at other times are irrelevant to the force it exerts on the charge.

Generation of the **H** field by a moving charge or current, however, involves creation of a magnetomotive force (mmf) along any curve surrounding and linking the current, no matter how large the curve. We therefore have action at a distance, because the source current generates mmf along a far-away curve, in accordance with Ampère's law, and the existence of mmf requires the existence of an **H** field over at least some part of the distant curve.

Although we have action at a distance, we do not yet see a time delay in moving-charge interaction. Ampère's law does not entail any difference in time of evaluation of the mmf along a closed curve and the linked current: In

$$U(t) = \oint_C \mathbf{H}(\mathbf{r}, t) \cdot d\boldsymbol{l} = I(t), \tag{2.95}$$

the time t is the same in all parts of the equation. Even if the point on the curve C nearest to the most far-flung portion of the surrounded current I is light-years away from the source, the mmf at time t equals the source current at that time t.

There is something fundamentally wrong with Ampère's law: It is incomplete in the time-varying case. To correct this law, as well as to discover how time delay comes about, we need to account for time variation of sources and fields, beyond merely the motion of charges.

We will also find that the direct process of generation of magnetic fields by source currents is overshadowed by the far more important and ubiquitous one of generation by electric effects.

We considered a current element in a conductor as a dipole flow source, a superposition of nearby source and sink of equal strength. This is a convenient point source, useful for dealing with various configurations by superposition.

We have again deliberately avoided introducing the notion of a magnetic (scalar or vector) potential, which would simplify the superposition of fields of combinations of source currents. The magnetic scalar potential loses its validity and utility not only in the time-varying case but also in the presence of currents (the latter make it multivalued). The magnetic vector potential has more to recommend it, as it does not lose its relation to the magnetic field in the time-varying case, but its connection to the electric field is more complicated and we argue that it is preferable to relate the time-varying electric and magnetic fields to each other directly, without the intervention of the potentials. We relegate our discussion of potentials to the end (Chapter 10), as part of our review of electromagnetics.

PROBLEMS

Ionized atom in magnetic field

2.1 A neutral atom traveling horizontally at speed v_0 enters a region with a uniform, downward magnetic field B. At time $t = 0$ it passes through the origin and is suddenly ionized.
  (a) Describe the subsequent motions of the ion (mass M, charge e) and the freed electron (mass m, charge $-e$).
  (b) What is the farthest distance apart the ion and electron ever get?
  Note: Assume the magnetic field is strong enough and the speed high enough for the electrical attraction between ion and electron to be negligible compared to the magnetic forces.

Arbitrary initial motion of charge in uniform magnetic field

2.2 An electron in a region with a uniform, downward magnetic field B has an initial velocity $\mathbf{v}(0)$ that has both a horizontal component $\mathbf{v}_0$ and a vertical component $\mathbf{v}_1$.
  (a) Obtain the orbit $\mathbf{r}(t)$ of the electron.
  (b) Find the motion of the *guiding center*, the average location of the electron, found upon averaging out the oscillatory part of the motion.

Motion in parallel electric and magnetic fields

2.3 Uniform (in space) and constant (in time), parallel, downward electric and magnetic fields $\mathbf{E}(\mathbf{r}, t) = \mathbf{E}_0$ and $\mathbf{B}(\mathbf{r}, t) = \mathbf{B}_0$ have been set up within a region of space. An electron enters that space at time $t = 0$ at position $\mathbf{r}_0$, with horizontal velocity $\mathbf{v}_0$.
  (a) Obtain the orbit $\mathbf{r}(t)$ of the electron.

(b) Verify that the position and velocity of the electron are not periodic but that its acceleration is periodic. What is the period of the acceleration?

(c) Where is the electron after one-half period (of the acceleration)?

Drift in crossed uniform electric and magnetic fields

2.4 An electron is at rest at the origin at time $t = 0$, in the presence of constant, uniform, perpendicular electric and magnetic fields $\mathbf{E} = E\hat{\mathbf{x}}$ and $\mathbf{B} = B\hat{\mathbf{y}}$. Let $\alpha = (e/m)E$ and $\omega = (e/m)B$.

(a) Obtain the orbit of the electron. Note that the orbit is spatially periodic.

(b) How far does the electron get along the x-direction, within one spatial period? Along the y-direction? Along the z-direction? How long does it take to complete one period? What is the drift velocity (averaged over a period)?

(c) The orbit is a cycloid, the same as that of a point on the edge of a circle that rolls steadily along a straight line. Give the radius, angular velocity, and direction of rolling of the "equivalent" circle, in terms of $\mathbf{E}$, $\mathbf{B}$. *Note:* For this part, just compare the equations of motion of the point on the circle and of the electron.

Open-loop mmf along helix

2.5 The magnetic intensity field $\mathbf{H}(\mathbf{r})$ is given by $\mathbf{H}(x, y, z) = J_0(x\hat{\mathbf{z}} + z\hat{\mathbf{x}})$, where J_0 is a constant. The curve C is the helix $\mathbf{r}(\xi) = a\cos\xi\,\hat{\mathbf{x}} + a\sin\xi\,\hat{\mathbf{y}} + b\xi\,\hat{\mathbf{z}}$ of radius a and pitch $2\pi b$. Find the mmf $U\{C\}$ along one turn of the helix, from $\xi = 0$ to $\xi = 2\pi$.

Closed-loop mmf around semiellipse

2.6 The tilted semiellipse in Figure P2-1 is described by

$$\mathbf{r}(\xi) = b[1 - \sin\xi]\hat{\mathbf{x}} + b\cos\xi\,\hat{\mathbf{y}} + h\sin\xi\,\hat{\mathbf{z}}$$

for $0 \leq \xi \leq \pi$; it is closed by its minor axis and is oriented as shown. Let the magnetic field in space be given by the "whirlwind" field $\mathbf{H}(\mathbf{r}) = J_0\hat{\mathbf{z}} \times \mathbf{r}$, where J_0 is a constant. Calculate the mmf U around the indicated closed curve.

(*Comment:* Take particular note of the dependence of U on the height h of the tilted semiellipse; see the next problem.)

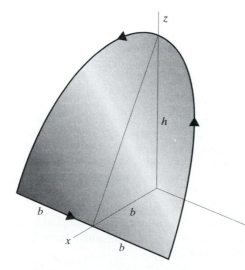

Figure P2-1 Semiellipse closed by its minor axis.

Closed-loop mmf by Stokes's theorem

2.7   Explain the lack of dependence of the closed-loop mmf around the semiellipse in the previous problem on the height h of its major axis by performing the calculation by use of Stokes's theorem.

Earth's dipole magnetic field

2.8   As a rough approximation, the earth's magnetic field outside the earth itself can be expressed as a dipole field: $\mathbf{B}(r, \theta) = B_0(R/r)^3[\hat{\mathbf{r}}\, 2\cos\theta + \hat{\boldsymbol{\theta}}\sin\theta]$, where r is the radius from the earth's center and θ is the elevation angle measured from the direction of the south magnetic pole (SMP) (which, incidentally, is not coincident with the geographic south pole). The relevant parameters are R = earth's radius = $(6.37)10^6$ m and, roughly, $B_0 = (0.5)10^{-4}$ T; the dipole field assumes $r \geq R$.

     (a) The field is vertically downward at the magnetic north pole (NMP). How strong is it there, according to this dipole approximation?

     (b) Obtain the mmf along a field line, beginning at the earth's surface at some magnetic elevation angle $\theta = \theta_0$ (in the southern hemisphere) and ending at elevation angle $\theta = \pi - \theta_0$ (in the northern hemisphere).

     Hint: By the properties of the dipole field, we know that the mmf is the same along the meridian path on the earth's surface between the same two points; the mmf calculation is simpler along that path. See Figure P2-2.

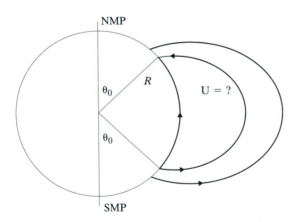

Figure P2-2   Dipole field line of the geomagnetic field.

Magnetic field of coaxial wire and shell

2.9   A wire of radius a carries a uniform current I; a coaxial cylindrical shell of inner and outer radii b, c carries the return current, also uniformly distributed. Find the magnetic intensity $\mathbf{H}(\mathbf{r})$, in magnitude and direction, at all points

     (a) inside the wire;   **(b)** between the wire and shell;

     (c) inside the shell;   **(d)** outside the shell.

Field of nonuniform cylindrical current

2.10   An infinitely long cylindrical region of radius a carries a total current I, distributed nonuniformly with density $J(\rho) = [(\lambda + 2)/(2\pi a^2)]I(\rho/a)^\lambda$, where λ is some constant (not necessarily an integer, subject to $\lambda > -2$).

(a) Verify that the total current is I.

(b) Determine the $\mathbf{H}(\mathbf{r})$ field, both inside and outside the cylinder.

Field of uniform planar current

2.11 (a) An infinite planar sheet that faces the x-direction carries a uniform surface current of density $\mathbf{K} = K\hat{\mathbf{z}}$. Find the magnetic field $\mathbf{H}(\mathbf{r})$ in the vacuum on both sides of the sheet.

(b) A second sheet, parallel to the first, carries an equal but opposite uniform surface current. Explain why the magnetic field now exists only between the planes and find its strength.

Field of uniform current on the surface of a sphere

2.12 Current I emerges from the "north pole" of a hollow spherical surface of radius R, flows uniformly in all directions along the "meridians" to the "south pole," where it is collected and pumped back up to the north pole along a thin wire that coincides with the south-to-north diameter of the sphere.

(a) What is the surface current $\mathbf{K}(\theta, \varphi)$ on the surface of the sphere?

(b) Find the magnetic field $\mathbf{H}(r, \theta, \varphi)$ inside and outside the sphere.

Hypothetical charge density of hailstorm in viscous air

2.13 As a somewhat less unrealistic version of the hailstorm in the text, let the cloud of area A and altitude h again drop N hailstones per second, steadily, releasing them with negligible initial velocity. The same unspecified mechanism causes each hailstone to emerge with one electron's charge. But this time, air resistance is not to be neglected, so that the hailstones attain a terminal velocity v_1, a constant that depends on the weight and size of the stones, assumed all identical, and on the viscosity of the air. The downward speed $v(t)$ of the hailstones obeys the differential equation $dv/dt + (g/v_1)v(t) = g$. This is easily solved for $v(t)$ and readily yields the height $z(t)$ of a stone released at $t = 0$, but this is not easily inverted to give $t(z)$ and hence $v(z)$. Nevertheless, we can obtain the inverses of the desired functions, specifically $z(v)$, $z(\rho)$, and $z(E)$.

(a) At what height $z = z(v)$ above the ground do the hailstones have an attainable downward velocity v (in the range $0 < v < v_1$)?

(b) In this steady state, at what height $z = z(\rho)$ is the charge density ρ (a negative value)?

(c) At what height $z = z(E)$ is the electric field E in the space between the ground and the cloud (assuming there is no field in the cloud itself)?

Transverse mmf from current-carrying coil

2.14 A thin wire is coiled into a helix of radius a and pitch h. The coil's axis is the z-axis, it is infinitely long, and it carries steady current I_0. The closed curve C_1 is a circle centered on the z-axis, of radius $b > a$, lying in a plane parallel to the xy-plane at height z, where $-h < z < h$. Obtain the mmf $U(z)$ around the curve C_1, as a function of its height z.

Longitudinal mmf from current-carrying coil

2.15 A thin wire is coiled into a helix of radius a and pitch h. The coil's axis is the z-axis, it is infinitely long, and it carries steady current I_0. The closed curve C_2 is a rectangle of width b (from the z-axis to a parallel line at distance $b > a$) and height z (extending from the xy-plane to a parallel line at level z). Obtain the mmf $U(z)$ around the rectangle C_2, as a function of the rectangle's height z.

Ideal solenoid as limit of helical coil

2.16 A solenoid, a closely wound coil of thin, current-carrying wire, is often modeled as an ideal cylindrical current sheet. The coil is actually a helix, of radius a and pitch h, but the pitch is taken to approach zero in the limit of an infinitesimally close winding. In an alternative view of this limit, the turns of the coil approach a series of closely spaced, parallel, transverse annular wires, each carrying current I_0, with $n = 1/h$ such circles per unit axial length (see Figure P2-3). In the first view of the limit, the current I_0 in the helix approaches zero along with h but the ratio I_0/h remains finite. In the second view, the current I_0 is finite when there is only a finite number n of turns per unit length but, in the limit, I_0 approaches zero and n becomes infinite, with a fixed product of the two. With either view, the limit is a continuous azimuthal current sheet $\mathbf{K} = \hat{\boldsymbol{\varphi}} I_0/h = \hat{\boldsymbol{\varphi}} n I_0 = \hat{\boldsymbol{\varphi}} K \,(\text{amp/m})$ flowing transversely around the cylindrical surface of radius a.

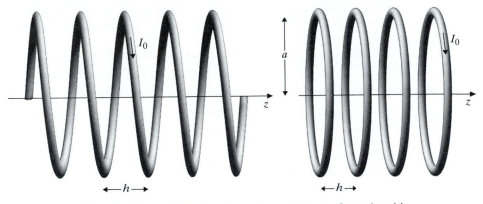

Figure P2-3   Helical and annular versions of a solenoid.

Assume the solenoid is infinitely long and that we are therefore dealing with the limiting case of an infinitely long cylinder of radius a, coaxial with the z-axis, with a uniform azimuthal surface current $\mathbf{K} = \hat{\boldsymbol{\varphi}} K$ flowing on it.

(a) The closed curve C_1 is a circle centered at the origin, of radius $b > a$, lying in the xy-plane. What is the mmf $U\{C_1\}$ around the curve C_1?

(b) The closed curve C_2 is a rectangle of width b (from the z-axis to a parallel line at distance $b > a$) and height z (extending from the xy-plane to a parallel line at level z). Obtain the mmf $U(z)$ around the rectangle C_2, as a function of the rectangle's height z.

(c) Obtain the magnetic intensity $\mathbf{H}$, both inside and outside the solenoid. Assume there is no field infinitely far from the axis.

Dissolving point charge

2.17 A point charge is located in a medium of permittivity ε and conductivity σ. At time $t = 0$, the amount of charge is q_0.

(a) From Gauss's law, obtain the field patterns $\mathbf{D}(\mathbf{r}, t)$ and $\mathbf{J}(\mathbf{r}, t)$ in terms of the amount of charge $q(t)$ remaining at time t.

(b) From conservation of charge, obtain a differential equation for $q(t)$.

(c) Solve the differential equation for $q(t)$.

(d) In sea water $(\varepsilon = 80\varepsilon_0, \sigma = 5$ siemens/m$)$, how long does it take to deplete the charge to 1% of its initial value q_0?

Redistribution of charge on two concentric spheres

2.18 A conducting spherical surface of radius a carries a charge q_0 at time $t = 0$ while surrounded by another, concentric spherical surface of radius b, which carries an initial charge Q_0, of the same sign, with $Q_0 > q_0$. The medium between the spheres has permittivity ε, permeability μ, and conductivity σ. After $t = 0$, a transfer of charge between the spherical surfaces will occur through the intervening medium, so that at time t the charges on the outer and inner spheres will have changed to $Q(t)$ and $q(t)$, respectively.

(a) Find the electric field $\mathbf{E}(\mathbf{r}, t)$ between the spheres, in terms of the charges $Q(t)$ and $q(t)$ at time t.

(b) Apply the law of conservation of charge to appropriate surfaces and volumes to determine two differential equations, one for $q(t)$ and one for $Q(t)$.

(c) Write the solutions to the two differential equations for $q(t)$ and $Q(t)$, in terms of the initial values q_0 and Q_0.

(d) On the basis of the solution, do the original charges on the two spheres attract, repel, or ignore each other?

Alternative calculations of the field of a semi-infinite current filament

2.19 Obtain the magnetic intensity $\mathbf{H}(r, \theta, \varphi)$ of a semi-infinite current filament I_0 along the positive z-axis feeding a point source in a conducting medium by applying Ampère's law to the circular field line C at any specified (r_0, θ_0), with each of the following surfaces that are bounded by C:

(a) planar disk perpendicular to the filament, for the case $z_0 = r_0 \cos \theta_0 > 0$;

(b) planar disk perpendicular to the filament, for the case $z_0 = r_0 \cos \theta_0 < 0$;

(c) spherical cap, centered at the origin, but on the far side of the filament (not pierced by the filament).

Design of Helmholtz coils

2.20 We need uniformity of magnetic field strength along a line 16 cm long, to within 1% of the strength at the center point. Design a pair of Helmholtz coils to achieve this specification. How far apart should the coils be? What should their diameter be?

Magnetic field of rectangular current loop

2.21 A rectangular area in the xy-plane extends from $x = -a$ to $x = a$ and from $y = -b$ to $y = b$. Current I_0 flows around the perimeter of the rectangle.

(a) Find the magnetic field $\mathbf{H}(z)$ along the z-axis.

(b) What is the magnetic field $\mathbf{H}(0)$ at the origin?

(c) Specialize $\mathbf{H}(0)$ to the limit $a \to \infty$, which leaves two parallel wires with equal and opposite currents I_0, separated by distance $2b$. Does your result agree with the field of two such parallel current filaments?

Combination of uniform and dipole field

2.22 We are in a region filled with a uniform flow field of current density $\mathbf{J} = J_0\hat{\mathbf{z}}$ and we introduce a dipole flow source $Idl\,\hat{\mathbf{z}}$ at some point in that region. Consider an imaginary spherical surface of radius a centered at the dipole.

   (a) Find the dipole moment Idl of the source that will result in a combined flow field that, at every point of the sphere's surface, is perpendicular to this surface.

   (b) Find the dipole moment Idl of the source that will result in a combined flow field that, at every point of the sphere's surface, is parallel to the surface.

   Note: The dipole moment will need to be finite, not infinitesimal, in both cases.

CHAPTER 3

Emf, Field Dynamics, and Maxwell's Equations

Starting with the concept of a field, we developed that of integrals of fields, over surfaces and along curves. We found the surface integral of the **D** field and the line integral of the **H** field to be especially significant; the former is the electric flux and the latter is the magnetomotive force (mmf). We will now discover the equal, or even greater, significance of the line integral of the electric field and the surface integral of the magnetic field.

ELECTROMOTIVE FORCE

With an electric field $\mathbf{E}(\mathbf{r})$ and a collection of points in space that form a curve C, we can associate a significant scalar quantity $V\{C\}$, defined as the line integral of **E** along C and called the *electromotive force*, abbreviated emf, or the *voltage* along the oriented curve:

$$V\{C\} = \int_C \mathbf{E} \cdot d\boldsymbol{l}. \tag{3.1}$$

This quantity is associated with both the electric field and the path of integration. Electrical measurements most often yield voltage readings along certain paths, rather than point-by-point electric field values.

   What is this quantity? We can arrive at a tentative significance of the emf on the basis of the electric field's definition as the force per unit charge and of the definition of mechanical work as the sum of dot products of a force with a displacement, along a path followed by an object subjected to that force. If we choose C to be the path followed by test charge q in electric field **E**, the quantity $qV\{C\}$ becomes the line integral of the force $q\mathbf{E}$ on the charge along its path; this would then be the work done on that charge and we could understand the emf to be the work per unit charge. This interpretation is of only limited utility and validity, however, because even when the field, or the path, is dependent on time, the definition of the voltage is given instantaneously. That is,

$$V(t) = \int_C \mathbf{E}(\mathbf{r}, t) \cdot d\mathbf{l} \tag{3.2}$$

is the emf along C at one, fixed, instant of time t. By contrast, the evaluation of the work done in moving charge q along the path C would involve all the times from departure at t_1 to arrival at t_2, yielding the constant W as

$$W = \int_C q\mathbf{E}(\mathbf{r}(t), t) \cdot d\mathbf{l} = \int_{t_1}^{t_2} q\mathbf{E}(\mathbf{r}(t), t) \cdot \left[d\mathbf{r}(t)/dt\right] dt, \tag{3.3}$$

in which $\left[d\mathbf{r}/dt\right] dt$ is the same as $d\mathbf{l}$ but the field $\mathbf{E}$ at any typical point $\mathbf{r}$ on C is evaluaed only at the time that point is reached by the charge. Compare this with the correct calculation of the voltage along this same path C, parametrized by a variable ξ unrelated to the time of arrival at $\mathbf{r}(\xi)$:

$$V(t) = \int_C q\mathbf{E}(\mathbf{r}, t) \cdot d\mathbf{l} = \int_{\xi_1}^{\xi_2} q\mathbf{E}(\mathbf{r}(\xi), t) \cdot \left[d\mathbf{r}(\xi)/d\xi\right] d\xi, \tag{3.4}$$

in which the field $\mathbf{E}$ at all the points $\mathbf{r}$ on C is evaluated at only the observation time t at which we evaluate the emf $V(t)$. Unless the field $\mathbf{E}$ does not change in time, it is therefore not correct that $qV\{C\}$ is W, the work done on q in traversing the path C. The significance of $V\{C\}$ is deeper.

The units of electric field, on the basis of its definition as force per unit charge, are N/C, so that the units of emf must be Nm/C, but this quantity is called a volt (V); in turn, the units of electric field are usually quoted as V/m.

Example 3.1

To emphasize the crucial concept of the difference between voltage and work per unit charge, let's calculate both and compare them, for the following case.

Suppose the electric field in space is $\mathbf{E}(\mathbf{r}, t) = E_0(x/a) \cos \omega t \, \hat{\mathbf{z}}$, where E_0, a, and ω are constants. Consider a charge Q that follows the parabolic arc C given by $\mathbf{r} = \mathbf{r}(t) = ut\,\hat{\mathbf{x}} + (vt - \frac{1}{2}gt^2)\hat{\mathbf{z}}$, from time $t = 0$ to time $t_0 = v/g$, where u, v, and g are constants.

(a) *Find the emf $V(t)$ along the curve C, at any specific observation time t.*

(b) *Find the work W done on the charge Q as it traverses the path C.* ◈

For the line integral that defines the emf, the parabolic arc can be parametrized in a variety of ways, but we may as well use the time of arrival at any of its points as a convenient parameter. We must, however, carefully distinguish between values of this parameter and the observation time t. We therefore use the time along the orbit but rename it λ, so that the curve is expressed as $\mathbf{r} = \mathbf{r}(\lambda) = u\lambda\hat{\mathbf{x}} + (v\lambda - \frac{1}{2}g\lambda^2)\hat{\mathbf{z}}$, with $0 < \lambda < t_0$.

(a) Noting that $x(\lambda) = u\lambda$ and that $d\mathbf{r}/d\lambda = u\hat{\mathbf{x}} + (v - g\lambda)\hat{\mathbf{z}}$, we have

$$\mathbf{E}(\mathbf{r}(\lambda), t) \cdot d\mathbf{r}(\lambda)/d\lambda = E_0\left[x(\lambda)/a\right] \cos \omega t \, \hat{\mathbf{z}} \cdot \{u\hat{\mathbf{x}} + (v - g\lambda)\hat{\mathbf{z}}\}$$

$$= E_0[u\lambda/a] \cos \omega t \, [v - g\lambda]$$

and the voltage is

$$V(t) = \int_0^{t_0} E_0 \cos \omega t \{[uv/a]\lambda - [ug/a]\lambda^2\} \, d\lambda$$

$$= E_0 \cos \omega t \{[uv/a](t_0^2/2) - [ug/a](t_0^3/3)\}$$

$$= E_0[uv^3/6ag^2] \cos \omega t$$

after substituting v/g for t_0.

(b) For the work done on charge Q as it moves along the arc, we must evaluate $\mathbf{E}(\mathbf{r}, t)$ at $\mathbf{r}(t)$ at time t, so that the calculation is the same, except that t and λ are the same variable (we can use λ or t for the integration; let's use λ):

$$Q\mathbf{E}(\mathbf{r}(\lambda), \lambda) \cdot d\mathbf{r}(\lambda)/d\lambda = QE_0[x(\lambda)/a] \cos \omega \lambda \, \hat{\mathbf{z}} \cdot \{u\hat{\mathbf{x}} + (v - g\lambda)\hat{\mathbf{z}}\}$$

$$= QE_0[u\lambda/a] \cos \omega \lambda [v - g\lambda]$$

and the work done from time 0 to t_0 is

$$W = \int_0^{t_0} QE_0 \cos \omega \lambda \{[uv/a]\lambda - [ug/a]\lambda^2\} \, d\lambda$$

$$= [QE_0uv/a] \int_0^{t_0} \lambda \cos \omega \lambda \, d\lambda - [QE_0ug/a] \int_0^{t_0} \lambda^2 \cos \omega \lambda \, d\lambda$$

$$= [QE_0uv/\omega^2 a][\omega t_0 \sin \omega t_0 - (1 - \cos \omega t_0)]$$

$$- [2QE_0ug/\omega^3 a][\omega t_0 \cos \omega t_0 - (1 - \tfrac{1}{2}\omega^2 t_0^2) \sin \omega t_0]$$

$$= QE_0[uv^3/6ag^2][12 \sin \theta / \theta - 6(1 + \cos \theta)]/\theta^2,$$

where $\theta = \omega t_0 = \omega v/g$.

Comparing W with $QV(t) = QE_0[uv^3/6ag^2] \cos \omega t$, we see that the fixed amount of work W and the function of time $QV(t)$ are not at all the same.

Note, however, that if ω were zero, the electric field would be static and the difference between the results for the work W and the quantity QV should disappear. It is in fact easy to verify, by expanding $\sin \theta$ and $\cos \theta$ in power series, that the factor $[12 \sin \theta / \theta - 6(1 + \cos \theta)]/\theta^2$ becomes 1 in the limit $\theta \to 0$, which corresponds to $\omega \to 0$, so that $QV = W$ in that limit.

An especially significant emf value is one associated with a closed curve, denoted $V\{C\} = \oint_C \mathbf{E} \cdot d\mathbf{l}$. An important property of Coulomb fields, which include the electric fields of point charges and of all superpositions of point charges, is that they have a zero closed-loop emf. To see why, consider a single point charge at the origin and, at first, any open curve, as in Figure 3-1. The Coulomb field of the point charge is proportional to $\mathbf{r}/r^3$, so that the emf along the curve is proportional to $\oint_C [\mathbf{r}/r^3] \cdot d\mathbf{l}$. If we parametrize the curve as $\mathbf{r} = r(\xi)$, then $d\mathbf{l} = [d\mathbf{r}/d\xi] \, d\xi$ and $\mathbf{r} \cdot d\mathbf{l}$ is $[\mathbf{r} \cdot (d\mathbf{r}/d\xi)] \, d\xi$. Now the distance r from the origin to any point is just the magnitude of the position vector: $\mathbf{r} \cdot \mathbf{r} = r^2$. Along the curve, both $\mathbf{r}$ and r are functions of the

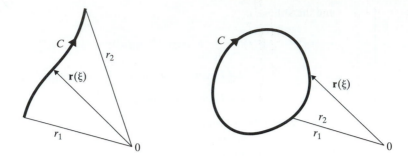

Figure 3-1   Integration of Coulomb field along open and closed curves.

parameter ξ, and we find on differentiating the relation $\mathbf{r} \cdot \mathbf{r} = r^2$ with respect to ξ that $2\mathbf{r} \cdot d\mathbf{r}/d\xi = 2r\,dr/d\xi$. We therefore have $\mathbf{r} \cdot d\mathbf{r} = r\,dr$, so that we can reduce the integrand from $[\mathbf{r}/r^3] \cdot d\mathbf{l} = [(\mathbf{r}/r^3) \cdot (d\mathbf{r}/d\xi)]\,d\xi$ to $[(r/r^3)(dr/d\xi)]\,d\xi$, which is just dr/r^2, and we find

$$\int_C \frac{\mathbf{r} \cdot d\mathbf{l}}{r^3} = \int_{\xi_1}^{\xi_2} \frac{\mathbf{r} \cdot d\mathbf{r}/d\xi}{r^3}\,d\xi = \int_{r(\xi_1)}^{r(\xi_2)} \frac{dr}{r^2} = \frac{1}{r(\xi_1)} - \frac{1}{r(\xi_2)}. \qquad (3.5)$$

But $r_1 = r(\xi_1)$ and $r_2 = r(\xi_2)$ are the distances of the end points of the curve C from the origin. If the curve is not open but closed, the end points coincide and the difference $1/r_1 - 1/r_2$ becomes zero: The closed line integral of a Coulomb field, and hence also of a superposition of many Coulomb fields, vanishes for any closed curve. The result means that *the electric field of any fixed distribution of charges must have zero emf along any closed curve.*

However, there are other ways of generating an electric field than by sprinkling charges here and there. For the **E** field of these other sources, the closed-loop emf need not vanish, as we will see.

Example 3.2

We want to compare the emf for one electric field, along different curves but with the same end points, and verify that the results are independent of the path only in special cases.

The electric field in space is $\mathbf{E}(\mathbf{r}) = \alpha x\hat{\mathbf{z}} + \beta z\hat{\mathbf{x}}$, where α and β are constants. Find the emf $V\{C\}$ along the specified curve C as a function of α, β and x_0, z_0,

  (a) *if C is the straight line from the origin to $\mathbf{r}_0 = x_0\hat{\mathbf{x}} + z_0\hat{\mathbf{z}}$;*

  (b) *if C is the parabola $z/z_0 = (x/x_0)^2$ from the origin to $\mathbf{r}_0 = x_0\hat{\mathbf{x}} + z_0\hat{\mathbf{z}}$;*

  (c) *if C is the elliptical path $x(\theta) = x_0 \sin\theta$, $z(\theta) = z_0[1 - \cos\theta]$, from the origin to $\mathbf{r}_0 = x_0\hat{\mathbf{x}} + z_0\hat{\mathbf{z}}$.*

  (d) *How can α and β be related to make all three of these answers the same?* ◇

We need only evaluate the given electric field along each of the specified curves and perform the line integral in each case.

  (a) The straight line from the origin to $\mathbf{r}_0$ is $\mathbf{r}(\lambda) = \lambda\mathbf{r}_0$ with $0 < \lambda < 1$. The field along this path is $\mathbf{E}(\lambda) = \lambda[\alpha x_0\hat{\mathbf{z}} + \beta z_0\hat{\mathbf{x}}]$ and $d\mathbf{r}/d\lambda = \mathbf{r}_0$, so that

$\mathbf{E} \cdot d\mathbf{r}/d\lambda = \lambda\left[\alpha x_0 z_0 + \beta z_0 x_0\right] = (\alpha + \beta) x_0 z_0 \lambda$ and $V = (\alpha + \beta) x_0 z_0 \int_0^1 \lambda \, d\lambda = x_0 z_0 (\alpha + \beta)/2.$

(b) Let $x/x_0 = \xi$. Then, on C, $\mathbf{r}(\xi) = x_0 \xi \hat{\mathbf{z}} + z_0 \xi^2 \hat{\mathbf{z}}$ and $d\mathbf{r}/d\xi = x_0 \hat{\mathbf{x}} + z_0 2\xi \hat{\mathbf{z}}$. The field along C is $\mathbf{E}(\xi) = \alpha x_0 \xi \hat{\mathbf{z}} + \beta z_0 \xi^2 \hat{\mathbf{x}}$ and $\mathbf{E} \cdot d\mathbf{r}/d\xi = \alpha x_0 z_0 2\xi^2 + \beta z_0 x_0 \xi^2 = (2\alpha + \beta) x_0 z_0 \xi^2$. Hence, $V = (2\alpha + \beta) x_0 z_0 \int_0^1 \xi^2 \, d\xi = x_0 z_0 (2\alpha + \beta)/3.$

(c) On C, $\mathbf{r}(\theta) = x_0 \sin\theta \, \hat{\mathbf{x}} + z_0 (1 - \cos\theta) \hat{\mathbf{z}}$ and $d\mathbf{r}/d\theta = x_0 \cos\theta \, \hat{\mathbf{x}} + z_0 \sin\theta \hat{\mathbf{z}}$. The range of θ is 0 to $\pi/2$; the field along C is $\mathbf{E}(\theta) = \alpha x_0 \sin\theta \, \hat{\mathbf{z}} + \beta z_0 (1 - \cos\theta) \hat{\mathbf{x}}$ and $\mathbf{E} \cdot d\mathbf{r}/d\theta = \beta z_0 x_0 \cos\theta (1 - \cos\theta) + \alpha x_0 z_0 \sin^2\theta$, so that the emf is $V = x_0 z_0 \int_0^{\pi/2} \left[\beta \cos\theta + \alpha \sin^2\theta - \beta \cos^2\theta\right] d\theta = x_0 z_0 [\beta + (\alpha + \beta)\pi/4].$

(d) The three emfs are $V_1 = x_0 z_0 (\alpha + \beta)/2$, $V_2 = x_0 z_0 (2\alpha + \beta)/3$, and $V_3 = x_0 z_0 [\beta + (\alpha - \beta)\pi/4]$. We need $(\alpha + \beta)/2 = (2\alpha + \beta)/3 = \beta + (\alpha - \beta)\pi/4$. These equations are identically satisfied if $\alpha = \beta$.

One conclusion we can reach is that, unless $\alpha = \beta$, the specified field cannot be generated by any distribution of fixed charges.

OHM'S LAW

If we subject a region containing charge to an electric field, the charges will experience forces and will be set in motion. If such motion is not precluded by the atomic structure, currents will flow. Here, the cause of the phenomenon would be the electric field $\mathbf{E}$ and the effect, the flow, would be described by the current density $\mathbf{J}$. Hence the $\mathbf{E}$ and $\mathbf{J}$ fields should be related.

In some materials called *conductors*, particularly metals and ionic solutions, charges are available and mobile, although neutralized by other charges. An electric field applied to such materials should set these charges in motion. The immediate effect of the electric field is to accelerate the charges, but the motion is impeded because of the presence of other particles or of the entire lattice structure of the material. The accelerated charges collide with these impediments and are stopped or slowed, then accelerated again by the applied electric field. On an average, macroscopic basis that blurs the details of this sort of stop-and-go behavior, there results an overall steady motion when the applied field is steady. The stronger the applied field, the stronger the resulting average drift of charge. A material in which this effect occurs is called an *Ohmic medium*, and the consequent proportionality between the applied field and the current density is expressed as

$$\mathbf{J} = \sigma\mathbf{E}. \tag{3.6}$$

This is termed *Ohm's law*, although it does not have the status of the other, usually inviolate laws of physics in that this one either holds or does not, depending on the nature of the material and the strength of the field. When it does hold, it is a pointwise relation and the current density then represents induced flow of preexisting, available, mobile charges that are inaccessible to observation by ordinary means.

Such flow, induced by the applied electric field in the conducting medium, is called *conduction current* and the proportionality coefficient σ is a property of the material, its *conductivity*.

Most materials are at least somewhat conducting, because their atoms comprise electric charges and some of these can drift between atoms. There is an enormous range of values of conductivity among common materials; metals are highly conducting, while glass, rubber, and ceramics are insulators. Vacuum has zero conductivity, because there are no preexisting, available charges in truly empty space. For reasonably low electric field strengths, air can be considered to have zero conductivity as well. The units of conductivity are based on that of the ratio of voltage to current, the ohm (Ω), or its reciprocal, the siemens (S): Since E is in V/m and J is in A/m^2, the conductivity must be in S/m. Metals have conductivities of the order of 10^8 S/m; quartz glass is a good insulator, with a conductivity of the order of only 10^{-12} S/m, or even considerably less, depending on impurity levels.

ELECTRIC POWER DENSITY

We wish now to show that the fundamental concepts of mechanical energy and power carry over to electrical phenomena.

We immerse ourselves in a region of space in which electric charges are streaming, in the presence of electric and magnetic fields. We consider an elementary area $d\mathbf{S}$ across which charges are flowing, forming an electric current, and follow the charges for an infinitesimal time dt.

The current crossing the area $d\mathbf{S}$ is $\mathbf{J} \cdot d\mathbf{S}$, by the definition of the current density $\mathbf{J}$. In time dt, an amount of charge $dq = \mathbf{J} \cdot d\mathbf{S}\, dt$ crosses the area $d\mathbf{S}$ and is conveyed along displacement $d\mathbf{l}$, which is parallel to the flow vector $\mathbf{J}$.

The motion of this charge across $d\mathbf{S}$ and along $d\mathbf{l}$ defines a volume element dV that is swept out in the time dt. As shown in Figure 3-2, this volume is generally an oblique cylinder, since there is no need for the direction in which the base area faces (that of $d\mathbf{S}$) to be the same as the one along which the cylinder's height stretches (that of $d\mathbf{l}$); it is only necessary that $\mathbf{J}$ and $d\mathbf{l}$ be parallel, since both are along the direction of flow in this river of moving charge. The volume element swept out in time

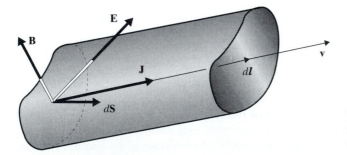

Figure 3-2   Volume element swept out along $\mathbf{J}$ or $\mathbf{v}$ or $d\mathbf{l}$ in time dt.

dt is given by $dV = d\mathbf{S} \cdot d\boldsymbol{l}$, not merely $dS \, dl$, because the cosine factor in the dot product accounts for the obliqueness of the cylinder.

The charge that crosses dS in time dt is subjected to a force during its displacement; this is given by the Lorentz force as

$$d\mathbf{F} = dq \, \mathbf{E} + dq \, \mathbf{v} \times \mathbf{B} = dq \, \mathbf{E} + dq (d\boldsymbol{l}/dt) \times \mathbf{B}, \qquad (3.7)$$

where $\mathbf{v} = d\boldsymbol{l}/dt$ is the velocity of streaming of the charge.

The work done on the charge by the fields during the time dt that the charge undergoes displacement $d\boldsymbol{l}$ is $dW = d\mathbf{F} \cdot d\boldsymbol{l}$ or

$$dW = d\mathbf{F} \cdot d\boldsymbol{l} = dq \, \mathbf{E} \cdot d\boldsymbol{l} + dq (d\boldsymbol{l}/dt) \times \mathbf{B} \cdot d\boldsymbol{l}$$

$$= \mathbf{J} \cdot d\mathbf{S} \, dt \, \mathbf{E} \cdot d\boldsymbol{l} + \mathbf{J} \cdot d\mathbf{S} \, d\boldsymbol{l} \times \mathbf{B} \cdot d\boldsymbol{l}$$

$$= \mathbf{J} \cdot d\mathbf{S} \, \mathbf{E} \cdot d\boldsymbol{l} \, dt + 0, \qquad (3.8)$$

since $dq = \mathbf{J} \cdot d\mathbf{S} \, dt$ and the triple product $d\boldsymbol{l} \times \mathbf{B} \cdot d\boldsymbol{l}$ is identically zero. The magnetic field does not contribute to the work done, because the force it exerts is at right angles to the motion of the charge.

In the last expression, $\mathbf{J}$ and $d\boldsymbol{l}$ can be interchanged without changing the overall product, because $\mathbf{J}$ and $d\boldsymbol{l}$ are parallel to each other, so that

$$\mathbf{J} \cdot d\mathbf{S} \, \mathbf{E} \cdot d\boldsymbol{l} = d\boldsymbol{l} \cdot d\mathbf{S} \, \mathbf{E} \cdot \mathbf{J}, \qquad (3.9)$$

by virtue of the fact that both expressions involve the same four vector magnitudes and the same cosines of the two angles between the pairs of vectors. Consequently, the work done in the process is

$$dW = \mathbf{E} \cdot \mathbf{J} \, d\boldsymbol{l} \cdot d\mathbf{S} \, dt = \mathbf{E} \cdot \mathbf{J} \, dV \, dt, \qquad (3.10)$$

since the element of volume is $dV = d\boldsymbol{l} \cdot d\mathbf{S}$. The power, or rate of expenditure of energy, is dW/dt, so that the coefficient of the infinitesimals,

$$p = \mathbf{E} \cdot \mathbf{J}, \qquad (3.11)$$

has the interpretation of power density: the power expended per unit volume, in watts per cubic meter (W/m^3), at the point where $\mathbf{E}$ and $\mathbf{J}$ both exist.

In ohmic media, the energy that is dissipated appears as heat; this process, called Joule heating, manifests macroscopically the impeded motion of the internal charges in such media.

At any point where $\mathbf{E}$ and $\mathbf{J}$ coexist (and are not perpendicular to each other), energy is being expended (by whatever agency is responsible for the presence of both $\mathbf{E}$ and $\mathbf{J}$) at the rate of $p = \mathbf{E} \cdot \mathbf{J}$ joules per second per cubic meter. This power density is a (scalar) field, being a function of position.

The total power expended in some region is the volume integral of the power density:

$$P = \int_V \mathbf{E} \cdot \mathbf{J} \, dV. \qquad (3.12)$$

The total energy expended within the region during a time interval T is the time integral of the total power:

$$W = \int_T P\, dt = \int_T \int_V \mathbf{E} \cdot \mathbf{J}\, dV\, dt. \tag{3.13}$$

Example 3.3

As a simple example, consider a wire of length l, cross-sectional area A, and conductivity σ that is bent around into a circle and the ends joined to form a complete circuit. Suppose the wire is somehow made to carry current I. What power is expended in this resistive circuit? ◈

Assuming that the current is uniformly distributed across the cross section of the wire, we can deduce that there is current density $J = I/A$ along the wire at every one of its points. This demands that there be an electric field $E = J/\sigma = I/A\sigma$ along the wire at every point in the conducting wire. The simultaneous existence of the parallel $\mathbf{E}$ and $\mathbf{J}$ fields entails a power density $p = EJ = I^2/A^2\sigma$ at every point of the wire. There is then a power expenditure within the wire, of amount

$$P = pV = pAl = I^2 l/A\sigma = I^2 R, \quad \text{with} \quad R = l/A\sigma, \tag{3.14}$$

where R is the resistance of the wire. The electrical energy is converted to heat throughout the wire. Whatever causes the current I must supply this much power to the wire.

We note a perhaps surprising result in this situation when we consider the electric field along the circuit, $E = I/A\sigma$. Since this is the same all along the wire, we readily find its integral along the full length of the wire to be $V = El = Il/A\sigma = IR$. What may be surprising about this mundane result is that it exhibits a nonzero closed-loop emf:

$$\oint \mathbf{E} \cdot d\boldsymbol{l} = V = IR \neq 0. \tag{3.15}$$

Evidently, whatever agency causes the current I to flow in this resistive circuit must be capable of creating a nonzero closed-loop emf. Since we saw that Coulomb fields of any distribution of charges have only zero closed-loop emfs, we conclude that the source of this emf cannot be merely some combination of suitably-disposed charges.

Various means that involve energy conversion from other types to electrical form can achieve both the closed-loop emf and the required provision of the power that gets converted to heat, as when the chemical energy of a battery is converted to electrical energy.

Whenever we know the field in a region of space and over an interval of time, we are in a position to calculate the power and energy expended in whatever process is taking place in that region and interval. We should deem a problem incompletely solved if the energetic aspect of the field configuration has not been assessed.

Example 3.4

A sphere of radius a has initial charge q_0 and is separated from a larger, concentric sphere of radius b with initial charge Q_0 by a medium with parameters ε, μ, σ. We considered the *ensuing charge transfer in a problem of the previous chapter and found that the redistribution of the charges as time elapses is expressed by*

$$q(t) = q_0 e^{-(\sigma/\varepsilon)t} \quad \text{and} \quad Q(t) = Q_0 + q_0\big[1 - e^{-(\sigma/\varepsilon)t}\big].$$

(a) *Find the power density $p(\mathbf{r}, t)$ between the two spheres.*

(b) *Calculate the total energy W expended in the process of transferring the charge.* ◈

We need the fields between the spheres, as functions of time, from which we form the power density. This can then be integrated, in both space and time, to get the energy expended.

(a) The spherical symmetry gives, as we know, $D(r, t) = q(t)/4\pi r^2$ and we get $E = q(t)/4\pi\varepsilon r^2$ and $J = \sigma q(t)/4\pi\varepsilon r^2$, giving us $p = EJ = \sigma q^2(t)/16\pi^2\varepsilon^2 r^4$ or $p(r, t) = \big[\sigma q_0^2/16\pi^2\varepsilon^2 r^4\big]e^{-2(\sigma/\varepsilon)t}$.

(b) The total power expended between the spheres is $P(t) = \displaystyle\int_a^b p(r, t)4\pi r^2 \, dr =$

$\big[\sigma q_0^2/4\pi\varepsilon^2\big]e^{-2(\sigma/\varepsilon)t}\displaystyle\int_a^b r^{-2}\,dr = \big[\sigma q_0^2/4\pi\varepsilon^2\big][1/a - 1/b]e^{-2(\sigma/\varepsilon)t}$. The total energy

expended is $W = \displaystyle\int_0^\infty P(t)\,dt = \big[q_0^2/8\pi\varepsilon\big][1/a - 1/b]$.

We may note that the conductivity of the medium between the spheres affects the rate at which energy is expended but not the total energy. We also note that the initial charge Q_0 on the *outer* sphere is irrelevant to what is happening *between* the spheres, as we expect from Gauss's law, which relates the electric flux to only the enclosed charge, not to any external charges.

We have noted that nonzero emfs around closed circuits cannot be generated by any distribution of fixed charges; they require some energy conversion from non-electric form, as from a chemical battery. However, the sovereign process for creating such emf is magnetic and dynamic, a process we will explore next.

FARADAY-MAXWELL LAW

The purely electromagnetic means for generating an electric field along a closed loop is a dynamic one, involving time-varying magnetic flux. For any open surface S whose edge is the closed curve C, with S and C mutually oriented by the right-hand rule, the emf $V\{C\}$ around the closed curve equals the rate of decrease of the magnetic flux $\Phi\{S\}$ across the surface:

$$V\{C\} = -\frac{d\Phi\{S\}}{dt}. \tag{3.16}$$

This is the Faraday-Maxwell law. In detail, it appears as

$$\oint_C \mathbf{E} \cdot d\mathbf{l} = -\frac{d}{dt}\int_S \mathbf{B} \cdot d\mathbf{S}. \tag{3.17}$$

This law holds for any surface S and its edge C, regardless of how the electric and magnetic fields may have been generated and no matter what causes the time variation of the magnetic flux across the surface. The negative sign is correct when we adhere to the convention of orienting the curve and surface by the right-hand rule. The law is valid instant by instant, in the sense that $V(t) = -d\Phi(t)/dt$ for all time t.

Example 3.5

Suppose a thin wire of length l, cross-sectional area A, and conductivity σ is bent around into a circle with the ends joined to form a complete circuit. Suppose this wire is immersed in a uniform magnetic field whose direction is perpendicular to the circle and whose strength is collapsing exponentially, with time constant T, from an initial value B_0:

$$B(t) = B_0 \exp(-t/T). \tag{3.18}$$

There is magnetic flux across the circular disk bounded by the wire, and this magnetic flux is a function of time. Consequently, there should be an emf along the wire. We seek the current induced in the wire and the energy expenditure that this entails. Figure 3-3 depicts the configuration and the decay of the applied magnetic field. ◈

Since the circumference of the circular wire is l, its radius is $l/2\pi$ and the area of the circle is $l^2/4\pi$. For this uniform field, the magnetic flux across the circular disk is just the field strength times the area, so that $\Phi(t) = B(t)\, l^2/4\pi$. There is a rate of decrease of this exponentially decaying flux:

$$-d\Phi/dt = -[dB/dt]l^2/4\pi = [B(t)/T]l^2/4\pi \tag{3.19}$$

and this equals the emf along the wire, $V(t)$.

Since the configuration is completely symmetrical, the electric field that integrates to this emf is the same all around the wire and, for the stipulated thin wire, we can assume there is no variation of the electric field strength across the wire's cross section. Hence $V = El$ and the electric field in the wire is $E(t) = B(t)l/4\pi T$. In the conducting wire, this implies a current density $J = \sigma E = \sigma Bl/4\pi T$ that is also uniform across the wire and integrates to a current

$$I(t) = JA = \frac{\sigma B(t)Al}{4\pi T} = \frac{\sigma B_0 Al}{4\pi T} e^{-t/T}. \tag{3.20}$$

To obtain the energy expenditure, we first integrate the power density $p = EJ = \sigma E^2$, which is uniform, over the volume Al of the wire to get the power expended,

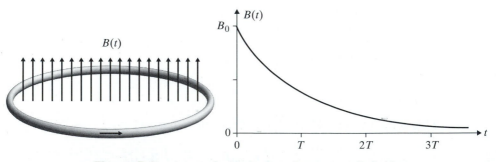

Figure 3-3   Loop of wire in decaying magnetic field.

$P(t) = \sigma A l E^2$. Since $I = \sigma E A$ and $V = E l$ for these uniform fields, the power is here simply $P = VI$. The total energy expended in heating the wire as the applied magnetic field collapses is the time integral of the power:

$$
\begin{aligned}
W &= \int_0^\infty P(t)\, dt = \int_0^\infty V(t) I(t)\, dt = \int_0^\infty \frac{B l^2}{4\pi T}\, \frac{\sigma B A l}{4\pi T}\, dt \\
&= \frac{\sigma A l^3}{16\pi^2 T^2} \int_0^\infty B^2(t)\, dt = \frac{\sigma A l^3 B_0^2}{16\pi^2 T^2} \int_0^\infty \exp(-2t/T)\, dt \\
&= \frac{\sigma A l^3 B_0^2}{32\pi^2 T}.
\end{aligned}
\tag{3.21}
$$

This is the amount of energy expended by the agency that causes the magnetic field to decay to zero, from its initial value B_0.

It is a bit early, yet important, to note that the result calculated in this example cannot be the entire story of what takes place in that situation. A current is induced in the closed circuit by the collapsing applied magnetic field, but that loop current, in turn, generates a magnetic field of its own. This additional field affects the total magnetic flux that is present and its time variation, so that the induced current is not quite what we calculated. Qualitatively, it is an illustration of *Lenz's law* that the additional field so generated acts to reduce or oppose the effect of the primary applied field. We will systematize later the calculation of the quantitative effects of this reaction back on the circuit, by calculating successive corrections to the approximate calculations we can make when we ignore such effects. For now, however, we need to complete our development of field dynamics.

The time variation of the applied magnetic field that we assumed in the last example is not the only process that can induce an emf in the loop. Any sort of time variation of the magnetic flux is equally effective, and a constant applied field can serve as well if the closed loop itself changes in time.

Example 3.6

Consider two parallel conducting rails, each with resistivity r_0 (in Ω/m), separated by distance l, connected at one end by a cross bar of resistance R_0 (in Ω) and with a sliding cross bar of resistance R_0 that can travel along the rails. A uniform, constant magnetic field B_0 fills the region, directed perpendicular to the plane of the rails. There is magnetic flux across the circuit formed by the rails and cross bars, and this magnetic flux is a function of time because of the variation of the area bounded by the circuit as the cross bar moves. There should therefore be an emf around the circuit. Suppose that the movable cross bar is made to slide back and forth along the rails, first forward at speed v for time T, then backward at speed v for the same time: The bar is at distance $x(t) = vt$ from the stationary one at time $t < T$ and then at $x(t) = v(2T - t)$ during the interval $T < t < 2T$. We seek the current induced in the rails and the resultant energy expenditure in each such cycle. Figure 3-4 shows the configuration. ◈◈

At time t, the area of the circuit is $lx(t)$, so that the magnetic flux across it is $\Phi(t) = B_0 l x(t)$. The rate of change of this magnetic flux is

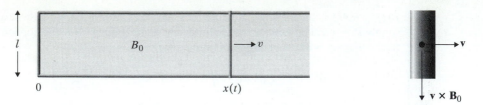

Figure 3-4   Sliding cross bar on parallel rails in magnetic field.

$$d\Phi/dt = B_0 l \, dx/dt = \pm B_0 lv, \qquad (3.22)$$

with the sign depending on the time interval. This must be the induced emf around the time-varying circuit. The direction of the emf is determined by the right-hand rule: If we orient the rectangular surface formed by the rails and cross bars in the direction of the applied field, then the flux is positive and the rectangle's perimeter must be orient-ed by the right-hand rule. The voltage along that orientation is then $V(t) = -d\Phi/dt = \mp B_0 lv$. The circuit has a total resistance of $R = 2R_0 + 2r_0 x(t)$, so that the current around the rectangle is

$$I(t) = \frac{-B_0 lv}{2R_0 + 2r_0 vt} \, (0 < t < T) \quad \text{and} \quad I(t) = \frac{+B_0 lv}{2R_0 + 2r_0 v(2T - t)} \, (T < t < 2T);$$

$$(3.23)$$

the signs indicate the direction of the current in relation to the positive orientation of the closed curve. Figure 3-5 is a plot of the induced current, labeled by the initial current $I_0 = B_0 lv/2R_0$, drawn for the case $r_0 vT/R_0 = 6$.

The electrical power expended in the process is the integral of EJ over the vol-ume of the rails and cross bars, or simply $P(t) = I(t)V(t)$, because E integrates to V along the length of the circuit and J integrates separately to I over the cross-sectional area of the conducting rails and cross bars. The energy expenditure over one cycle is the time integral

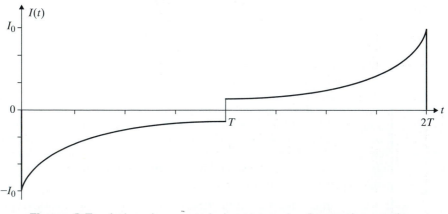

Figure 3-5   Induced current during one cycle of cross bar motion.

$$W = \int_0^{2T} P(t)\, dt = \int_0^{2T} V(t) I(t)\, dt$$

$$= \int_0^T \frac{B_0^2 l^2 v^2}{2R_0 + 2r_0 vt}\, dt + \int_T^{2T} \frac{B_0^2 l^2 v^2}{2R_0 + 2r_0 v[2T - t]}\, dt$$

$$= \frac{B_0^2 l^2 v}{r_0} \ln\left(1 + \frac{r_0 vT}{R_0}\right). \tag{3.24}$$

This is the amount of energy expended by the agency that causes the moving cross bar to slide back and forth. Half this total amount is dissipated in each half of the cycle, of course.

In this example, it was the motion of the cross bar that caused the magnetic flux to vary in time, even with the magnetic field kept steady. In this case, we can see the underlying mechanism responsible for the induced emf. Any mobile charge q within the cross bar moves with it and is therefore subjected to a magnetic force $q\mathbf{v} \times \mathbf{B}_0$ that is directed along the length of the cross bar, as indicated in Figure 3-4. This force pumps the charge around the closed circuit, giving rise to the induced current that we have attributed to the induced emf. We can see that this qualitative picture is also valid quantitatively by noting that the force per unit charge $\mathbf{v} \times \mathbf{B}_0$ exists only over the length l of the moving cross bar, not along the stationary part of the circuit ($\mathbf{v} = 0$), so that this effective electric field integrates around the circuit to the voltage $B_0 lv$ that we calculated from the Faraday-Maxwell law to be the induced emf.

For cases that feature a time-varying magnetic field as well as a changing surface or boundary curve, we calculate the time variation of the flux through the surface, from all its causes, and then take the time derivative of this flux to obtain the emf around the edge of the surface.

Example 3.7

Two parallel conducting rails of resistivity r_0 (ohm/m) are separated by distance l and connected by one stationary and one sliding cross bar, each of resistance R_0. A nonuniform, nonconstant magnetic field perpendicular to the plane of the rails fills the region. The field is $B(x, t) = B_0(x/L)(t/T)$. The sliding cross bar moves along the rails, from rest at the location of the fixed one (at $x = 0, t = 0$), at constant acceleration a, for a time T. Figure 3-4 in the previous example applies, except that the magnetic field is now not constant.

(a) *What current $I(t)$ is induced in the rails?*

(b) *What is the energy expended in this process?* ◈

We have here a nonuniform field that is time varying, and also accelerated motion of the cross bar. There are therefore multiple causes of the time variation of the magnetic flux. But whatever the causes, we need only calculate the flux $\Phi(t)$ as a function of time and then get the induced emf as $V = -d\Phi/dt$. The minus sign actually has no special significance in this context, since all we want is the emf around the circuit.

(a) At time t, the circuit extends from $x = 0$ to $w(t) = \frac{1}{2}at^2$ and the magnetic flux surrounded by the circuit is $\Phi = \int_0^w B_0(x/L)(t/T)l\, dx = B_0(l/L)(t/T)\frac{1}{2}w^2 =$

$\frac{1}{2}B_0(l/L)(t/T)(\frac{1}{2}at^2)^2 = (1/8)B_0(l/L)(a^2/T)t^5$. The emf induced around the circuit at time t, appropriately oriented, is $V(t) = (5/8)B_0(l/L)(a^2/T)t^4$. The total resistance of the circuit at time t is $R(t) = 2R_0 + 2r_0w(t) = 2R_0 + r_0at^2$, so the induced current is $I(t) = (5/8)B_0(l/L)(a^2/T)t^4/[2R_0 + r_0at^2]$.

(b) The energy expended is

$$W = \int_0^T V(t)I(t)\,dt = \int_0^T \left[(25/64)B_0^2(l/L)^2(a^4/T^2)t^8/(2R_0 + r_0at^2)\right]dt$$

$$= \left[(25/8)(B_0^2l^2R_0^3)/(TL^2r_0^4\theta)\right]\int_0^\theta \left[u^8/(1 + u^2)\right]du$$

$$= \left[(25/8)B_0^2l^2R_0^3/(TL^2r_0^4)\right]f(\theta),$$

where $\theta = T\sqrt{r_0a/2R_0}$ and $f(\theta) = \theta^6/7 - \theta^4/5 + \theta^2/3 - 1 + (\tan^{-1}\theta)/\theta$. The integral is elementary after the long division is performed.

Having seen how a time-varying magnetic flux can generate an emf around a closed loop, and therefore an electric field, we now consider how a time-varying electric flux can similarly generate an mmf, and therefore a magnetic field.

HOW AMPÈRE'S LAW FAILS

Ampère's law is capable of relating the generated magnetic field **H** to the source current I, provided that the current flows in a closed circuit, as must be the case for steady currents. In the time-varying case, the flow of charge need not be in a closed circuit; the current can end at some point and the charge can then simply pile up at that terminus. In that case, there is immediately a fatal ambiguity in Ampère's law, in that we cannot be certain that the current does or does not cross the open surface S along whose edge we want to compute the mmf. Figure 3-6 shows that we can get an unambiguous value for the current across any surface whose edge is a given closed curve C if the current flows in a closed loop, but that we can find different values for the current across different surfaces, for the same boundary curve, if the current ends somewhere.

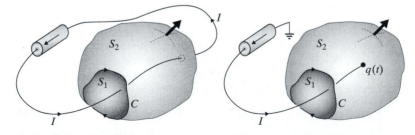

Figure 3-6   Mmf around a continuous and a discontinuous current.

In the figure, the current source pumps current I, which pierces the small surface S_1 bounded by the closed curve C, whether the current is continuous around a closed circuit or is discontinuous. If we prefer to use the large, ballooning surface S_2 that has the same closed curve C as its edge, then we find the same current I piercing that surface in the continuous case, but no current at all crossing S_2 in the discontinuous case. The mmf $U\{C\}$ is ambiguous in the discontinuous case, being $I\{S_1\} = I$ if we choose S_1 as the open surface, but being $I\{S_2\} = 0$ if we choose S_2 instead.

We can see more directly that Ampère's law is not consistent with the law of conservation of charge by considering a small, closed curve C with a large, ballooning, open surface S bounded by C. According to Ampère's law, $U\{C\} = I\{S\}$ and, if we go to the limit of C shrinking down to zero size, this mmf must vanish as we integrate whatever magnetic field may exist at the location of the curve C over a vanishingly small closed curve. However, while $U\{C\}$ necessarily vanishes in this limit, the current $I\{S\}$ need not vanish, because the surface S need not shrink along with the closed curve C: The balloon can remain large but simply close as we tie off its mouth with the shrinking curve C so that, in the limit, we deal with a closed surface S_0 instead of the open one S; see Figure 3-7. But the net current across a closed surface S_0 may not be zero, if the charge q enclosed by the surface changes in time, in accordance with the law of conservation of charge:

$$U\{C\} \to 0 \quad \text{but} \quad I\{S\} \to I\{S_0\} = -dq/dt \neq 0. \qquad (3.25)$$

Clearly, there is an inconsistency in Ampère's law $U\{C\} = I\{S\}$, at least in the time-varying case. While Ampère's law is applicable in the static case, it must be modified to allow for time variation.

We can get a clue to the resolution of the inconsistency from the problem of the magnetic field generated by a semi-infinite wire in a conducting medium. We were able to find this magnetic field because the spherically symmetric outflow from the tip of the wire could not possibly generate a magnetic field of its own. The result we found for the azimuthal magnetostatic field, for the case that the current I_0 is directed down the positive z-axis to the tip at the origin, was

$$H_\varphi(r, \theta) = \frac{-I_0}{4\pi r} \cot \frac{\theta}{2}. \qquad (3.26)$$

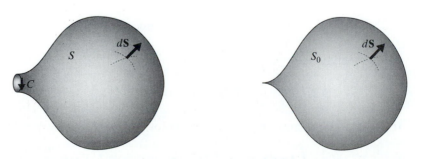

Figure 3-7 Balloon-type large surface with vanishingly small edge.

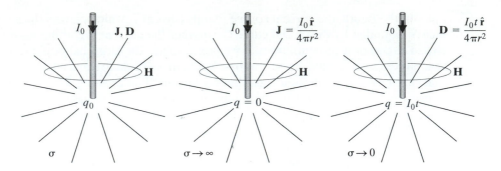

Figure 3-8   Semi-infinite current in general, in conductor, in insulator.

The clue furnished by this expression lies in the fact that the conductivity σ of the medium does not appear in the result. The conductivity seems to be irrelevant to the strength of the generated magnetic field. Let us therefore compare the configuration of currents in two extreme cases, of very high and of very low conductivity, as in Figure 3-8.

How can we reconcile the observable fact that the magnetic field is identical for all values of the medium's conductivity σ, even in the extreme cases of infinite and of zero conductivity, with the fact that the spherical outflow $\mathbf{J} = I_0\hat{\mathbf{r}}/4\pi r^2$, which we integrated over a spherical cap to get the mmf around a circle, becomes zero in the case of an insulator?

The difference between the two extreme cases is that, for a good conductor, the current that reaches the tip continues outward, distributed as the radial, inverse-square current density field $\mathbf{J}$ while, for a good insulator, the same current results in an accumulation of charge at the tip, which is accompanied instead by a radial, inverse-square electric flux density field $\mathbf{D}$. Since, by conservation of charge, $I_0 = dq/dt$, we find that the conducting case has field $\mathbf{J} = I_0\hat{\mathbf{r}}/4\pi r^2$ while the insulating case features a field $\mathbf{D} = q(t)\hat{\mathbf{r}}/4\pi r^2$ whose time derivative is $\partial\mathbf{D}/\partial t = I_0\hat{\mathbf{r}}/4\pi r^2$, exactly the same as the $\mathbf{J}$ field of the conductor. For both cases, the generated magnetic field is the same; the fields $\mathbf{J}$ and $\partial\mathbf{D}/\partial t$ seem to be interchangeable.

Following this clue, we resolve the inconsistency by combining Gauss's law with the law of conservation of charge: For any *closed* surface S_0, the two laws demand

$$\oint_{S_0} \mathbf{J} \cdot d\mathbf{S} = -dq/dt \quad \text{and} \quad \oint_{S_0} \mathbf{D} \cdot d\mathbf{S} = q, \tag{3.27}$$

which combine into

$$\oint_{S_0} \mathbf{J} \cdot d\mathbf{S} + \frac{d}{dt} \oint_{S_0} \mathbf{D} \cdot d\mathbf{S} = 0, \quad \text{or} \quad I\{S_0\} + \frac{d\Psi\{S_0\}}{dt} = 0. \tag{3.28}$$

If we form the corresponding expression for any *open* surface S, we get a quantity

$$\int_S \mathbf{J} \cdot d\mathbf{S} + \frac{d}{dt} \int_S \mathbf{D} \cdot d\mathbf{S} \quad \text{or} \quad I\{S\} + \frac{d\Psi\{S\}}{dt} \tag{3.29}$$

that *must* become zero if the surface S closes *and* that reduces to just $I\{S\}$ in the static case. Maxwell recognized that this quantity has all the features needed to extend Ampère's law consistently to the time-varying case. We therefore proclaim the Ampère-Maxwell law to extend and replace Ampère's law, as follows.

AMPÈRE-MAXWELL LAW

For any open surface S whose edge is the closed curve C, and if S and C are mutually oriented by the right-hand rule, the mmf $U\{C\}$ around the closed curve equals the sum of the current $I\{S\}$ and the rate of increase of the electric (or displacement) flux $\Psi\{S\}$ across the surface:

$$\oint_C \mathbf{H} \cdot d\boldsymbol{l} = \int_S \mathbf{J} \cdot d\mathbf{S} + \frac{d}{dt} \int_S \mathbf{D} \cdot d\mathbf{S}, \tag{3.30}$$

or

$$U\{C\} = I\{S\} + \frac{d\Psi\{S\}}{dt}. \tag{3.31}$$

This law is consistent with the static version and allows the closed curve C to shrink to a point and still leave an equality, with zero on both sides, by virtue of the law of conservation of charge and Gauss's law. Experiments confirm that this extension of the static law is valid generally. It states that the rate of change of electric flux across any surface S contributes to the mmf along the edge C of the surface, just as the current through the surface does. Charge flow, measured by $\mathbf{J}$, and charge accumulation, measured by $\partial \mathbf{D}/\partial t$, are equally effective in generating mmf and magnetic fields.

The quantity $d\Psi/dt$, which in the law adds to the current I across the surface, itself behaves as a current insofar as it generates mmf; it is called *displacement current*. The displacement current density is $\partial \mathbf{D}/\partial t$. Since the total electric flux, or displacement, Ψ_0 across a closed surface equals the enclosed charge, the displacement current $d\Psi_0/dt$ equals the rate of increase of the enclosed charge, dq/dt, and hence equals the rate at which charge enters the enclosed region by actual charge movement or current I_0. The displacement current thereby serves to complement conduction current, to make the total current continuous. Total current, which includes displacement current, always flows in closed circuits, so that the linking of a closed path C with the circuit of total current has a clear meaning and the Ampère-Maxwell law is unambiguous.

Example 3.8

We are now in a position to ascertain the full truth about the electric and magnetic fields generated by the semi-infinite current filament, in a general medium with constitutive parameters μ, ε, and σ. Let the current fed by the remote current source to the insulated filament be a step of height I_0, meaning that the current is zero before time $t = 0$ and becomes

I_0 after $t = 0$. Does a charge develop at the tip? What fields are generated? How do they vary in space and time? ◈

Once the current is fed to the filament, charge is conveyed to the tip, where it can accumulate or flow away in all directions. Both processes should occur, because the presence of charge results in a **D** field around the tip and hence an **E** field in the medium, by $\mathbf{D} = \varepsilon\mathbf{E}$, and hence a **J** flow field in the medium, by $\mathbf{J} = \sigma\mathbf{E}$. We need to know the outcome of the competition between charge accumulation and charge outflow. The law of conservation of charge tells us what happens to the charge q at the tip.

If the charge at the tip is $q(t)$ then there is the spherical displacement field $\mathbf{D} = q(t)\hat{\mathbf{r}}/4\pi r^2$, the spherical electric field $\mathbf{E} = \mathbf{D}/\varepsilon$, and the spherical current density $\mathbf{J} = (\sigma/\varepsilon)\mathbf{D}$. If we surround the tip with a closed surface, then the electric flux $\oint\mathbf{D}\cdot d\mathbf{S}$ over that surface is just $q(t)$ by Gauss's law. Hence the total conduction current is $\oint\mathbf{J}\cdot d\mathbf{S} = (\sigma/\varepsilon)\oint\mathbf{D}\cdot d\mathbf{S} = (\sigma/\varepsilon)q(t)$ in distributed form, plus the inward filamentary current I_0 (after $t = 0$) in concentrated form, for a total outward conduction current of $(\sigma/\varepsilon)q(t) - I_0$. By charge conservation, this must equal $-dq(t)/dt$, the rate of decrease of the charge enclosed by the surface. Hence the charge $q(t)$ satisfies the first-order differential equation

$$\frac{dq(t)}{dt} + \frac{\sigma}{\varepsilon}q(t) = I_0 \qquad (3.32)$$

for $t > 0$.

The solution to this differential equation, with zero charge as the initial condition at $t = 0$, is a steady-state value and an exponential transient:

$$q(t) = (\varepsilon/\sigma)I_0\left[1 - e^{-(\sigma/\varepsilon)t}\right]. \qquad (3.33)$$

The charge at the tip increases from zero, rapidly at first but then more slowly, finally settling down to an asymptotic value $q_0 = (\varepsilon/\sigma)I_0$. The time constant for this process is ε/σ.

The displacement density and current density fields are therefore

$$\mathbf{D}(\mathbf{r}, t) = \frac{\varepsilon I_0\hat{\mathbf{r}}}{\sigma 4\pi r^2}\left[1 - e^{-(\sigma/\varepsilon)t}\right], \qquad (3.34)$$

$$\mathbf{J}(\mathbf{r}, t) = \frac{I_0\hat{\mathbf{r}}}{4\pi r^2}\left[1 - e^{-(\sigma/\varepsilon)t}\right]. \qquad (3.35)$$

Both these fields approach their steady-state values gradually, with time constant ε/σ.

We now ask what magnetic field is generated in this time-varying situation. Applying the Ampère-Maxwell law, we choose as our closed curve through any observation point at r_0, θ_0 the horizontal circle $r = r_0$, $\theta = \theta_0$, along which the magnetic field amplitude is the unknown $H(r_0, \theta_0, t)$. The azimuthally directed mmf along this circle is

$$U\{C\} = \oint_C \mathbf{H}\cdot d\mathbf{l} = \int H(r_0, \theta_0, t)\hat{\boldsymbol{\varphi}}\cdot\hat{\boldsymbol{\varphi}}\,dl = H(r_0, \theta_0, t)2\pi\,r_0\sin\theta_0, \qquad (3.36)$$

where we find ourselves integrating a constant along the periphery, and where $2\pi r_0\sin\theta_0$ is the circumference of the circle, as in Figure 3-9.

We have the mmf around C, in terms of the unknown $H(r_0, \theta_0, t)$; we choose for the surface S the spherical cap of radius r_0 and extending only to elevation angle θ_0, so

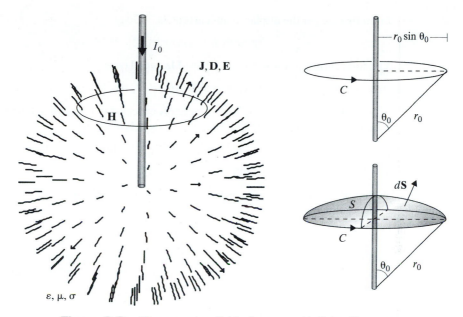

Figure 3-9 Time-varying fields from semi-infinite filament.

that its edge is the circle C, and we orient S radially to conform to the right-hand rule for the azimuthally directed circle, as indicated in the figure.

The total current $I\{S\}$ that crosses this surface in the radial direction includes the concentrated current I_0 in the filament, which is directed negatively (toward the origin), as well as the distributed outflow of the current density $\mathbf{J}(\mathbf{r}, t)$. These total

$$
\begin{aligned}
I\{S\} &= -I_0 + \int_S \mathbf{J}(\mathbf{r}, t) \cdot d\mathbf{S} \\
&= -I_0 + \int_0^{2\pi} \int_0^{\theta_0} \frac{I_0}{4\pi}\left[1 - e^{-(\sigma/\varepsilon)t}\right]\sin\theta \, d\theta \, d\varphi \\
&= -I_0 + \frac{I_0}{4\pi}\left[1 - e^{-(\sigma/\varepsilon)t}\right]2\pi\left(1 - \cos\theta_0\right) \\
&= -\tfrac{1}{2}I_0\left[(1 + \cos\theta_0) + (1 - \cos\theta_0)e^{-(\sigma/\varepsilon)t}\right] \\
&= -I_0\left[\cos^2(\theta_0/2) + \sin^2(\theta_0/2)e^{-(\sigma/\varepsilon)t}\right].
\end{aligned}
\tag{3.37}
$$

The electric flux over this same surface is

$$
\begin{aligned}
\Psi\{S\} &= \int_S \mathbf{D}(\mathbf{r}, t) \cdot d\mathbf{S} \\
&= \int_0^{2\pi} \int_0^{\theta_0} \frac{\varepsilon I_0}{\sigma 4\pi}\left[1 - e^{-(\sigma/\varepsilon)t}\right]\sin\theta \, d\theta \, d\varphi \\
&= \frac{\varepsilon I_0}{\sigma 4\pi}\left[1 - e^{-(\sigma/\varepsilon)t}\right]2\pi\left(1 - \cos\theta_0\right) \\
&= (\varepsilon/\sigma)I_0 \sin^2(\theta_0/2)\left[1 - e^{-(\sigma/\varepsilon)t}\right].
\end{aligned}
\tag{3.38}
$$

From this, we get the displacement current as

$$d\Psi/dt = I_0 \sin^2(\theta_0/2)e^{-(\sigma/\varepsilon)t}. \tag{3.39}$$

The total current for use in the Ampère-Maxwell law is

$$I\{S\} + d\Psi\{S\}/dt = -I_0\cos^2(\theta_0/2), \tag{3.40}$$

in which the time dependence has completely canceled out. Equating the mmf and total current, in accordance with the Ampère-Maxwell law, we find

$$U\{C\} = H(r_0, \theta_0, t)2\pi r_0 \sin\theta_0 = H(r_0, \theta_0, t)4\pi r_0 \sin(\theta_0/2)\cos(\theta_0/2)$$
$$= I\{S\} + d\Psi\{S\}/dt = -I_0\cos^2(\theta_0/2), \tag{3.41}$$

so that the unknown amplitude $H(r_0, \theta_0, t)$ is now known to be

$$H(r_0, \theta_0, t) = -\frac{I_0}{4\pi r_0}\cot\frac{\theta_0}{2}. \tag{3.42}$$

This result is valid for all r_0 and θ_0 (except for $\theta_0 = 0$, which is on the filament, or $r_0 = 0$, which is at its tip). We conclude that the magnetic field pattern for the semi-infinite filament is, after $t = 0$,

$$\mathbf{H}(\mathbf{r}, t) = H(r, \theta, t)\hat{\boldsymbol{\varphi}} = -\hat{\boldsymbol{\varphi}}\frac{I_0}{4\pi r}\cot\frac{\theta}{2}. \tag{3.43}$$

This result is independent of time (after the step of current has occurred), meaning that the magnetic field follows the current step instantly, without the gradual, exponential buildup that we found for the $\mathbf{J}, \mathbf{E}$, and $\mathbf{D}$ fields in the space around the tip of the semi-infinite filament. While these three fields depend on the parameters ε and σ of the medium, the magnetic field is oblivious to those values, confirming again that this $\mathbf{H}$ field is generated entirely by the semi-infinite filament, with the radial, spherically symmetric conduction and displacement currents making no contribution of their own. Figure 3-10 shows the time variation of all the fields generated in the space. The ultimate

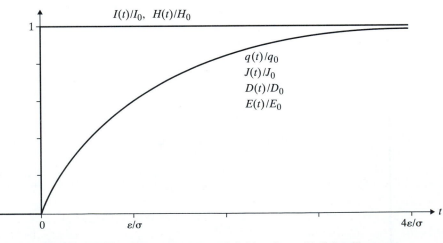

Figure 3-10   Step response of fields of semi-infinite filament.

(saturation) values of the various quantities are related to the size I_0 of the current step, and to the observation point, by

$$q_0 = (\varepsilon/\sigma)I_0, \qquad J_0 = I_0/4\pi r^2, \qquad D_0 = (\varepsilon/\sigma)J_0,$$

$$E_0 = J_0/\sigma, \qquad H_0 = -[I_0 \cot(\theta/2)]/4\pi r. \qquad (3.44)$$

The permittivity and conductivity of the medium determine the rate at which charge delivered to the tip of the filament is drained away. Whatever current may be pumped into the filament, the charge at the tip builds up to the ultimate level that just balances the rates of delivery and of outflow. The spherical fields $\mathbf{E}$, $\mathbf{D}$, and $\mathbf{J}$ build up in proportion to the charge, with time constant ε/σ, but the magnetic fields $\mathbf{H}$ and $\mathbf{B} = \mu\mathbf{H}$ have no contribution from the spherically symmetric $\mathbf{J}$ field, are generated by only the semi-infinite filament current, and can therefore follow the input current instantly.

In the limiting case of infinite conductivity, the time constant is zero and the charge at the tip is drained immediately; there is no charge buildup, no $\mathbf{D}$ or $\mathbf{E}$ field, but the $\mathbf{J}$ field appears at once at its full saturation value to drain the full filament current.

In the opposite extreme case of zero conductivity, there is no means for draining the charge delivered to the tip by the filamentary current; the charge accumulates steadily, as $q(t) = I_0 t$. There is a steadily rising $\mathbf{D}$ and $\mathbf{E}$ field, but no $\mathbf{J}$ field. The resultant constant displacement current completes the circuit for the constant filament current. In all cases, even in the two extremes, the magnetic field remains the same.

TYPES OF CURRENT

Now that we are acquainted with displacement current, it may be worthwhile to review the distinctions among the various kinds of current we may encounter.

Current being charge motion, the basic flow is a stream of charged particles. If a region contains charge distributed with density ρ and moving with velocity $\mathbf{v}$, the *convection current density* is $J = \rho\mathbf{v}$. Both ρ and $\mathbf{v}$ are fields and so is $\mathbf{J}$.

In an ohmic medium, *conduction current* consists of motion of inaccessible charges whose density and velocity are unknown but whose aggregate motion is manifested by an overall proportionality to the applied electric field. Its current density is $\mathbf{J} = \sigma\mathbf{E}$.

Whether we describe the flow of charge by giving ρ and $\mathbf{v}$ or by specifying the conductivity σ of the medium depends on our state of knowledge or ignorance of the underlying mechanism responsible for the motion. If we model a medium with some assumed charge density ρ and calculate the charge motion in response to applied fields, we usually take the background region to be a vacuum and use $\sigma = 0$ for it; effectively, we are calculating the medium's conductivity if the response turns out to be simply proportional to the applied field. On the other hand, if we specify a conductivity σ for some medium, then we usually concede ignorance of the properties ρ and $\mathbf{v}$ of the atomic charges and use $\rho = 0$ in the equations while letting $\mathbf{J} = \sigma\mathbf{E}$. We could even take a perverse attitude and claim knowledge, or a model, of the density and motion of one charge species while relegating the properties of another type of charge in the region to oblivion by attributing a conductivity σ to the background medium in which the interesting species responds to applied fields. In that case, our equations of motion would involve ρ and $\mathbf{v}$ but also feature the constitutive parameter σ to describe the effects of the other species.

Displacement current, as we saw, does not consist of a flow of charge but rather manifests changes in an accumulation of charge. Such variations of stored charges are accompanied by alterations in the resultant $\mathbf{D}$ field and generate the same magnetic effects as do currents formed by the flow of free charges. The density of displacement current is the field $\partial \mathbf{D}/\partial t$.

In material media, part of the displacement current, the excess over the vacuum displacement current, does reflect some motion of charge, but that motion (displacement) is only of limited extent. In response to an applied field, the *polarization current* in a dielectric manifests the slight displacement of bound charges, as opposed to the long-distance drift of free charge carriers in a conductor. The atomic configuration in a dielectric may get stretched or distorted in the force field, so that the atoms get *polarized*. That slight stretching motion of the charged atomic constituents contributes to the macroscopic displacement current.

MAGNETIC FIELD OF A MOVING CHARGE

As another application of the Ampère-Maxwell law, let us find the magnetic field generated by a moving point charge, the most rudimentary of currents.

Example 3.9

Point charge q drifts uniformly at speed v along a straight line in vacuum. Its motion constitutes a current, so that a magnetic field, as well as an electric one, should be generated by this source. We seek $\mathbf{H}(\mathbf{r}, t)$ *at any observation point and time.* ◈

We set the z-axis of our coordinate system along the path of the charge, with the origin at its initial location. From the azimuthal symmetry, we see at once that the magnetic field will be directed in circles around this axis, with the circles lying in planes perpendicular to the motion. The field strength at any one instant will be the same all around any such circle. Hence the mmf along one such circle, of radius ρ, will be $U = \oint \mathbf{H} \cdot d\mathbf{l} = 2\pi\rho H$, in terms of the unknown azimuthal field amplitude $H = H(\rho, z, t)$ on that circle. The Ampère-Maxwell law will tell us the value of U and hence of H.

To apply the law, we need to choose an open surface over which to evaluate the electric flux, one whose edge is the circle of radius ρ. Since the medium is a vacuum, there is no conduction current and we need to consider only the $\mathbf{D}$ field. At time t, the point charge is at location $\mathbf{v}t$ and, at that instant, it should be surrounded by its spherically symmetric inverse-square Coulomb field, centered at $\mathbf{v}t$:

$$\mathbf{D}(\mathbf{r}, t) = \frac{q(\mathbf{r} - \mathbf{v}t)}{4\pi|\mathbf{r} - \mathbf{v}t|^3}, \tag{3.45}$$

where $\mathbf{r}$ is the fixed observation point and $\mathbf{v}t$ is the moving source point at the observation time t. We are here making the perhaps rash assumption that the Coulomb field is not affected by the motion of the source charge, other than being convected along with it; we will reconsider that assumption later.

With respect to the source point at $\mathbf{v}t$ as center, this is a radial field and it will be convenient to select a surface whose normal at time t is also radial, with that center point. We therefore choose surface S to be the spherical cap centered at $\mathbf{v}t$ and extending only to the circle of radius ρ perpendicular to and centered on the path of the charge and passing through the observation point at $\mathbf{r}$. We can conveniently perform the surface integration in terms of spherical coordinates centered at $\mathbf{v}t$ at the instant t, as indicated in Figure 3-11. The spherical cap has radius $r_0 = |\mathbf{r} - \mathbf{v}t|$ and extends to elevation angle θ_0, the angle between the z-axis and the relative position vector $\mathbf{r} - \mathbf{v}t$ of the observation point at $\mathbf{r}$. The flux across the spherical cap is

$$\Psi\{S\} = \int_S \mathbf{D} \cdot d\mathbf{S} = \int_0^{2\pi} \int_0^{\theta_0} \frac{q}{4\pi r_0^2} \, \hat{\mathbf{r}} \cdot \hat{\mathbf{r}} r_0^2 \sin\theta \, d\theta \, d\varphi$$

$$= \tfrac{1}{2} q(1 - \cos\theta_0), \tag{3.46}$$

where $\hat{\mathbf{r}}$ is the radial unit vector at θ, φ, on the spherical cap whose center is at $\mathbf{v}t$. There is no current $I\{S\}$ in the vacuum, so that we need to equate the mmf $U\{C\}$ to only the displacement current $d\Psi\{S\}/dt$. The next step is therefore to obtain the time derivative of this flux.

But is the displacement flux $\Psi\{S\} = \tfrac{1}{2} q(1 - \cos\theta_0)$ we just calculated a function of time? Yes, it is, because θ_0 increases in time as the charge moves. The displacement current is therefore

$$d\Psi/dt = \tfrac{1}{2} q \sin\theta_0 \big[d\theta_0/dt \big] \tag{3.47}$$

and we need the rate of change of the elevation angle to the fixed observation point as the source point moves. Since $\theta_0(t)$ is the angle between the vectors $\hat{\mathbf{z}}$ and $\mathbf{r} - \mathbf{v}t$, their dot product gives $r_0 \cos\theta_0 = \hat{\mathbf{z}} \cdot (\mathbf{r} - \mathbf{v}t) = z - vt$, from which we could get $d\theta_0/dt$ easily, but for the fact that $r_0 = r_0(t)$ also changes in time. We prefer to express $\theta_0(t)$ in terms of quantities that do not change in time; among these is ρ, the radial cylindrical coordinate of the observation point, which is related to the spherical coordinates by $\rho = r_0(t) \sin\theta_0(t)$. We therefore write

$$z - vt = \rho \cot\theta_0(t) \tag{3.48}$$

and differentiate with respect to time t, noting that both ρ and z are fixed coordinates of the observation point, to find

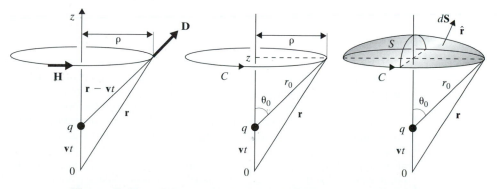

Figure 3-11   Calculation of magnetic field of moving charge.

$$-v = -\rho \csc^2 \theta_0 [d\theta_0/dt] \quad \text{or} \quad d\theta_0/dt = [v/\rho]\sin^2 \theta_0. \tag{3.49}$$

The displacement current is therefore

$$d\Psi/dt = \tfrac{1}{2}q[v/\rho]\sin^3 \theta_0 \tag{3.50}$$

and this equals the mmf $U = 2\pi\rho H$, giving

$$H_\varphi = \frac{qv \sin^3 \theta_0}{4\pi\rho^2}. \tag{3.51}$$

This is the answer we sought, but it is expressed in terms of mixed cylindrical and spherical coordinates, the latter changing as the charge moves. Let us look at this in unmingled coordinates. We can reexpress it in terms of only the moving spherical coordinate system by substituting $\rho = r_0 \sin \theta_0$, so that

$$H_\varphi = \frac{qv \sin \theta_0}{4\pi r_0^2}. \tag{3.52}$$

This looks familiar! It is identical to the field of a current element, but with $I dl$ replaced with qv.

In this version of the expression for the field, we must recall that both r_0 and θ_0 vary in time as the charge moves. We can dispense with specific coordinate systems by expressing the **H** field in terms of the fixed observation position vector **r** and the explicit observation time t. We need only recognize that θ_0 is the angle between the charge velocity vector **v** and the relative position vector $\mathbf{r} - \mathbf{v}t$, as well as the fact that the azimuthal direction of the **H** field is perpendicular to both, so that the quantity $[v \sin \theta_0 r_0]\hat{\boldsymbol{\varphi}}$ can be expressed as the vector $\mathbf{v} \times (\mathbf{r} - \mathbf{v}t)$. This gives us the magnetic field of the moving point charge explicitly in terms of the observation point **r** and time t, as

$$\mathbf{H}(\mathbf{r}, t) = \frac{q\mathbf{v} \times (\mathbf{r} - \mathbf{v}t)}{4\pi|\mathbf{r} - \mathbf{v}t|^3}. \tag{3.53}$$

In this form, we can recognize the property of a uniformly moving point charge that the electric and magnetic fields it generates are simply related by

$$\mathbf{H}(\mathbf{r}, t) = \mathbf{v} \times \mathbf{D}(\mathbf{r}, t). \tag{3.54}$$

Also, because $\mathbf{v} \times \mathbf{v} = 0$, a slightly simpler form is

$$\mathbf{H}(\mathbf{r}, t) = \frac{q\mathbf{v} \times \mathbf{r}}{4\pi|\mathbf{r} - \mathbf{v}t|^3}. \tag{3.55}$$

We again recognize these expressions to be of the same form as the magnetic field of a current element, whose moment $I dl$ is replaced by $q\mathbf{v}$ and whose field moves with the point charge. The magnetic field pattern in space is therefore just that of a current element, but the pattern as a whole is convected along with the point charge.

The magnetic field strength as a function of time, as it might be sensed at a fixed observation point as the point charge moves by, is plotted in Figure 3-12. The curve is drawn for an observation point such that $\rho/z = 5/6$. Note that the magnetic field peaks, at the strength $H_0 = qv/4\pi\rho^2$, when the point charge comes abreast of the observation point, which is located at (ρ, z). Note also that the time scale ρ/v is an indication of how long the observed field remains significantly strong. The curve is sharp for observation

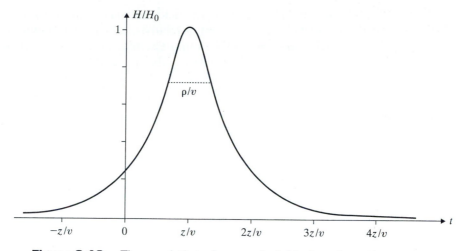

Figure 3-12 Time variation of magnetic field of moving point charge.

points near the path of the charge (small ρ); it is broader if observed at more remote points (large ρ).

Comment. One difficulty with the solution we have just found for the magnetic field generated by a uniformly moving point charge is that it is wrong!

However, it is only slightly wrong, to an insignificant extent under normal circumstances, but the correct result departs from this one when the speed and energy of the moving point charge become relativistic, which is when the speed is a significant fraction of the speed of light or the energy is not negligible compared to the particle's rest mass energy. This may be of concern to us for both a practical and a theoretical reason. Practically, it is quite feasible to bring electrons, and even protons, to highly relativistic speeds, and we should be careful not to apply low-speed formulas to such cases. Theoretically, we should be disappointed with any nonrelativistic result from electromagnetics, because one basis for relativity theory is that Maxwell's equations remain valid as the speed of light is approached. The theory of relativity teaches that it is not electromagnetics but Newtonian mechanics that must be corrected (for example, by distinguishing observation time from proper time).

What is wrong with our application of the Ampère-Maxwell law when speeds do become relativistic is that we assumed that the Coulomb field of the charge is undistorted when the charge moves, merely moving with it. This is essentially true at ordinary, low speeds but not when v^2/c^2 is no longer negligible. What we find from a relativistically correct analysis is that the moving charge's **D** field is no longer isotropic: Its strength is different in different directions, the spherical symmetry being broken by the existence of a preferred direction, that of the motion. The field strength is diminished in the direction of motion and enhanced at right angles to the motion. Interestingly, it is still true that the **D** field emanates from the present location of the charge and that the magnetic field is related to it by $\mathbf{H} = \mathbf{v} \times \mathbf{D}$, but this last relation also indicates that as the Coulomb electric field distorts at high energies, so does the accompanying magnetic field. At extremely relativistic energies, both the electric field and the magnetic field become concentrated essentially in a plane perpendicular to the motion, moving with the charge at nearly the speed of light.

The integral-law technique we used to get the magnetic field of a moving charge is not limited to uniform motion. Consider the accelerated charge in the next example.

Example 3.10

A metal ball of mass M, radius a carries charge Q. Starting from rest at time $t = 0$ at the origin, the ball falls (acceleration of gravity $= g$) along the negative z-axis, in air. Refer to Figure 3-13.

(a) *Obtain the mmf $U(t)$ along the circumference of a circle of radius ρ_0 in the plane $z = z_0$, centered on the z-axis and oriented azimuthally, with $\rho_0 > a$.*

(b) *What magnetic field $\mathbf{H}(\rho, \varphi, z, t)$ is generated by the falling ball?* «»

We approach this the same way we did the uniformly moving charge, but the differentiation of the electric flux must now account for the accelerated motion.

(a) At time $t > 0$, the ball is at $z = -\frac{1}{2}gt^2$. The Coulomb field is radial, of strength $D = Q/4\pi r^2$, centered on the moving ball. The flux Ψ on the cap of a sphere centered at the ball, out to elevation angle $\theta_0 = \tan^{-1}[\rho_0/(z_0 + \frac{1}{2}gt^2)]$, is $\Psi = \int \mathbf{D} \cdot d\mathbf{S} = \iint (Q/4\pi) \sin\theta \, d\theta \, d\varphi = \frac{1}{2}Q[1 - \cos\theta_0]$. The mmf is $U = d\Psi/dt$, where $\cot\theta_0(t) = (z_0 + \frac{1}{2}gt^2)/\rho_0$ differentiates to $-\csc^2\theta_0[d\theta_0/dt] = gt/\rho_0$. Hence, $U(t) = \frac{1}{2}Q\sin\theta_0[d\theta_0/dt] = -\frac{1}{2}Q\sin^3\theta_0[gt/\rho_0]$. Expressing $\sin\theta_0$ as an explicit function of t, we have $\sin\theta_0 = \rho_0/[\rho_0^2 + (z_0 + \frac{1}{2}gt^2)^2]^{1/2}$ and therefore $U(t) = -\frac{1}{2}Qgt\rho_0^2/[\rho_0^2 + (z_0 + \frac{1}{2}gt^2)^2]^{3/2}$.

(b) By azimuthal symmetry, $U = 2\pi\rho_0 H_\varphi$ and the result is valid for any ρ_0, z_0 outside the ball, so that $\mathbf{H}(\rho, \varphi, z, t) = -\hat{\varphi}[Qg/4\pi]\rho t/[\rho^2 + (z + \frac{1}{2}gt^2)^2]^{3/2}$.

Figure 3-14 shows the magnetic field strength as a function of time, as it might be sensed at a fixed observation point as the point charge falls nearby. The curve is drawn for an observation point such that $\rho/(-z) = 5/6$ and the time scale t_0 expresses the height of the observer as $z = -\frac{1}{2}gt_0^2$, below the origin. Note that the peak of the field oc-

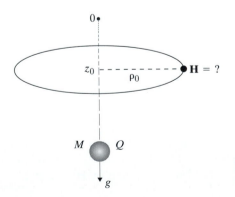

Figure 3-13   A magnetic field is generated by a falling charge.

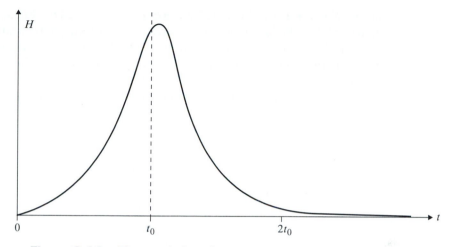

Figure 3-14   Time variation of magnetic field of falling point charge.

curs slightly after the ball comes abreast of the observer, because the ball's speed keeps increasing, which tends to make the field stronger. Despite this effect, the increasing distance of the charge from the observer eventually reduces the field strength toward zero, as shown.

GAUSS'S MAGNETIC LAW

We amended Ampère's law, to make it applicable in the time-varying case as the Ampère-Maxwell law, by recognizing the inconsistency of the static version with the law of conservation of charge when we allow the open surface S in the law to close on itself by shrinking the edge curve C to a point. We should now ask whether a similar amendment is needed for the Faraday-Maxwell law. What happens to that law when we allow the open surface S to become closed?

The Faraday-Maxwell law equates the emf $V\{C\}$ around the edge curve C of the open surface S to the rate of decrease $-d\Phi\{S\}/dt$ of the magnetic flux through S. If the curve C shrinks to a point, the emf becomes the line integral of the electric field at that point over a path of ultimately zero length, so that the emf $V\{C\}$ approaches zero. Simultaneously, the surface S closes and the Faraday-Maxwell law can remain in effect only if we can guarantee that the rate of change of the magnetic flux over a closed surface is also zero.

We find experimentally that the magnetic flux over any closed surface is in fact always zero; this is the magnetic version of Gauss's law:

$$\Phi\{S\} = 0 \qquad \text{or} \qquad \oint_S \mathbf{B} \cdot d\mathbf{S} = 0, \qquad (3.56)$$

for any closed surface S. As much flux leaves a region enclosed by the surface as enters it. This fact allows no time variation of the closed surface integral and leaves the Faraday-Maxwell law self-consistent and in no need of amendment.

Gauss's law states that the closed surface integral of the **D** field equals the enclosed charge $Q\{V\}$; the magnetic version asserts that no "magnetic charge" is ever enclosed. That this is so reflects the fact that no one has ever found a concentrated source of magnetic flux unaccompanied by a corresponding sink of flux. There is no magnetic version of electric charge.

That does not mean that none could ever be found and, in fact, investigators are still looking for magnetic charges, in cosmic ray debris, for example. Several "sightings" have been reported but later discredited. There is strong motivation to find magnetic charge, because one consequence of quantum theory is that if there exists magnetic charge, then electric charge must be quantized. Of course, we know electric charge to be quantized by the charge on an electron, so that it would be supportive and comforting to discover some magnetic charge to "explain" this.

If magnetic charge were ever found, Gauss's magnetic law would have to be modified, by simply changing the zero to the amount of magnetic charge enclosed by the surface. Additionally, the Faraday-Maxwell law would become inconsistent with the magnetic version of conservation of charge, until we add to $d\Phi/dt$, as an additional source of closed-loop emf, the magnetic current that may cross the open surface S whose edge is the closed curve C. As long as no confirmed discovery of magnetic charges and currents is made, however, we retain Gauss's magnetic law and the Faraday-Maxwell law intact, asserting that magnetic flux is always formed in closed circuits.

The option of amending these two laws to allow for magnetic charges and currents is nevertheless of more than academic interest, because inventing such fictitious sources of magnetic flux and emf, particularly at boundaries between regions, can often be helpful in solving realistic field problems. Boundaries between regions are our next concern.

BOUNDARY CONDITIONS

Until now, we have examined electric and magnetic fields in infinitely extended, uniform media; that is, the medium had the same, constant constitutive parameters μ, ε, σ throughout the universe. We now wish to allow these parameters to be different at different points. The general case can be extremely difficult to deal with. The simplest case is that of just *two* uniform regions, with an *abrupt* change of parameters from one to the other, across an interface, as illustrated in Figure 3-15. Each region forms a half-space; the unit normal at the interface is $\hat{n}$.

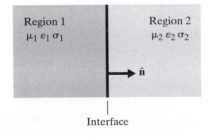

Region 1
$\mu_1 \, \varepsilon_1 \, \sigma_1$

Region 2
$\mu_2 \, \varepsilon_2 \, \sigma_2$

$\hat{n}$

Interface

Figure 3-15   Interface between two uniform regions.

Finding the fields in the two regions entails solving Maxwell's equations in all of space. Our ability to solve the equations in either of the two regions does not entitle us to claim we have the solution everywhere. We still need to satisfy Maxwell's equations at the points on the interface itself. What does it mean to satisfy Maxwell's equations at a collection of points at which the constitutive parameters that enter into the equations suffer discontinuities? To answer this question, we need to examine the field discontinuities permitted by Maxwell's equations at the boundary between two dissimilar regions of space.

To satisfy Maxwell's equations at the points of a surface across which the constitutive parameters suffer an abrupt discontinuity, we use the Maxwell equations in integral form, $V = -d\Phi/dt$ and $U = I + d\Psi/dt$, which involve integrations in space and hence are not troubled by spatial discontinuities. Applying these equations at the points of the interface results in *boundary conditions* that can be used to relate the fields on one side of the surface of discontinuity to those on the other side.

To obtain these boundary conditions, we first ask the question, Must fields vary continuously, or can they jump suddenly from one value at one point to another value at an immediately neighboring point? This question requires us to compare the fields, say $\mathbf{E}_1$ and $\mathbf{E}_2$, at two adjacent points.

Now adjacent points automatically define a direction in space and also a bisecting plane that separates the two points, as in Figure 3-16. The direction is that from one point to the other; the plane is the one that is perpendicular to and bisects the line that joins the two points. In the figure, (a) shows two adjacent points and (b) depicts the direction from point 1 to point 2. The bisecting plane is shown on edge in (c) and obliquely in (d). In (e), two hypothetical values of an electric field, associated with the two neighboring points, are shown with a hypothetically finite difference between them; that is, the field suddenly jumps in its vector value from one point to the immediately adjacent one, instead of varying continuously from one to the other. Our goal is to determine what restrictions Maxwell's equations place on the possible discontinuity in the field. In (f), the two fields are shown decomposed into their components perpendicular to the bisecting plane and tangential to it; we will find that Maxwell's equations dictate the sorts of discontinuities these field components may

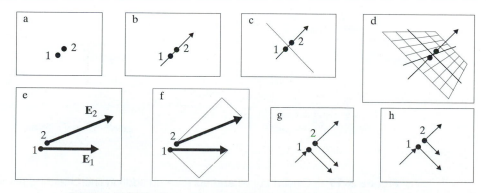

Figure 3-16   Two adjacent points and their bisecting plane.

undergo. A hypothetical case of fields at the two points that have the same tangential component but dissimilar normal components is given in (g), and another case, with equal normal components but with a discontinuity in tangential component, is shown in (h). The discontinuities allowed by Maxwell's equations can be determined by comparing the fields at adjacent points, as they enter into those equations.

Now what the equations speak of are closed line and surface integrals of the various electromagnetic fields, so we should specify the curves or surfaces for the integrations. In order to learn something about the allowed discontinuities at adjacent points, we must see to it that the elements of length or area of these curves or surfaces pass through the two neighboring points. This puts one part of the curve or surface on one side of the bisecting plane and the rest on the other side. The curves and surfaces should hence straddle, and tightly hug, the bisecting plane, as suggested in Figure 3-17, which shows both the upper side of the plane (on the left of the figure) and the lower side (on the right).

As a consequence of the hugging of the bisecting plane by the curve or surface of integration, the line elements dl along the curve have directions tangential to the bisecting plane and the surface elements dS are directed perpendicular to that plane; at least that becomes the case in the limit that the two neighboring points, 1 and 2, nearly merge, with the upper and lower portions of either the closed curve or the closed surface being then separated only infinitesimally. The most important consequence of the merging of the upper and lower portions of the curve or surface is that, in the limit, the surface that is surrounded by the closed curve can be merely a thin sliver, ultimately of zero area, and that the volume enclosed by the closed surface approaches a flat area on the bisecting plane, ultimately of zero volume.

Turning our attention first to the closed curve that straddles the plane, we have been led to choose curves of integration for the emf V and the mmf U such that each curve passes through the two points 1 and 2 and such that the surfaces of integration bounded by those curves, on which the electric and magnetic fluxes, Ψ and Φ, are to be evaluated, are infinitesimally thin. In the limit of zero separation between the points, we deal with the electric and magnetic fields just at the interface, and on its two sides. For this infinitesimal separation, however, the surface bounded by the curve

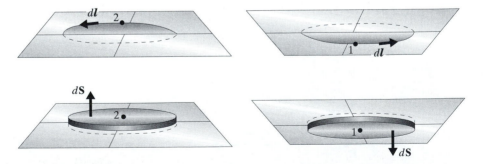

Figure 3-17   Closed curve and surface straddling the bisecting plane.

can be merely a thin sliver, with zero area in the limit. As a result, the flux or current across that thin surface must become zero in the limit, unless the current density were infinite.

We postpone discussion of the extraordinary case of infinite current density for a moment and consider first the normal situation of finite densities, leading to zero total flux and current across the thin sliver. Thus, Φ or Ψ or I must be zero in the limit of zero separation between the two halves of the closed curve. Then so are $d\Phi/dt$ and $d\Psi/dt$ zero and Maxwell's equations then require V and U to be zero along the closed curve. This means that the line integral of $\mathbf{E}$, and also of $\mathbf{H}$, along the curve must vanish, which implies that the integral along one side of the bisecting plane and that on the other side must be equal and opposite, to cancel to zero. This can happen for *every* closed curve that straddles and hugs the plane only if the *tangential* field components are the *same* on both sides. That only the tangential components are so restricted follows from the $V = \oint \mathbf{E} \cdot d\mathbf{l}$ or $U = \oint \mathbf{H} \cdot d\mathbf{l}$ integrals; these involve only the field components along $d\mathbf{l}$, which is itself tangential to the plane. The components of $\mathbf{E}$ and $\mathbf{H}$ normal to the plane do not enter into the line integrals and are not affected by the $V = 0$ and $U = 0$ requirements in the limit.

The result just obtained applies to any two neighboring points; they mandate *continuity* of the $\mathbf{E}$ and $\mathbf{H}$ field components *tangential* to the bisecting plane between the two adjacent points. When the bisecting plane is the interface between two media of dissimilar constitutive parameters, this continuity requirement becomes a boundary condition that restricts the discontinuities that the fields may suffer. Figure 3-18 shows allowable jumps in $\mathbf{E}$ and $\mathbf{H}$ field vectors across a discontinuity plane; the tangential components must match across the plane.

To summarize, in the ordinary case of finite fields and current densities, the boundary conditions at an interface between two regions are, for the field components tangential to the interface,

$$\text{continuity of } \mathbf{E}_{\text{tan}}: \qquad (\mathbf{E}_1)_{\text{tan}} = (\mathbf{E}_2)_{\text{tan}}, \qquad (3.57)$$

$$\text{continuity of } \mathbf{H}_{\text{tan}}: \qquad (\mathbf{H}_1)_{\text{tan}} = (\mathbf{H}_2)_{\text{tan}}. \qquad (3.58)$$

Imposing these two conditions is equivalent to satisfying the two Maxwell equations $V = -d\Phi/dt$ and $U = I + d\Psi/dt$ at the points of the interface.

Turning next to the closed surface that straddles the plane, we have chosen surfaces of integration for the electric and magnetic fluxes Ψ and Φ such that each closed surface passes through the two points 1 and 2 and such that the volumes of integration

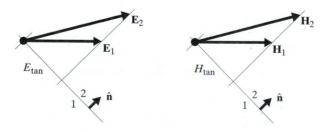

Figure 3-18 Permissible discontinuities of $\mathbf{E}$ and $\mathbf{H}$ fields.

bounded by those surfaces are infinitesimally thin, with zero volume in the limit. As a result, the total charge enclosed by the surface must become zero in the limit, unless the charge density were infinite.

We put off discussion of the extraordinary case of infinite charge density for a moment and consider first the normal situation of finite charge density, leading to zero total charge Q enclosed by the surface. Thus, $\Psi_0 = Q$ must be zero in the limit of zero separation between the two halves of the closed surface. Maxwell's equations also require Φ_0 to be zero over this, or any, closed surface. This means that the closed surface integral of $\mathbf{D}$, and also of $\mathbf{B}$, must vanish, which implies that the integral along one side of the bisecting plane and that on the other side must be equal and opposite, to cancel to zero. This can happen for *every* closed surface that straddles and hugs the plane only if the *normal* field components are the same on both sides. Only the normal components are so restricted, as follows from the $\Psi_0 = \oint \mathbf{D} \cdot d\mathbf{S}$ or $\Phi_0 = \oint \mathbf{B} \cdot d\mathbf{S}$ integrals; these involve only the field components along $d\mathbf{S}$, which is itself normal to the plane. The components of $\mathbf{D}$ and $\mathbf{B}$ tangential to the plane do not enter into the surface integrals and are not affected by the requirements that $\Psi_0 = 0$ and $\Phi_0 = 0$.

This result applies to any two neighboring points; they require *continuity* of the $\mathbf{D}$ and $\mathbf{B}$ field components *normal* to the bisecting plane between the two adjacent points. When the bisecting plane is the interface between two media of dissimilar constitutive parameters, this continuity requirement becomes a boundary condition that restricts the discontinuities that the fields may suffer. Figure 3-19 shows allowable jumps in $\mathbf{D}$ and $\mathbf{B}$ field vectors across a discontinuity plane; the normal components must match across the plane.

To summarize, in the ordinary case of finite charge density, the boundary conditions at an interface between two regions are, for the field components perpendicular to the interface,

$$\text{continuity of } \mathbf{D}_{\text{nor}}: \qquad \left(\mathbf{D}_1\right)_{\text{nor}} = \left(\mathbf{D}_2\right)_{\text{nor}}, \tag{3.59}$$

$$\text{continuity of } \mathbf{B}_{\text{nor}}: \qquad \left(\mathbf{B}_1\right)_{\text{nor}} = \left(\mathbf{B}_2\right)_{\text{nor}}. \tag{3.60}$$

Imposing these two conditions is equivalent to satisfying the two subsidiary requirements of the Maxwell equations, $\Psi_0 = Q$ and $\Phi_0 = 0$ for closed surfaces, at the points of the interface.

Satisfying the continuity requirements on both the tangential components of $\mathbf{E}$ and the normal components of $\mathbf{D}$ at an interface between media of given, differ-

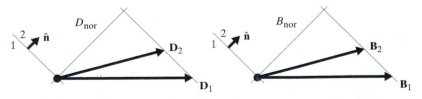

Figure 3-19   Permissible discontinuities of **D** and **B** fields.

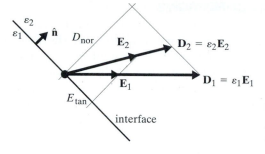

Figure 3-20 Unique E_{tan} and D_{nor} fix all of $\mathbf{E}_1$, $\mathbf{E}_2$, $\mathbf{D}_1$, $\mathbf{D}_2$ from any one.

ent permittivities leads to a determination of the full discontinuity of each of the vectors, as illustrated in Figure 3-20. This shows the $\mathbf{E}$ and $\mathbf{D}$ fields on the two sides of the interface between two dielectrics of given permittivities ε_1 and ε_2, with $\varepsilon_1 > \varepsilon_2$. Given any one of the four vectors $\mathbf{E}_1$, $\mathbf{E}_2$, $\mathbf{D}_1$, $\mathbf{D}_2$ and the values of the two permittivities, the other three vectors are fully determined by the requirements of continuity of tangential $\mathbf{E}$ and of normal $\mathbf{D}$. As a result, the discontinuities in both $\mathbf{E}$ and $\mathbf{D}$ are determined by these boundary conditions. The same point is illustrated also in the following example.

Example 3.11

We want to verify that a field that obliquely traverses the boundary between dissimilar media generally bends and changes magnitude.
Let the interface between two dielectrics with $\varepsilon_1 = 2\varepsilon_0$ and $\varepsilon_2 = 6\varepsilon_0$ coincide with the xy-plane. Suppose that the electric field within side $1(z < 0)$ were uniform, with a magnitude $E_1 = 12$ V/m and a direction inclined at $\theta_1 = 30°$ to the z-axis. Can the electric field be uniform on the other side of the plane? If so, what are the strength and direction of $\mathbf{E}_2$? ◈

Since the field components tangential to the interface plane must be continuous, those of $\mathbf{E}_2$ must be the same as those of $\mathbf{E}_1$, whose tangential part has magnitude $E_1 \sin\theta_1 = 6$ V/m, with some given orientation in the xy-plane. Thus $E_2 \sin\theta_2 = 6$ V/m and $\mathbf{E}_2$ is in the same plane as $\mathbf{E}_1$ and $\hat{\mathbf{n}}$. The z-component, which is normal to the plane, must be found. Now $\mathbf{D}_1 = \varepsilon_1\mathbf{E}_1 = 2\varepsilon_0\mathbf{E}_1$ has as its normal component, $D_{1z} = 2\varepsilon_0 E_1 \cos\theta_1$, which must be the same on the other side: If $\mathbf{E}_2$ is inclined at θ_2 to the normal, $D_{2z} = \varepsilon_2 E_2 \cos\theta_2 = 6\varepsilon_0 E_2 \cos\theta_2 = 2\varepsilon_0 E_1 \cos\theta_1$ or $E_2 \cos\theta_2 = (E_1 \cos\theta_1)/3 = 3.4641$ V/m. The two results give at once that $\tan\theta_2 = 6/3.4641 = 1.7321$ or $\theta_2 = 60°$ and $E_2 = 6.9282$ V/m. The electric field jumps in strength and bends in direction, while maintaining the same tangential orientation in the xy-plane.
Note that the $\mathbf{D}$ field's discontinuity is now also known: It jumps from $\mathbf{D}_1 = 2\varepsilon_0\mathbf{E}_1$, with a strength of $24\varepsilon_0$ C/m^2, to $\mathbf{D}_2 = 6\varepsilon_0\mathbf{E}_2$, with a magnitude of $41.569\varepsilon_0$ C/m^2, and bends from $30°$ to $60°$ from one side of the plane to the other. The field is uniform within the upper region as well, but with a new strength and direction.

EXTRAORDINARY BOUNDARY CONDITIONS

To complete the development of boundary conditions, we should consider the extraordinary, but important, cases in which the limit of zero area or volume does *not* necessarily imply that the area must have zero current traversing it, or that the volume must enclose zero charge. We still exclude finite electric or magnetic flux Ψ or Φ over a sliver of infinitesimal area, because that would require an infinite **D** or **B** field, a physical possibility. However, we must not hastily exclude the possibility of infinite density of charge or current because, from a macroscopic viewpoint, charge *can* be concentrated to such an extent that a finite amount of charge can be smeared over a surface, occupying negligible volume. If that charge is moving, then a finite current flows in a sheet along the surface, which has negligible thickness. In those circumstances in which a surface charge or current sheet has been caused to appear on a surface of zero or infinitesimal thickness, the total charge Q within a volume that straddles and hugs that surface can be finite even within zero volume and the total current I crossing a sliver of area that straddles and hugs the surface can be finite even across zero area.

In those cases, the appropriate densities that describe the amount of charge or current on the surface are the surface charge density ρ_S (C/m^2) instead of volume charge density ρ (C/m^3) and the surface current density vector **K** in amperes per unit width (A/m) instead of the volume current density vector **J** (A/m^2). In Figure 3-21, the area A on the plane within the thin volume has surface charge of density ρ_S, so that the total enclosed charge is the finite amount $Q = \rho_S A$ despite the lack of finite volume in the limit. The width w on the plane within the thin sliver of area straddling the plane carries current as a sheet with density K, so that the sliver has a finite current $I = Kw$ crossing it, along the plane and at right angles to the width w, despite the lack of finite thickness in the limit.

What happens to the boundary conditions when surface sources can be present? Gauss's law requires $\Psi_0 = Q$ and now $Q = \rho_S A$ remains finite even as the volume enclosed by the surface that passes through the neighboring points on the two sides of the charged surface loses all thickness. The electric flux Ψ_0 over the closed surface then becomes $\Psi_0 = \oint \mathbf{D} \cdot d\mathbf{S} = \mathbf{D}_2 \cdot \hat{\mathbf{n}} A + \mathbf{D}_1 \cdot (-\hat{\mathbf{n}})A = (\mathbf{D}_2 - \mathbf{D}_1) \cdot \hat{\mathbf{n}} A$ if, as shown, $d\mathbf{S}$ is along $+\hat{\mathbf{n}}$ on the side of point 2 but along $-\hat{\mathbf{n}}$ on the other side. (The area A has, for simplicity, been taken small enough for the field **D** not to vary significantly over either the top or the bottom part of the closed surface. Rigorously, we should consider the area A to be a first-order infinitesimal and the thickness to be a higher-order one.) Canceling the arbitrary area, from both sides of $\Psi_0 = Q$ leaves

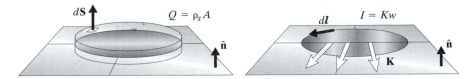

Figure 3-21 Surface charge and surface current sheet.

$$\left(\mathbf{D}_2 - \mathbf{D}_1\right) \cdot \hat{\mathbf{n}} = \rho_S \qquad (3.61)$$

as the more general boundary condition to replace the continuity of normal $\mathbf{D}$ at neighboring points. This relation states that the component of $\mathbf{D}$ normal to the bisecting plane between neighboring points is discontinuous by just the surface charge density on that plane, if any has been placed there. This boundary condition is equivalent to satisfying Gauss's law on the plane.

The Ampère-Maxwell law $U = I + d\Psi/dt$, applied to the closed curve that surrounds the thin sliver of area that can now carry a current sheet K, requires that $U = I + 0$ ($\Psi = 0$ in the limit, since $\mathbf{D}$ will not be infinite), and $I = Kw$ remains finite even as the two parts of the closed curve that passes through the neighboring points on the two sides of the current-carrying surface become only infinitesimally separated. The mmf $U = \oint \mathbf{H} \cdot d\boldsymbol{l}$ along the closed curve becomes $(H_2 - H_1)w$ if, as shown, $d\boldsymbol{l}$ is along the direction of $\mathbf{K} \times \hat{\mathbf{n}}$ on the side of point 2 but along $-\mathbf{K} \times \hat{\mathbf{n}}$ on the other side. (The width w has, for simplicity, been taken small enough for the field $\mathbf{H}$ not to vary significantly over either the top or the bottom part of the closed curve. Rigorously, we should consider the width w to be a first-order infinitesimal and the thickness of the sliver to be a higher-order one.) Canceling the arbitrary width w from both sides of $U = I$ leaves

$$H_2 - H_1 = K \qquad (3.62)$$

as the more general boundary condition to replace the continuity of tangential $\mathbf{H}$ at neighboring points. In this boundary condition, the component of $\mathbf{H}$ along the direction of $\mathbf{K} \times \hat{\mathbf{n}}$ is meant; note that the right-hand rule continues to relate the direction of $\mathbf{H}$ and of the current in the Ampère law. In vector form, the boundary condition can be stated as

$$\hat{\mathbf{n}} \times \left(\mathbf{H}_2 - \mathbf{H}_1\right) = \mathbf{K}, \qquad (3.63)$$

since both the direction and magnitude of the discontinuity are thereby properly (i.e., by the right-hand rule) related to the surface current. This relation states that the component of $\mathbf{H}$ tangential to the bisecting plane between neighboring points and at right angles to the direction of the current sheet is discontinuous by just the surface current density on that plane, if any has been placed there. This boundary condition is equivalent to satisfying the Ampère-Maxwell law on the plane.

Satisfying the continuity requirements (or, if there is a current sheet, the discontinuity requirements) on the tangential components of $\mathbf{H}$ and the continuity requirements on normal components of $\mathbf{B}$ at an interface between media of given, different permeabilities leads to a determination of the full discontinuity of each of the vectors, as illustrated in the following example.

Example 3.12

The interface between two permeable media with $\mu_1 = 2\mu_0$ and $\mu_2 = 6\mu_0$ coincides with the xy-plane. Suppose that the magnetic field within side $1(z < 0)$ were uniform, with a magnitude $H_1 = 12$ A/m and a direction inclined at $\theta_1 = 30°$ to the z-axis, at zero azimuth. Suppose also that a current sheet of strength $\mathbf{K} = (3\hat{\mathbf{x}} + 4\hat{\mathbf{y}})$ A/m were known to *be flowing along the xy-plane. Could the magnetic field be uniform on the other side of the plane? If so, what would be the strength and direction of $\mathbf{H}_2$?* ◈

Since the field component of $\mathbf{B}$ normal to the interface plane must be continuous, that of $\mathbf{B}_2$ must be the same as that of $\mathbf{B}_1$. The latter has normal component $B_1 \cos\theta_1 = \mu_1 H_1 \cos\theta_1 = \mu_0(20.7846 \text{ A/m})$, and $B_{2z} = B_2 \cos\theta_2$ must have this same value, if $\mathbf{B}_2$ is inclined at θ_2 to the normal. We need to find the tangential components of $\mathbf{H}_2$.

The components of $\mathbf{H}$ tangential to the interface plane jump discontinuously, in accordance with $\hat{\mathbf{n}} \times \mathbf{H}_2 = \hat{\mathbf{n}} \times \mathbf{H}_1 + \mathbf{K}$, with $\hat{\mathbf{n}} = \hat{\mathbf{z}}$. This yields the tangential components of $\mathbf{H}_2$, but not the normal component, as follows. $\hat{\mathbf{z}} \times \mathbf{H}_2 = \hat{\mathbf{z}} \times \mathbf{H}_1 + \mathbf{K} = \hat{\mathbf{z}} \times \left[H_1 \sin\theta_1 \hat{\mathbf{x}} + H_1 \cos\theta_1 \hat{\mathbf{z}} \right] + (3\hat{\mathbf{x}} + 4\hat{\mathbf{y}})$, which is $\hat{\mathbf{z}} \times \mathbf{H}_2 = 6\hat{\mathbf{y}} + (3\hat{\mathbf{x}} + 4\hat{\mathbf{y}}) = (3\hat{\mathbf{x}} + 10\hat{\mathbf{y}})$, expressible as $\hat{\mathbf{z}} \times (10\hat{\mathbf{x}} - 3\hat{\mathbf{y}} + H_{2z}\hat{\mathbf{z}})$. We get the missing H_{2z}, the normal component of $\mathbf{H}_2$, from $B_{2z} = \mu_2 H_{2z} = \mu_0(6 H_{2z}) = \mu_0(20.7846 \text{ A/m})$, which yields $H_{2z} = 3.4641$ A/m. Thus, $\mathbf{H}_2 = (10\hat{\mathbf{x}} - 3\hat{\mathbf{y}} + 3.4641\hat{\mathbf{z}})$ A/m.

The result can be visualized as a vector of magnitude $H_2 = 11$ A/m, at azimuth $\varphi_2 = \tan^{-1}(-3/10) = -16.70°$, at angle $\theta_2 = \cos^{-1}(3.4641/11) = 71.64°$ to the z-axis. The magnetic field jumps in strength and bends in direction in crossing the current-carrying xy-plane, because of both the change in permeability and the presence of the surface current.

Note that the $\mathbf{B}$ field's discontinuity is now also known: It jumps from $\mathbf{B}_1 = 2\mu_0\mathbf{H}_1$, with a strength of $(24\mu_0)$ T, to $\mathbf{B}_2 = 6\mu_0\mathbf{H}_2$, with a magnitude of $(66\mu_0)$ T, and bends from 30° to the z-axis, in the xz-plane, to 71.64° to the z-axis, at azimuth $-16.70°$, in crossing from one side of the xy-plane to the other. The field is uniform within the upper region as well as below, but with a new strength and direction.

Finally, the equation of conservation of charge, $I_0 = -dQ/dt$, relating the outflow of current I_0 across any closed surface to the enclosed charge Q, yields one further boundary condition, restricting discontinuities in electric current density. As indicated in Figure 3-22, the outflow I_0 of current from the closed surface that straddles the plane has three contributions: $\mathbf{J}_2 \cdot \hat{\mathbf{n}} A$ on the upper surface, $\mathbf{J}_1 \cdot (-\hat{\mathbf{n}}) A$ on the lower one, and $\oint \mathbf{K} \cdot d\mathbf{s}$ around the periphery, where $d\mathbf{s}$ has the length of an element of the curve along the edge but is directed outward, in the plane. Let us exclude the possibility that a net surface current $\oint \mathbf{K} \cdot d\mathbf{s}$ may spread out of any region along the surface, as it might if the surface has its own conducting properties, independent of those of the bulk media above and below it. It follows then that the total outflow is just $I_0 = A\{(\mathbf{J}_2 - \mathbf{J}_1) \cdot \hat{\mathbf{n}}\}$ and this must equal $-dQ/dt = -A\, d\rho_S/dt$. Canceling the small but arbitrary area A from both sides of the relation $I_0 = -dQ/dt$ leaves

$$(\mathbf{J}_2 - \mathbf{J}_1) \cdot \hat{\mathbf{n}} = -d\rho_S/dt. \tag{3.64}$$

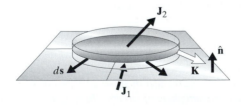

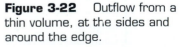

Figure 3-22   Outflow from a thin volume, at the sides and around the edge.

If there is no surface charge distribution, we have here the ordinary boundary condition that mandates *continuity* of the normal component of the electric current density at neighboring points. If, however, the bisecting plane carries a surface charge distribution (or a surface current), this becomes instead the extraordinary boundary condition that specifies the discontinuity in the normal component of the bulk current density **J** across the surface.

Let's pause and note that we have developed both the ordinary and the extraordinary boundary condition that apply at the interface between different media. How do we know when we are dealing with the ordinary cases that demand continuity of tangential **E** and **H** and of normal **D** and **B** and **J** at an interface between two media, rather than with the extraordinary cases that allow specific discontinuities in the presence of surface sources? It all depends on whether such surface sources are present, by design or of necessity, as follows.

For insulating media, unless we are told that we are to conceive of some given surface charge or current sheet that has deliberately been "painted" or smeared on some given surface, no such surface source should be present and continuity of the appropriate field components will reign. For conducting media, however, any charges that may initially have been introduced will tend to redistribute themselves; they may or may not thereafter accumulate on a boundary between conductors, depending on the rate at which they can move in the two media. If they do pile up or flow along the boundary, we have to deal with the surface charge and surface current and the resultant discontinuities.

Above all, we must ensure that Maxwell's equations be satisfied at all points, including the points that form an interface between two media, and we must also arrange for all the constitutive relations to be satisfied on both sides of the interface. If either or both of the two media has nonzero conductivity, therefore, the opportunity exists for charges to be conveyed within the conductors, by whatever electric field is present, from the sources to the interface. These charges may accumulate or move along the boundary, forming a surface charge or a surface current and causing discontinuities in normal **D** and **J** and in tangential **H**, in accordance with the boundary conditions just discussed.

Example 3.13

We can see time-varying fields and the appearance of surface charges in the following situation. Consider a spherical region of radius a, which we call the core, comprised of a conducting dielectric of conductivity σ_1 and permittivity ε_1. This is surrounded by a spherical shell extending between radii a and b, which we refer to as the mantle, formed of another conductor of parameters σ_2 and ε_2. Beyond the mantle is air ($\sigma = 0$ and $\varepsilon = \varepsilon_0$), as in Figure 3-23. Suppose that, somehow, a charge has been caused to fill the core and mantle, uniformly distributed throughout both at time t = 0, with charge density ρ_0, so that the total charge is $Q_0 = (4/3)\pi b^3 \rho_0$. We wish to determine the fields throughout space for all time and ask what happens to the charge distribution. ◇

From the symmetry, the electric field clearly is, and will remain, radial and dependent on only the radial coordinate r and time t. The **D** field at radius r is determined by the total charge $Q(r, t)$ present at time t up to radius r, by Gauss's law:

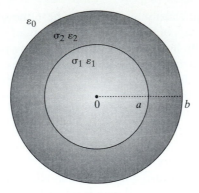

Figure 3-23   Concentric spherical core and mantle in air.

$$\oint \mathbf{D} \cdot d\mathbf{S} = 4\pi r^2 D(r,t) = Q(r,t). \tag{3.65}$$

We need to determine the total charge $Q(r,t)$ within a sphere of radius r. This will change in time, because the $\mathbf{D}$ field at radius r is accompanied by an electric field $\mathbf{E} = \mathbf{D}/\varepsilon$ and a current density $\mathbf{J} = \sigma\mathbf{E} = (\sigma/\varepsilon)\mathbf{D}$, where ε and σ take on the values appropriate to radius r. The law of conservation of charge tells us how $Q(r,t)$ will change in time as a result of the current formed by $\mathbf{J}$:

$$\oint \mathbf{J} \cdot d\mathbf{S} = -\frac{\partial Q(r,t)}{\partial t}. \tag{3.66}$$

Combining the two laws (Gauss's and charge conservation) gives us a differential equation for how $Q(r,t)$ varies in time.

$$-\frac{\partial Q(r,t)}{\partial t} = \oint \mathbf{J} \cdot d\mathbf{S} = \frac{\sigma}{\varepsilon} \oint \mathbf{D} \cdot d\mathbf{S} = \frac{\sigma}{\varepsilon} Q(r,t). \tag{3.67}$$

In this differential equation, σ/ε is a (piecewise constant) function of r but not a function of t, since the material properties do not change in time, so that the differential equation is of the constant-coefficient type, solved by an exponential:

$$Q(r,t) = Q(r,0)e^{-[\sigma(r)/\varepsilon(r)]t}. \tag{3.68}$$

In this result, the exponent varies as follows:

$$
\begin{aligned}
\sigma(r)/\varepsilon(r) = \sigma_1/\varepsilon_1 &= 1/T_1 && (r < a) \\
= \sigma_2/\varepsilon_2 &= 1/T_2 && (a < r < b) \\
&= 0 && (b < r). \tag{3.69}
\end{aligned}
$$

The parameters T_1 and T_2 represent the (generally unequal) exponential decay times in the core and mantle. The initial values $Q(r,0)$ are known for all r from the initially uniform charge distribution of density ρ_0:

$$
\begin{aligned}
Q(r,0) = 4\pi r^3 \rho_0/3 &= Q_0(r^3/b^3) && (r < b); \\
= 4\pi b^3 \rho_0/3 &= Q_0 && (b < r). \tag{3.70}
\end{aligned}
$$

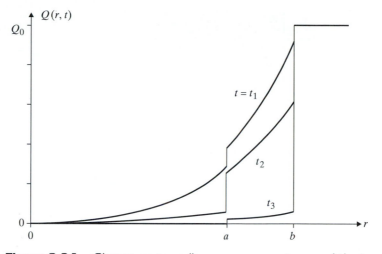

Figure 3-24   Charge up to radius r, versus r, at several times.

The complete radial and time dependence of the charge enclosed up to radius r is therefore

$$Q(r,t) = Q_0(r^3/b^3)e^{-t/T_1} \qquad (r < a) \qquad \text{core}$$

$$= Q_0(r^3/b^3)e^{-t/T_2} \qquad (a < r < b) \qquad \text{mantle}$$

$$= Q_0 \qquad (b < r) \qquad \text{air} \qquad (3.71)$$

Figure 3-24 shows a plot of the $Q(r,t)$ function, as a function of r, for several times t, one before T_1, another between T_1 and T_2, and one after T_2. The plot applies to the case $a/b = 3/4$ and $T_1/T_2 = 1/4$.

The feature of this result that we wish to stress is that, not initially but after $t = 0$, it is a discontinuous function of r. Assuming $T_1 \neq T_2$, there is a jump in the $Q(r,t)$ function both at the core–mantle interface at $r = a$ and at the mantle–air boundary at $r = b$. The amounts of the discontinuities are, at time t,

$$\Delta Q(a,t) = Q(a^+,t) - Q(a^-,t) = Q_0(a/b)^3[e^{-t/T_2} - e^{-t/T_1}], \qquad (3.72)$$

$$\Delta Q(b,t) = Q(b^+,t) - Q(b^-,t) = Q_0[1 - e^{-t/T_2}], \qquad (3.73)$$

These finite amounts of charge evidently accumulate at the two spherical surfaces at $r = a$ and $r = b$, each as a *surface* charge distribution smeared uniformly over the surface, of area $4\pi a^2$ or $4\pi b^2$. The amounts of these surface charge distributions $\rho_S(t)$ at $r = a$ and $r = b$ are given by

$$\rho_{Sa}(t) = \frac{\Delta Q(a,t)}{4\pi a^2} = \frac{Q_0}{4\pi b^2}\frac{a}{b}(e^{-t/T_2} - e^{-t/T_1}), \qquad (3.74)$$

$$\rho_{Sb}(t) = \frac{\Delta Q(b,t)}{4\pi b^2} = \frac{Q_0}{4\pi b^2}(1 - e^{-t/T_2}). \qquad (3.75)$$

Figure 3-25 shows plots of the two surface charge densities, as functions of time, again for the case $a/b = 3/4$ and $T_1/T_2 = 1/4$.

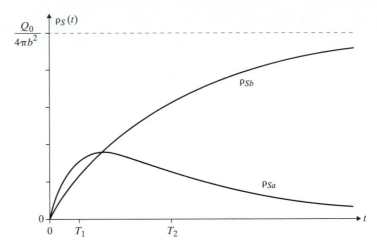

Figure 3-25   Surface charge densities at the interfaces, versus time.

The surface charge at $r = a$ is initially and finally zero but in the interim builds up to an amount that depends on the difference between the time constants T_1 and T_2. These time constants represent the rates at which charges can be swept away by the conduction currents in the two regions. The surface charge $\rho_{sa}(t)$ is positive (for $Q_0 > 0$) if $T_2 > T_1$, meaning that it takes longer to clear away the mantle than the core, so that the charge temporarily piles up at the boundary between them; it is negative if $T_2 < T_1$, meaning that charges are swept from the mantle to the outer boundary faster than from the core, so that there is a temporary depletion of charge at the core–mantle interface. At $r = b$, the surface charge is initially zero but builds up until all the charge Q_0 has piled up at the outer surface and remains there, unable to flow into the nonconducting air region.

Finally, we can obtain the electric field everywhere, directly from Gauss's law and the function $Q(r, t)$, by use of $D(r, t) = Q(r, t)/4\pi r^2$ to yield the radial **D** field throughout space. This is plotted in Figure 3-26, for the same set of parameters, for several times.

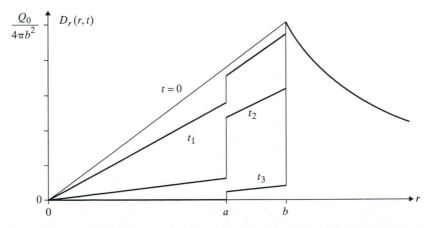

Figure 3-26   Radial electric flux density field versus radius, at several times.

> The field is a piecewise linear function of r in the core and mantle and an inverse-square field outside. The discontinuities in the radial **D** field are exactly equal to the surface charge densities at the two interfaces.

SUMMARY OF BOUNDARY CONDITIONS AND EFFECTS OF CONDUCTIVITY

To summarize the boundary conditions that apply at the interface between two media, they are expressed in vector form, in terms of the unit vector $\hat{\mathbf{n}}$ normal to the boundary and pointing toward medium 2, as

$$\hat{\mathbf{n}} \times \mathbf{E}_2 - \hat{\mathbf{n}} \times \mathbf{E}_1 = 0, \tag{3.76}$$

$$\hat{\mathbf{n}} \times \mathbf{H}_2 - \hat{\mathbf{n}} \times \mathbf{H}_1 = \mathbf{K}. \tag{3.77}$$

We also have the subsidiary relations

$$\hat{\mathbf{n}} \cdot \mathbf{D}_2 - \hat{\mathbf{n}} \cdot \mathbf{D}_1 = \rho_S, \tag{3.78}$$

$$\hat{\mathbf{n}} \cdot \mathbf{B}_2 - \hat{\mathbf{n}} \cdot \mathbf{B}_1 = 0, \tag{3.79}$$

$$\hat{\mathbf{n}} \cdot \mathbf{J}_2 - \hat{\mathbf{n}} \cdot \mathbf{J}_1 = -d\rho_S/dt, \tag{3.80}$$

and the fields in each region are related by the constitutive parameters ε, μ, σ for each of the two media.

The boundary conditions are not all independent. *The two boundary conditions that deal with the tangential components provide sufficient conditions to determine the fields at a boundary.* The additional ones that involve the normal components are subsidiary conditions that can determine the surface sources that make all the boundary conditions consistent with one another.

We should take careful note of how the *conductivity* of a medium affects the behavior of electric fields within it and on its surface. Within the realm of *electrostatics*, which considers only charges that are at rest, there can be no electric field inside the conductor and no tangential component of an electric field along the surface. If there were such a field, the nonzero conductivity σ would demand a corresponding current density $\mathbf{J} = \sigma\mathbf{E}$ within the material or tangential to its surface, which means that the charges in the conductor would be moving, contradicting the assumption of static conditions. That would require us to abandon the equations of electrostatics and allow for time variation.

We can consider a variant of the static situation, one in which there is flow of charge, but only steady flows. We then have charge motion but only as steady currents; the situation then does not change in time and appears static; we speak then of a stationary state. In that case, nonzero, steady electric fields accompany the steady currents in the conductor, as $\mathbf{J} = \sigma\mathbf{E}$; such currents flow in closed circuits, so that the electric field can follow a closed loop, which it cannot do in electrostatics. There must then be a source other than electric charges for such fields and flows. The steady currents also generate steady magnetic fields, and this is the realm of *magnetostatics*.

Once we allow time variation of the fields, a conductor can and does allow the existence of an electric field within it, accompanied by the current density $\mathbf{J} = \sigma\mathbf{E}$.

The corresponding flow tends to reduce the charge within the material, sweeping it to the outer surface and thereby reducing the strength of the internal electric field (unless the charge is replenished by some source). The time scale for this reduction of an initial internal field is given by $T = \varepsilon/\sigma$; eventually, the field inside the material and along its surface is reduced to zero.

Because there is so vast a range of values of conductivity in common materials, we consider extremes of good insulators and good conductors. We speak of good insulators, for which the time constant T is long (compared to other times of interest in a problem), and their limiting case of the perfect insulator, for which the time constant is infinite (as for a vacuum and, for most purposes, air). We also have good conductors, for which the time constant T is short (compared to other times of interest in the system), and we adopt the opposite limit of a *perfect conductor*, for which the time constant T is zero. A perfect conductor never tolerates an electric field, whether static or time varying, because the infinite conductivity that makes $T = 0$ would then call for an infinite current and infinitely strong heating. We often model metals as perfect conductors, because their time constant T is typically shorter by far than any other time of interest in the system. The difference between an ordinary conductor and a perfect conductor is that for a conductor, an unreplenished electric field *eventually* becomes zero (the higher the conductivity, the sooner the field is reduced to zero). For a perfect conductor, modeled as having infinite conductivity, there is *immediately* no electric field internal to the material or tangential at its surface, even in the time-varying case.

SUPERCONDUCTORS, PLASMAS, AND SPACE CHARGE

We deal also with more exotic media that cannot be described simply by parameters such as ε and σ. One important instance is the *superconductor*. A superconducting material can be in one of two states, the superconducting state and the normal state. The material is characterized by a critical temperature T_c and a critical magnetic field H_c that jointly determine the state of the material. Within the body, if the temperature is less than critical $(T < T_c)$ *and* the magnetic field is less than the critical field strength $(H < H_c)$, then the medium is in the superconducting state, which has three properties:

1. The conductivity is infinite: $\sigma = \infty$ (so that $E = 0$ for any J).
2. The permeability is zero: $\mu = 0$ (so that $B = 0$ for any H).
3. The current density **J** is of the electrostatic type ($\oint \mathbf{J} \cdot d\mathbf{l} = 0$).

Wherever $T > T_c$ or $H > H_c$, the medium is in the normal state, with finite values of conductivity and permeability: $\mathbf{J} = \sigma\mathbf{E}$ and $\mathbf{B} = \mu\mathbf{H}$ (typically, $\mu = \mu_0$). While in the superconducting state, the bulk material exhibits no resistance to the flow of electric current and tolerates no magnetic flux within it.

Another medium whose description defies the simplifications inherent in simple parameters like ε and σ is a *plasma*, which is, typically, an electrified gas comprised of free electrons neutralized by ions. To describe the relation between an applied electric field, say, and the resulting currents, we can often assume the heavy ions are nearly immobile, in comparison to the much lighter electrons. If there are n electrons (and the same number of ions, for charge neutrality) per unit volume and these electrons move with velocity $\mathbf{v}$, then the presence of an electric field $\mathbf{E}$ causes an acceleration of the electron motion:

$$m \, \partial \mathbf{v}/\partial t = -e\mathbf{E}, \qquad (3.81)$$

where m and $-e$ are the mass and charge of an electron. This acceleration affects the current $\mathbf{J} = -ne\mathbf{v}$, which is thereby related to the electric field in the plasma. For example, if we assume the electron density to remain constant and consider the simple case in which no magnetic field at all exists in the plasma, then the consequent relation $\mathbf{J} + \partial \mathbf{D}/\partial t = 0$ entails (for a plasma in a vacuum)

$$-ne\mathbf{v} + \varepsilon_0 \, \partial \mathbf{E}/\partial t = 0. \qquad (3.82)$$

Combining these two relations indicates that the electric field can exist in the plasma, without a magnetic field, provided it obeys the differential equation

$$\partial^2 \mathbf{E}/\partial t^2 + \left[ne^2/m\varepsilon_0\right]\mathbf{E} = 0. \qquad (3.83)$$

This differential equation implies sinusoidal oscillation of the electric field, at the *plasma frequency*

$$\omega_p = \left[ne^2/m\varepsilon_0\right]^{1/2}. \qquad (3.84)$$

This oversimplified development barely hints at the properties of a plasma, which exhibits both electromagnetic and fluid-flow phenomena.

One last example of a medium that requires combined electrical and mechanical analysis for its description is the *space charge* that appears in the space between the cathode and anode of a vacuum diode. When operated with the cathode grounded and a positive voltage applied to the anode, the resultant electric field enhances the emission of electrons from the hot cathode, but the cloud of electrons in the vacuum space tends to repel other electrons and retards further emission. An equilibrium is reached between these opposed effects, wherein the electrons form a space charge that modifies the electric field that would otherwise be present between the electrodes and thereby affects the electron current that can be collected at the anode. The relation between the applied voltage and the resultant current can be derived as in the following example. That relation is the Langmuir-Child law.

Example 3.14

A straight tungsten filament of length l and negligible thickness emits electrons radially, uniformly, and steadily at the rate I_0 C/s into the surrounding vacuum. The electrons emerge from this cathode at negligible speed but are accelerated by the radial electric field $E(r)$ and cross the interelectrode space until they are collected by a coaxial cylindrical anode at radius b. We want to determine the electric field distribution $E(r)$, the space charge density $\rho(r)$, and the voltage-current relation of the diode. The configuration is illustrated in Figure 3-27. ◇

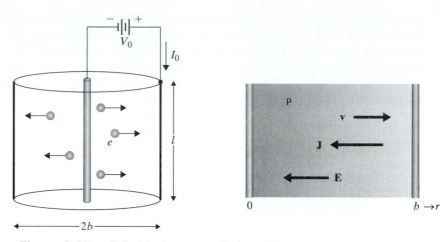

Figure 3-27   Cylindrical vacuum diode and its space charge and fields.

In the steady state represented by the equilibrium, the current crossing a cylinder of any radius is the same, so that the convection current density, directed radially inward toward the cathode, has the distribution

$$J(r) = I_0/2\pi rl. \tag{3.85}$$

We do not know the radial variation $v(r)$ of the electron velocity, but let us assume it to depend on radius as some as-yet-unknown power α of r, with an also-unknown proportionality coefficient v_0:

$$v(r) = v_0(r/b)^\alpha. \tag{3.86}$$

Both the exponent α and the factor v_0 (which is evidently the velocity at which the electrons reach the anode at $r = b$) are to be found. If this is really the correct form of the velocity distribution, then the electron density $\rho(r)$ must have a corresponding distribution found from $\mathbf{J} = \rho\mathbf{v}$ (noting that $\mathbf{J}$ and $\mathbf{v}$ are oppositely directed) as

$$\rho(r) = -[I_0/2\pi blv_0](b/r)^{\alpha + 1}. \tag{3.87}$$

To obtain the electric field distribution, we apply Gauss's law on a cylindrical surface of length l and radius r, so that we will need the total charge $Q(r)$ in the enclosed cylindrical space. This is

$$Q(r) = \int_0^r \rho(r)2\pi rl\, dr = \frac{-I_0 b^\alpha}{v_0} \int_0^r r^{-\alpha}\, dr = \frac{-I_0 b^\alpha}{v_0(1 - \alpha)} r^{1-\alpha} \tag{3.88}$$

and gives the radial electric field, from $2\pi rl\varepsilon_0 E(r) = Q(r)$, as

$$E(r) = \frac{-I_0 b^\alpha}{2\pi \varepsilon_0 v_0 l(1 - \alpha)} r^{-\alpha}. \tag{3.89}$$

The voltage applied to the anode is the emf

$$V_0 = \int_b^0 E(r)\, dr = \frac{I_0 b^\alpha}{2\pi \varepsilon_0 v_0 l(1 - \alpha)} \int_0^b r^{-\alpha}\, dr = \frac{I_0 b}{2\pi \varepsilon_0 v_0 l(1 - \alpha)^2}. \tag{3.90}$$

To determine the unknown constants v_0 and α, we need to invoke the mechanical relation between the electron motion and the electric field in the interelectrode space. The force law $m\, dv/dt = -eE$ becomes

$$m\frac{dv(r)}{dt} = m\frac{dv(r)}{dr}\frac{dr}{dt} = m\frac{dv}{dr}v = -eE(r). \tag{3.91}$$

Combining

$$v(r) = v_0[r/b]^\alpha \quad \text{and hence} \quad dv/dr = [\alpha v_0/b][r/b]^{\alpha-1} \tag{3.92}$$

with

$$E(r) = -[I_0/2\pi\varepsilon_0 v_0 l(1-\alpha)][r/b]^{-\alpha} \tag{3.93}$$

yields

$$\frac{m\alpha v_0^2}{b}\frac{r^{2\alpha-1}}{b^{2\alpha-1}} = \frac{eI_0}{2\pi\varepsilon_0 v_0 l(1-\alpha)}\frac{r^{-\alpha}}{b^{-\alpha}}, \tag{3.94}$$

which can be satisfied for all r if

$$2\alpha - 1 = -\alpha \quad \text{or} \quad \alpha = 1/3 \tag{3.95}$$

and if also

$$\frac{m\alpha v_0^2}{b} = \frac{eI_0}{2\pi\varepsilon_0 v_0 l(1-\alpha)} \quad \text{or} \quad v_0^3 = \frac{eI_0 b}{2\pi\varepsilon_0 m l\alpha(1-\alpha)}. \tag{3.96}$$

That we can find α and v_0 justifies the assumption of a power law for the radial variation of the fields. We have, therefore, that the voltage cubed is

$$V_0^3 = \frac{I_0^3 b^3}{[2\pi\varepsilon_0 l(1-\alpha)^2]^3 v_0^3} = \frac{I_0^2 b^2 m\alpha}{l^2(2\pi\varepsilon_0)^2 e(1-\alpha)^5} \tag{3.97}$$

or, finally, substituting $1/3$ for α,

$$I_0 = GV_0^{3/2} \tag{3.98}$$

with

$$G = 2\pi\varepsilon_0\frac{l}{b}\frac{(1-\alpha)^{5/2}}{\alpha^{1/2}}\left(\frac{e}{m}\right)^{1/2} = \frac{2\sqrt{2}}{9}4\pi\varepsilon_0\left(\frac{e}{m}\right)^{1/2}\frac{l}{b}. \tag{3.99}$$

The three-halves power law relating current to voltage in the space-charge limited diode is the Langmuir-Child law; the coefficient G is the *perveance* of the diode. As an example, if $l = 1$ cm and $b = 2$ mm, applying $V_0 = 100$ V to the diode will draw $I_0 = 73$ mA under space-charge limited operation.

The voltage-current characteristic of the diode is plotted in Figure 3-28, along with the distributions of the charge density, electron velocity, current density, and electric field in the interelectrode space. The normalizing field values are those at the anode; they are given by

$$\tfrac{1}{2}mv_0^2 = eV_0, \qquad \rho_0 = -I_0/2\pi blv_0,$$

$$J_0 = -I_0/2\pi bl, \qquad E_0 = -3I_0/4\pi\varepsilon_0 lv_0. \tag{3.100}$$

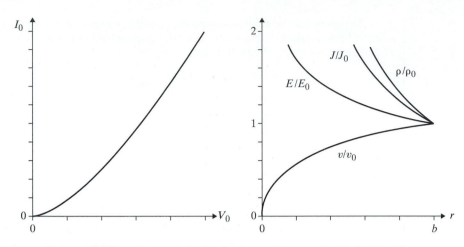

Figure 3-28　Vacuum diode characteristic and field distributions.

Note that the current density **J** is not simply proportional to the electric field **E**, as it would be in an ohmic conductor. The space charge behaves as a medium of a different sort; it is a nonneutral version of a plasma.

SUMMARY

We have defined the concept of emf, or voltage, as the line integral of the electric field along a specified path. It is the dual of the mmf, in the same sense that current serves as the dual of voltage. We explored currents and Ohm's law and introduced conductivity as a property of the medium. We demonstrated that Coulomb fields, those electric fields that are generated only by charge distributions, cannot have a nonzero emf around any closed curve. Yet, in examining currents and electric power expenditure in circuital flows along conducting loops, we realized that there do exist nonzero emfs around such closed circuits. This pointed to alternative means, other than distributions of charge, to generate electric fields and circuital currents.

The electric field generation process of surpassing importance is the one based on time-varying magnetic flux, as dictated by the Faraday-Maxwell law,

$$V\{C\} = -d\Phi\{S\}/dt, \tag{3.101}$$

applicable to any open surface S bounded by closed curve C, with S and C mutually oriented by the right-hand rule: *The emf around the closed curve C is the rate of decrease of the magnetic flux across the open surface S.* Although we have illustrated this law for selected applications, we have only hinted at its far-reaching consequences, which remain to be explored.

We have also discussed the failure of Ampère's law to be consistent with the law of conservation of charge and the need to modify and extend it into the Ampère-Maxwell law,

$$U\{C\} = d\Psi\{S\}/dt + I\{S\}, \tag{3.102}$$

also applicable to any open surface S bounded by closed curve C, mutually oriented by the right-hand rule. This states that *the mmf around the closed curve C is the sum of the rate of increase of the electric flux and the current across the open surface S*. This also furnishes an alternative to generation of magnetic fields by currents, allowing for electric generation of magnetic fields through the displacement current. Again, the consequences are supremely important and need to be investigated.

We also stated the magnetic counterpart of Gauss's law. While the electric version states that the electric flux across any closed surface equals the enclosed electric charge, the magnetic one asserts that there are no magnetic charges, so that the magnetic flux across any closed surface is zero.

We categorized various types of current, including convection, conduction, displacement, and polarization currents. We applied Maxwell's equations in integral form to obtain boundary conditions at interfaces between different media, where the fields can be discontinuous. This is to enable us to conjoin field patterns that may exist on the two sides of the interface, in a way that allows Maxwell equations to be satisfied everywhere, even at the boundary points between media. We found that, ordinarily, *the tangential components of the* **E** *and* **H** *fields at the interface must be continuous*. Correspondingly, the normal components of **D** and **B** must be continuous. We also discussed the extraordinary cases of boundaries, such as those between different conducting media, at which distributions of surface charges and surface currents may appear. In those cases, the tangential H field will be discontinuous by the amount of the surface current, the normal D field will be discontinuous by the amount of the surface charge, and the normal current density J will be discontinuous by the rate of decrease of the surface charge.

Finally, we glanced at a few media that are not ohmic, in that the relation between electric field and current density is not a mere proportionality. These media included superconductors, plasmas, and space charge. A material in the superconducting state tolerates no electric field and no magnetic flux, while allowing a current density. We found the simplest description of a plasma, entailing a plasma frequency as an inherent property. We explored space charge in a vacuum diode, obtaining the Langmuir-Child three-halves power law for the relation between current and voltage.

All these considerations are merely preparatory to the main business at hand, which is to obtain and study electromagnetic field patterns that are consistent with *both* of Maxwell's equations simultaneously, with the time-varying electric fields sustaining the magnetic ones and the time-varying magnetic fields maintaining the electric ones. The consequences will be seen to be action at a distance, with a delay.

PROBLEMS

Emf of given electric field

3.1 The electric field in space is $\mathbf{E}(\mathbf{r}) = F(r)[\hat{\mathbf{r}} + \hat{\boldsymbol{\theta}} + \hat{\boldsymbol{\varphi}}]\sin\theta$, where $F(r)$ depends only on the radial coordinate and is therefore constant on the surface of the sphere $r = R$. Find

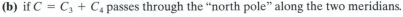

the emf $V\{C\}$ along the specified curve C on the surface of this sphere, from the point A at $(R, \pi/2, 0)$ to the point B at $(R, \pi/4, \pi/2)$,

(a) if $C = C_1 + C_2$ follows the "equator" and then the meridian, as in Figure P3-1;

(b) if $C = C_3 + C_4$ passes through the "north pole" along the two meridians.

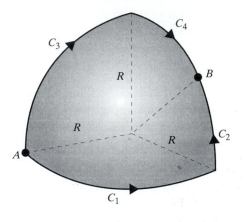

Figure P3-1   Two paths for emf calculation from point A to point B.

Capacitance of conducting sphere

3.2   The *capacitance* C of a single conductor is defined by $C = Q/V$, where V is the voltage from the conductor to infinity when Q is the charge on the conductor. Find the capacitance of a conducting sphere of radius a in a vacuum. Give a numerical answer, in farads, for a sphere of radius $a = 10$ cm and for a sphere the size of the earth, for which $a = (6.37)10^6$ m.

Resistance in regions of nonuniform flow

3.3   A region of space bounded by a closed surface S comprised of a section S_1 across which a nonuniformly distributed current enters the region and which is perpendicular to the flow, a section S_2, also perpendicular to the flow, across which the same current exits the region, and the remainder, S_0, joining S_1 and S_2 such that there is no flow across S_0 is called a *tube of flow*. If the enclosed region is conducting, the nonuniform flow is accompanied by an electric field and there is a voltage from surface S_1 to surface S_2. The ratio of this voltage to the total current in the tube is the *resistance* $R = V/I$ of the tube of flow.

For a spherically symmetric source of current, the tubes of flow are conical sections, capped by spherical segments. For a medium of conductivity σ, calculate the resistance of a conical tube of angular opening α, between radial distances a and b from the source point, as in Figure P3-2.

Approximate skin-effect resistance of a wire

3.4   Although the current distribution in a cylindrical wire may be uniform when it is unvarying, it tends to be concentrated near the surface of the wire when it is rapidly time varying. In that case, a reasonable approximation to the nonuniform distribution of current in the wire may be $J(\rho) = J_0 e^{-(a-\rho)/\delta}$, where a is the radius of the wire and δ is the "skin depth," so called because $\delta \ll a$, which implies that the current flows primarily in a thin layer just within the outer surface of the wire. The wire has bulk conductivity σ.

(a) What is the total current I_0 carried by the wire?

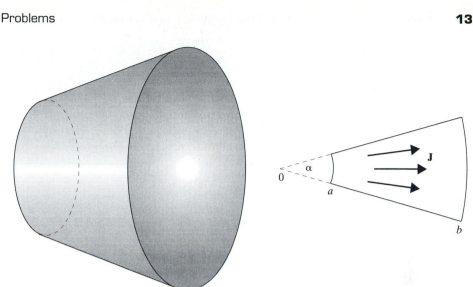

Figure P3-2   Section of conical tube of flow carrying a nonuniform current.

   (b) What is the voltage V_0 along a length l of the wire, measured along its surface, when it carries the total current I_0?

   (c) Compare the resistance R of a length l of the wire when the "skin effect" is present to its resistance R_0 when it carries the same total current I_0 in a uniform distribution.

   (d) If δ is so small that we wish to describe the flow as a surface current K, how is the value of K related to J_0?

   (e) What is the surface resistivity r_s (measured in ohms) in the relation $E = r_s K$ between the electric field E at the surface and the surface current K?

Power dissipated by current from submerged sphere

3.5   A copper sphere of radius a is deeply submerged in sea water of conductivity σ and an insulated wire carries current I_0 to it. Assume the wire interferes negligibly with the outflow; assume infinite depth.

   (a) Calculate the power P dissipated in the water.

   (b) If P is written as $P = I_0^2 R = I_0^2/G$, what are the resistance R and the conductance G of the sea water in this flow?

Fields of current injected and extracted from a sphere

3.6   A conducting sphere of radius a has current I_1 injected at its "south pole" from a straight wire along the negative z-axis, while current I_2 is extracted at its "north pole" through another straight wire along the positive z-axis (Figure P3-3). As a result, if the two steady currents are not the same, charge either accumulates or is depleted from the sphere. Assume that this charge spreads uniformly over the sphere's surface and that the wires interfere negligibly with the field pattern. The sphere is neutral at time $t = 0$. In the air space outside the sphere and wires (i.e., for all r, θ such that $a < r$ and $0 < \theta < \pi$), obtain in terms of I_1 and I_2

   (a) the electric field $\mathbf{E}(r, \theta, \varphi, t)$;

   (b) the magnetic field $\mathbf{H}(r, \theta, \varphi, t)$.

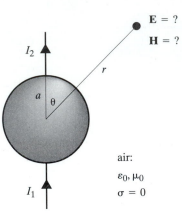

E = ?

H = ?

I_2

r

a θ

air:

ε_0, μ_0

I_1 $\sigma = 0$

Figure P3-3   Current injected and extracted from a sphere.

Power and energy expended in charge transfer

3.7 A cylinder of radius a and length l has initial charge q_0 and is separated from a wider, coaxial cylinder of radius b, also of length l, by a medium with parameters ε, μ, σ. Neglect fringing at the ends of the cylinder.

(a) The ensuing charge transfer will leave charge $q(t)$ on the inner cylinder at time t. Find $q(t)$ for $t > 0$.

(b) Obtain the power density $p(\mathbf{r}, t)$ between the two cylinders.

(c) Find the total energy W expended in the process of transferring the charge.

Induced emf in nonuniform magnetic field

3.8 A thin wire of length l, cross sectional area A, conductivity σ is bent around into a circle and the ends joined. The wire is immersed in a nonuniform, decaying magnetic field $B(\rho, t) = B_0\left[1 - (2\pi\rho/l)^2\right]e^{-t/T}$ for $t > 0$, in the plane of the wire, directed perpendicularly to that plane, where ρ is the radial distance from the center of the circle.

(a) Note that the magnetic field is zero at the wire, for all time. Is there nevertheless an emf around the wire? If so, what is this emf $V(t)$?

(b) What energy W is expended in heating the wire during the field's decay?

Sliding cross bar on parallel rails in nonuniform magnetic field

3.9 Two parallel conducting rails of resistivity r_0 (ohm/m) are separated by distance l and connected by one stationary and one sliding cross bar, each of resistance R_0. A constant but nonuniform magnetic field perpendicular to the plane of the rails fills the region. The field is $B(x) = B_0x/l$, where x is the distance from the stationary cross bar. The sliding cross bar moves at constant speed v along the rails, for a time T (Figure P3-4, left).

(a) What current $I(t)$ is induced in the rails?

(b) What is the energy expended in this process?

Sliding cross bar on parallel rails in nonconstant magnetic field

3.10 Two parallel conducting rails of resistivity r_0 (ohm/m) are separated by distance l and connected by one stationary and one sliding cross bar, each of resistance R_0. A uniform but nonconstant magnetic field perpendicular to the plane of the rails fills the region. The field is $B(t) = B_0t/T$. The sliding cross bar moves at constant speed v along the rails, for a time T (Figure P3-4, left).

Figure P3-4   Cross bar sliding on rails in perpendicular magnetic field.

(a) What current $I(t)$ is induced in the rails?
(b) What is the energy expended in this process?

Sliding cross bar on nonparallel rails in magnetic field

3.11 Two conducting rails of resistivity r_0 (ohm/m) meet at angle α and are connected by a sliding cross bar of the same resistivity (Figure P3-4, right). A uniform, constant magnetic field B_0 perpendicular to the plane of the rails fills the region. The sliding cross bar starts at the apex and moves perpendicularly to itself, in the direction that bisects the angle between the rails, at constant speed v, until the spot where the separation between the rails is l.
(a) What current $I(t)$ is induced in the rails?
(b) What is the energy expended in this process?

Accelerated cross bar on parallel rails in magnetic field

3.12 Two parallel conducting rails of resistivity r_0 (ohm/m) are separated by distance l and connected by one stationary and one sliding cross bar, each of resistance R_0. A uniform, constant magnetic field B_0 perpendicular to the plane of the rails fills the region. The sliding cross bar moves along the rails, from rest at the location of the fixed one, at constant acceleration a, for a time T (Figure P3-4, left).
(a) What current $I(t)$ is induced in the rails?
(b) What is the energy expended in this process?

Accelerated motion, nonparallel rails, nonuniform and nonconstant field

3.13 Two conducting rails of resistivity r_0 (ohm/m) meet at angle α and are connected by a sliding cross bar of the same resistivity (Figure P3-4, right). A nonuniform, nonconstant magnetic field perpendicular to the plane of the rails fills the region. The field is $B(x, t) = B_0(x/L)(t/T)$. The sliding cross bar starts at the apex and moves perpendicularly to itself, in the x-direction that bisects the angle between the rails, at constant acceleration a, until the spot where the separation between the rails is l.
(a) What current $I(t)$ is induced in the rails?
(b) What is the energy expended in this process?

Decaying current fed to point source in conductor

3.14 Starting at time $t = 0$, current $I(t) = I_0 e^{-(\sigma/\varepsilon)t}$ is fed along an insulated wire to a point source in a medium of conductivity σ and permittivity ε. There is no charge at the tip initially.

(a) Find the charge $q(t)$ at the tip after $t = 0$.

(b) When does the charge attain a maximum? What is that maximum charge?

Conductivity of semiconductor

3.15 Electrons (charge $-e$) and holes (charge $+e$) in a semiconductor respond to an applied electric field $\mathbf{E}$ and to the crystal lattice that impedes accelerated motion by attaining an average velocity proportional to the electric field. The proportionality is the *mobility* μ and we have, for the electrons, $\mathbf{v}_n = -\mu_n \mathbf{E}$ and, for the holes, $\mathbf{v}_p = \mu_p \mathbf{E}$. If there are n electrons/m^3 and p holes/m^3 in the crystal, what is the conductivity σ of the semiconductor?

Magnetic field of decelerating charge

3.16 A metal ball of mass M, radius a carries charge Q (Figure P3-5). It is shot upward with initial velocity v_0, so that it will reach its peak height $z_0 = v_0^2/2g$ at time v_0/g.

(a) Obtain the mmf $U(t)$ along the circumference of a circle of radius ρ_0 in the plane $z = z_0$, centered on the z-axis and oriented azimuthally, with $\rho_0 > a$.

(b) What magnetic field $\mathbf{H}(\rho, \varphi, z_0, t)$ is generated by the ball, in the plane $z = z_0$ of its peak height and for $\rho > a$, during its deceleration?

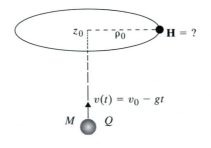

Figure P3-5   A magnetic field is generated by a decelerating charge.

Refraction of current at interface between conductors

3.17 Uniform current density $\mathbf{J}_1$ in a medium of conductivity σ_1 bends to a new direction when it flows across a planar boundary into another medium of conductivity σ_2, continuing as another uniform flow $\mathbf{J}_2$ (Figure P3-6). If the given $\mathbf{J}_1$ makes angle θ_1 with the normal to the boundary plane, determine the angle θ_2 of $\mathbf{J}_2$ with the normal, in the steady state.

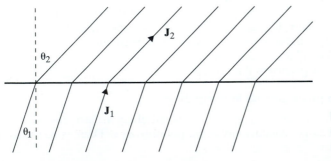

Figure P3-6   Refraction of current at interface between conductors.

Refraction of current and charge accumulation at interface between dielectrics

3.18   Uniform current density $\mathbf{J}_1$ in a medium of conductivity σ_1 and permittivity ε_1 bends to a new direction when it flows across a planar boundary into another medium of conductivity σ_2 and permittivity ε_2, continuing as another uniform flow $\mathbf{J}_2$. Charge may accumulate on the boundary plane. If the given $\mathbf{J}_1$ makes angle θ_1 with the normal to the boundary plane, find the angle θ_2 of $\mathbf{J}_2$ with the normal, in the steady state. Determine what surface charge density q_s must be deposited along the boundary plane by the incident current to maintain the steady-state flow.

Point source of magnetic flux

3.19   Total magnetic flux Φ_0 emerges at the origin from an infinitely long and infinitesimally thin solenoid that lies along the positive z-axis. The tip acts as a point source and the flux spreads outward in all directions into the infinite medium of permeability μ. Assume the thin solenoid interferes negligibly with this outflux.
   (a) Find the magnetic field $\mathbf{B}(\mathbf{r})$ in the medium.
   (b) If the total flux Φ_0 supplied by the solenoid varies (slowly) with time, find the induced electric field $\mathbf{E}(\mathbf{r}, t)$ in the medium.

Cylindrical shell in narrow magnetic field

3.20   A conducting cylindrical shell (occupying the region $a < \rho < b, 0 < z < h$ and with parameters μ, ε, σ) is subjected to an externally imposed, uniform but time-varying magnetic field $B(t)\hat{\mathbf{z}}$ that occupies the cylindrical space $0 \leq \rho < R$. Take $R < a$ and neglect fringing of the field. The magnetic field was $B = B_0$ for $t < 0$ but then is caused to decay with time constant T, as $B(t) = B_0 \exp(-t/T)$ for $t > 0$. Neglect corrections to this field that may arise from induced currents.
   (a) Obtain the emf $V(t)$ along the circumference of a circle centered on the z-axis, of radius ρ_0 at height z_0, within the shell $(a < \rho_0 < b)$, oriented azimuthally.
   (b) Obtain the azimuthal electric field $E_\varphi(\rho, \varphi, z, t)$ and the current density $J_\varphi(\rho, \varphi, z, t)$ within the shell.
   (c) Find the total current $I_0(t)$ that circulates around and within the shell (across the area $(b - a)h$ of a vertical section of the shell).
   (d) Find the power density $p(\rho, \varphi, z, t)$ that heats the shell.
   (e) Obtain the total energy expended within the shell as the magnetic field decays to zero.

Cylindrical shell in wide magnetic field

3.21   A conducting cylindrical shell (occupying the region $a < \rho < b, 0 < z < h$ and with parameters μ, ε, σ) is subjected to an externally imposed, uniform but time-varying magnetic field $B(t)\hat{\mathbf{z}}$ that occupies the cylindrical space $0 \leq \rho < R$. Take $R > b$ and neglect fringing of the field. The magnetic field was $B = B_0$ for $t < 0$ but then is caused to decay with time constant T, as $B(t) = B_0 \exp(-t/T)$ for $t > 0$. Neglect corrections to this field that may arise from induced currents.
   (a) Obtain the emf $V(t)$ along the circumference of a circle centered on the z-axis, of radius ρ_0 at height z_0, within the shell $(a < \rho_0 < b)$, oriented azimuthally.
   (b) Obtain the azimuthal electric field $E_\varphi(\rho, \varphi, z, t)$ and the current density $J_\varphi(\rho, \varphi, z, t)$ within the shell.
   (c) Find the total current $I_0(t)$ that circulates around and within the shell (across the area $(b - a)h$ of a vertical section of the shell).

(d) Find the power density $p(\rho, \varphi, z, t)$ that heats the shell.

(e) Obtain the total energy expended within the shell as the magnetic field decays to zero.

Cylindrical shell in magnetic field of comparable extent

3.22 A conducting cylindrical shell (occupying the region $a < \rho < b, 0 < z < h$ and with parameters μ, ε, σ) is subjected to an externally imposed, uniform but time-varying magnetic field $B(t)\hat{\mathbf{z}}$ that occupies the cylindrical space $0 \leq \rho < R$. Take $a < R < b$ and neglect fringing of the field. The magnetic field was $B = B_0$ for $t < 0$ but then is caused to decay with time constant T, as $B(t) = B_0 e^{-t/T}$ for $t > 0$. Neglect corrections to this field that may arise from induced currents.

(a) Obtain the emf $V(t)$ along the circumference of a circle centered on the z-axis, of radius ρ_0 at height z_0, within the shell $(a < \rho_0 < b)$, oriented azimuthally.

(b) Obtain the azimuthal electric field $E_\varphi(\rho, \varphi, z, t)$ and the current density $J_\varphi(\rho, \varphi, z, t)$ within the shell.

(c) Find the total current $I_0(t)$ that circulates around and within the shell (across the area $(b - a)h$ of a vertical section of the shell).

(d) Find the power density $p(\rho, \varphi, z, t)$ that heats the shell.

(e) Obtain the total energy expended within the shell as the magnetic field decays to zero.

Magnetic field of point charge in simple harmonic motion

3.23 A point charge q is caused to execute a low-frequency simple harmonic motion along the z-axis, in air, so that its location at time t is $z(t) = a \sin \omega t$.

(a) Obtain the mmf $U(t)$ along the circumference of a circle of radius ρ_0 in the xy-plane, centered at the origin and oriented azimuthally.

(b) Find the magnetic field strength $H(\rho, t)$ anywhere in the $z = 0$ plane.

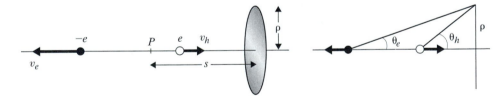

Figure P3-7   Electron-hole pair's birth and motion create a charge pulse.

Charge pulse from motion of electron and hole pair after their creation

3.24 An electron-hole pair is created (by externally incident radiation) at time $t = 0$ at one point P in a semiconducting material. Thereafter, the electron and hole, having different mobilities, travel in opposite directions at different speeds, v_e and v_h, when a voltage is applied along the axis of the material. The result is a current pulse in the circuit of which the semiconductor is a part. We are interested in the shape of the current pulse and in the total charge that circulates around the circuit.

(a) Evaluate the mmf $U(t)$ around a circular path of radius ρ (perpendicular to and centered on the axis) at a distance s from the birthplace P of the pair. [*Note*: This is most easily expressed in terms of the time-varying angles $\theta_e(t)$ and $\theta_h(t)$ shown in Figure P3-7, before getting $U(t)$ explicitly as a function of t.]

(b) Why is this mmf $U(t)$ the same as the waveform of the current pulse that is intercepted by the disk whose boundary is the circle?

(c) Find the total charge transfer by the current pulse in the circuit (i.e., find the integral of the current pulse, from $t = 0$ to $t = \infty$). [*Note:* Again, this is most easily calculated in terms of the two angles θ_e, θ_h, (rather than t) as the integration variables.]

Naive model of critical current in superconducting wire

3.25 A long cylindrical superconducting wire of radius a is maintained at temperature $T < T_c$ (T_c = critical temperature; recall that superconductivity requires both $T < T_c$ and $H < H_c$, otherwise the normal state prevails).

(a) Find a critical current I_c, the highest value such that the wire can carry any steady, uniformly distributed current $I < I_c$ while remaining superconducting throughout its volume.

(b) If the current exceeds I_c, then only a part of the wire can be in the superconducting state, with the rest in the normal state. A naive model of this situation assumes that the current density is still uniformly distributed across the wire and that the superconducting portion of the wire extends radially only from the axis to radius R (that is, for radius ρ such that $0 \le \rho < R$, with the normal state prevailing in $R < \rho < a$). Find the limiting radius R.

(c) Why must this naive model be wrong? What law of electromagnetics does it violate?

Dipole as image of uniform field in insulating sphere

3.26 An otherwise uniform flow field $\mathbf{J}_0$ in a conducting fluid of conductivity σ_0 encounters an impenetrable spherical bubble of radius a and zero conductivity; see Figure P3-8. The distortion of the current density $\mathbf{J}$ as it flows around the obstacle can be expressed as the superposition of the original uniform flow and the field of an "image" dipole located at the center of the sphere.

(a) What boundary condition applies to the current density field $\mathbf{J}$ at the surface of the bubble?

(b) What is the strength $I d\mathbf{l}$ of the image dipole that gives the correction to the uniform field outside the bubble?

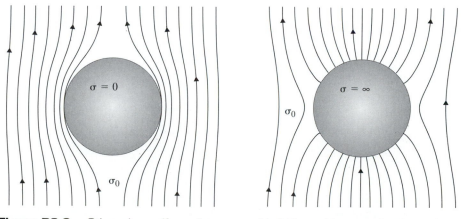

Figure P3-8 Otherwise uniform flow around bubble and into conducting sphere.

Dipole as image of uniform field in conducting sphere

3.27 An otherwise uniform flow field $\mathbf{J}_0$ in a conducting fluid of conductivity σ_0 encounters a perfectly conducting sphere of radius a; see Figure P3-8. The distortion of the current density $\mathbf{J}$ as it flows into the obstacle can be expressed as the superposition of the original uniform flow and the field of an "image" dipole located at the center of the sphere.

(a) What boundary condition applies to the current density field $\mathbf{J}$ at the surface of the sphere?

(b) What is the strength $I\,dl$ of the image dipole that gives the correction to the uniform field outside the sphere?

CHAPTER 4

Maxwell's Equations and Quasistatic Analysis

We have now reached the point where we know the rules of the game for time-varying fields and we have to start playing the game with full dynamics. We have Maxwell's equations, which govern the spatial and temporal behavior of electromagnetic fields, but we need to develop means for solving them for specified configurations of conductors and insulators and for given sources of the fields.

FIELD EQUATIONS

Let us review the field equations we have developed. They may seem to be numerous and bewildering, but we will see that the crucial ones are only the two Maxwell equations.

Fields. We have four electromagnetic fields, vector functions of position $\mathbf{r}$ and time t.

$$\mathbf{E}(\mathbf{r}, t) \qquad \mathbf{B}(\mathbf{r}, t) \qquad \mathbf{D}(\mathbf{r}, t) \qquad \mathbf{H}(\mathbf{r}, t) \tag{4.1}$$

Sources. The electric current density and electric charge density fields act as the sources that generate the electromagnetic fields.

$$\mathbf{J}(\mathbf{r}, t) \qquad \rho(\mathbf{r}, t) \tag{4.2}$$

Line integrals. Line integrals of the electric and magnetic field intensities $\mathbf{E}, \mathbf{H}$ along a specified, oriented curve C define electromotive force (emf, in volts) $V\{C\}$ and magnetomotive force (mmf, in amperes) $U\{C\}$, along that curve.

$$\text{emf} \qquad V\{C\} = \int_C \mathbf{E} \cdot d\mathbf{l} \tag{4.3}$$

$$\text{mmf} \qquad U\{C\} = \int_C \mathbf{H} \cdot d\mathbf{l} \tag{4.4}$$

Surface integrals. Surface integrals of the electric and magnetic flux densities $\mathbf{D}, \mathbf{B}$ along a specified, oriented surface S define electric flux (also called displacement flux, in coulombs), $\Psi(S)$, and magnetic flux (in webers), $\Phi\{S\}$, over that surface. The surface integral of the current density $\mathbf{J}$ gives the total current (in amperes), $I\{S\}$, crossing the surface.

$$\text{electric flux} \qquad \Psi\{S\} = \int_S \mathbf{D} \cdot d\mathbf{S} \tag{4.5}$$

$$\text{magnetic flux} \qquad \Phi\{S\} = \int_S \mathbf{B} \cdot d\mathbf{S} \tag{4.6}$$

$$\text{current} \qquad I\{S\} = \int_S \mathbf{J} \cdot d\mathbf{S} \tag{4.7}$$

Volume integrals. The volume integral of the charge density ρ over a specified volume V gives the total charge (in coulombs), Q, within that volume.

$$\text{electric charge} \qquad Q\{V\} = \int_V \rho \, dV \tag{4.8}$$

We also integrate mechanical quantities like force density and power density over volumes, to get total force on the contents or the total power expended within.

Integrals on open surfaces. If open surface S is bounded by closed curve C and the two are mutually oriented by the right-hand rule, then *Maxwell's equations* apply.

Faraday-Maxwell law.

$$V\{C\} = -d\Phi\{S\}/dt \tag{4.9}$$

or

$$\oint_C \mathbf{E} \cdot d\mathbf{l} = -\frac{d}{dt} \int_S \mathbf{B} \cdot d\mathbf{S} \tag{4.10}$$

Ampère-Maxwell law.

$$U\{C\} = I\{S\} + d\Psi\{S\}/dt \tag{4.11}$$

or

$$\oint_C \mathbf{H} \cdot d\mathbf{l} = \int_S \mathbf{J} \cdot d\mathbf{S} + \frac{d}{dt} \int_S \mathbf{D} \cdot d\mathbf{S} \tag{4.12}$$

Integrals on closed surfaces. If volume V is bounded by closed surface S, oriented outward, then the following *constraints* apply. These are necessary for consistency with, and as a prerequisite for, Maxwell's equations.

Conservation of charge.

$$I\{S\} = -dQ\{V\}/dt \tag{4.13}$$

or

$$\oint_S \mathbf{J} \cdot d\mathbf{S} = -\frac{d}{dt} \int_V \rho \, dV \tag{4.14}$$

Gauss's law.

$$\Psi\{S\} = Q\{V\} \tag{4.15}$$

or

$$\oint_S \mathbf{D} \cdot d\mathbf{S} = \int_V \rho \, dV \tag{4.16}$$

Continuity of total current.

$$I\{S\} + d\Psi\{S\}/dt = 0 \tag{4.17}$$

or

$$\oint_S \mathbf{J} \cdot d\mathbf{S} + \frac{d}{dt} \oint_S \mathbf{D} \cdot d\mathbf{S} = 0 \tag{4.18}$$

Nonexistence of magnetic charge.

$$\Phi\{S\} = 0 \tag{4.19}$$

or

$$\oint_S \mathbf{B} \cdot d\mathbf{S} = 0 \tag{4.20}$$

We may be tempted to view the conservation of charge equation and the continuity of total current equation as equivalent. They surely are not. One deals with a volume integral (of the enclosed charge); the other involves a surface integral over the bounding surface (the emerging flux). One expresses conservation of charge; the other declares that conduction and displacement currents are complementary (one takes over for the other, so that the sum is zero). The two concepts should not be blurred. Only Gauss's law makes one equation imply the other, but they are decidedly not the same.

Connection to mechanical concepts. Force on a charge q that moves with velocity $\mathbf{v}$:

Lorentz force $\qquad\qquad \mathbf{F} = q\mathbf{E} + q\mathbf{v} \times \mathbf{B}. \tag{4.21}$

Force density wherever there is charge density ρ and current density $\mathbf{J}$:

Lorentz force density $\qquad \mathbf{f} = \rho\mathbf{E} + \mathbf{J} \times \mathbf{B} \tag{4.22}$

power density $\qquad\qquad\qquad p = \mathbf{E} \cdot \mathbf{J} \tag{4.23}$

convection current density $\quad \mathbf{J} = \rho\mathbf{v}. \tag{4.24}$

Constitutive relations. Description of media, via conductivity σ, permittivity ε, and permeability μ:

ohmic conductor	$\mathbf{J} = \sigma\mathbf{E}$	(4.25)
dielectric medium	$\mathbf{D} = \varepsilon\mathbf{E}$	(4.26)
permeable medium	$\mathbf{B} = \mu\mathbf{H}$	(4.27)

In a vacuum,
$$\sigma = 0, \quad \varepsilon = \varepsilon_0, \quad \mu = \mu_0. \tag{4.28}$$

MAXWELL'S EQUATIONS

The primary formulas in the foregoing compendium are the two Maxwell equations, which we repeat and stress here.

$$V\{C\} = -\frac{d\Phi\{S\}}{dt} \tag{4.29}$$

$$U\{C\} = I\{S\} + \frac{d\Psi\{S\}}{dt} \tag{4.30}$$

The power of these laws comes from the fact that they apply to *any* closed curve C and *any* open surface S, provided that the curve C is the edge of surface S and that C and S are oriented in accordance with the right-hand rule.

The curve and surface need not be fixed; they can vary in time. For this reason, it is important that the time derivative applies to the flux, not merely to the flux density. The flux can be a function of time not only because the flux density, $\mathbf{B}$ or $\mathbf{D}$, varies in time but possibly also because the surface S can change in time. In the explicit form expressed in terms of line and surface integrals, the time derivatives stand outside of the flux integrals and cannot be taken inside, unless the surface of integration is unchanging.

A differential form of Maxwell's equations can be derived from the preceding integral form, provided that the surfaces of integration remain fixed. Recall that Stokes's theorem converts any line integral of a vector field $\mathbf{F}$ around a closed curve into a surface integral of the curl of the field over any surface whose edge is the curve, provided that the curve and surface are oriented by the right-hand rule:

$$\oint_C \mathbf{F} \cdot d\mathbf{l} = \int_S \nabla \times \mathbf{F} \cdot d\mathbf{S}. \tag{4.31}$$

This allows us to express the two Maxwell equations entirely in terms of surface integrals, as

$$\int_S \nabla \times \mathbf{E} \cdot d\mathbf{S} = -\frac{d}{dt} \int_S \mathbf{B} \cdot d\mathbf{S} \tag{4.32}$$

$$\int_S \nabla \times \mathbf{H} \cdot d\mathbf{S} = \int_S \mathbf{J} \cdot d\mathbf{S} + \frac{d}{dt} \int_S \mathbf{D} \cdot d\mathbf{S}. \tag{4.33}$$

If the surface S is unchanging, then the flux integrals are functions of time only because their integrands, the flux densities, are time varying. Under these conditions, the time derivative can be transferred inside the integral, instead of remaining outside in the most general case. Of course, since the integrand depends on position as well as time, the time derivative becomes a partial derivative when applied to the integrand, rather than to the integral. We then have

$$\int_S \mathbf{\nabla} \times \mathbf{E} \cdot d\mathbf{S} = -\int_S \partial\mathbf{B}/\partial t \cdot d\mathbf{S} \tag{4.34}$$

$$\int_S \mathbf{\nabla} \times \mathbf{H} \cdot d\mathbf{S} = \int_S \mathbf{J} \cdot d\mathbf{S} + \int_S \partial\mathbf{D}/\partial t \cdot d\mathbf{S} \tag{4.35}$$

and these equations must hold for *any* (fixed) surface S. The only way these relations can hold for all such surfaces is for the integrands to be equal. The resulting equations are Maxwell's equations in differential form.

$$\mathbf{\nabla} \times \mathbf{E} = -\partial\mathbf{B}/\partial t \tag{4.36}$$

$$\mathbf{\nabla} \times \mathbf{H} = \mathbf{J} + \partial\mathbf{D}/\partial t \tag{4.37}$$

These are vector partial differential equations, because the curl is a combination of partial derivatives of the components of the vector fields with respect to the coordinates. Note, however, that these differential equations are less general than the equations in integral form, since they require that the surfaces of integration in the latter form remain independent of time. Situations in which we cannot rely on the differential form include problems involving moving media, for which, without appropriate adjustments to the equations, the surfaces of integration may undergo time variation and vitiate the differential equations. The integral laws hold in all cases, however, and are arguably easier to deal with than the differential form.

Example 4.1

If we do want to work with the differential version of Maxwell's equations, we need to express the curl operation in terms of the components of the field in the coordinate system that we adopt for the problem. While the expressions for the curl are fairly simple for a rectangular coordinate system, they become rather complicated in other systems. The integral form of the equations is readily applied in any coordinate system, however, and if we choose a surface of infinitesimal size and conforming to the coordinate system, we can arrive at the differential form with little pain and no prodigious feats of memory.

To illustrate this, let us obtain the radial component of the curl of a field $\mathbf{F}$ in the spherical coordinate system. Refer to Figure 4-1.

We choose an infinitesimal surface at a point (r, θ, φ), facing the radial direction. The surface is formed by letting θ change by $d\theta$ and φ by $d\varphi$, while keeping r fixed (i.e., we remain on the sphere of radius r centered at the origin). The bounding curve for this surface is made up of four segments of infinitesimal lengths. We perform the closed line integral of $\mathbf{F}$ around this bounding curve, noting that the four segments are already infinitesimal and therefore do not require breaking up into tinier pieces.

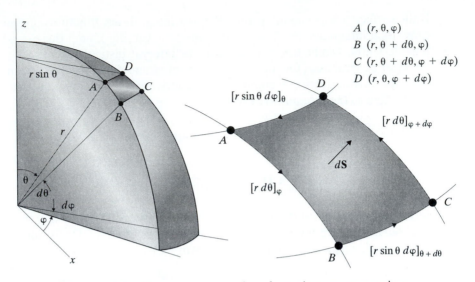

Figure 4-1   Boundary curve of surface element on a sphere.

On the segment from A to B, $\mathbf{F} \cdot d\mathbf{l} = F_\theta r \, d\theta$
On the segment from B to C, $\mathbf{F} \cdot d\mathbf{l} = F_\varphi r \sin\theta \, d\varphi$
On the segment from C to D, $\mathbf{F} \cdot d\mathbf{l} = -F_\theta r \, d\theta$
On the segment from D to A, $\mathbf{F} \cdot d\mathbf{l} = -F_\varphi r \sin\theta \, d\varphi$

Summing the four contributions seems to result in total cancellation, until we realize that the first and third segments require evaluation at different values of φ—namely, at φ and at $\varphi + d\varphi$, respectively—and the second and fourth involve different values of θ, at $\theta + d\theta$ and at θ, respectively. Consequently, the pairwise differences become differentials. On the segments from A to B and from C to D, the sum is

$$\left[F_\theta r \, d\theta\right]_\varphi + \left[-F_\theta r \, d\theta\right]_{\varphi + d\varphi} = -\frac{\partial\left[F_\theta r \, d\theta\right]}{\partial\varphi} \, d\varphi = -\frac{\partial F_\theta}{\partial\varphi} r \, d\theta \, d\varphi \qquad (4.38)$$

and on the segments from B to C and from D to A, the sum is

$$\left[F_\varphi r \sin\theta \, d\varphi\right]_{\theta + d\theta} + \left[-F_\varphi r \sin\theta \, d\varphi\right]_\theta = \frac{\partial\left[F_\varphi r \sin\theta \, d\varphi\right]}{\partial\theta} \, d\theta$$

$$= \frac{\partial\left[F_\varphi \sin\theta\right]}{\partial\theta} r \, d\theta \, d\varphi. \qquad (4.39)$$

The closed line integral around the infinitesimal surface is therefore

$$\oint \mathbf{F} \cdot d\mathbf{l} = \left\{\partial\left[F_\varphi \sin\theta\right]/\partial\theta - \partial F_\theta/\partial\varphi\right\} r \, d\theta \, d\varphi \qquad (4.40)$$

and, by Stokes's theorem, this should be $\boldsymbol{\nabla} \times \mathbf{F} \cdot d\mathbf{S}$, with no further integration, since the surface is already infinitesimal. But $d\mathbf{S} = \hat{\mathbf{r}} r \, d\theta \, r \sin\theta \, d\varphi$, so that the radial component of the curl of $\mathbf{F}(r, \theta, \varphi)$ must be

$$\hat{\mathbf{r}} \cdot \nabla \times \mathbf{F} = \{\partial[\sin\theta\, F_\varphi]/\partial\theta - \partial F_\theta/\partial\varphi\}/r\sin\theta. \tag{4.41}$$

This is the expression sought. The other two components of the curl can be found in a similar way, by choosing infinitesimal surfaces that face along the directions of increasing θ and of increasing φ.

The divergence theorem, which converts any closed surface integral of a vector field into an integral of the divergence of the field over the enclosed volume, can similarly be used to express the constraint equations in differential form. We have therefore

$$\text{Gauss's law (electric)} \qquad \nabla \cdot \mathbf{D} = \rho \tag{4.42}$$

$$\text{Gauss's law (magnetic)} \qquad \nabla \cdot \mathbf{B} = 0 \tag{4.43}$$

and, provided we are not in a moving medium,

$$\text{conservation of charge} \qquad \nabla \cdot \mathbf{J} + \partial\rho/\partial t = 0 \tag{4.44}$$

$$\text{continuity of current} \qquad \nabla \cdot (\mathbf{J} + \partial\mathbf{D}/\partial t) = 0. \tag{4.45}$$

There is an elaborate mathematical framework for the solution of partial differential equations, at least in a small number of coordinate systems for which the technique of separation of variables works. This can be applied successfully when the material configuration fits one of these coordinate systems. However, these techniques usually offer little insight into the nature of the solutions. We prefer to use Maxwell's equations in integral form and rely at first on quasistatic analysis, which assumes that the fields vary slowly. This approach offers opportunities for physical reasoning, for understanding the behavior of the fields, and for analysis of more complicated configurations that do not necessarily conform to a simple coordinate system. Later, we will remove the restriction to slow time variation.

Quasistatic analysis furnishes a more user-friendly, gentle transition from the more intuitive results of low-frequency circuit theory to those of faster circuitry. Quasistatics imparts insights difficult to gain from formal solutions to differential equations, which typically require one to recall or learn advanced, special functions of mathematical analysis. Quasistatics emphasizes deviations from the results of circuit theory that are due to interactions between the electric and the magnetic fields in the configuration. It can be applied to geometries that defy analytic solutions. It is particularly helpful in readily furnishing approximate answers in complicated cases. Most important, it provides the engineer with estimates of the frequency range of applicability of possible designs that are ultimately based on static analysis.

QUASISTATIC FIELDS

One feature of Maxwell's equations that complicates their solution is that they are coupled. In a medium characterized by constitutive parameters μ, ε, σ, there are just two fields to be found, the electric field $\mathbf{E}(\mathbf{r}, t)$ and the magnetic field $\mathbf{H}(\mathbf{r}, t)$. But each of the two Maxwell equations involves both these fields, so that the two equations must be solved simultaneously:

$$\oint_C \mathbf{E} \cdot d\boldsymbol{l} = -\frac{d}{dt} \int_S \mu \mathbf{H} \cdot d\mathbf{S} \tag{4.46}$$

with

$$\oint_C \mathbf{H} \cdot d\boldsymbol{l} = \int_S \sigma \mathbf{E} \cdot d\mathbf{S} + \frac{d}{dt} \int_S \varepsilon \mathbf{E} \cdot d\mathbf{S}. \tag{4.47}$$

To extract, say, the **E** field from the emf in the first of these equations, we need to know the magnetic field **H** in order to evaluate the magnetic flux. In the same way, to extract the **H** field from the mmf in the second one, we need to know the electric field **E** in order to evaluate the electric flux and the electric current. It seems we cannot attack either equation individually. Any field configuration we can arrive at that satisfies one of the two equations must also be consistent with the other. We need a solution procedure that yields self-consistent fields that satisfy both equations simultaneously.

There exists a special condition that uncouples the two equations and allows us to deal with one at a time. This is the condition of *statics*, wherein there is no time variation of the fields. For that case, the equations reduce to

$$\oint_C \mathbf{E} \cdot d\boldsymbol{l} = 0, \tag{4.48}$$

which involves only **E** and asks no questions about **H**, together with

$$\oint_C \mathbf{H} \cdot d\boldsymbol{l} = \int_S \sigma \mathbf{E} \cdot d\mathbf{S}, \tag{4.49}$$

which either involves only **H** if the medium is a perfect insulator ($\sigma = 0$) or else does require knowledge of **E** if the medium is conductive, but then we can use the **E** field previously obtained from the emf equation, without **H**, to evaluate the current in the mmf equation. In the static case, then, the two equations are uncoupled and we face a considerably simpler problem than when they are coupled.

The two uncoupled equations in the static case are, in fact, too easy to solve! They suffer from having an infinite variety of solutions. For example, the emf equation by itself is solvable by the Coulomb field of any charge configuration whatsoever. The problem is that knowing only the line integral of a field does not determine it uniquely; information about its flux must be provided as well, in order to fix the field pattern specifically. In the time-varying case, the Maxwell equations have both the line and surface integrals for each field, but in the static case, we lose the magnetic and electric fluxes in the equations.

The resolution of this difficulty in the static case is to recall the constraint equations, which relate fluxes to their sources, and append them to the two equations for the emf and mmf. For a static field configuration, we then have equations that are uncoupled, yet sufficient to yield a field pattern:

electrostatics $\quad \oint_C \mathbf{E} \cdot d\mathbf{l} = 0, \qquad\qquad\qquad \oint_S \varepsilon \mathbf{E} \cdot d\mathbf{S} = Q\{V\};$ (4.50)

magnetostatics $\quad \oint_C \mathbf{H} \cdot d\mathbf{l} = I\{S\} + \int_S \sigma \mathbf{E} \cdot d\mathbf{S}, \qquad \oint_S \mu \mathbf{H} \cdot d\mathbf{S} = 0,$ (4.51)

where $Q\{V\}$ is the source charge in the volume V enclosed by closed surface S of the electrostatics equations and $I\{S\}$ is the source current that crosses the open surface S in the magnetostatics case. In the case of a nonconducting medium, the two equations are entirely separate; when there is conductivity, we merely solve the electrostatic equations first and then use the known $\mathbf{E}$ field in the magnetostatic equations to solve for the $\mathbf{H}$ field.

We already have experience solving these static equations, if necessary by superposition of the fields of point sources (point charges and current elements). What we find is field configurations of a typical static type, with field strengths proportional to the strength of the sources. That is, for electrostatics, the field *pattern* is of the type that has no closed-loop emf and the field *strength* is proportional to the source charge; for magnetostatics, the field *pattern* is of the type that has no closed-surface flux and the field *strength* is proportional to the source current. We are here using the word *pattern* to connote the spatial variation of the field configuration, say inverse-square in the distance from the source, without regard to the field magnitude, and the word *strength* to denote the field amplitude at a typical field point.

We now invoke some physical reasoning and ask what may happen if the source strength, instead of remaining static, were to undergo some slow, gradual change. Since the field strength is proportional to that of the source, it seems reasonable to expect that the field amplitude would follow the slow variations in the source strength, with no expectation that the field pattern would change. For example, suppose the charge in some region is 3 coulombs and that it produced some field pattern, perhaps inverse square. If now the amount of charge is changed to 6 coulombs, we may expect the field pattern to still be inverse square, but now twice as strong. We also expect that during the transition from 3 to 6 coulombs, the field pattern remains the same but with an amplitude that follows the increase in the source strength slavishly and synchronously.

Under this reasoning, when the sources of the electrostatic or magnetostatic field configuration become slowly varying time functions $Q(t)$ and $I(t)$, we expect the fields to retain their fully static field pattern but become functions of time that follow the source variations proportionately. We call such field configurations *quasistatic*, that is, "as if they were static," although they do vary in time, synchronously with the sources. Such fields satisfy the same laws of electrostatics and magnetostatics as far as their pattern is concerned, so that they still have no closed-loop emf, for example, but they carry the time variation of their sources. Let us append a zero subscript to such quasistatic fields; they satisfy

$$\oint_C \mathbf{E}_0(\mathbf{r}, t) \cdot d\mathbf{l} = 0, \qquad\qquad \oint_S \varepsilon \mathbf{E}_0(\mathbf{r}, t) \cdot d\mathbf{S} = Q(t); \qquad (4.52)$$

$$\oint_C \mathbf{H}_0(\mathbf{r}, t) \cdot d\mathbf{l} = I(t) + \int_S \sigma \mathbf{E}_0(\mathbf{r}, t) \cdot d\mathbf{S}, \qquad \oint_S \mu \mathbf{H}_0(\mathbf{r}, t) \cdot d\mathbf{S} = 0. \qquad (4.53)$$

The quasistatic fields are no more difficult to find than are the perfectly static ones; they are physically reasonable field configurations for the case of sources that vary. They cannot be correct, however. Before pointing out just what is wrong with these fields and how to correct their deficiency, let us examine an example of such fields.

Example 4.2

Two parallel circular conducting plates are separated by a nonconducting dielectric with permittivity ε and permeability μ. The plates are separated by height h and are each of area $A = \pi a^2$. This capacitor is connected to a voltage source, as indicated in Figure 4-2. If the voltage source is slowly varying, as $V = V(t)$, find the quasistatic fields inside the capacitor. ◇

What the voltage source does is to push charges around the circuit, piling them up on the upper plate and draining them from the lower one. These charges cannot bridge the gap between the plates, because the dielectric is nonconducting. The opposite charges on the two plates attract each other, so that they tend to remain on the inner surface of the two plates, but they repel their like kind on each plate, so that they spread out over the expanse of each plate. The charges spread out as evenly as possible, since any charges in clumps would repel each other, except that at the outer edge of the circular plates the charges cannot push any farther outward and can be expected to pile up to a higher density than in the middle region of the plate.

If we stay away from the capacitor's outer edge, the resultant electric field that accompanies the charge distribution on the two plates will be seen to be uniform. We will make the approximation that the field is uniformly distributed throughout the region between the plates, even out to the edge. This neglects the slight buildup of charges at the edge and the consequent *fringing* of the electric field, which actually bulges somewhat into the air region beyond the plates. For large plates with small separation, and especially if the dielectric has a high permittivity relative to the air beyond it, the region of fringing is a negligible fraction of the volume that carries the uniform field we are as-

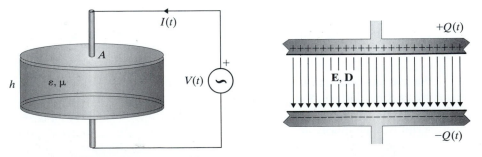

Figure 4-2   Circular-plate capacitor and its uniform field.

suming. One of the important advantages of the quasistatic analysis we are pursuing is that we can readily incorporate inexact field configurations and other approximations into our analysis.

With this assumption of uniformity of the charge distribution and neglecting the very weak fringing field on the outside of the plates, we can easily obtain the electric field strength inside the capacitor in terms of the specified applied voltage $V(t)$ and also find the relation among the circuit variables $V(t), Q(t), I(t)$. If we surround just the upper plate with a closed surface S, for convenience one that just hugs that plate, then the enclosed charge is $Q(t)$ and the electric flux is just the integral of the uniform $\mathbf{D}_0$ field over the part of the surface that lies inside the capacitor. Since we are integrating a constant field strength D_0 over the area A of the plate, the flux is just $D_0 A$. By Gauss's law, then, the quasistatic electric field is

$$\mathbf{D}_0(t) = [Q(t)/A]\hat{\mathbf{e}}, \qquad \mathbf{E}_0(t) = [Q(t)/\varepsilon A]\hat{\mathbf{e}}, \qquad (4.54)$$

where $\hat{\mathbf{e}}$ is a unit vector directed from the upper plate to the lower one.

We can calculate the emf around the entire circuit, including the voltage source, the wires that connect it to the capacitor, and continuing through the capacitor. We get no contribution from the conducting wires, where there is no tangential electric field, so that the entire voltage provided by the source appears across the capacitor and equals the integral of the $\mathbf{E}_0$ field from the upper plate to the lower one. Since we are integrating a constant field strength E_0 over a distance h along $\hat{\mathbf{e}}$, we get

$$V(t) = E_0(t)h, \qquad V(t) = [h/\varepsilon A]Q(t), \qquad (4.55)$$

or

$$Q(t) = CV(t), \qquad C = \varepsilon A/h \qquad (4.56)$$

as the familiar defining circuit relation for a capacitor of capacitance C and the well-known capacitance $C = \varepsilon A/h$ of a parallel-plate capacitor.

Finally, we can get the current-voltage circuit relation for this configuration by applying the law of charge conservation to the same closed surface that hugs the upper plate. The enclosed charge is just $Q(t)$ again and the outwardly directed current is the negative of the current flowing around the circuit and into the upper plate, or $-I(t)$, so that the law requires

$$I(t) = dQ(t)/dt \qquad \text{or} \qquad I(t) = C\, dV(t)/dt, \qquad (4.57)$$

as the familiar circuit relation for a capacitor.

Since there is no current flowing inside the perfect insulator between the plates, there is no quasistatic magnetic field inside the capacitor. We are about to find, however, that the preceding quasistatic fields and circuit relations are not correct, and that there is a magnetic field inside the capacitor, in the time-varying case.

What is wrong with the quasistatic fields we have just calculated is that they satisfy the equations of electrostatics but with the fields varying in time in synchronism with the time variation of the source. As such, the fields do vary in time and hence produce time-varying flux, which is neglected in the static equations.

What was neglected in the full Maxwell equations was really the time derivatives of the electric and magnetic fluxes. If the fields vary slowly, because the sources themselves vary slowly, then the rates of change of the fluxes are small quantities, which we can either neglect or treat as small corrections to the results we have found. The slower the time variation of the sources, the smaller the time derivatives of the fluxes, even if the fluxes are themselves substantially large. In the limit of unvarying sources, the quasistatic results become exact, but as the sources begin to vary in time, corrections to the quasistatic fields can be calculated from the full Maxwell equations. Let us first continue the foregoing example, to find the magnetic field that must accompany the electric one inside the capacitor, in the time-varying case.

Example 4.3

Find the magnetic field inside the parallel circular plate capacitor excited by a slowly varying voltage source of the previous example. ◈

We have already found a $\mathbf{D}$ field inside the capacitor, directed from the upper to the lower plate and proportional to the applied voltage:

$$\mathbf{D}_0(t) = \left[\varepsilon V(t)/h\right]\hat{\mathbf{e}}. \tag{4.58}$$

Since this is time varying, there is a displacement current inside the capacitor, of density

$$\partial \mathbf{D}_0(t)/\partial t = \left[(\varepsilon/h)\,dV(t)/dt\right]\hat{\mathbf{e}}. \tag{4.59}$$

This generates a magnetic field surrounding it. This displacement current density is uniform and, because of the circular symmetry, the magnetic field lines will be circular, centered on the axis of the plates and parallel to them. The circular symmetry also dictates that the magnetic field strength will depend only on the radial distance from the axis, not on the azimuthal angle; it will also not vary with vertical distance from a plate, because of the axial uniformity of the electric flux density. Hence, the magnetic field is of the form $\mathbf{H}(\mathbf{r}) = H(\rho)\hat{\mathbf{h}}$, where $\hat{\mathbf{h}}$ is directed azimuthally, perpendicular to $\hat{\mathbf{e}}$ and oriented with respect to $\hat{\mathbf{e}}$ by the right-hand rule; see Figure 4-3.

To discover the radial dependence of the magnetic field strength, we calculate the mmf around a horizontal circle of typical radius ρ, centered on the axis, inside the capacitor. Since $\mathbf{H} \cdot d\mathbf{l} = H(\rho)\hat{\mathbf{h}} \cdot \hat{\mathbf{h}}\,dl$ and $H(\rho)$ is constant on the circle, the mmf is just that constant $H(\rho)$ times the circumference of the circle:

$$U(\rho) = 2\pi\rho H(\rho). \tag{4.60}$$

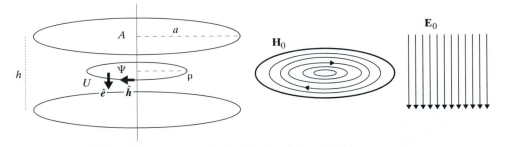

Figure 4-3   Calculation of magnetic field inside a capacitor.

On the horizontal circular disk of radius ρ, oriented by $\hat{\mathbf{e}}$ to conform to the right-hand rule, the electric flux of the uniform $\mathbf{D}$ field is

$$\Psi(\rho) = D_0\pi\rho^2 = (\varepsilon/h)\pi\rho^2 V(t). \tag{4.61}$$

Equating the mmf around the edge of the disk to the displacement current through it, we find

$$U = d\Psi/dt = 2\pi\rho H(\rho) = (\varepsilon/h)\pi\rho^2\, dV(t)/dt, \tag{4.62}$$

so that the magnetic field intensity at radius ρ is

$$H(\rho) = (\varepsilon/2h)\big[dV(t)/dt\big]\rho. \tag{4.63}$$

Note that this magnetic field increases linearly, from zero on the axis to a maximum of $H(a) = (\varepsilon a/2h)\big[dV(t)/dt\big]$ at the edge of the dielectric, where the radius of the plates is a, with $A = \pi a^2$. Note also that the magnetic field is absent in the perfectly static case; it requires time variation of the voltage source for its existence.

Having found that, in the time-varying case, a magnetic field appears inside the capacitor, we recognize the existence of a magnetic flux between the plates, circulating around the axis of the capacitor. This flux is also time varying, so that a nonzero emf must surround any portion of that magnetic flux. But this is inconsistent with the uniform electric field we assumed at the outset, because a uniform field has no closed-loop emf. We conclude that, once the applied voltage begins to vary, the electric field inside the capacitor cannot remain uniform, as it is in the perfectly static case. The electric field can still be directed vertically and there is still azimuthal symmetry, so that the field must develop a nonuniformity in the radial direction. However, this nonuniformity should be slight, since it evolves from the time variation of a magnetic flux, which itself arises from the slow time variation of the electric flux that follows the slow variations in the voltage source. We can apply the Faraday-Maxwell law to determine the radial distribution of the electric field in the capacitor, as follows.

Example 4.4

Find the nonuniformity of the electric field inside the parallel circular plate capacitor excited by a slowly varying voltage source of the prior examples. ◈

We think of the nonuniformity as a small correction to the uniform electric field we assumed at the outset, brought on by the slow time variation of the applied voltage. In the static limit, that nonuniformity should disappear, leaving the uniform field $E = V/h$. We need to evaluate the emf induced by the time-varying magnetic flux that circulates around the axis; we can then extract the radial variation of the electric field, $E(\rho)$.

Since the magnetic field is directed azimuthally, we will choose a surface that faces that direction to calculate the magnetic flux. This surface will be vertical, and we want its edge to be at a typical radius ρ_1, so that the resultant emf will involve the electric field at any such radius. Let us therefore choose as our surface for the evaluation of the magnetic flux the rectangular area of height h (from one plate to the other) and extending radially from the axis to the typical radius ρ_1, at any convenient azimuth, as in Figure 4-4.

We have the magnetic field $H(\rho)$ and therefore the magnetic flux density $B(\rho)$ inside the capacitor:

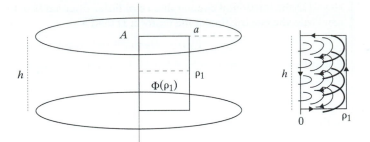

Figure 4-4   Calculation of magnetic flux and emf inside capacitor.

$$H(\rho) = (\varepsilon/2h)[dV(t)/dt]\rho, \qquad B(\rho) = (\mu\varepsilon/2h)[dV(t)/dt]\rho. \tag{4.64}$$

We integrate this flux density over the rectangular area; the surface element is $d\mathbf{S} = \hat{\mathbf{h}}\,d\rho\,dz$ and $\mathbf{B} = \hat{\mathbf{h}}B(\rho)$, so that the surface integration entails a factor h from the vertical integration and the following radial integration:

$$\Phi(\rho_1) = \int_0^{\rho_1}(\mu\varepsilon/2)[dV(t)/dt]\rho\,d\rho = (\mu\varepsilon/4)[dV(t)/dt]\rho_1^2. \tag{4.65}$$

Thus, the magnetic flux out to some radius increases quadratically with that radius.

Because this magnetic flux is a function of time, an emf is induced around the edge of the rectangle. Since there can be no horizontal electric field tangential to the conducting plates, this emf must come from a nonuniformity of the vertical electric field; that is, the electric field must be a function of the radius if it is a function of time. However, there is no cause for any field variation vertically. The emf around the rectangle is therefore

$$V\{C\} = E(0)h - E(\rho_1)h \qquad \text{and} \qquad V\{C\} = -d\Phi\{S\}/dt, \tag{4.66}$$

so that

$$E(\rho_1) = E(0) + [d\Phi/dt]/h = E(0) + (\mu\varepsilon/4h)[d^2V(t)/dt^2]\rho_1^2. \tag{4.67}$$

We do not yet know what the electric field strength is at $\rho = 0$, on axis, but we see already that the field varies parabolically with radius, instead of being uniform. The nonuniformity appears only when there is time variation; in fact, it arises when there is a second derivative of the applied voltage.

How shall we determine the field on axis, $E(0)$, so that we can complete the calculation? This unknown is not determined by Maxwell's equations alone; physically, we must furnish information about just how the source is impressed upon the capacitor; mathematically, we must provide a boundary condition. In our problem, the voltage is applied externally and, considering the emf around the entire circuit and continuing to neglect the very weak magnetic field outside the capacitor (corresponding to the inductance of the wires connected to it), then the full source voltage is applied to the outer surface of the dielectric, from the edge of the upper plate to that of the lower one. Still neglecting any fringing near the edge of the dielectric, this means that the electric field is fully specified at the outer edge of the capacitor:

$$E(a) = V(t)/h. \tag{4.68}$$

This boundary condition enables us to find the missing value of the electric field on the axis. We set $\rho_1 = a$ in the previous result and set the field equal to $V(t)/h$, which yields

$$V(t)/h = E(0) + (\mu\varepsilon/4h)[d^2V(t)/dt^2]a^2, \qquad (4.69)$$

or

$$E(0) = V(t)/h - (\mu\varepsilon/4h)[d^2V(t)/dt^2]a^2. \qquad (4.70)$$

The field on axis differs from that at the edge by the term proportional to the second derivative of the applied voltage.

Finally, since ρ_1 could have been any radius from 0 to a, we find that the electric field distribution inside the capacitor is the nonuniform field

$$E(\rho) = V(t)/h - (\mu\varepsilon/4h)[d^2V(t)/dt^2][a^2 - \rho^2], \qquad (4.71)$$

instead of the uniform one $E(\rho) = E_0 = V(t)/h$. This is the result we sought.

SURPRISES FOR CIRCUIT ANALYSTS

There are several surprising consequences to the above deviations of the electromagnetic field inside the capacitor from the purely electrostatic configuration. One result that surprises anyone used to circuit analysis is that the voltage across the two plates varies from one point to another of the plates. The reason is, of course, that there is a time-varying magnetic flux inside the capacitor, which is itself due to the displacement current inside it. The voltage along any vertical line from the upper to the lower plate at radius ρ is just h times the electric field at that radius. That emf varies quadratically with radius ρ, from the axis ($\rho = 0$) to the outer edge ($\rho = a$) of the capacitor:

$$\text{emf}(\rho) = V(t) - (\mu\varepsilon/4)[d^2V(t)/dt^2][a^2 - \rho^2]. \qquad (4.72)$$

The emf does equal the applied voltage at the outer edge $\rho = a$ but, at the axis $\rho = 0$, it differs from the applied voltage by the amount $(\mu\varepsilon a^2/4)d^2V(t)/dt^2$.

Example 4.5

Find the difference between the voltages at the axis and at the outer edge of the parallel circular plate capacitor excited by a sinusoidal voltage source. Let the source be $V(t) = V_0 \cos(\omega t + \theta)$, with amplitude $V_0 = 100$ V and frequency $f = \omega/2\pi = 100$ MHz. The plates are 1 mm apart and their radius is 10 cm. The dielectric is nonmagnetic $(\mu = \mu_0)$ and has relative permittivity $\varepsilon/\varepsilon_0 = 3$. ◇

Since $d^2V(t)/dt^2 = -\omega^2V(t)$ for a sinusoidal voltage at frequency ω, we have

$$\text{emf}(0) - \text{emf}(a) = [\mu\varepsilon\omega^2a^2/4]V_0 \cos(\omega t + \theta), \qquad (4.73)$$

which is itself a sinusoid, in phase with the source, of amplitude

$$[\mu_0 3\varepsilon_0 \pi^2 f^2 a^2]V_0 = 3.29 \text{ volts}. \qquad (4.74)$$

We see that the voltage across the plates is nonuniform to the extent of about 3% for sinusoidal time variation at 100 MHz; we get a sense that we can usually consider our sources to be "slowly varying," and circuit theory to apply well enough, for frequencies up into the megahertz range, if the structure involved is of a size not larger than a few centimeters.

Another surprise for the circuit analyst is the result that even the ideal capacitor we have considered fails to behave as the circuit-theoretical capacitor is supposed to act. We have postulated a perfectly insulating dielectric, perfectly conducting plates, and perfect geometry, together with perfect wires and an ideal voltage source. We have even neglected fringing of the fields at the edges. Yet the relation between the current into the capacitor and the voltage across it is not the ideal $I(t) = C\,dV(t)/dt$ that we might have expected. To see this, we need to calculate the current delivered to the capacitor, for the given applied voltage $V(t)$. Using the constraint of continuity of total current over the closed surface that surrounds and hugs the upper plate, we find the outward current $I\{S\}$ to be $-I(t)$, concentrated in the wire, and the outward electric flux $\Psi\{S\}$ of the fringe-free electric field pattern to be the integral of the electric flux density $\varepsilon E(\rho)$ across the plate:

$$\Psi\{S\} = \int_0^{2\pi} \int_0^a \varepsilon\{V/h - (\mu\varepsilon/4h)[d^2V/dt^2][a^2 - \rho^2]\}\rho\,d\rho\,d\varphi$$

$$= [\varepsilon\pi a^2/h]\{V - [\mu\varepsilon a^2/8]d^2V/dt^2\}. \tag{4.75}$$

Hence, the continuity relation $I\{S\} + d\Psi\{S\}/dt = 0$ yields for the current-voltage relation of our capacitor,

$$I(t) = [\varepsilon\pi a^2/h]\{dV/dt - [\mu\varepsilon a^2/8]d^3V/dt^3\}$$

$$= C\{dV/dt - [\mu\varepsilon a^2/8]d^3V/dt^3\}, \tag{4.76}$$

instead of the $I(t) = C\,dV(t)/dt$ relation of pure circuit theory, where $C = \varepsilon\pi a^2/h$ is the static capacitance we calculated before. The structure, which we made as ideal as possible, does not quite behave as a capacitor. That is, we now see that the most ideal parallel-plate capacitor structure we can conceive of is inherently incapable of behaving like an ideal circuit-theoretic capacitor, because it must generate time-varying magnetic flux along with its electric flux.

Example 4.6

Based on the preceding calculation of the relation between the current into the parallel-plate capacitor structure and the voltage across it, which yielded

$$I(t) = C\{dV/dt - [\mu\varepsilon a^2/8]d^3V/dt^3\},$$

obtain an equivalent circuit for the structure. ◈

Because the last term in the expression for the current has a negative sign, it is easier to obtain an equivalent circuit after first bringing the negative term to the other side of the equation:

$$C\,dV/dt = I(t) + C[\mu\varepsilon a^2/8]d^3V/dt^3.$$

This can be considered the time derivative of a simpler expression:

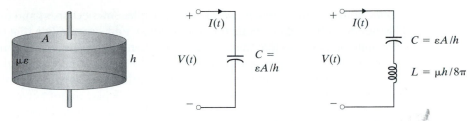

Figure 4-5 Elementary and corrected equivalent circuit of ideal capacitor.

$$V(t) = (1/C) \int I(t')dt' + [\mu\varepsilon a^2/8]d^2V/dt^2.$$

The last term is a small correction to the previous term; to this order of approximation, the d^2V/dt^2 term can be thought of as the derivative of dV/dt, which is approximately $I(t)/C$, so that d^2V/dt^2 can be replaced by $(1/C)dI/dt$. This expresses the voltage-current relation as

$$V(t) = (1/C) \int I(t') dt' + [\mu\varepsilon a^2/8](1/C) dI/dt.$$

This looks just like the sum of the voltages across a capacitor and an inductor, both with the same current, so that this version of the relation can be expressed by a capacitor and an inductor in series, as in Figure 4-5. The equivalent circuit relation is simply

$$V(t) = (1/C) \int I \, dt + L \, dI/dt$$

and the two relations agree if the capacitor has the value C already given and the inductor value is such that $L = [\mu\varepsilon a^2/8](1/C)$. Since $C = \varepsilon\pi a^2/h$, the equivalent inductance in series with the static capacitance is $L = \mu h/8\pi$.

Another observation to make about the calculation of the nonuniformity of the electric field in the capacitor is that it cannot be the end of the story. When we calculated the magnetic field inside the capacitor, we used a displacement current based on a uniform electric field. Since the electric field is actually not uniform, the displacement current we used is not the correct one. Consequently, the magnetic field we found is not quite right and hence the magnetic flux we used is also faulty. But this flux is what we used to find the nonuniformity of the electric field, so that even this is in doubt.

It appears that we must go back to our calculation of the displacement current, this time using the nonuniform electric field distribution, and get a corrected version of the magnetic field, which we can then use to get a better expression for the magnetic flux and then correct our version of the nonuniform electric field. Seemingly, there will be no end to this recursive process of correcting our corrections. The saving grace is that the corrections to our calculations involve successively higher derivatives of the applied voltage and each such derivative entails an ever smaller correction if the original applied voltage is slowly varying. Even if the many volts of the applied source were to result in amperes of displacement current and even if this would generate webers of magnetic flux, the resultant emf surrounding that flux might be in the millivolt range, so that the accompanying nonuniform electric field might correspond to merely milliamperes of additional displacement current, which would then generate milliwebers of additional magnetic flux and only microvolts of surrounding emf, and so on to microamperes of further displacement current and ever smaller corrections to the electric and magnetic fields and fluxes. For a sufficiently slowly varying source, then, we wind up with an infinite series of correction terms that we can expect to converge to a field configuration that is consistent with both Maxwell equations.

This approach ultimately yields solutions to Maxwell's equations in the form of infinite series, or at least the first few terms of such series. The alternative is to attack Maxwell's equations directly, which we will do later. If we do choose to carry out quasistatic calculations, and whether or not we will attempt to sum the infinite series of corrections, we will need to be highly systematic about our scheme of computation. A simple, tabular scheme is particularly effective and easy to apply.

SYSTEMATIC QUASISTATIC ANALYSIS

Let us return to the two Maxwell equations and see how we can systematize the successive approximation procedure we have just outlined to arrive at self-consistent electromagnetic fields that satisfy both equations simultaneously.

The equations are

$$V\{C\} = -d\Phi\{S\}/dt \quad \text{and} \quad U\{C\} = I\{S\} + d\Psi\{S\}/dt, \quad (4.77)$$

where closed curve C is the edge of open surface S for each equation separately; the surface need not be the same one for the two equations. The medium is described by parameters ε, μ, σ, so that we need to find just the two fields $\mathbf{E}(\mathbf{r}, t)$ and $\mathbf{H}(\mathbf{r}, t)$. The magnetic flux $\Phi\{S\}$ is the surface integral of $\mu\mathbf{H}$, the electric flux $\Psi\{S\}$ is the integral of $\varepsilon\mathbf{E}$, and the current $I\{S\}$ is the integral of $\sigma\mathbf{E}$.

We start by pretending the sources are static and we find electric and magnetic field configurations that are consistent with the electrostatic and magnetostatic equations, supplemented by Gauss's laws or other constraints or symmetry conditions. The static fields must also be consistent with the specified source. We then allow the source to be time varying and let the static fields acquire the same time variation, following the source synchronously but retaining the static spatial field pattern. These are the quasistatic fields $\mathbf{E}_0(\mathbf{r}, t)$ and $\mathbf{H}_0(\mathbf{r}, t)$; they satisfy the static version of the Maxwell equations

$$V_0\{C\} = 0 \quad \text{and} \quad U_0\{C\} = I_0\{S\}, \quad (4.78)$$

with the time derivatives of the fluxes absent, despite the fact that the fields and their fluxes do vary in time.

These quasistatic fields are not consistent with the Maxwell equations because, although they exhibit time-varying magnetic flux $\Phi_0(t) = \int_S \mu \mathbf{H}_0 \cdot d\mathbf{S}$, electric flux $\Psi_0(t) = \int_S \varepsilon \mathbf{E}_0 \cdot d\mathbf{S}$, and current $I_0(t) = \int_S \sigma \mathbf{E}_0 \cdot d\mathbf{S}$ that we can calculate, the emf around the edge of the surface S is zero instead of $-d\Phi_0/dt$ and the mmf is only $I_0(t)$ instead of $I_0 + d\Psi_0/dt$. We conclude that some change in the spatial pattern must accompany the onset of time variation.

For sufficiently slow time variation, however, the static equations are very nearly exact, because the time derivatives of the slowly varying fluxes are then very small. For slow changes in time, only a slight correction to the quasistatic fields may be needed:

$$\mathbf{E} = \mathbf{E}_0 + \mathbf{E}' \quad \text{and} \quad \mathbf{H} = \mathbf{H}_0 + \mathbf{H}' , \tag{4.79}$$

to which will correspond the integrals along curves and surfaces

$$V = V_0 + V' \qquad U = U_0 + U'$$

$$\Phi = \Phi_0 + \Phi' \qquad \Psi = \Psi_0 + \Psi' \qquad I = I_0 + I'. \tag{4.80}$$

These will have to satisfy

$$V_0 + V' = -d\Phi_0/dt - d\Phi'/dt \tag{4.81}$$

and

$$U_0 + U' = I_0 + I' + d\Psi_0/dt + d\Psi'/dt. \tag{4.82}$$

But the quasistatic equations already guarantee that $U_0 = I_0$ and $V_0 = 0$, so that the correction (primed) fields will satisfy

$$V' = -d\Phi_0/dt - d\Phi'/dt \tag{4.83}$$

and

$$U' = I' + d\Psi_0/dt + d\Psi'/dt. \tag{4.84}$$

On the right side of the equation for the correction V' to the emf V_0, only $d\Phi_0/dt$ is known. However, we argue that this known amount is nearly all there is, because the remaining term is the time derivative of a slowly varying quantity that is itself a small correction term Φ'. For example, $d\Phi_0/dt$ might be of the order of volts but $d\Phi'/dt$ may amount to only millivolts. We can therefore get almost the entire correction to the quasistatic electric field by retaining only the $d\Phi_0/dt$ term. This will yield, say, only $\mathbf{E}_1$ instead of the full correction $\mathbf{E}'$; a still-smaller correction $\mathbf{E}''$ will have been neglected.

Let us simplify the situation, temporarily, by restricting ourselves to a nonconducting medium ($\sigma = 0$), so that the equation for the correction to the quasistatic mmf reduces to

$$U' = d\Psi_0/dt + d\Psi'/dt. \tag{4.85}$$

In this case, on the right side of the equation, only $d\Psi_0/dt$ is known. Once again, we argue that this known amount is nearly all there is, because the remaining term is the time derivative of a slowly varying quantity that is itself a small correction term Ψ'. For example, $d\Psi_0/dt$ might be on the order of amperes but $d\Psi'/dt$ may amount to only milliamperes. We can therefore get almost the entire correction to the quasistatic magnetic field by retaining only the $d\Psi_0/dt$ term. This will yield, say, only $\mathbf{H}_1$ instead of the full correction $\mathbf{H}'$; a still-smaller correction $\mathbf{H}''$ will have been neglected.

What we have to this stage is then

$$\mathbf{E} = \mathbf{E}_0 + \mathbf{E}_1 + \mathbf{E}'' \quad \text{and} \quad \mathbf{H} = \mathbf{H}_0 + \mathbf{H}_1 + \mathbf{H}'', \tag{4.86}$$

with their line and surface integrals satisfying

$$V_0 = 0 \qquad V_1 = -d\Phi_0/dt \qquad V'' = -d\Phi_1/dt - d\Phi''/dt \tag{4.87}$$

$$U_0 = 0 \qquad U_1 = d\Psi_0/dt \qquad U'' = d\Psi_1/dt + d\Psi''/dt. \tag{4.88}$$

The equation for V_1 has a fully known right side and can yield $\mathbf{E}_1$; the equation for U_1 has a fully known right side and can yield $\mathbf{H}_1$. From $\mathbf{H}_1$ we can get $d\Phi_1/dt$, which is most of V'' but not all of it; from $\mathbf{E}_1$ we can get $d\Psi_1/dt$, which is most of U'' but not all of it. We can continue the successive approximations by setting

$$\mathbf{E}'' = \mathbf{E}_2 + \mathbf{E}''' \quad \text{and} \quad \mathbf{H}'' = \mathbf{H}_2 + \mathbf{H}''', \tag{4.89}$$

so that

$$V'' = V_2 + V''' \quad \text{with} \quad V_2 = -d\Phi_1/dt \quad \text{and} \quad V''' = -d\Phi''/dt, \tag{4.90}$$

in which $d\Phi_1/dt$ is known and yields V_2 and $\mathbf{E}_2$, and

$$U'' = U_2 + U''' \quad \text{with} \quad U_2 = d\Psi_1/dt \quad \text{and} \quad U''' = d\Psi''/dt, \tag{4.91}$$

in which $d\Psi_1/dt$ is known and yields U_2 and $\mathbf{H}_2$.

The successive approximations are generating successive terms of an infinite series of correction terms. Each such term in the series for the fields is obtainable from the fluxes of the previous ones:

$$\mathbf{E} = \mathbf{E}_0 + \mathbf{E}_1 + \mathbf{E}_2 + \mathbf{E}_3 + \cdots \quad \text{and} \quad \mathbf{H} = \mathbf{H}_0 + \mathbf{H}_1 + \mathbf{H}_2 + \mathbf{H}_3 + \cdots, \tag{4.92}$$

with line and surface integrals satisfying

$$\begin{aligned} V &= V_0 + V_1 + V_2 + V_3 + \cdots \\ &= -d\Phi/dt = 0 - d\Phi_0/dt - d\Phi_1/dt - d\Phi_2/dt - \cdots \end{aligned} \tag{4.93}$$

and

$$\begin{aligned} U &= U_0 + U_1 + U_2 + U_3 + \cdots \\ &= d\Psi/dt = 0 + d\Psi_0/dt + d\Psi_1/dt + d\Psi_2/dt + \cdots \end{aligned} \tag{4.94}$$

We equate these in pairs, obtaining successively

$$V_0 = 0 \qquad V_1 = -d\Phi_0/dt \qquad V_2 = -d\Phi_1/dt \qquad V_3 = -d\Phi_2/dt \tag{4.95}$$

$$U_0 = 0 \qquad U_1 = d\Psi_0/dt \qquad U_2 = d\Psi_1/dt \qquad U_3 = d\Psi_2/dt. \tag{4.96}$$

The general form is the set of infinite series

$$\mathbf{E} = \sum_{m=0}^{\infty} \mathbf{E}_m \qquad \mathbf{H} = \sum_{m=0}^{\infty} \mathbf{H}_m$$

and

$$V = \sum_{m=0}^{\infty} V_m \qquad U = \sum_{m=0}^{\infty} U_m \qquad \Phi = \sum_{m=0}^{\infty} \Phi_m \qquad \Psi = \sum_{m=0}^{\infty} \Psi_m \qquad (4.97)$$

with the successive terms related by

$$V_0 = 0 \qquad\qquad U_0 = 0$$
$$V_{m+1} = -d\Phi_m/dt \qquad U_{m+1} = d\Psi_m/dt \qquad (4.98)$$

for $m = 0, 1, 2, \cdots$.

The sums of all the corrections will clearly satisfy both Maxwell equations, provided that the sums converge. Convergence may reasonably be expected if the time variation of the source is sufficiently slow, since successive terms involve ever smaller time derivatives of ever smaller prior correction terms.

We have that Φ_0 and Ψ_0 vary in synchronism with the source; hence $V_1, U_1, \Phi_1,$ Ψ_1 vary as the first time derivative of the source; V_2, U_2, Φ_2, Ψ_2 vary as the second time derivative of the source, and so on.

If we now reinstate the conductivity of the medium ($\sigma \neq 0$), then the mmf equations acquire an additional conductive current. The mth-order field $\mathbf{E}_m$ produces not only $\Psi_m = \displaystyle\int_S \varepsilon \mathbf{E}_m \cdot d\mathbf{S}$ but also $I_m = \displaystyle\int_S \sigma \mathbf{E}_m \cdot d\mathbf{S}$, both of which vary as the mth time derivative of the source. We maintain the identification of the order of the correction term with the order of time derivative of the source by imposing the following equations on the corrections:

$$V_0 = 0 \qquad\qquad U_0 = I_0$$
$$V_{m+1} = -d\Phi_m/dt \qquad U_{m+1} = I_{m+1} + d\Psi_m/dt \qquad (4.99)$$

In words, taking a time derivative of a flux labeled with index m produces a quantity that is to be balanced against others labeled $m + 1$. Quantities of successive orders are expected to be successively smaller, and time derivatives of slowly varying quantities result in weaker terms, comparable to quantities of the next order.

We can truncate the infinite series of corrections if we can accept that correction terms beyond a certain stage are negligible, or else we may be able to sum the entire infinite series to get an exact solution to Maxwell's equations.

We recommend a tabular scheme to systematize the calculations of successive approximations to the fields. We begin with the field pattern that would apply if the source were static and choose two suitable open surfaces, one, S_m, that will have a magnetic flux through it, another, S_e, that will carry the electric flux and current. The quantities we can calculate in sequence are, typically, the emf $V\{C_m\}$, from which we may extract $\mathbf{E}$ if we have chosen the closed curve C_m judiciously; we get $\mathbf{D}$ as $\varepsilon\mathbf{E}$, then $\Psi\{S_e\}$, as well as $\mathbf{J}$ as $\sigma\mathbf{E}$ and $I\{S_e\}$ for the surface S_e. We can calculate the mmf $U\{C_e\}$ from I and Ψ; we may extract $\mathbf{H}$ from $U\{C_e\}$ and calculate $\mathbf{B}$ as $\mu\mathbf{H}$ and then $\Phi\{S_m\}$. We complete the circle by obtaining $V\{C_m\}$ again from $\Phi\{S_m\}$.

We carry this scheme out repeatedly, for successive orders of time derivatives of the source; the zeroth derivative corresponds to time variation synchronous with the source and the zero-order quantities are the quasistatic ones. Thereafter, each time we perform a differentiation in time, we move to a higher order of correction term. The table should therefore set out the preceding quantities in successive rows, say, and the orders of time variation in successive columns. Ultimately, each field or flux or other quantity is the sum of all the terms in its row, to all orders of time variation.

Let us illustrate the procedure first for the circular capacitor we have already looked at.

Example 4.7

Tabulate the successive approximations to the electromagnetic quantities associated with the space inside the circular-plate capacitor of the previous examples. For this case, the dielectric is nonconducting. ◈

The static field pattern has only the vertical electric field, but the resulting displacement current generates a magnetic field in horizontal circles around that field. We therefore choose a horizontal circular disk as the "electric surface" S_e and a vertical rectangular area for the "magnetic surface" S_m, as previously outlined. Both these surfaces extend to a variable radius ρ: The disk is centered on the axis and has radius ρ; the rectangle extends the full height h of the capacitor but its width is from the axis to radius ρ; see Figure 4-6. It would actually have been more expedient to place the other side of the rectangle at radius a rather than radius zero, because we know the field there; it is always best to have the surface extend from a location where the field is known to another at a variable location. However, we will continue with the less judicious choice of rectangle we made earlier.

Because of the symmetry, we will be able to extract the azimuthal magnetic field $H(\rho)$ from the mmf around the edge of the disk as

$$H(\rho) = U\{C_e\}/2\pi\rho \tag{4.100}$$

and the vertical electric field $E(\rho)$ from the emf around the rectangle by way of an undetermined constant $E(0)$ as

$$V\{C_m\} = E(0)h - E(\rho)h \qquad \text{or} \qquad E(\rho) = E(0) - V\{C_m\}/h. \tag{4.101}$$

The boundary condition that the applied voltage $V(t)$ appears at the edge of the capacitor will determine the unknown constant $E(0)$.

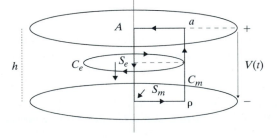

Figure 4-6   Electric and magnetic surfaces for quasistatic analysis.

The table we want has the following appearance, at an early stage.

	$0 \sim V(t)$	$1 \sim dV/dt$	$2 \sim d^2V/dt^2 \cdots$
$V\{C_m\} = -d\Phi/dt$	0		
$E(\rho) = E(0) - V\{C_m\}/h$	$E(0) = V(t)/h$		
$D = \varepsilon E$			
$\Psi\{S_e\} = \displaystyle\int D\, dS$			
$U\{C_e\} = d\Psi/dt$	0		
$H = U/2\pi\rho$	0		
$B = \mu H$	0		
$\Phi\{S_m\} = \displaystyle\int B\, dS$	0		

In the column headed "$0 \sim V(t)$," corresponding to quasistatic entries, both $V\{C_m\}$ and $U\{C_e\}$ are automatically set to zero, since the time rate of change of the fluxes is absent in the static case. Since $U = 0$, H, B, and Φ are also zero. The entry for E to zero order is just the unknown constant $E_0(0)$, since $V\{C_m\} = 0$, but we immediately evaluate this $E(\rho)$ at $\rho = a$ and set the result to $V(t)/h$, since the source voltage $V(t)$ is applied at $\rho = a$. Since $E_0(0)$ is a constant independent of ρ, we replace $E_0(0)$ with $E_0(a) = V(t)/h$ in this column. We can now continue to fill the table, in terms of the applied $V(t)$.

	$0 \sim V(t)$	$1 \sim dV/dt$	$2 \sim d^2V/dt^2 \cdots$
$V\{C_m\} = -d\Phi/dt$	0		
$E(\rho) = E(0) - V\{C_m\}/h$	$E_0(0) = V(t)/h$		
$D = \varepsilon E$	$\varepsilon V(t)/h$		
$\Psi\{S_e\} = \displaystyle\int D\, dS$	$\varepsilon V(t)\pi\rho^2/h$		
$U\{C_e\} = d\Psi/dt$	0		
$H = U/2\pi\rho$	0		
$B = \mu H$	0		
$\Phi\{S_m\} = \displaystyle\int B\, dS$	0		

At this point, we have all the quasistatic quantities and we proceed to the next column, the one headed 1, whose entries must vary as dV/dt, and we continue. The top entry is the negative derivative of Φ_0, which was zero, giving zero for $V_1\{C_m\}$. The next entry is $E_1(\rho) = E_1(0) - V_1\{C_m\}/h$, which is just $E_1(0)$, where $E_1(0)$ means that part of the constant $E(0)$ that varies (in time, not in ρ) as dV/dt. It should be understood that the full field on axis, $E(0)$, has contributions to all orders:

$E(0) = E_0(0) + E_1(0) + E_2(0) + \cdots$, where $E_m(0)$ varies in time as $d^m V(t)/dt^m$. To get $E_1(0)$, we note that the boundary condition is that the full field at the edge, $E(a)$, must be exactly $V(t)/h$, because of the voltage applied there. The quasistatic field $E_0(\rho)$ was uniform and equal to $E_0(0) = E_0(a) = V(t)/h$; hence, in the sum $E(0) = E_0(0) + E_1(0) + E_2(0) + \cdots$, the term $E_0(0)$ was already equal to $V(t)/h$, so that there should be no corrections to this quantity. That is, *all the correction terms for $E(a)$, of order higher than zero, must vanish.* Thus, we set $E_1(\rho) = E_1(0)$ equal to zero at $\rho = a$, which leaves $E_1(0) = 0$ as the entry for $E_1(\rho)$. Consequently, all four upper entries in this column are zeros. This reflects the fact that there is no quasistatic magnetic field in the perfectly insulating dielectric.

	$0 \sim V(t)$	$1 \sim dV/dt$	$2 \sim d^2V/dt^2$
$V\{C_m\} = -d\Phi/dt$	0	0	
$E(\rho) = E(0) - V\{C_m\}/h$	$E_0(0) = V(t)/h$	$E_1(0) = 0$	
$D = \varepsilon E$	$\varepsilon V(t)/h$	0	
$\Psi\{S_e\} = \displaystyle\int D\, dS$	$\varepsilon V(t)\pi\rho^2/h$	0	
$U\{C_e\} = d\Psi/dt$	0	$\varepsilon[dV/dt]\pi\rho^2/h$	
$H = U/2\pi\rho$	0	$\varepsilon[dV/dt]\rho/2h$	
$B = \mu H$	0	$\mu\varepsilon[dV/dt]\rho/2h$	
$\Phi\{S_m\} = \displaystyle\int B\, dS$	0	$\mu\varepsilon[dV/dt]\rho^2/4$	

The next row, for $U\{C_e\}$, calls for the time derivative of Ψ, which refers to the Ψ_0 of the previous column, because taking a time derivative always moves us to the next column. We fill in $d\Psi_0/dt$ for U_1 and continue down the column; the last entry, for Φ_1, is obtained by integration of $B(\rho)$ over the rectangle S_m.

To proceed from the bottom entry of column "1," we need the negative derivative of Φ_1 as the entry for V_2, since the additional time derivative moves us to the next column. Now comes the only tricky part. The entry for $E_2(\rho)$ calls for $E_2(0) - V_2\{C_m\}/h$, where $E_2(0)$ means the part of the constant $E(0)$ that varies (in time, not in ρ) as d^2V/dt^2. How shall we get $E_2(0)$?

	$0 \sim V(t)$	$1 \sim dV/dt$	$2 \sim d^2V/dt^2$
$V\{C_m\} = -d\Phi/dt$	0	0	$-\mu\varepsilon[d^2V/dt^2]\rho^2/4$
$E(\rho) = E(0) - V\{C_m\}/h$	$E_0(0) = V(t)/h$	0	$E_2(0) + \mu\varepsilon[d^2V/dt^2]\rho^2/4h$
$D = \varepsilon E$	$\varepsilon V(t)/h$	0	
$\Psi\{S_e\} = \displaystyle\int D\, dS$	$\varepsilon V(t)\pi\rho^2/h$	0	
$U\{C_e\} = d\Psi/dt$	0	$\varepsilon[dV/dt]\pi\rho^2/h$	
$H = U/2\pi\rho$	0	$\varepsilon[dV/dt]\rho/2h$	
$B = \mu H$	0	$\mu\varepsilon[dV/dt]\rho/2h$	
$\Phi\{S_m\} = \displaystyle\int B\, dS$	0	$\mu\varepsilon[dV/dt]\rho^2/4$	

Once again, all the correction terms for $E(a)$ of order higher than zero must vanish, to ensure that $E(a) = V(t)/h$ have no corrections, as required by the boundary condition. We therefore set

$$E_2(a) = E_2(0) + \mu\varepsilon[d^2V/dt^2]a^2/4h = 0 \qquad (4.102)$$

to evaluate the undetermined constant $E_2(0)$ and then replace the entry for $E_2(\rho) = E_2(0) + \mu\varepsilon[d^2V/dt^2]\rho^2/4h$ with

$$E_2(\rho) = -\mu\varepsilon[d^2V/dt^2][a^2 - \rho^2]/4h. \qquad (4.103)$$

The remainder of the column is straightforward to fill in, with $U_2 = d\Psi_1/dt$ being the derivative of a zero entry in the previous column, so that the last four entries in the "2" column are all zeros.

	$0 \sim V(t)$	$1 \sim dV/dt$	$2 \sim d^2V/dt^2$	$3 \sim d^3V/dt^3$
$V\{C_m\}$	0	0	$-\mu\varepsilon[d^2V/dt^2]\rho^2/4$	0
E	$V(t)/h$	0	$-\mu\varepsilon[d^2V/dt^2][a^2 - \rho^2]/4h$	0
D	$\varepsilon V(t)/h$	0	$-\mu\varepsilon^2[d^2V/dt^2][a^2 - \rho^2]/4h$	0
$\Psi\{S_e\}$	$\dfrac{\varepsilon V(t)\pi\rho^2}{h}$	0	$-\mu\varepsilon^2\dfrac{\pi a^2}{4h}\dfrac{d^2V}{dt^2}\left(\rho^2 - \dfrac{\rho^4}{2a^2}\right)$	0
$U\{C_e\}$	0	$\varepsilon\dfrac{dV}{dt}\dfrac{\pi\rho^2}{h}$	0	$-\mu\varepsilon^2\dfrac{\pi a^2}{4h}\dfrac{d^3V}{dt^3}\left(\rho^2 - \dfrac{\rho^4}{2a^2}\right)$
H	0	$\varepsilon\dfrac{dV}{dt}\dfrac{\rho}{2h}$	0	$-\mu\varepsilon^2\dfrac{a^2}{8h}\dfrac{d^3V}{dt^3}\left(\rho - \dfrac{\rho^3}{2a^2}\right)$
B	0	$\mu\varepsilon\dfrac{dV}{dt}\dfrac{\rho}{2h}$	0	$-\mu^2\varepsilon^2\dfrac{a^2}{8h}\dfrac{d^3V}{dt^3}\left(\rho - \dfrac{\rho^3}{2a^2}\right)$
$\Phi\{S_m\}$	0	$\mu\varepsilon\dfrac{dV}{dt}\dfrac{\rho^2}{4}$	0	$-\mu^2\varepsilon^2\dfrac{a^2}{16}\dfrac{d^3V}{dt^3}\left(\rho^2 - \dfrac{\rho^4}{4a^2}\right)$

In the next column, the entry for $E_3(\rho)$ is $E_3(0) - V_3\{C_m\}/h$ but $V_3\{C_m\} = 0$, which leaves $E_3(\rho) = E_3(0)$. Setting the correction $E_3(a)$ at the edge equal to zero, we get $E_3(0) = 0$, leaving a zero entry for $E_3(\rho)$, and hence for D_3 and Ψ_3 too. After this, U_3 is set to $d\Psi_2/dt$ and the rest of the column is easily filled in.

Until we get tired, there is nothing to stop us from continuing the table to as many correction terms as we may want, forming the infinite series for the fields, fluxes, and emf and mmf that satisfy Maxwell's equations. To third order in time variation, we have found the fields in the capacitor to be

$$E(\rho, t) = \frac{V(t)}{h} - \frac{\mu\varepsilon a^2}{4h}\frac{d^2V}{dt^2}\left(1 - \frac{\rho^2}{a^2}\right), \qquad (4.104)$$

$$H(\rho, t) = \frac{\varepsilon a}{2h}\frac{dV}{dt}\frac{\rho}{a} - \frac{\mu\varepsilon^2 a^3}{8h}\frac{d^3V}{dt^3}\left(\frac{\rho}{a} - \frac{1}{2}\frac{\rho^3}{a^3}\right). \qquad (4.105)$$

NORMAL MODES

We have examined a parallel-plate capacitor and found the fields generated by a voltage applied externally across the two plates. The source of the fields is the applied voltage and their strength is proportional to that voltage. In particular, if the source voltage were reduced to zero (i.e., if the plates were shorted out), the fields in the capacitor would become zero. This is elementary, but it is useful to examine a given structure for possible sourceless fields.

Sourceless fields, or *normal modes*, of a structure are field patterns that satisfy Maxwell's equations but are unemcumbered by actual sources to generate them, such as voltage sources, with their accompanying wires that carry a current that affects the field pattern. Such source-free solutions solve the equations for the case that the source is zero. If any such field pattern is found, its strength could be anything we please and the source would remain zero. Consequently, the sourceless fields can have undetermined amplitudes and, if more than one such field pattern were found, they could be superimposed, forming highly useful building blocks for field patterns generated by actual sources. Mathematically, the normal modes correspond to the homogeneous solutions of, say, a differential equation. Let's look for the normal modes of the capacitor that we have already studied.

Example 4.8

A parallel-plate capacitor has a sourceless internal electric field parallel to the axis, with field amplitude $E(0, t) = F(t)$ on axis, as in Figure 4-7. If $F(t)$ were static, the electric field would be uniform but the time-varying field is not; we neglect fringing at the edge of the capacitor. Assume a perfectly insulating dielectric between the plates (parameters μ, ε; area $A = \pi a^2$, height h). Find the electric and magnetic fields $E(\rho, t)$ and $H(\rho, t)$, to fourth order in $F(t)$. ◇

The process of filling in the table is the same as before, except that it is now $E(0, t) = F(t)$ that is the given quantity and the electric field at $\rho = 0$ is therefore the quantity that must remain uncorrected. The table appears as follows.

	$0 \sim F(t)$	$1 \sim dF/dt$	$2 \sim d^2F/dt^2$	$3 \sim d^3F/dt^3$	$4 \sim d^4F/dt^4$
$V\{C_m\}$		0	$\mu\varepsilon h[d^2F/dt^2]\rho^2/4$	0	$\dfrac{\mu^2\varepsilon^2 h}{64}\dfrac{d^4F}{dt^4}\rho^4$
E	$F(t)$	0	$\mu\varepsilon[d^2F/dt^2]\rho^2/4$	0	$\dfrac{\mu^2\varepsilon^2}{64}\dfrac{d^4F}{dt^4}\rho^4$
D	$\varepsilon F(t)$	0	$\mu\varepsilon^2[d^2F/dt^2]\rho^2/4$	0	
$\Psi\{S_e\}$	$\varepsilon F(t)\pi\rho^2$	0	$\pi\mu\varepsilon^2[d^2F/dt^2]\rho^4/8$	0	
$U\{C_e\}$	0	$\varepsilon\dfrac{dF}{dt}\pi\rho^2$	0	$\dfrac{\pi\mu\varepsilon^2}{8}\dfrac{d^3F}{dt^3}\rho^4$	
H	0	$\dfrac{\varepsilon}{2}\dfrac{dF}{dt}\rho$	0	$\dfrac{\mu\varepsilon^2}{16}\dfrac{d^3F}{dt^3}\rho^3$	
B	0	$\dfrac{\mu\varepsilon}{2}\dfrac{dF}{dt}\rho$	0	$\dfrac{\mu^2\varepsilon^2}{16}\dfrac{d^3F}{dt^3}\rho^3$	
$\Phi\{S_m\}$	0	$-\mu\varepsilon h\dfrac{dF}{dt}\dfrac{\rho^2}{4}$	0	$-\dfrac{\mu^2\varepsilon^2 h}{64}\dfrac{d^3F}{dt^3}\rho^4$	

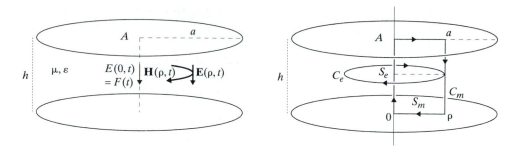

Figure 4-7   Open capacitor with internal sourceless field pattern.

Consequently, the electric and magnetic fields are, to fourth order,

$$E(\rho,t) = F(t) + \mu\varepsilon[d^2F/dt^2]\rho^2/4 + \mu^2\varepsilon^2[d^4F/dt^4]\rho^4/64, \tag{4.106}$$

$$H(\rho,t) = \varepsilon[dF/dt]\rho/2 + \mu\varepsilon^2[d^3F/dt^3]\rho^3/16. \tag{4.107}$$

For the sake of those readers who may tend to get frustrated by a technique that seems to be limited practically to yielding only the first few terms of an infinite series, let's demonstrate one technique that can furnish the full infinite series of corrections.

Example 4.9

If the sourceless field on axis in the capacitor of the previous example is sinusoidal, as $E(0,t) = F(t) = E_0 \sin \omega t$, then the fields we found are, to fourth order,

$$E(\rho,t) = E_0\big[1 - \omega^2\mu\varepsilon\rho^2/4 + \omega^4\mu^2\varepsilon^2\rho^4/64 - \cdots\big]\sin\omega t, \tag{4.108}$$

$$H(\rho,t) = \omega\varepsilon E_0\rho\big[1/2 - \omega^2\mu\varepsilon\rho^2/16 + \cdots\big]\cos\omega t. \tag{4.109}$$

It appears that the two infinite series in the brackets are Taylor series for functions $f(x)$ and $g(x)$, with $x^2 = \omega^2\mu\varepsilon\rho^2$ and with

$$f(x) = 1 - x^2/4 + x^4/64 - \cdots; \tag{4.110}$$

$$g(x) = 1/2 - x^2/16 + \cdots. \tag{4.111}$$

We want expressions for the numerical coefficients e_n and h_n in

$$f(x) = \sum_n e_n x^{2n}, \qquad g(x) = \sum_n h_n x^{2n}.$$

We already have $e_0 = 1$, $e_1 = -1/4$, $e_2 = 1/64$ and $h_0 = 1/2$, $h_1 = -1/16$. To get all the coefficients in one fell swoop, obtain the entries in the quasistatic analysis table, starting at the field E of order $2n$ and cycling through to order $2n + 2$, thereby determining e_{n+1} in terms of e_n, as well as h_n in terms of e_n. From these relations, obtain the nth coefficients e_n and h_n as functions of n. ◈

If $E(\rho) = E_0 f(k\rho)\sin\omega t$ and $H(\rho) = \omega\varepsilon E_0\rho g(k\rho)\cos\omega t$, where $k = \omega\sqrt{\mu\varepsilon}$, then at $2n$th order we have as successive entries to the table

$$E = E_0 \sin \omega t \, e_n \big(\omega^2 \mu \varepsilon \rho^2\big)^n,$$

$$D = \varepsilon E_0 \sin \omega t \, e_n \big(\omega^2 \mu \varepsilon \rho^2\big)^n,$$

$$\Psi = 2\pi \varepsilon E_0 \sin \omega t \, e_n \int_0^\rho \big(\omega^2 \mu \varepsilon \rho^2\big)^n \rho \, d\rho$$

$$= 2\pi \varepsilon E_0 \sin \omega t \, e_n \big(\omega^2 \mu \varepsilon\big)^n \rho^{2n+2}/(2n + 2).$$

At order $2n + 1$, we have

$$U = 2\pi \omega \varepsilon E_0 \cos \omega t \, e_n \big(\omega^2 \mu \varepsilon\big)^n \rho^{2n+2}/(2n + 2),$$

$$H = \omega \varepsilon E_0 \cos \omega t \, e_n \big(\omega^2 \mu \varepsilon\big)^n \rho^{2n+1}/(2n + 2),$$

$$B = \omega \mu \varepsilon E_0 \cos \omega t \, e_n \big(\omega^2 \mu \varepsilon\big)^n \rho^{2n+1}/(2n + 2),$$

$$\Phi = -\omega \mu \varepsilon E_0 h \cos \omega t \, e_n \big(\omega^2 \mu \varepsilon\big)^n \int_0^\rho \rho^{2n+1} \, d\rho/(2n + 2)$$

$$= -\omega \mu \varepsilon E_0 h \cos \omega t \, e_n \big(\omega^2 \mu \varepsilon\big)^n \rho^{2n+2}/(2n + 2)^2.$$

At order $2n + 2$, we have

$$V = -\omega^2 \mu \varepsilon E_0 h \sin \omega t \, e_n \big(\omega^2 \mu \varepsilon\big)^n \rho^{2n+2}/(2n + 2)^2,$$

$$E = -\omega^2 \mu \varepsilon E_0 \sin \omega t \, e_n \big(\omega^2 \mu \varepsilon\big)^n \rho^{2n+2}/(2n + 2)^2$$

$$= -E_0 \sin \omega t \, e_n \big(\omega^2 \mu \varepsilon \rho^2\big)^{n+1}/(2n + 2)^2.$$

But we have gone through a full cycle of the table and this last quantity should be $E_0 \sin \omega t \, e_{n+1}\big(\omega^2 \mu \varepsilon \rho^2\big)^{n+1}$. This requires that

$$e_{n+1} = -e_n/(2n + 2)^2. \tag{4.112}$$

Also, since

$$H = \omega \varepsilon E_0 \cos \omega t \, e_n \big(\omega^2 \mu \varepsilon\big)^n \rho^{2n+1}/(2n + 2)$$

should be

$$H = \omega \varepsilon E_0 \cos \omega t \, \rho h_n \big(\omega^2 \mu \varepsilon \rho^2\big)^n,$$

we have

$$h_n = e_n/(2n + 2). \tag{4.113}$$

Starting from $e_0 = 1$, we get successively

$$e_1 = -1/2^2, \; e_2 = 1/2^2 4^2, \; e_3 = -1/2^2 4^2 6^2, \; e_4 = 1/2^2 4^2 6^2 8^2, \quad \text{etc.,}$$

or

$$e_n = (-1)^n/\big[(2 \cdot 1)^2(2 \cdot 2)^2(2 \cdot 3)^2 \cdots (2 \cdot n)^2\big]$$

$$= (-1)^n/\big[2^n n!\big]^2 \tag{4.114}$$

and

$$h_n = (-1)^n/\big[2^{2n+1} n!(n + 1)!\big]. \tag{4.115}$$

Comment: The functions so obtained as Taylor series are *Bessel functions*. More particularly, the function that gives the sourceless electric field

$$f(x) = \sum e_n x^{2n} = \sum_{n=0}^{\infty} (-1)^n x^{2n}/\big[2^n n!\big]^2 = J_0(x)$$

is called the Bessel function of the first kind, of order zero, and denoted $J_0(x)$. The function representing the sourceless magnetic field is

$$g(x) = \sum h_n x^{2n} = \sum_{n=0}^{\infty} (-1)^n x^{2n}/[2^{2n+1} n!(n+1)!] = J_1(x)/x,$$

where $J_1(x)$ denotes the Bessel function of the first kind, of order one.

CYLINDRICAL CAVITY

A *cavity* is a hollow space enclosed by a conducting surface. Electric and magnetic field patterns may exist in a cavity and oscillate in time, but only at certain natural resonant frequencies. If the boundary surface were perfectly conducting and the enclosed dielectric were perfectly insulating, the oscillation, once started, could persist indefinitely.

Example 4.10

A cylindrical cavity is formed when we close a parallel-plate capacitor (circular plates of radius a, separated by height h) with a cylindrical conducting wall at its edge, as in Figure 4-8. The sourceless field pattern of the last two examples can oscillate sinusoidally within this cavity if the field satisfies the appropriate boundary condition at the wall, but this is possible only at certain frequencies. Up to eighth order, the axial electric field and azimuthal magnetic field are, as found previously,

$$E(\rho,t) = E_0[1 - x^2/4 + x^4/64 - x^6/2304 + x^8/147456]\sin \omega t,$$

$$H(\rho,t) = \omega \varepsilon E_0 \rho[1/2 - x^2/16 + x^4/384 - x^6/18432]\cos \omega t,$$

where $x^2 = \omega^2 \mu \varepsilon \rho^2$.

(a) *What boundary condition must the field satisfy at $\rho = a$?*

(b) *Find the lowest resonant frequency ω_0 of this cavity, for this field pattern, in terms of the smallest $x_0 = \omega_0 a \sqrt{\mu \varepsilon}$ that satisfies the boundary condition, when the field expansion is retained to second order, to fourth order, to sixth order, and to eighth order.*

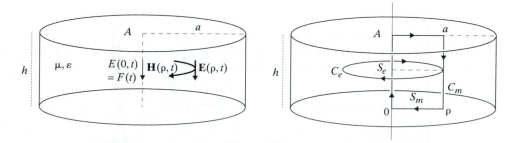

Figure 4-8   Cylindrical cavity with internal sourceless field pattern.

Comment: *We may use any root-finding algorithm that may be available on a calculator or computer, or else apply Newton's method, for the cubic and quartic equations to be solved; only the lowest, real root is needed. For comparison, the exact value (i.e., to infinite order) is $x_0 = 2.4048$, to five significant figures, as determined from the properties of Bessel functions.*

   (c) *What is the lowest resonant frequency $f_0 = \omega_0/2\pi$, in Hz, if the cavity radius is $a = 2.54$ cm and the dielectric inside is a vacuum?* ◈

(a) The axial electric field is tangential to the wall at $\rho = a$ and must vanish there, so that the boundary condition is that $E(a, t) = 0$.

(b) The equations whose roots are needed to ensure the boundary condition is satisfied are set out in the following table.

order	equation for x_0	lowest root
2	$1 - x^2/4 = 0$	$x_0 = 2.00000$
4	$1 - x^2/4 + x^4/64 = 0$	$x_0 = 2.82843$
6	$1 - x^2/4 + x^4/64 - x^6/2304 = 0$	$x_0 = 2.39165$
8	$1 - x^2/4 + x^4/64 - x^6/2304 + x^8/147456 = 0$	$x_0 = 2.40564$

(c) $x_0 = \omega a \sqrt{\mu_0 \varepsilon_0}$ or $f_0 = x_0 / \left[2\pi a \sqrt{\mu_0 \varepsilon_0} \right] = 4.52$ GHz.

FIELDS IN A WIRE

Having found the fields in a capacitor and demonstrated that Maxwell's equations prevent even an ideal capacitive structure from behaving like a circuit-theoretic capacitor should, let us now examine the fields in a segment of wire and discover that this prototypical resistor cannot act as a pure resistor. This example will also afford us an opportunity to include conductivity in the scheme of calculation by successive approximation.

Example 4.11

Tabulate the successive approximations to the electromagnetic quantities associated with the space inside a segment of cylindrical wire made of imperfectly conducting material. The wire has radius a and length l and its constitutive parameters are μ, ε, σ. The wire carries current I(t). ◈

    We first determine what the field structure may be in the static case, whereupon we can choose appropriate electric and magnetic surfaces for our calculations. The direct current fed by a static source will be distributed uniformly over the cross section of the wire. The quasistatic current density will therefore be

$$J = I(t)/\pi a^2, \tag{4.116}$$

so that the electric field, parallel to the wire axis, is

$$E = I(t)/\sigma \pi a^2. \tag{4.117}$$

The current distribution will surround itself with an azimuthal magnetic field, in circles around the axis. Around a typical circle of radius ρ transverse to and centered on the axis, the mmf will be the magnetic field at radius ρ times the circumference of this circle,

$$U = 2\pi\rho H(\rho). \tag{4.118}$$

The current linked by this mmf is the fraction $\pi\rho^2/\pi a^2$ of the total current $I(t)$ carried by the wire, so that Ampère's law

$$2\pi\rho H(\rho) = U(\rho) = I(t)\rho^2/a^2 \tag{4.119}$$

gives as the quasistatic magnetic field

$$H(\rho) = \left[I(t)/2\pi a\right](\rho/a). \tag{4.120}$$

Based on this elementary calculation of the quasistatic fields, a reasonable choice of the two surfaces for the successive approximation analysis is a flat, transverse, circular disk of radius ρ for the electric surface S_e and a rectangular area extending from radius ρ to radius a and of length l for the magnetic surface S_m; see Figure 4-9. The reason for choosing the rectangle to end at radius a rather than on axis is that we will shortly be interested in the field at the surface of the wire. The boundary condition is that the total current in the wire be the specified current $I(t)$.

We can begin to fill in a table for all the relevant quantities to several orders of time variation, as follows. In this instance, with finite conductivity, we include rows for the current density field J and for the conduction current $I\{S_e\}$ through surface S_e. In addition, we find it convenient to insert a row for the special entry of the total current through the wire, which is $I\{S_e(a)\}$, the value of $I\{S_e(\rho)\}$ at $\rho = a$. This allows us to incorporate the boundary condition in the table itself; the specification that the total current be $I(t)$ is already satisfied by just the quasistatic flow and therefore demands that there be zero corrections to that current. For simplicity, we also decide to neglect the displacement current inside the wire; we will justify this after the calculation. We therefore omit D and Ψ from the table.

	$0 \sim I(t)$	$1 \sim dI/dt \cdots$
$V\{C_m\} = -d\Phi/dt$	0	
$E(\rho) = V\{C_m\}/l + E(a)$	$E_0(a) = I(t)/\sigma\pi a^2$	
$J = \sigma E$	$I(t)/\pi a^2$	
$I\{S_e\} = \displaystyle\int J\, dS$	$I(t)(\rho^2/a^2)$	
$I\{S_e(a)\} = I(t)$	$I(t)$	0
$U\{C_e\} = I\{S_e\}$	$I(t)(\rho^2/a^2)$	
$H = U/2\pi\rho$	$I(t)\rho/2\pi a^2$	
$B = \mu H$	$\mu I(t)\rho/2\pi a^2$	
$\Phi\{S_m\} = \displaystyle\int B\, dS$	$(\mu l/4\pi)I(t)(1 - \rho^2/a^2)$	

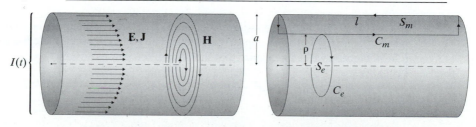

Figure 4-9   Current-carrying wire's quasistatic fields.

We have already filled in zero for the higher-order correction to the total current up to radius a and we have made entries in the quasistatic column by working upward from $I\{S_e(a)\} = I(t)$, on the basis of a uniform current distribution, as calculated earlier. Then the magnetic part of the column is easily filled in; note that we neglect displacement current in arriving at U.

Moving along to the first-order column, we encounter the first-order contribution to the undetermined constant $E(a)$; this $E_1(a)$ is to be found from the boundary condition. Carrying it along as we go down the column, we reach the entry for the first-order $I\{S_e(a)\}$, which is to be zero. This determines the unknown constant $E_1(a)$ as

$$\sigma E_1(a)\pi a^2 - (\mu\sigma/4)[dI/dt](a^2/2) = 0$$

or
$$E_1 = (\mu/8\pi)dI/dt, \tag{4.121}$$

which then allows us to go back and make all the entries in the column depend explicitly on dI/dt. We then readily fill in the entire column.

	$0 \sim I(t)$	$1 \sim dI/dt \cdots$
$V\{C_m\} = -d\Phi/dt$	0	$-\left(\dfrac{\mu l}{4\pi}\right)\left[\dfrac{dI}{dt}\right]\left(1 - \dfrac{\rho^2}{a^2}\right)$
$E(\rho) = V\{C_m\}/l + E(a)$	$E_0(a) = I(t)/\sigma\pi a^2$	$E_1(a) - \left(\dfrac{\mu}{4\pi}\right)\left[\dfrac{dI}{dt}\right]\left(1 - \dfrac{\rho^2}{a^2}\right)$
$J = \sigma E$	$I(t)/\pi a^2$	$\sigma E_1 - \left(\dfrac{\mu\sigma}{4\pi}\right)\left[\dfrac{dI}{dt}\right]\left(1 - \dfrac{\rho^2}{a^2}\right)$
$I\{S_e\} = \displaystyle\int J\, dS$	$I(t)(\rho^2/a^2)$	$\sigma E_1\pi\rho^2 - \left(\dfrac{\mu\sigma}{4}\right)\left[\dfrac{dI}{dt}\right]\left(\rho^2 - \dfrac{\rho^4}{2a^2}\right)$
$I\{S_e(a)\} = I(t)$	$I(t)$	$\sigma E_1\pi a^2 - \left(\dfrac{\mu\sigma}{4}\right)\left[\dfrac{dI}{dt}\right]\left(\dfrac{a^2}{2}\right) = 0$
$U\{C_e\} = I\{S_e\}$	$I(t)(\rho^2/a^2)$	
$H = U/2\pi\rho$	$I(t)\rho/2\pi a^2$	
$B = \mu H$	$\mu I(t)\rho/2\pi a^2$	
$\Phi\{S_m\} = \displaystyle\int B\, dS$	$\left(\dfrac{\mu l}{4\pi}\right)I(t)\left(1 - \dfrac{\rho^2}{a^2}\right)$	

The table with the constant $E_1(a)$ in the first-order column replaced by its newly found value $(\mu/8\pi)\,dI/dt$ appears as follows. The last entry in this table has a simplified form because we can recognize that the integration entails

$$\int_\rho^a [-\rho + \rho^3/a^2]\, d\rho = -(a^2 - \rho^2)/2 + (a^4 - \rho^4)/4a^2 = -(a^2 - \rho^2)^2/4a^2. \tag{4.122}$$

	$0 \sim I(t)$	$1 \sim dI/dt \cdots$
$V\{C_m\} = -d\Phi/dt$	0	$-\left(\dfrac{\mu l}{4\pi}\right)\left[\dfrac{dI}{dt}\right]\left(1 - \dfrac{\rho^2}{a^2}\right)$
$E(\rho) = V\{C_m\}/l + E(a)$	$I(t)/\sigma\pi a^2$	$\left(\dfrac{\mu}{8\pi}\right)\left[\dfrac{dI}{dt}\right]\left(-1 + \dfrac{2\rho^2}{a^2}\right)$
$J = \sigma E$	$I(t)/\pi a^2$	$\left(\dfrac{\mu\sigma}{8\pi}\right)\left[\dfrac{dI}{dt}\right]\left(-1 + \dfrac{2\rho^2}{a^2}\right)$
$I\{S_e\} = \displaystyle\int J \, dS$	$I(t)(\rho^2/a^2)$	$\left(\dfrac{\mu\sigma}{8}\right)\left[\dfrac{dI}{dt}\right]\left(-\rho^2 + \dfrac{\rho^4}{a^2}\right)$
$I\{S_e(a)\} = I(t)$	$I(t)$	0
$U\{C_e\} = I\{S_e\}$	$I(t)(\rho^2/a^2)$	$\left(\dfrac{\mu\sigma}{8}\right)\left[\dfrac{dI}{dt}\right]\left(-\rho^2 + \dfrac{\rho^4}{a^2}\right)$
$H = U/2\pi\rho$	$I(t)\rho/2\pi a^2$	$\left(\dfrac{\mu\sigma}{16\pi}\right)\left[\dfrac{dI}{dt}\right]\left(-\rho + \dfrac{\rho^3}{a^2}\right)$
$B = \mu H$	$\mu I(t)\rho/2\pi a^2$	$\left(\dfrac{\mu^2\sigma}{16\pi}\right)\left[\dfrac{dI}{dt}\right]\left(-\rho + \dfrac{\rho^3}{a^2}\right)$
$\Phi\{S_m\} = \displaystyle\int B \, dS$	$\left(\dfrac{\mu l}{4\pi}\right)I(t)\left(1 - \dfrac{\rho^2}{a^2}\right)$	$-\left(\dfrac{\mu^2\sigma l}{64\pi a^2}\right)\left[\dfrac{dI}{dt}\right](a^2 - \rho^2)^2$

At this point, we should justify our neglect of the displacement current $d\Psi/dt$ inside the conducting wire, compared to the conduction current. This is what impelled us to write $U\{C_e\} = I\{S_e\}$ instead of $U\{C_e\} = I\{S_e\} + d\Psi\{S_e\}/dt$. Since $J = \sigma E$ and $D = \varepsilon E$, we have that $\Psi\{S_e\} = (\varepsilon/\sigma)I\{S_e\}$ and the first-order conduction and displacement currents are

$$I_1\{S_e\} = (\mu\sigma/8)[dI/dt](-\rho^2 + \rho^4/a^2) \qquad (4.123)$$

and

$$d\Psi_0\{S_e\}/dt = (\varepsilon/\sigma)[dI/dt](\rho^2/a^2). \qquad (4.124)$$

The two currents peak at $\rho^2/a^2 = 1/2$ and at $\rho = a$, respectively; the ratio of the peak strengths of the conduction and displacement currents is

$$I/[d\Psi/dt] = [\mu\sigma a^2/32]/(\varepsilon/\sigma) = \mu\sigma^2 a^2/32\varepsilon. \qquad (4.125)$$

For a copper wire of radius $a = 1$ mm, using $\mu = \mu_0$ for the nonmagnetic metal, $\varepsilon = \varepsilon_0$ for lack of better data on copper, and $\sigma = 6 \cdot 10^7$ S/m, we find that the conduction current is $(1.6)10^{13}$ times stronger than the displacement current, which explains the lack of better data on the metal's permittivity and confirms that our neglect of the displacement current is amply justified.

Let us now examine the current-voltage relation for this segment of wire. We evaluate the voltage along the wire at its surface, which is

$$V = E(a)l = [l/\sigma\pi a^2]I(t) + (\mu l/8\pi)[dI/dt] \tag{4.126}$$

and has the form of a series combination of a resistor and inductor:

$$V = RI + L\,dI/dt \quad \text{with} \quad R = l/\sigma\pi a^2 \quad \text{and} \quad L = \mu l/8\pi. \tag{4.127}$$

Evidently, even keeping terms to only first order in time variation, the piece of wire exhibits not just the familiar resistance $R = l/\sigma\pi q^2$ but an inductance $L = \mu l/8\pi$ as well. This inductance is referred to as the internal inductance, because it arises from the magnetic field inside the wire; we have not examined the magnetic field outside the wire, which also contributes to the overall inductance.

Once we are adept at visualizing fluxes and their associated emfs and mmfs, the entire calculation can be streamlined somewhat, as in the next example.

Example 4.12

A perfectly conducting U-shaped bracket, of dimensions l, w, h, with air around it ($\mu = \mu_0$, $\varepsilon = \varepsilon_0, \sigma = 0$), is fed a total current I_0 at one leg (with an equal current extracted from the other leg) at z = 0; see Figure 4-10. Neglecting fringing, this results in a uniform magnetic field $H_0 = I_0/w$ in the air space within the bracket, when the current is static. If the input current $I_0(t)$ now varies slowly with time, obtain

    (a) *the first-order electric field $E_1(z, t)$;*

    (b) *the second-order magnetic field $H_2(z, t)$;*

    (c) *the voltage $V_0(t)$ across the input end of the bracket (up to second order);*

    (d) *the equivalent inductance L of the bracket (up to second order);*

    (e) *the current $I^0(t)$ along the end of the bracket, at z = l (up to second order).* ◈

The static magnetic field is directed along the width of the bracket, in accordance with the right-hand rule. This suggests use of the rectangular surface marked S_m in Figure 4-11 for calculations of the magnetic flux. There is no static electric field, but the

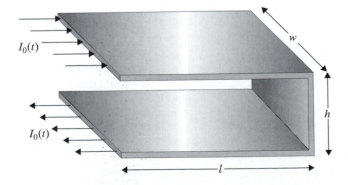

$I_0(t)$

$I_0(t)$

w

h

l

Figure 4-10   U-shaped bracket with input current $I_0(t)$.

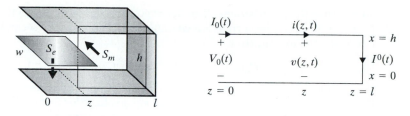

Figure 4-11   Flux surfaces and simplified circuit for bracket.

induced electric field should be vertical, along the bracket's height, which suggests the horizontal rectangular area, marked S_e in the figure, for the electric flux. Since there is no finite conductivity in this problem, only the displacement current is relevant for that surface. Note that both surfaces have one edge at a variable point z and that another edge has been placed at two opposite ends of the bracket, because we know the edge field there: The edge at $z = l$ must have zero electric field, being tangential to the conducting end wall of the bracket; the edge at $z = 0$ has a known magnetic field $H_0 = I_0/w$, with no corrections, in conformity with the given input current $I_0(t)$. The figure also shows the bracket reduced to a simplified "circuit," showing the current $i(z, t)$ and voltage $v(z, t)$ in relation to the input current I_0 and input voltage V_0, as well as the wall current I^0.

Starting with the quasistatic magnetic field $H_0 = I_0/w$, we get the zero-order magnetic flux $\Phi_0\{S_m\}$ as its integral from z to l and over height h as

$$\Phi_0\{S_m\} = \mu_0(h/w)I_0(t)[l - z]. \tag{4.128}$$

This implies an emf around the edge of S_m amounting to

$$V_1(t) = -d\Phi_0/dt = -\mu_0(h/w)[dI_0(t)/dt][l - z]. \tag{4.129}$$

But this emf is the integral of the vertical, downward electric field, or

$$V_1(t) = E_1(l, t)h - E_1(z, t)h, \tag{4.130}$$

in which $E_1(l, t) = 0$ because of the conducting wall at $z = l$. This gives the first-order electric field $E_1(z, t)$ for Part (a) as

$$E_1(z, t) = (\mu_0/w)[dI_0(t)/dt][l - z]. \tag{4.131}$$

We now integrate this first-order electric field over the electric flux surface S_e to get the first-order electric flux

$$\Psi_1\{S_e\} = \int_0^z \varepsilon_0 E_1(z', t)w \, dz' = \varepsilon_0 \mu_0[dI_0(t)/dt] \int_0^z (l - z') \, dz'$$

$$= \varepsilon_0 \mu_0[dI_0(t)/dt][lz - \tfrac{1}{2}z^2]. \tag{4.132}$$

This implies an mmf around the edge of S_e amounting to

$$U_2(t) = d\Psi_1/dt = \varepsilon_0 \mu_0[d^2I_0(t)/dt^2][lz - \tfrac{1}{2}z^2]. \tag{4.133}$$

But this mmf is the integral of the horizontal, rearward magnetic field, or

$$U_2(t) = H_2(0, t)w - H_2(z, t)w, \tag{4.134}$$

in which $H_2(0, t) = 0$ because there can be no correction to the magnetic field at the input end $z = 0$, where the current is a *given* quantity. This gives the second-order magnetic field $H_2(z, t)$ for Part (b) as

$$H_2(z, t) = -(\varepsilon_0 \mu_0/w)[d^2 I_0(t)/dt^2][lz - \tfrac{1}{2} z^2]. \tag{4.135}$$

Since we now have the electric and magnetic fields up to second order, we can answer the remaining questions directly. The input voltage for Part (c) is the integral of just $E_1(0, t)$ over the height h, since there is no zero-order electric field and the next correction to E would be third order:

$$V_0(t) = E_1(0, t)h = (\mu_0 h/w)[dI_0(t)/dt]l. \tag{4.136}$$

Because this result has the form $V_0 = L \, dI_0/dt$, we simply read off the value of the inductance of the bracket, correct to second order, for Part (d):

$$L = \mu_0 hl/w. \tag{4.137}$$

Finally, for the current in the end wall of the bracket, $I^0(t)$, to second order, we need the integral across the width of the bracket of the magnetic field at the wall, up to second order:

$$I^0(t) = H(l, t)w = [H_0(l, t) + H_2(l, t)]w$$
$$= I_0(t) - \varepsilon_0 \mu_0 [d^2 I_0(t)/dt^2][\tfrac{1}{2} l^2], \tag{4.138}$$

which answers Part (e).

Note that, along the way, we have obtained the voltage and current distributions, up to second order in time variation, as functions of z and t, in terms of the input current $I_0(t)$ and its derivatives:

$$v(z, t) = E_1(z, t)h = \mu_0(h/w)[dI_0/dt][l - z], \tag{4.139}$$

$$i(z, t) = [H_0(t) + H_2(z, t)]w$$
$$= I_0(t) - \mu_0 \varepsilon_0 [d^2 I_0/dt^2][lz - \tfrac{1}{2} z^2]. \tag{4.140}$$

The second-order correction term is small as long as the time variation of the input current is sufficiently slow to make $|\mu_0 \varepsilon_0 l^2 d^2 I_0(t)/dt^2| \ll |I_0(t)|$. For a sinusoidal input current at frequency ω, this criterion for usefulness of the truncated series of correction terms becomes that the product of the frequency and the length of the bracket must be small enough so that $\mu_0 \varepsilon_0 \omega^2 l^2 \ll 1$. For example, if the frequency can get as high as $\omega = 2\pi f = 2\pi (10 \text{ MHz})$, then this criterion requires that the length of the bracket be small enough that $l^2 \ll (4.77)^2 \text{ m}^2$ or $l \ll 4.77$ m. Apparently, for centimeter-sized structures, we are safe in using the quasistatic analysis for operating frequencies up to the megahertz range.

MICROSTRIP LINES

As an illustration of the approximate answers to complex problems obtainable from rough estimates of unknown field patterns, consider the capacitance associated with a conducting strip deposited on a dielectric substrate, itself resting on a conducting layer, as in Figure 4-12. This is a microstrip line, a microwave structure of great prac-

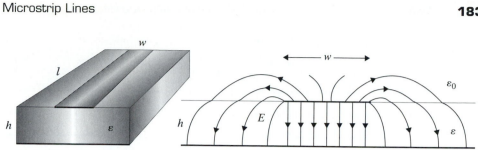

Figure 4-12   Microstrip line and a sketch of its electric field pattern.

tical importance because it can be easily and inexpensively fabricated by photolithographic means and is readily integrated with other circuitry and microwave devices.

An evaluation of the microstrip line's capacitance is crucial to its design and operation as a microwave element. To calculate the capacitance, we need the electric field pattern that results when the line is charged. Unfortunately, as is the case for most structures that are not fully enclosed, the electric field that satisfies Maxwell's equations and the boundary conditions is rather difficult to calculate, even if we assume ideally thin and perfectly shaped electrodes. Recourse is usually had to numerical analysis of such configurations, but even with such aids, the fringing fields depend on unknown details of the thickness and precise shape of the strip in realistic devices.

Figure 4-12 shows a sketch of the cross-sectional field pattern, as revealed by advanced, numerical analysis techniques. It confirms an essentially uniform field region directly under the strip, as might be expected, but also features field lines that emerge into the air from the upper surface of the strip, bend around toward the dielectric layer, cross obliquely into the layer with appropriate bending to satisfy the boundary conditions, and end up normal to the conducting ground plane.

Rather than attempt to calculate the exact field pattern, we propose to illustrate how a crude estimate of that pattern, based on physical reasoning, can be used to obtain at least a rough approximation of the required capacitance parameter. This could be used for initial engineering designs, to be refined by computer-aided numerical analysis.

The strip has length l and width w; we assume zero thickness of the strip. The substrate has thickness h and permittivity ε. The region above the strip is air. We conjecture that the field is uniform under the strip. For the air region, we crudely replace what are really more complicated field lines with circular ones. We even allow these hypothetical field lines to cross into the substrate perpendicularly, although we realize that they really ought to strike the dielectric interface at an angle and then bend. We extend the circular field lines with straight ones within the dielectric layer, so as to meet the conducting ground plane normally, as required. Figure 4-13 depicts the conjectured field pattern. Note that it is symmetric about the midplane, of course.

Still intent on simplifying our calculations as much as is reasonable, we add to our hypothetical model that the field strength be uniform along a particular semicircular arc; it then must change strength to a weaker value inside the dielectric, to maintain continuity of normal D field. Each semicircular field line has its own field

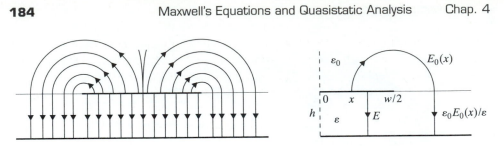

Figure 4-13   Crude estimate of field pattern of a microstrip line.

strength, $E_0(x)$, where we measure x from the midplane. We have placed the centers of the semicircles at the edge of the strip, at $x = w/2$, to guarantee continuity of the substrate electric field under that edge, from the region directly under the strip to the space beyond it.

We need the voltage V from the strip to the ground plane when the strip carries charge Q (and the ground plane has charge $-Q$), in order to find the capacitance C in $Q = CV$. The assumed electric field pattern allows us to estimate the ratio of Q to V. Under the strip, the voltage is $V = Eh$, so $E = V/h$. Along a particular semicircular segment of an external field line, of radius $(w/2 - x)$, the emf is just $E_0(x)\pi(w/2 - x)$. The additional emf along the straight segment that extends this field line within the dielectric layer is $[\varepsilon_0 E_0(x)/\varepsilon]h$. Quasistatically, the sum of these two voltages is also V, so we have

$$E_0(x) = V/[\pi(w/2 - x) + \varepsilon_0 h/\varepsilon], \qquad \text{for } 0 \le x \le w/2. \qquad (4.141)$$

We use Gauss's law to get the charge on the half of the strip corresponding to this range of x; this charge is $Q/2$. We need to integrate the D field over both the bottom and top of the half-strip:

$$Q/2 = \int_0^{w/2} \varepsilon E l\, dx + \int_0^{w/2} \varepsilon_0 E_0(x) l\, dx$$

$$= \int_0^{w/2} (\varepsilon V/h)l\, dx + \int_0^{w/2} \frac{\varepsilon_0 V}{\pi(w/2 - x) + \varepsilon_0 h/\varepsilon}\, l\, dx$$

$$= (\varepsilon l w/2h)V + (\varepsilon_0 l V/\pi)\ln[(w/2 + \varepsilon_0 h/\pi\varepsilon)/(\varepsilon_0 h/\pi\varepsilon)] \qquad (4.142)$$

and we get the estimate of the capacitance of the microstrip line as

$$C = \varepsilon l w/h + \varepsilon_0 l(2/\pi)\ln[1 + (\pi/2)(\varepsilon/\varepsilon_0)(w/h)]. \qquad (4.143)$$

We do realize that this expression can only offer rough guidance as to how the capacitance varies with the permittivity and the width-to-height ratio of the microstrip line, because it has been derived from a quite crude conjecture as to the shape of the field lines and the field strength along them. We know the assumed field configuration cannot be right; it has some important flaws. For one thing, the assumed vertical electric field within the substrate ends abruptly at the end of the largest semicircle. This entails a discontinuity of tangential electric field. Another flaw is the variation of the substrate field with distance from the midplane, as expressed by Eq. (4.141) but modified by the factor $\varepsilon_0/\varepsilon$. This implies that the emf around a rectangle with two vertical sides at different distances from the midplane and two horizontal sides at the upper and lower planes of the dielectric layer will have a nonzero emf, because the field strength is different at the two vertical sides and we have allowed no horizontal field component. These flaws become less severe when the w/h ratio gets larger. For a typical case, the capacitance expression we derived is about 10% in error, compared to the result of a full numerical analysis.

We could improve our capacitance estimate by postulating field lines that are not merely semicircular and reach the substrate obliquely. We could also obtain quasistatic corrections to the field pattern when there is time variation. We will not pursue these pathways here, however, as we wanted only to illustrate how to get useful approximate circuit parameters from even crude assumptions about the field pattern.

SUMMARY

We have developed and illustrated a technique for converting static field configurations into field distributions that satisfy the full Maxwell equations when the sources are slowly varying. It is a successive approximation technique, whereby the static fields are first considered to acquire the slow time variation of the source in their amplitude, without changing the static field pattern. These are termed the quasistatic fields.

For example, an inverse-square field satisfies the laws of electrostatics and is proportional to the charge at its center. The quasistatic version of this Coulomb field would still be inverse square but its amplitude would now vary in time exactly as the source charge may have been specified to do. The entire field pattern grows and wanes in synchronism with the time variation of the source. We then have action at a distance but the synchronous time variation entails no delay.

If there are several sources, each with its own time variation, we simply superimpose the responses to each source individually, so that the resultant field strength becomes a linear combination of the several source time functions.

The quasistatic fields cannot be correct, however, because they are accompanied by electric and magnetic fluxes that also vary synchronously with the source. Such time-varying fluxes induce closed-loop mmfs and emfs, which are absent in the static field pattern. The quasistatic fields are therefore not consistent with the full Maxwell equations.

The successive approximation technique calls for the calculation of the electric and magnetic fluxes of the quasistatic fields over some open surfaces, whereupon the time derivatives of these fluxes contribute to correction fields that do possess a closed-loop mmf or emf around the edges of the surfaces. These correction fields vary in time, as does the time derivative of the source. Because we assume slow time variation of the source, this time derivative represents only a small correction to the quasistatic field pattern.

The process does not end there, however, because the correction fields are also accompanied by their own fluxes, which also vary in time as the first derivative of the source. The rate of change of these correction fluxes contributes to the corresponding line integrals around the edges of the surfaces, so that a further correction field appears, varying as the second derivative of the source. Further corrections can be calculated ad infinitum; the process is useful if the source variation is slow enough to make successive terms of the infinite series of corrections decline in amplitude sufficiently to achieve convergence of the sums.

How slow the source variation needs to be for this process to give useful results depends on the size of the physical structure involved and the constitutive parameters of the materials used, as well as on how many correction terms we are willing to calculate and retain in the sum. The technique is especially well suited to approximate analysis, wherein we may neglect fields in regions where those fields are weak (for example, in the case of fringing fields).

The correction terms are classified according to the order of the time derivative of the source time function: for example, the third-order terms vary as the third derivative of the source. Not only the fields but also their line and surface integrals, hence all the electromagnetic quantities, can be written as infinite series of such corrections to the static version. We have set forth a systematic tabular method of keeping track of all the quantities to be calculated, to all orders.

We first determine what the field pattern may look like in the static case, then select an open surface for the electric field and one for the magnetic one. Each surface should have part of its edge at a variable location, the rest of the edge, whenever possible, at locations where the field is absent or known. A logical structure for the table has a row for each of

the emf around the edge of the magnetic surface

the electric field

the electric flux density

the electric flux

the current density

the current across the electric surface

the mmf around the the edge of the magnetic surface

the magnetic field

the magnetic flux density

the magnetic flux

It may also be desirable to include a row for whatever quantity, perhaps an applied voltage or current or charge or magnetic flux, is most directly related to the specified source. Imposing zero corrections to that quantity allows constants of integration to be evaluated for the specific configuration being treated. Such constants of integration are constant in space but still vary in time. The columns of the table correspond to successive orders of time variation of the source. Ultimately, each field quantity, including the constants of integration, becomes the sum of the entries in its row.

As we fill in successive rows of the table, we encounter quantities that involve time derivatives of previously calculated fluxes. Any time derivative moves us to the next column, to maintain consistency with our classification of terms of the series according to the order of the time derivatives of the source. In particular, if the last row of the table is the magnetic flux, then the entry that would follow logically is its time derivative, which brings us to the emf at the top of the table. The table therefore really wraps around and continues in the next column. Note also that mmf entries beyond the quasistatic one combine conduction currents from the same column with displacement currents derived from electric fluxes in the previous column.

An alternative to truncating the infinite series after a few terms is to carry the series out far enough for its sum to be recognizable. Summing the series furnishes an exact solution to Maxwell's equations. We will perform this summation in our next example, for a structure of great engineering importance, a transmission line.

We applied the quasistatic analysis technique to a parallel-plate capacitor, finding that there are magnetic fields as well as electric fields inside that structure. We also introduced the notion of sourceless fields, or normal modes, of a structure. These are fields whose patterns satisfy Maxwell's equations but that have no sources in the region considered, so that their amplitude is an undetermined constant factor. Superpositions of such sourceless fields can serve as homogeneous solutions for a case that has a source only at the boundary of a region (such cases are called *boundary value problems*). Superpositions of the sourceless solutions can also describe the behavior of the fields after a source has been turned off (*initial value problems*).

We applied quasistatic analysis to an idealized parallel-plate capacitor and discovered that Maxwell's equations do not permit even an ideal such structure to behave as a pure capacitance. We applied it also to a piece of wire of finite conductivity and found that Maxwell's equations do not allow it to behave as a pure resistance. We will apply the same technique next to a parallel-wire transmission line and discover that a short circuit is not really a short circuit, nor is an open circuit really open, and that most circuit concepts are only approximations. Along the way, we will also encounter, at long last, action at a distance with a delay.

PROBLEMS

Component of curl in spherical coordinates

4.1 Obtain the θ-component of $\nabla \times \mathbf{F}$ for $\mathbf{F}(r, \theta, \varphi)$ expressed in spherical coordinates, by applying Stokes's theorem to an infinitesimal surface that faces the direction of $\hat{\theta}$ and is formed by a change dr in r and $d\varphi$ in φ, at fixed θ.

Component of curl in spherical coordinates

4.2 Obtain the φ-component of $\nabla \times \mathbf{F}$ for $\mathbf{F}(r, \theta, \varphi)$ expressed in spherical coordinates, by applying Stokes's theorem to an infinitesimal surface that faces the direction of $\hat{\varphi}$ and is formed by a change dr in r and $d\theta$ in θ, at fixed φ.

Component of curl in cylindrical coordinates

4.3   Obtain the ρ-component of $\nabla \times \mathbf{F}$ for $\mathbf{F}(\rho, \varphi, z)$ expressed in cylindrical coordinates, by applying Stokes's theorem to an infinitesimal surface that faces the direction of $\hat{\boldsymbol{\rho}}$ and is formed by a change dz in z and $d\varphi$ in φ, at fixed ρ.

Component of curl in cylindrical coordinates

4.4   Obtain the z-component of $\nabla \times \mathbf{F}$ for $\mathbf{F}(\rho, \varphi, z)$ expressed in cylindrical coordinates, by applying Stokes's theorem to an infinitesimal surface that faces the direction of $\hat{\mathbf{z}}$ and is formed by a change $d\rho$ in ρ and $d\varphi$ in φ, at fixed z.

Alternative calculation of fields in capacitor with external source

4.5   Since the parallel circular-plate capacitor in the examples has its applied voltage $V(t)$ impressed from the exterior, at radius a, it is more convenient to choose the rectangular surface for the calculation of magnetic flux and emf to extend from radius ρ to a, as in Figure P4-1, rather than from 0 to ρ, as was done in the text. Assume a perfectly insulating dielectric between the plates (parameters μ, ε; area $A = \pi a^2$, height h).
   (a) How is $E(\rho)$ related to the emf $V\{C_m\}$ quasistatically (order 0) and for all higher orders?
   (b) How do the two series for $\Phi\{S_m\}$ and $V\{C_m\}$ begin (up to fourth order)?
   (c) Verify that the series for the E and H fields, using the alternative surface S_m, are the same as those calculated in the text.

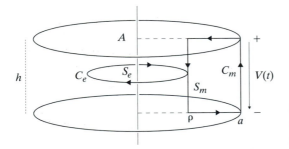
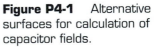

Figure P4-1   Alternative surfaces for calculation of capacitor fields.

Equivalent circuit for capacitor with external current source

4.6   A parallel circular-plate capacitor is fed current $I(t)$ externally, to and from the center points outside the plates, as in Figure P4-2. The dielectric between the plates is perfectly insulating (parameters μ, ε; area $A = \pi a^2$, height h). Neglect fringing beyond the edge of the capacitor.
   Since a static current accumulates charge and results in a nonstatic electric field, it is convenient to express the zero-order quasistatic field pattern in terms of the total charge $Q(t)$ on the upper plate, where $I(t) = dQ/dt$, rather than $I(t)$. The higher-order contributions can be given in terms of $I(t)$ and its derivatives.
   (a) Express $E(\rho)$, $\Psi(\rho) = \Psi\{S_e\}$, and $\Psi(a)$ in terms of the electric field at the edge, $E(a)$, to zero order. Reexpress $\Psi(a)$, $E(\rho)$, and $\Psi(\rho)$ in terms of $Q(t)$.
   (b) Give $U(\rho) = U\{C_e\}$, $H(\rho)$, and $\Phi(\rho) = \Phi\{S_m\}$ to first order [i.e., proportional to $dQ(t)/dt$, expressed as $I(t)$].
   (c) Find $V\{C_m\}$ and give the second-order ($\sim d^2Q/dt^2 = dI/dt$) values of $E(\rho)$, $\Psi(\rho)$, and $\Psi(a)$ in terms of the second-order electric field at the edge $E(a)$. After deciding what $\Psi(a)$ must be to this order, reexpress $E(a)$, $E(\rho)$, and $\Psi(\rho)$ in terms of dI/dt.

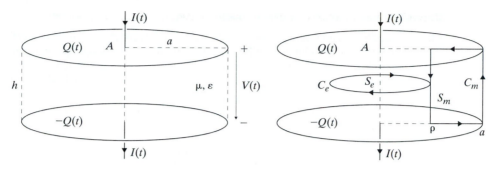

Figure P4-2   Parallel-plate capacitor with external current source.

(d) Obtain the third-order $(\sim d^3Q/dt^3 = d^2I/dt^2)$ values of $U(\rho)$ and $H(\rho)$, in terms of d^2I/dt^2.

(e) Write the electric and magnetic fields $E(\rho,t)$ and $H(\rho,t)$, including only terms proportional to $Q(t)$, $I(t)$, dI/dt, and d^2I/dt^2.

(f) Find the voltage $V(a)$ across the edge of the capacitor, up to d^2I/dt^2; verify that this corresponds to the voltage across a series combination of a capacitor C and inductor L when current $I(t)$ traverses both. What are C and L?

Parallel-plate resistor with external voltage source

4.7   The parallel circular-plate geometry can model a resistor if the plates are perfect conductors but the material between the plates is an imperfect conductor. We neglect displacement current, compared to the conduction current from one plate to the other, so that the medium between the plates is sufficiently described by the parameters μ, σ and the area $A = \pi a^2$ and height h. The resistor has voltage $V(t)$ applied externally to the center points outside the plates, as in Figure P4-3, so that the voltage appears symmetrically across the resistor at its outer edge. Neglect fringing beyond the edge of the resistor.

(a) Obtain the electric and magnetic fields $E(\rho,t)$ and $H(\rho,t)$, up to first order in the time variation of $V(t)$.

(b) Find the total current $I(t)$ in the resistor, to first order in $V(t)$. Verify that, if terms of order higher than the first are neglected, $I(t)$ is related to $V(t)$ as if $V(t)$ were applied to the series combination of a conductance G and an inductance L. What are G and L?

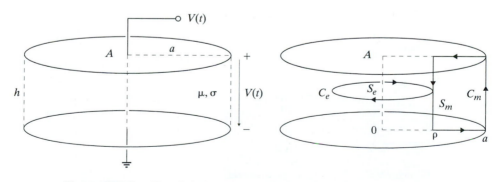

Figure P4-3   Parallel-plate resistor with external voltage source.

Capacitor with external current source and imperfect insulator

4.8   A parallel circular-plate capacitor is fed current $I(t)$ externally, to and from the center points outside the plates, as in Figure P4-4. The plates are perfectly conducting but the dielectric between the plates is imperfectly insulating (parameters μ, ε, σ; area $A = \pi a^2$, height h). Neglect fringing beyond the edge of the capacitor; retain both conduction and displacement current.

  The zero-order electric and magnetic fields are $E_0(\rho, t) = I(t)/\sigma A$ and $H_0(\rho, t) = I(t)\rho/2A$.

(a) Obtain the first-order electric field, $E_1(\rho, t)$.

(b) Obtain the first-order magnetic field, $H_1(\rho, t)$.

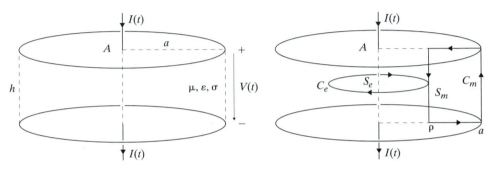

Figure P4-4   Current-fed capacitor with imperfect dielectric.

Fields in air gap between magnetic pole pieces

4.9   The "north" and "south" pole pieces of a cylindrical electromagnet are brought close to each other, leaving an air gap of height h, as in Figure P4-5. The parallel circular faces of the two pole pieces are of radius a. The electromagnet delivers magnetic flux $\Phi(t)$ to the air gap. Neglect fringing of the fields beyond the edge of the gap space. Obtain the electric and magnetic fields $E(\rho, t)$ and $H(\rho, t)$ in the air gap, up to second order in the time variation of $\Phi(t)$.

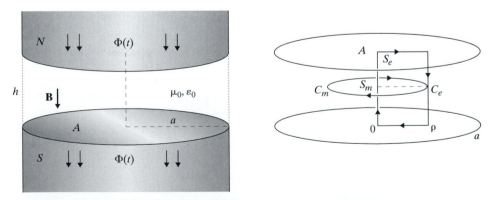

Figure P4-5   Cylindrical air gap between pole pieces with given flux.

Fields in air gap in magnetic circuit

4.10 A high-permeability core $(\mu_1/\mu_0 \gg 1)$ is wound with N turns of wire that carries current $I(t)$. The core has a cylindrical air gap (area $A = \pi a^2$, height h) as in Figure P4-6. Assume the permeability μ_1 is so high that H is negligible in the core; also assume fringing beyond the edge of the air gap can be neglected.

    (a) What is the mmf $U_1(t)$ around loop C_1, in terms of the applied current $I(t)$? What is the zero-order magnetic field $H(t)$ in the air gap?

    (b) Obtain the electric and magnetic fields $E(\rho, t)$ and $H(\rho, t)$ in the air gap, up to second order in the time variation of $I(t)$.

    (c) What is the emf $V\{C_2\}$ (up to order dI/dt) around closed loop C_2 that follows the (perfectly conducting) coil of N turns and then bridges the space between the input terminals? Note that the flux $\Phi(t)$ is encircled N times by the curve C_2. What is the inductance L of the magnetic circuit, at its terminals?

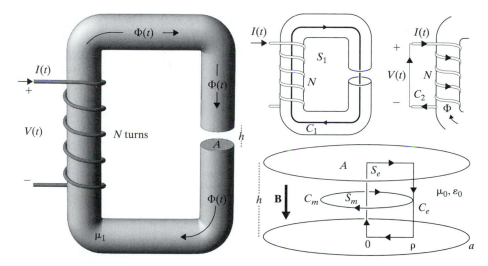

Figure P4-6   Cylindrical air gap in magnetic circuit with given coil current.

Fields of vertical wire ending on horizontal plane

4.11 Current $I_0(t)$ flows along the surface of a perfectly conducting, vertical, semi-infinite cylindrical wire (radius $= a$) until it reaches a perfectly conducting, infinitely extended, horizontal plane, whereupon the current spreads out along the surface of the plane, equally in all directions; see Figure P4-7. The space outside the wire and above the plane is air. Assume no spatial variation of the current along the length of the vertical wire. The static magnetic field in the air space is given, of course, by $\mathbf{H}(\rho, \varphi, z) = -\hat{\varphi} I_0/2\pi\rho$ and there is no static electric field.

    (a) What is the static surface current density vector $\mathbf{K}(\rho, \varphi)$ on the plane?

    (b) If the current varies slowly in time, as $I_0(t)$, what are the first-order corrections to the static electric and magnetic fields?

    (c) Find the second-order corrections to the electric and magnetic fields.

    (d) Is there a surface charge density ρ_s on the wire and on the plane? If so, give ρ_s in terms of $I_0(t)$, up to second order.

Note: The integral $\int x^{m-1}\ln x \, dx = (x^m/m)(\ln x - 1/m)$ may be useful.

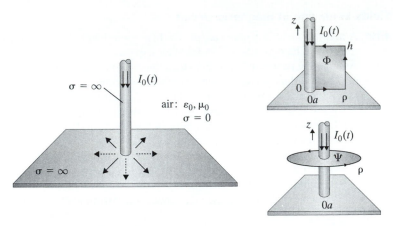

Figure P4-7   Vertical wire ending on horizontal plane.

Capacitor with internal voltage source

4.12 A parallel circular-plate capacitor has voltage $V(t)$ applied internally, along a wire coaxial with the plates, as in Figure P4-8. The dielectric between the plates is perfectly insulating (parameters μ, ε; area $A = \pi a^2$, height h); the wire radius is b. Neglect fringing beyond the edge of the capacitor. Assume the voltage source itself is tiny and disturbs the field pattern negligibly. Note that the open-circuit condition at the edge $\rho = a$ of the plates implies that the total conduction current into the top plate is balanced by the total displacement current in the dielectric, leaving no mmf at $\rho = a$.

(a) Find the electric and magnetic fields $E(\rho, t)$ and $H(\rho, t)$, up to second order in the time variation of $V(t)$.

(b) What is the current $I(t)$ supplied by the voltage source, to second order?

(c) Find the voltage $V(a)$ across the edge of the capacitor, to second order.

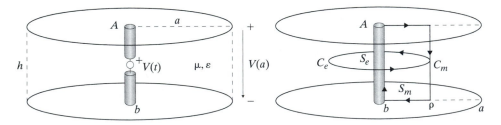

Figure P4-8   Parallel-plate capacitor with internal voltage source.

Capacitor with internal current source

4.13 A parallel circular-plate capacitor is fed current $I(t)$ internally, along a wire coaxial with the plates, as in Figure P4-9. The dielectric between the plates is perfectly insulating (parameters μ, ε; area $A = \pi a^2$, height h); the wire radius is b. Neglect fringing beyond the edge of the capacitor. Assume the current source itself is tiny and disturbs the field pattern negligibly. Note that the open-circuit condition at the edge $\rho = a$ of the plates im-

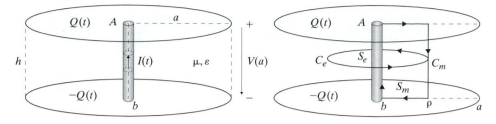

Figure P4-9 Parallel-plate capacitor with internal current source.

plies that the total conduction current into the top plate is balanced by the total displacement current in the dielectric, leaving no mmf at $\rho = a$.

Since a static current accumulates charge and results in a nonstatic electric field, it is convenient to express the zero-order quasistatic field pattern in terms of the total charge $Q(t)$ on the upper plate, where $I(t) = dQ/dt$, rather than $I(t)$. The higher-order contributions can be given in terms of $I(t)$ and its derivatives.

(a) Find the zero-order electric field $E_0(\rho, t)$, electric flux $\Psi_0(\rho) = \Psi\{S_e\}$, and voltage $V_0(t)$ developed across the current source, in terms of $Q(t)$.

(b) Find the first order (in Q) mmf $U_1(\rho) = U\{C_e\}$, magnetic field $H_1(\rho)$, and magnetic flux $\Phi_1(\rho) = \Phi\{S_m\}$, in terms of $I(t) = dQ/dt$.

(c) Express the second-order (in Q) electric field $E_2(\rho)$ in terms of the second-order voltage $V_2(t)$ developed across the current source. Without carrying out the tedious integrations, indicate how the calculation of the second-order electric flux $\Psi_2(\rho)$ would furnish $V_2(t)$ in terms of $dI/dt = d^2Q/dt^2$.

(d) Express the voltage $V(a)$ across the edge of the capacitor, up to second-order terms $(\sim dI/dt)$ and in terms of $V_2(t)$.

Fields in notch cut out of a wire

4.14 A cylindrical space is cut out of a thick wire (radius b) until only a radius a of the original wire is left, over a length h. The notch is filled with a perfect dielectric (permittivity ε high enough to allow fringing fields to be neglected). Current $I_0(t)$ flows along the surface of the perfectly conducting, cylindrical wire. Assume no spatial variation of the current along the periphery of the wire. The static magnetic field in the dielectric is given by $\mathbf{H}(\rho, \varphi, z) = \hat{\varphi} I_0/2\pi\rho$, as we know, and there is no static electric field.

(a) What is the static magnetic flux Φ_0 over the rectangular surface marked Φ in Figure P4-10?

(b) If the current varies slowly in time, as $I_0(t)$, what are the first-order corrections to the static electric and magnetic fields in the dielectric?

(c) What is the first-order electric flux Ψ_1 over the circular surface marked Ψ in Figure P4-10?

(d) In terms of $H(\rho)$, the azimuthal magnetic field, which of the following is the appropriate boundary condition to use, given that the total current fed to the wire is $I_0(t)$: $H_m(a) = 0$ for $m > 0$, or $H_m(b) = 0$ for $m > 0$? Why?

(e) What are the second-order corrections to the static electric and magnetic fields in the dielectric?

Note: The integral $\int x^{m-1}\ln x \, dx = (x^m/m)(\ln x - 1/m)$ may be useful.

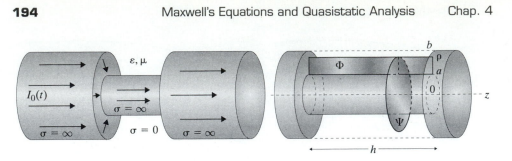

Figure P4-10   Cylindrical wire with notch cut out of it.

Fields of uniform surface current on perfect conductor

4.15 The xy-plane is the boundary between a perfectly conducting substrate (in $z < 0$) and an imperfect dielectric above it ($z > 0$). The dielectric has permittivity ε, permeability μ, and conductivity σ. Uniform surface current $K_0(t)$ is caused to flow along the surface of the perfectly conducting substrate, in the y-direction. Assume no spatial variation of the surface current density over the entire surface and no fringing of the resultant fields. The static magnetic field in the dielectric is $\mathbf{H}(x, y, z) = \hat{\mathbf{x}}K_0$, as we know, and there is no static electric field.

(a) Justify briefly the preceding statements that, statically, $H_{x0} = K_0$ and $E_0 = 0$.

(b) What is the static magnetic flux Φ_0 over the rectangular surface marked Φ in Figure P4-11?

(c) If the surface current varies slowly in time, as $K_0(t)$, what is the first-order correction E_{y1} to the static electric field in the dielectric?

(d) What are the first-order electric flux Ψ_1 and electric current I_1 over the rectangular surface marked Ψ in Figure P4-11?

(e) What is the first-order correction H_{x1} to the static magnetic field in the dielectric?

(f) What are the second-order corrections E_{y2} and H_{x2} to the static electric and magnetic fields in the dielectric?

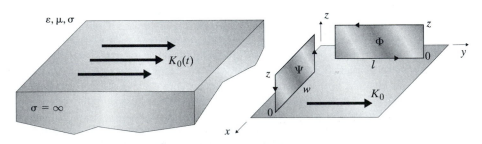

Figure P4-11   Surface current on perfectly conducting substrate.

Fields of ideal solenoid

4.16 An ideal solenoid is formed by a current sheet $K_0(t)$ that flows in the azimuthal direction on an infinitely long cylindrical surface of radius a. Quasistatically, this results in a uniform, axial, zero-order magnetic field $\mathbf{H}_0(\rho, \varphi, z, t) = \hat{\mathbf{z}}K_0(t)$ inside the cylindrical space and no magnetic field outside, with no zero-order electric field at all. The solenoid is air filled.

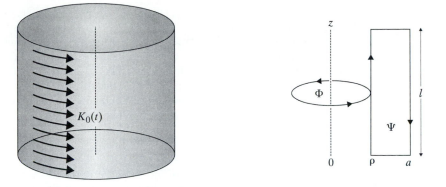

Figure P4-12   Surface current on section of ideal solenoid.

(a) Tabulate the values of mmf, magnetic field, magnetic flux and emf, electric field, electric flux $(U, H_z, \Phi, V, E_\varphi, \Psi)$, associated with the two surfaces indicated in Figure P4-12, up to third order in time variation.

(b) What is the magnetic field on the axis of the solenoid, to third order?

(c) What is the electric field at the edge $(\rho = a)$ of the solenoid, to third order?

Equivalent circuit of bracket excited by voltage source

4.17   A perfectly conducting U-shaped bracket, of dimensions l, w, h, in air is fed across the open end by a voltage source $V_0(t)$ at $z = 0$, as in Figure P4-13. Neglect fringing beyond the edges of the bracket.

A static voltage accumulates magnetic flux and makes a nonstatic magnetic field, so it is convenient to express the quasistatic field pattern in terms of the total flux $\Phi_0(t)$ inside the bracket, where $V_0(t) = d\Phi_0/dt$, rather than $V_0(t)$. The higher-order contributions can be given in terms of $V_0(t)$ and its derivatives.

(a) Express $H(z), \Phi(z) = \Phi\{S_m\}$, and $\Phi(0)$ in terms of the magnetic field at the edge, $H(0)$, to zero order. Reexpress $\Phi(0)$, $H(z)$, and $\Phi(z)$ in terms of $\Phi_0(t)$.

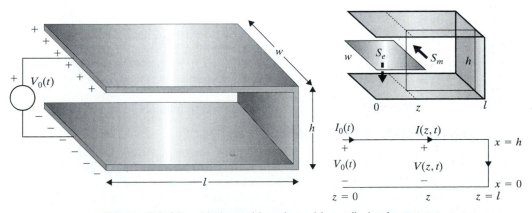

Figure P4-13   U-shaped bracket with applied voltage source.

(b) Give $V(z) = V\{C_m\}$, $E(z)$, and $\Psi(z) = \Psi\{S_e\}$ to first order [i.e., proportional to $d\Phi_0(t)/dt$, expressed as $V_0(t)$].

(c) Find $U\{C_e\}$ and give the second-order $(\sim d^2\Phi_0/dt^2 = dV_0/dt)$ values of $H(z), \Phi(z)$, and $\Phi(0)$ in terms of the second-order magnetic field at the edge $H_2(0)$. After deciding what $\Phi(0)$ must be to this order, reexpress $H_2(0)$ and $H_2(z)$ in terms of dV_0/dt.

(d) Find the current $I_0(t)$ into the upper leg of the bracket, up to dV_0/dt; verify that this corresponds to the current into a parallel combination of a capacitor C and inductor L when voltage $V_0(t)$ is applied to both. What are C and L?

Spherical capacitor with internal voltage source

4.18 Two concentric, perfectly conducting spheres of radii a, b are separated by air. A voltage source $V(t)$ is connected between the "north poles" of the two spheres, as in Figure P4-14. Statically, this results in uniform and opposite charge distributions on the two spheres, with an inwardly radial electric field. In the time-varying case, an azimuthal magnetic field also appears. To avoid infinitely strong magnetic fields from an infinitely thin wire at $\theta = 0$, allow the source structure to be a conical wire, with its surface at $\theta = \theta_0 > 0$. The suggested flux surfaces shown in Figure P4-14 are the spherical cap S_e at radius r, extending from angle 0 to θ, and the planar area S_m between radii a and b and extending from angle θ to π. In the space $a < r < b$, and $\theta_0 < \theta < \pi$, find

(a) the zero-order electric field $E_0(r, \theta, t)$, in terms of $V(t)$;

(b) the first-order azimuthal magnetic field $H_1(r, \theta, t)$;

(c) the second-order electric field $E_2(r, \theta, t)$;

(d) the source current $I(t)$ and "south pole" voltage $V(\pi, t)$, to second order.

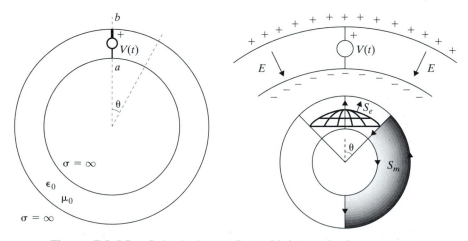

Figure P4-14   Spherical capacitor with internal voltage source.

Capacitor with nonparallel plates

4.19 A capacitor is formed of flat but nonparallel conducting plates and is excited at the narrower end by a voltage source $V(t)$, as in Figure P4-15. The plates' dimensions are l by w, they are in air, and they would meet at angle α if they were extended. Neglect fringing beyond the edges of the plates. Statically, we can expect electric field lines that are circular, perpendicular to the plates. In the time-varying case, we expect a magnetic field parallel to both plates. If we neglect fringing, the magnetic field must fall to zero at the wider end, $x = x_2$, since there is no current at the end of the plates.

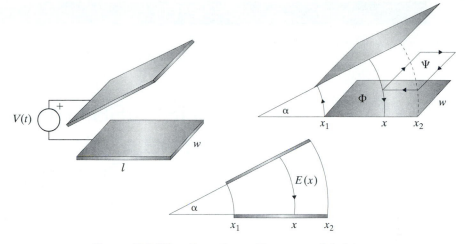

Figure P4-15 Capacitor with nonparallel plates.

(a) Find the zero-order electric field $E_0(x)$.
(b) Find the first-order magnetic field $H_1(x)$.
(c) Find the input current, to first order, and the capacitance of the structure.

Self and mutual capacitances of conducting strips on a substrate

4.20 Consider the capacitive coupling between two parallel conducting strips deposited on a dielectric substrate, itself resting on a conducting layer. Each strip has length l and width w; they are separated from each other by separation s and the substrate has thickness h and permittivity ε.

With voltage sources $V_1(t)$ and $V_2(t)$ applied from the ground plane to the two strips, charges $Q_1(t)$ and $Q_2(t)$ appear on the strips. They are related to the voltages by $Q_1 = C_{11}V_1 + C_{12}V_2$, $Q_2 = C_{21}V_1 + C_{22}V_2$, where C_{11} and C_{22} are the self-capacitances and $C_{12} = C_{21}$ is the mutual capacitance of the two strips.

To arrive at a good guess for the field pattern, we take the thickness of the strips to be zero and ignore fringing. We can then reasonably conjecture that the field is uniform under each strip and follows circular field lines from one strip to the other, as in Figure P4-16. We can also assume constant field strength along each semicircular field line.

Find C_{11}, C_{22}, and C_{12} on the basis of this estimated field pattern.

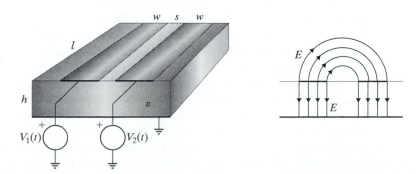

Figure P4-16 Capacitive coupling between conducting strips on a substrate.

CHAPTER 5

Transmission Lines, Time Delay, Wave Propagation

Let us now apply quasistatic analysis to a simple circuit formed by a voltage source connected to a load resistor through a pair of parallel wires. We will find that this simple circuit does not really obey the rules of circuit theory; instead, it behaves as a transmission line. This time, we will obtain not just a few but all the correction terms in the analysis and we will sum up the infinite series to get an exact solution.

Prior to developing the transmission line equations and to solving and interpreting them, we obtain the parameters of the parallel wires that will enable us to calculate the most fundamental property of the structure, its characteristic impedance. Although it is possible simply to assume some value of the characteristic impedance and solve the transmission line equations accordingly, an understanding of the principles involved and any possibility of designing or developing specifications for the line demand that we be capable of deducing the characteristic impedance from the geometry and constitution of the structure. We will see that we can get the characteristic impedance if we can deduce the field pattern associated with the transmission line.

TRANSMISSION LINE PARAMETERS

The voltage source varies in time as $V(t)$; it is connected to the load resistor R_l by a pair of parallel wires of length l, as in Figure 5-1. The medium surrounding the structure is air, which is nonconducting. We ask, What happens? What current flows?

Circuit theory furnishes one answer—namely, that current $I(t) = V(t)/R_l$ flows in the circuit, so that the voltage $V(t)$ is delivered at time t across the load resistor, without delay, and that this same voltage appears across the pair of wires at any location z between the ends $z = 0$ and $z = l$.

We realize at once that this cannot be correct, since the current must generate a magnetic field and the associated magnetic flux will vary in time as the source voltage varies. This must create an emf around the edge of any surface with which this flux is linked. This means that the voltage cannot be the same at every position along the

Figure 5-1   Voltage source applied to a load resistor via a pair of wires.

circuit, since any two positions z_1 and z_2 define a rectangular area, between them and between the wires, and this rectangle has time-varying magnetic flux across it. This demands that $V(z_1)$ differ from $V(z_2)$, which contradicts the circuit-theory result. Let us therefore analyze this structure more carefully.

We first need to ascertain what kind of field pattern we will be dealing with, on the basis of physical reasoning. The generator maintains a voltage $V(t)$ at $z = 0$; it therefore charges the line, so that a current can circulate. Because a time-varying magnetic field will accompany the current that will flow in the circuit, the voltage across the wires will vary from point to point along the line, as will the current, the charge distribution, the magnetic field, and all other electromagnetic quantities. Voltage $V = V(z, t)$, current $I = I(z, t)$, and charge per unit length $q(z, t)$ are the key quantities whose variation with position z and time t are needed.

In the charging process, the generator pushes charges into one wire and pulls them from the other. Unless special steps are taken to disrupt the symmetry, as much current enters as leaves the section to either side of any transverse plane. Hence, the currents in the upper and lower wires are equal and opposite at any location z. The same is true for the charge distribution: If an infinitesimal segment of the line from z to $z + dz$ carries charge $q(z, t)\,dz$ in the upper wire, then an equal and opposite charge $-q(z, t)\,dz$ is present in the lower wire.

This charge distribution is unknown; it is one of the quantities we want to calculate. But whatever this charge distribution may be, we can relate it to the electric field pattern it generates by superimposing the contributions from the upper and lower wires. In the static case, the charge distribution along the wires is uniform and we expect an electric field pattern in transverse planes, directed out of one wire and into the other. We are trying to maintain this transverse field pattern even in the quasistatic case, although we should no longer expect a uniform charge distribution.

We can use Gauss's law to determine the field configuration from the pair of equal and opposite charges $\pm q(z)\,dz$ at location z. By itself, the charge $q(z)\,dz$ in the upper wire will surround itself with a uniformly distributed radially outward electric field centered at the upper wire. In the lower wire, the corresponding charge $-q(z)\,dz$ at the same location z will generate a uniformly distributed, radially inward electric field centered on the lower wire. As illustrated in Figure 5-2, we superimpose the two fields vectorially to get the combined field pattern.

On a cylindrical surface of radius ρ_1 centered on the upper wire and of infinitesimal length dz, the electric flux generated by the upper wire is

$$d\Psi_1 = 2\pi\rho_1\varepsilon_0 E_1\,dz = q(z)\,dz \tag{5.1}$$

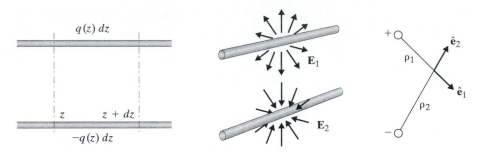

Figure 5-2   Section of parallel wires and the electric field it generates.

by Gauss's law, with the field E_1 in the direction of $\hat{\mathbf{e}}_1$. On a cylindrical surface of radius ρ_2 centered on the lower wire and of infinitesimal length dz, the electric flux generated by the lower wire is

$$d\Psi_2 = 2\pi\rho_2\varepsilon_0 E_2\, dz = -q(z)\, dz \qquad (5.2)$$

with the field E_2 in the direction of $\hat{\mathbf{e}}_2$ (but with sense reversed by the minus sign). The total field at the observation point at distances ρ_1 and ρ_2 from the two wires is the vector superposition

$$\mathbf{E} = E_1\hat{\mathbf{e}}_1 + E_2\hat{\mathbf{e}}_2 = \frac{q(z)}{2\pi\varepsilon_0}\left(\frac{\hat{\mathbf{e}}_1}{\rho_1} - \frac{\hat{\mathbf{e}}_2}{\rho_2}\right). \qquad (5.3)$$

Figure 5-3 shows electric field lines in a transverse plane. It is not difficult to confirm that the field lines are circles that pass through the two wires.

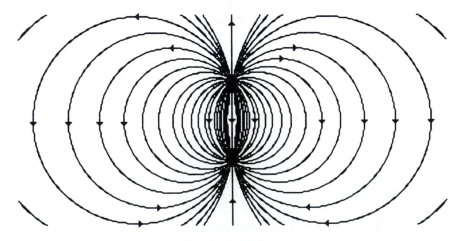

Figure 5-3   Electric field lines in a plane transverse to parallel wires.

We will need this field only in the xz-plane, the plane that passes through the two wires. In this plane, let us direct the x-axis from the upper wire to the lower one. Then $\hat{e}_1$ and $-\hat{e}_2$ are both along $\hat{x}$ and the electric field is along x; if s is the separation of the two parallel wires, then $\rho_1 = x$ and $\rho_2 = s - x$ in this plane and the electric field varies with x between the wires as

$$E(x) = \frac{q(z)}{2\pi\varepsilon_0}\left(\frac{1}{x} + \frac{1}{s-x}\right). \tag{5.4}$$

We wish to integrate this electric field from one wire to the other, in order to get the voltage across the pair of wires. This entails calculating

$$V(z) = \int_0^s E_x(x)\,dx = \int_0^s [q(z)/2\pi\varepsilon_0][1/x + 1/(s-x)]\,dx$$

$$= [q(z)/2\pi\varepsilon_0]\int_0^s [1/x + 1/(s-x)]\,dx \tag{5.5}$$

$$= [q(z)/2\pi\varepsilon_0][\ln x - \ln(s-x)]_0^s = ??$$

but we are immediately stuck with an improper integral: Both terms integrate to logarithms but each of these logarithms becomes infinite at one of the two limits. Why are we unable to complete this calculation of the voltage $V(z)$ between the wires at z for an assumed charge distribution $\pm q(z)$ on them?

What the recalcitrant mathematics is warning us about here is that we have made the physically untenable assumption that the wires are infinitely thin. If the wires really had no thickness, then a finite charge density on them would indeed require an infinite voltage between the wires. The realistic case, however, involves wires of some nonzero thickness, so let us specify the radius of each of the two wires as the small value a. Then the charge distribution $q(z)$ is spread over the surface of the wire and there is no electric field inside the perfectly conducting wires. The integration that gives the voltage from the upper wire to the lower one must therefore begin at the surface of the one and end at the surface of the other. This means that the appropriate limits for our integration in the xz-plane are $x = a$ and $x = s - a$, instead of $x = 0$ and $x = s$. Figure 5-4 shows the more precise geometry involved and a plot of the electric field in the plane between the wires.

The calculation of the voltage across the wires now presents no difficulty:

$$V(z) = \int_a^{s-a} E_x\,dx = \int_a^{s-a} [q(z)/2\pi\varepsilon_0][1/x + 1/(s-x)]\,dx$$

$$= [q(z)/2\pi\varepsilon_0]\ln[x/(s-x)]_a^{s-a} \tag{5.6}$$

$$= [q(z)/\pi\varepsilon_0]\ln[(s-a)/a],$$

which we may approximate by

$$V(z) = [q(z)/\pi\varepsilon_0]\ln[s/a] \tag{5.7}$$

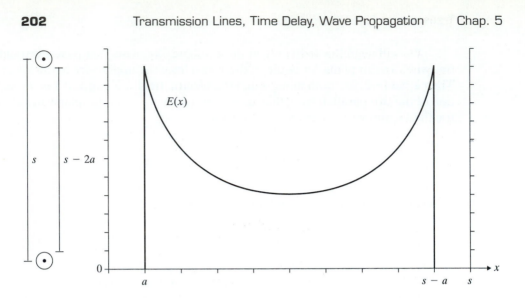

Figure 5-4 Electric field between parallel wires, in the plane of the wires.

because, for thin wires, a is negligible compared to s, although it cannot be replaced by zero in the argument of the logarithm.

What we have calculated here is the capacitance, per unit length, of the pair of parallel wires, because the charge per unit length $q(z)$ is associated with the voltage $V(z)$ between the wires. The ratio is the constant C in

$$q(z)/V(z) = C = \pi\varepsilon_0/\ln(s/a) \tag{5.8}$$

in farads/m. This capacitance involves only the geometry of the wires (separation, thickness) and the medium between them (here, ε_0 for air).

> **Example 5.1**
>
> *What is the capacitance per unit length of a pair of parallel wires separated by* 12 mm *and each of diameter* 2 mm, *in air?* ◈
>
> The ratio of separation to radius is $s/a = 12$ and
>
> $$C = \pi\varepsilon_0/\ln(12) = 11.2 \text{ pF/m.} \tag{5.9}$$
>
> A note to the purist: We have intended to analyze the case of thin wires with a substantial separation, so that $s/a = 12$ is supposed to be a large number for our expression for the capacitance to be valid. A more exact result, valid for thick wires of any separation, would replace $\ln(12) = 2.485$ in the formula with $\cosh^{-1}(12/2) = 2.478$; the details appear in Problem 5.3.

We calculated the voltage from the upper wire to the lower wire along the convenient straight-line path between them. We could have used any other path in a transverse plane, because our field pattern does not include an axial magnetic field, leaving us with an emf that is independent of the path in such planes and with a capacitive relation between charge and voltage that is unique.

The current distribution along the wires is unknown; it is another of the quantities we want to calculate. Whatever this current distribution may be, we can relate it to the magnetic field pattern it generates by superimposing the contributions from the upper and lower wires. In the static case, the current distribution along the wires is uniform, equal and opposite in the two wires, and we expect a magnetic field pattern in transverse planes, surrounding each of the wires. We are trying to maintain this transverse field pattern even in the quasistatic case, although we should no longer expect a uniform current.

We can use Ampère's law to determine the field configuration from the pair of equal and opposite currents $\pm I(z)dz$ at location z. By itself, the current $I(z)dz$ in the upper wire will surround itself with a uniformly distributed, azimuthal magnetic field centered at the upper wire. In the lower wire, the corresponding current $-I(z)dz$ at the same location z will generate a uniformly distributed, azimuthal magnetic field centered on the lower wire. As illustrated in Figure 5-5, we superimpose the two fields vectorially to get the combined field pattern.

Along a circle of radius ρ_1 centered on the upper wire, the mmf generated by the upper wire alone is

$$U_1 = 2\pi\rho_1 H_1 = I(z) \tag{5.10}$$

with the field H_1 in the direction of $\hat{\mathbf{h}}_1 = \hat{\mathbf{z}} \times \hat{\boldsymbol{\rho}}_1$. Along a circle of radius ρ_2 centered on the lower wire, the mmf generated by the lower wire is

$$U_2 = 2\pi\rho_2 H_2 = -I(z) \tag{5.11}$$

with the field H_2 in the direction of $\hat{\mathbf{h}}_2 = \hat{\mathbf{z}} \times \hat{\boldsymbol{\rho}}_2$ (but with sense reversed by the minus sign). The total field at the observation point at distances ρ_1 and ρ_2 from the two wires is the vector superposition

$$\mathbf{H} = H_1\hat{\mathbf{h}}_1 + H_2\hat{\mathbf{h}}_2 = \frac{I(z)}{2\pi}\left(\frac{\hat{\mathbf{h}}_1}{\rho_1} - \frac{\hat{\mathbf{h}}_2}{\rho_2}\right). \tag{5.12}$$

Figure 5-6 shows magnetic field lines in a transverse plane. It is readily confirmed that the field lines are circles surrounding, but not concentric with, each of the two wires.

We will need this field only in the xz-plane, the plane that passes through the two wires. In this plane, we have directed the x-axis from the upper wire to the lower

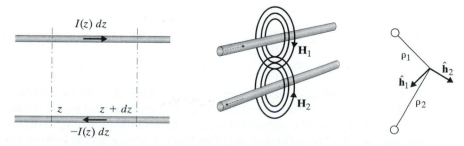

Figure 5-5   Section of parallel wires and the magnetic field it generates.

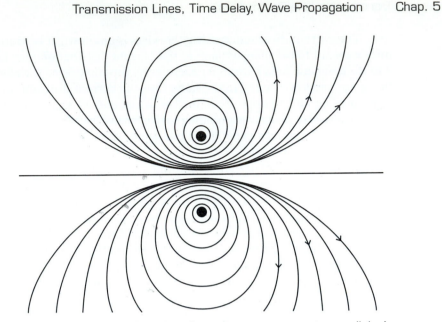

Figure 5-6   Magnetic field lines in a plane transverse to parallel wires.

one. Then $\hat{\mathbf{h}}_1$ and $-\hat{\mathbf{h}}_2$ are both along $\hat{\mathbf{y}}$ and the magnetic field is along y; once again $\rho_1 = x$ and $\rho_2 = s - x$ in this plane and the magnetic field varies with x between the wires as

$$H(x) = \frac{I(z)}{2\pi} \left(\frac{1}{x} + \frac{1}{s-x} \right). \tag{5.13}$$

The profile of this field strength is identical to that of the electric field, plotted earlier.

We wish to integrate this magnetic field over a rectangle that extends from one wire to the other and is infinitesimally thin along z, to get the magnetic flux per unit length $\varphi(z)$ between the wires. This entails calculating

$$\varphi(z) = \int_0^s \mu_0 H_y(x)\,dx = \int_0^s [\mu_0 I(z)/2\pi][1/x + 1/(s-x)]\,dx$$

$$= [\mu_0 I(z)/2\pi]\int_0^s [1/x + 1/(s-x)]\,dx \tag{5.14}$$

$$= [\mu_0 I(z)/2\pi][\ln x - \ln(s-x)]_0^s = ??$$

and we are again faced with the same improper integral as for the emf calculation. The cure is the same; we realize that the magnetic field is absent inside the perfectly conducting wires, since the current flows along their surfaces. The integration for the flux should therefore begin at $x = a$ and end at $x = s - a$, just as before. The integral is the same as before, giving for the magnetic flux per unit length

$$\varphi(z) = [\mu_0 I(z)/\pi]\ln[s/a], \tag{5.15}$$

with the same approximation that neglects a compared to s but not a compared to zero.

What we have calculated here is the inductance, per unit length, of the pair of parallel wires, because the magnetic flux per unit length $\varphi(z)$ is associated with the current $I(z)$ along the wires. The ratio is the constant L in

$$\varphi(z)/I(z) = L = [\mu_0/\pi]\ln(s/a), \tag{5.16}$$

in henrys/m. This inductance involves only the geometry of the wires (separation, thickness) and the medium between them (here, air).

Example 5.2

What is the inductance per unit length of a pair of parallel wires separated by 12 mm *and each of diameter* 2 mm, *in air?* «»

The ratio of separation to radius is $s/a = 12$ and

$$L = [\mu_0/\pi]\ln(12) = 0.994 \ \mu H/m. \tag{5.17}$$

Again, the result is an approximation valid for large ratios of s to a.

We calculated the magnetic flux between the wires over a planar rectangle between them. We would have found the same flux across any other surface (of the same length dz) that joins the two wires, because our field pattern does not include an axial magnetic flux, leaving us with a flux that is independent of such surfaces and with an inductive relation between current and flux that is unique.

The capacitance per unit length C and the inductance per unit length L are parameters of the transmission line. They depend only on the geometry and material constitution (here, air) of the medium where the fields exist. What we have at this point are the charge per unit length $q(z)$, or equivalently the electric flux per unit length $\psi(z)$, in relation to the voltage distribution $V(z)$ and the magnetic flux per unit length $\varphi(z)$ in relation to the current distribution $I(z)$; Figure 5-7 illustrates the flux surfaces involved. The parameters of the line establish the proportionalities between the electric flux (or charge) and the voltage and also between the magnetic flux and the current. We still need equations to determine the voltage and current distributions, given the time variation of the voltage source.

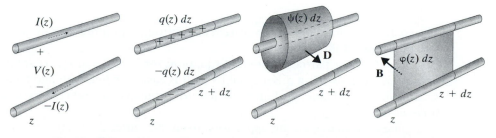

Figure 5-7   Infinitesimally wide electric and magnetic flux surfaces.

TRANSMISSION LINE EQUATIONS

The way the voltage distribution $V(z, t)$ varies from one point to another along the line is determined by the time variation of the magnetic flux between the two points; that magnetic flux depends on the current distribution. The way the current distribution $I(z, t)$ varies from one point to another along the line is determined by the time variation of the electric flux between the two points; that electric flux depends on the voltage distribution. What we need to do is to apply and solve the two Maxwell equations self-consistently.

We can apply the Faraday-Maxwell law to the rectangular area in Figure 5-7 that carries magnetic flux $\varphi(z, t) \, dz$. Since the electric field is zero along the wires, the emf around the edge of the rectangle is

$$\text{emf} = V(z + dz, t) - V(z, t) \tag{5.18}$$

and this must equal the rate of decrease of the magnetic flux through it:

$$\text{emf} = -\partial\big[\varphi(z, t) \, dz\big]/\partial t. \tag{5.19}$$

But $\varphi(z, t) = LI(z, t)$, so that

$$\big[V(z + dz, t) - V(z, t)\big]/dz = -L \, \partial I(z, t)/\partial t, \tag{5.20}$$

which becomes, in the limit of infinitesimal dz,

$$\frac{\partial V(z, t)}{\partial z} = -L \frac{\partial I(z, t)}{\partial t}. \tag{5.21}$$

This says that the spatial rate of change of the voltage distribution is proportional to the temporal rate of change of the current.

We can also apply the Ampère-Maxwell law to the cylindrical surface in Figure 5-7 or, even more simply, apply the law of conservation of charge to the segment of the upper wire from z to $z + dz$ in the figure. The total current that emerges from the segment must equal the rate of decrease of the enclosed charge:

$$I(z + dz, t) - I(z, t) = -\partial\big[q(z, t) \, dz\big]/\partial t. \tag{5.22}$$

But $q(z, t) = CV(z, t)$, so that in the limit of infinitesimal dz,

$$\frac{\partial I(z, t)}{\partial z} = -C \frac{\partial V(z, t)}{\partial t}. \tag{5.23}$$

The spatial rate of change of the current distribution is proportional to the temporal rate of change of the voltage.

Together, the partial differential equations for $V(z, t)$ and $I(z, t)$ are the *transmission line equations,* in differential form. It is easier to deal with and to apply our successive approximation technique to the corresponding integral version of the equations, as follows.

The total charge on the upper wire, from the source at $z = 0$ to a variable position z along the line, is related to the transverse voltage distribution by

$$Q(z, t) = \int_0^z q(z', t) \, dz' = \int_0^z CV(z', t) \, dz'. \tag{5.24}$$

The total magnetic flux between the wires, from the source at $z = 0$ to some variable position z along the line, is related to the current distribution by

$$\Phi(z, t) = \int_0^z \varphi(z', t) \, dz' = \int_0^z LI(z', t) \, dz'. \tag{5.25}$$

The current at z differs from the current entering the upper wire at $z = 0$ by the rate of decrease of the enclosed charge:

$$I(z, t) - I(0, t) = -\partial Q(z, t)/\partial t. \tag{5.26}$$

Since there is no voltage along the perfectly conducting wires, the Faraday-Maxwell law demands that the voltage across the wires at z differ from that at $z = 0$ (which in this case is the applied source voltage) by the rate of decrease of the total magnetic flux from $z = 0$ to z:

$$V(z, t) - V(0, t) = -\partial \Phi(z, t)/\partial t. \tag{5.27}$$

The *transmission line equations in integral form* are therefore

$$V(z, t) = V(0, t) - \partial \Phi(z, t)/\partial t, \qquad \Phi(z, t) = \int_0^z LI(z', t) \, dz', \tag{5.28}$$

$$I(z, t) = I(0, t) - \partial Q(z, t)/\partial t, \qquad Q(z, t) = \int_0^z CV(z' t) \, dz'. \tag{5.29}$$

For the transmission line we have analyzed, the two parameters are

$$L = \left[\mu_0/\pi\right]\ln(s/a), \qquad C = \left[\varepsilon_0\pi\right]/\ln(s/a). \tag{5.30}$$

For our example, the voltage at $z = 0$ is the specified source

$$V(0, t) = V(t). \tag{5.31}$$

Note carefully that we are using the letter V in two closely related but distinct senses. As a function with just one argument (time), it represents the given source voltage applied at the front end ($z = 0$) of the transmission line, as a function of time. As a function with two arguments (the first, distance; the second, time), it denotes the distribution of the voltage along the length of the transmission line, as well as in time. We are given $V(t)$ and we seek $V(z, t)$; it is imperative to take notice of the number of arguments of the V functions that appear in the equations we are developing.

We are given the generator voltage $V(0, t) = V(t)$ but the current at $z = 0$ is an unknown; this current $I(0, t)$ is the current supplied by the generator and is of particular interest to us, but we have no information about it yet. Instead, we have the alternative information that the load resistor at $z = l$ is a specified value R_l, so that we have the boundary condition

$$V(l, t) = R_l I(l, t). \tag{5.32}$$

What the transmission line equations are saying is that the voltage across the wires at any position along the line is related to the time derivative of the spatial integral of the current, while the current in the wires is related to the time derivative of the spatial integral of the voltage. How can we get both?

QUASISTATIC ANALYSIS OF TRANSMISSION LINE

The transmission line equations in integral form,

$$V(z, t) = V(0, t) - \partial\Phi(z, t)/\partial t, \qquad \Phi(z, t) = \int_0^z LI(z', t)\, dz', \qquad (5.33)$$

$$I(z, t) = I(0, t) - \partial Q(z, t)/\partial t, \qquad Q(z, t) = \int_0^z CV(z'\, t)\, dz'. \qquad (5.34)$$

are coupled, in that we need the current distribution to get the voltage and we need the voltage distribution to get the current, but they are in a form that is ideal for application of our successive approximation technique. This will give us an infinite series of correction terms for the voltage and current, and we plan to sum up the series to get an exact solution.

In the static case, the equations lack the $\partial Q/\partial t$ and $\partial\Phi/\partial t$ terms, so that to zeroth order in time variation, the voltage and current distributions along the transmission line are just

$$V_0(z, t) = V(0, t) = V(t), \qquad I_0(z, t) = I(0, t) = I(t), \qquad (5.35)$$

where $V(t)$ is the specified source voltage but $I(0, t)$ is the unknown source current, which we have simply designated $I(t)$ for convenience. Again, we must distinguish carefully between the function $I(t)$ of one argument (time), representing the generator current, and the function $I(z, t)$ of two arguments (distance, time), denoting the distribution of the current along the length of the transmission line, as well as in time; both these functions are unknown.

The preceding quasistatic results indicate a uniform distribution (independent of z) for both the voltage and current along the line. We can use these zero-order results to calculate the zero-order flux and charge, up to position z, as

$$\Phi_0(z, t) = \int_0^z LI_0(z', t)\, dz' = \int_0^z LI(t)\, dz' = LI(t)z, \qquad (5.36)$$

$$Q_0(z, t) = \int_0^z CV_0(z', t)\, dz' = \int_0^z CV(t)\, dz' = CV(t)z. \qquad (5.37)$$

We use these results for the time-varying flux and charge to get corrections to the zero-order voltage and current distributions. The time derivatives of the fluxes contribute to first-order correction terms. Note that $V(0, t)$ is the specified source voltage $V(t)$ and should therefore not suffer any corrections. The source current

$I(0, t)$ has been designated $I(t)$ and is unknown, but it too should not have corrections appended to it; it is whatever function of time $I(t)$ it will turn out to be after we satisfy all the conditions, so that nothing would be gained by writing it as an unknown zero-order term plus unknown corrections. At other locations $z > 0$, we will need the corrections to our zero-order approximations to the spatial distributions.

Consequently, the first-order corrections are just the time derivatives of the zero-order charge and flux we have just calculated.

$$I_1(z, t) = -\partial Q_0(z, t)/\partial t = -C[dV/dt]z, \tag{5.38}$$

$$V_1(z, t) = -\partial \Phi_0(z, t)/\partial t = -L[dI/dt]z. \tag{5.39}$$

Note that, since $V(t)$ is a specified function of time, its derivative dV/dt is also known, as a function of time. Note also that $I(t)$ is still unknown, but its derivative dI/dt is not a separate unknown function; it will be known once we solve for the $I(t)$ time function.

These first-order current and voltage results allow us to obtain the first-order flux and charge, up to location z:

$$\Phi_1(z, t) = \int_0^z LI_1(z', t)\, dz' = \int_0^z [-LC][dV(t)/dt]z'\, dz'$$

$$= -LC[dV(t)/dt]z^2/2, \tag{5.40}$$

$$Q_1(z, t) = \int_0^z CV_1(z', t)\, dz' = \int_0^z [-CL][dI(t)/dt]z'\, dz'$$

$$= -CL[dI(t)/dt]z^2/2. \tag{5.41}$$

In turn, these furnish corrections to the voltage and current distributions:

$$I_2(z, t) = -\partial Q_1(z, t)/\partial t = CL[d^2I/dt^2]z^2/2, \tag{5.42}$$

$$V_2(z, t) = -\partial \Phi_1(z, t)/\partial t = LC[d^2V/dt^2]z^2/2. \tag{5.43}$$

Again, we know d^2V/dt^2, since we are given $V(t)$, and d^2I/dt^2 will be known if and when we find the source current $I(t)$.

Let us go through one more cycle of the equations before collecting our results. The second-order current and voltage give us the second-order flux and charge:

$$\Phi_2(z, t) = \int_0^z LI_2(z', t)\, dz' = \int_0^z [L^2C][d^2I(t)/dt^2][z'^2/2]\, dz'$$

$$= L^2C[d^2I(t)/dt^2]z^3/3!, \tag{5.44}$$

$$Q_2(z, t) = \int_0^z CV_2(z', t)\, dz' = \int_0^z [LC^2][d^2V(t)/dt^2][z'^2/2]\, dz'$$

$$= LC^2[d^2V(t)/dt^2]z^3/3!. \tag{5.45}$$

Note that second-order quantities vary in time as the second derivative of the source voltage and source current, of which the former is given, the latter not yet known. The third-order corrections to the voltage and current are

$$I_3(z, t) = -\partial Q_2(z, t)/\partial t = -LC^2[d^3V/dt^3]z^3/3!, \qquad (5.46)$$

$$V_3(z, t) = -\partial \Phi_2(z, t)/\partial t = -L^2C[d^3I/dt^3]z^3/3!. \qquad (5.47)$$

By now, we see the pattern of successive corrections and we collect the terms of the series, up to fourth order, as

$$V(z, t) = V - L\frac{dI}{dt}z + LC\frac{d^2V}{dt^2}\frac{z^2}{2!} - L^2C\frac{d^3I}{dt^3}\frac{z^3}{3!} + L^2C^2\frac{d^4V}{dt^4}\frac{z^4}{4!}\cdots \qquad (5.48)$$

$$I(z, t) = I - C\frac{dV}{dt}z + LC\frac{d^2I}{dt^2}\frac{z^2}{2!} - LC^2\frac{d^3V}{dt^3}\frac{z^3}{3!} + L^2C^2\frac{d^4I}{dt^4}\frac{z^4}{4!}\cdots \qquad (5.49)$$

What remains to be done is to evaluate these at $z = l$ and impose the boundary condition that $V(l, t) = R_l I(l, t)$. This should yield the one remaining unknown function, $I(t)$.

We can do this after truncating the infinite series of corrections at some point, say after fourth-order terms. Equating $V(l, t)$ to $R_l I(l, t)$ then results in a fourth-order ordinary differential equation in the unknown function $I(t)$. If we truncate the series at the tenth-order term, to get a better approximation, we will get a tenth-order differential equation instead, and so on.

We can do better by not truncating the series of corrections at all. This yields a differential equation of infinite order for the unknown $I(t)$. This may sound frightening, but it could be handled by, say, Laplace transformation.

We prefer, however, to sum the two infinite series first and then apply the condition $V(l, t) = R_l I(l, t)$ to get a functional relationship between $V(t)$ and $I(t)$, then to find $I(t)$ and thereby evaluate $V(z, t)$ and $I(z, t)$.

SUMMATION OF SERIES

The key to summing the two infinite series is obscured by the alternation of terms involving $V(t)$ and $I(t)$ and their derivatives. The cure for this small complication is to combine the two series, as follows.

We form the combination $V(z, t) + \xi I(z, t)$, using an as yet unspecified constant coefficient ξ. We get

$$
\begin{aligned}
V(z, t) + \xi I(z, t) = &\, [V(t) + \xi I(t)] - (d/dt)[LI(t) + \xi CV(t)]z \\
&+ LC(d^2/dt^2)[V(t) + \xi I(t)]z^2/2! \\
&- LC(d^3/dt^3)[LI(t) + \xi CV(t)]z^3/3! \\
&+ L^2C^2(d^4/dt^4)[V(t) + \xi I(t)]z^4/4! \cdots \qquad (5.50)
\end{aligned}
$$

and since the value of the constant ξ is up to us, we choose it so as to simplify this sum. We see terms involving $[V(t) + \xi I(t)]$, alternating with terms that feature

$[LI(t) + \xi CV(t)]$. We can select the constant ξ to make the quantity $[LI(t) + \xi CV(t)]$, which appears in half the terms, the same as, or at least proportional to, $[V(t) + \xi I(t)]$, which appears in the other half. We rewrite the first bracketed expression as $\xi C[V(t) + \{L/\xi C\}I(t)]$ and set the latter combination of V and I equal to the second expression:

$$[LI(t) + \xi CV(t)] = \xi C[V(t) + \{L/\xi C\}I(t)] = \xi C[V(t) + \xi I(t)], \qquad (5.51)$$

which is satisfied if we choose ξ to obey the requirement

$$L/\xi C = \xi \quad \text{or} \quad \xi = \pm\sqrt{L/C}. \qquad (5.52)$$

Either value of ξ allows us to combine the two infinite series into a single one. The combined series appears as

$$\begin{aligned}
V(z,t) + \xi I(z,t) = {} & [V(t) + \xi I(t)] - \xi C(d/dt)[V(t) + \xi I(t)]z \\
& + LC(d^2/dt^2)[V(t) + \xi I(t)]z^2/2! \\
& - \xi CLC(d^3/dt^3)[V(t) + \xi I(t)]z^3/3! \\
& + L^2C^2(d^4/dt^4)[V(t) + \xi I(t)]z^4/4! \cdots
\end{aligned} \qquad (5.53)$$

We can simplify the appearance of this result by noting that, by our choice of ξ, we have $LC = (\xi C)^2$, which we rewrite in terms of a new parameter u that will aid us in interpreting the results as

$$\xi C = \sqrt{LC} = 1/u. \qquad (5.54)$$

We also abbreviate the combination of V and I in the brackets as W:

$$W(t) = V(t) + \xi I(t). \qquad (5.55)$$

This $W = W(t)$ is a function of one argument, time, alone. This abbreviation results in

$$V(z,t) + \xi I(z,t) = W(t) - \frac{dW(t)}{dt}(z/u) + \frac{d^2W(t)}{dt^2}\frac{(z/u)^2}{2!}$$

$$- \frac{d^3W(t)}{dt^3}\frac{(z/u)^3}{3!} + \frac{d^4W(t)}{dt^4}\frac{(z/u)^4}{4!} - \cdots \qquad (5.56)$$

We are now ready to sum this infinite series of successive correction terms. We can perform this magic because the form of this series, which may look vaguely familiar, should really be immediately recognizable. Yes, it is a Taylor series. It is a Taylor series expansion of the function W about the value t as center point (the point at which all the derivatives are evaluated), in powers of $[-z/u]$ (which must therefore be the deviation from the center point). Hence, wherever the infinite series converges, it sums to the W function, evaluated at the center point t plus the deviation $[-z/u]$:

$$V(z,t) + \xi I(z,t) = W(t - z/u). \qquad (5.57)$$

We have thereby succeeded in summing the infinite series that gives the combined function of two arguments, $V(z,t) + \xi I(z,t)$; the sum is a function of one argument, W, evaluated at the argument $t - z/u$.

We still need to separate the two quantities $V(z, t)$ and $I(z, t)$ out of this combined form. We can do even this additional bit of magic, because the coefficient ξ actually represents either one of two possible values, $\xi = \pm\sqrt{L/C}$. For notational convenience, let us define the (positive) quantity Z_0 as

$$Z_0 = \sqrt{L/C}, \tag{5.58}$$

so that $\xi = \pm Z_0$. Correspondingly, the parameter u also has the two values

$$u = 1/\xi C = \pm 1/Z_0 C = \pm 1/\sqrt{LC} = \pm u_0 \tag{5.59}$$

and we denote the positive value by $u_0 = 1/\sqrt{LC}$. We therefore have two versions of the combination of source voltage and current that we called $W(t)$; we will denote these two by $F(t)$ and $G(t)$, using the upper and the lower sign, respectively [i.e., $W(t)$ is $F(t)$ if we use $\xi = Z_0$ and $u = 1/\sqrt{LC} = u_0$, while $W(t)$ is $G(t)$ if we use $\xi = -Z_0$ and $u = -1/\sqrt{LC} = -u_0$]:

$$F(t) = V(t) + Z_0 I(t), \qquad G(t) = V(t) - Z_0 I(t). \tag{5.60}$$

The combined voltage and current distributions along the transmission line are therefore

$$V(z, t) + Z_0 I(z, t) = F(t - z/u_0)$$
$$= V(t - z/u_0) + Z_0 I(t - z/u_0), \tag{5.61}$$

$$V(z, t) - Z_0 I(z, t) = G(t + z/u_0)$$
$$= V(t + z/u_0) - Z_0 I(t + z/u_0). \tag{5.62}$$

Note that the V and I functions on the left are the functions of two arguments that we are seeking, the distributions of voltage and current along the line. The V and I functions on the right are, however, functions of a single argument (time), although this single argument is a combination of t and a term proportional to z. The equations are telling us that the distributions along the line are combinations of the generator voltage $V(t)$ (given) and the generator current $I(t)$ (sought), evaluated at arguments that are a combination of time t and distance z.

Let us pause to review what the known and the unknown quantities are. The parameters $Z_0 = \sqrt{L/C}$ and $u_0 = 1/\sqrt{LC}$ are known quantities derived from the known inductance L and capacitance C of the line. The source voltage $V(t)$ is given for all arguments t, so that $V(t \pm z/u_0)$ are both known. However, the source current $I(t)$ remains unknown for all arguments t, so that $I(t \pm z/u_0)$ are both unknown. As a result, the two functions $F(t)$ and $G(t)$ are not yet known; we will need them at their respective arguments $t - z/u_0$ and $t + z/u_0$. The two combinations of voltage and current distributions along the line—namely, $V(z, t) \pm Z_0 I(z, t)$—are correspondingly also still unknown, so that we are not ready to add and subtract the last two equations to extract the voltage $V(z, t)$ and current $I(z, t)$ separately. What is needed is one additional condition that can furnish the missing function $I(t)$.

That additional condition is provided by imposing the requirement that, at the end of the line, at $z = l$, there is a specified load resistor R_l that demands that the voltage and current at the end of the line be in just that proportion:

$$V(l, t) = R_l I(l, t). \tag{5.63}$$

The functions $F(t)$ and $G(t)$ are still unknown individually, although their sum is known for this problem as

$$F(t) + G(t) = 2V(t). \tag{5.64}$$

The second equation needed to determine both $F(t)$ and $G(t)$ comes from applying the boundary condition $V(l, t) = R_l I(l, t)$ at the loaded end of the line, at $z = l$. Substituting l for z, this requirement becomes

$$V(l, t) + Z_0 I(l, t) = F(t - l/u_0) = [R_l + Z_0]I(l, t) \tag{5.65}$$

from (5.61) and

$$V(l, t) - Z_0 I(l, t) = G(t + l/u_0) = [R_l - Z_0]I(l, t) \tag{5.66}$$

from (5.62). We can now eliminate the unknown current $I(l, t)$ in the load resistor, leaving two equations for the two unknown functions $F(t)$ and $G(t)$:

$$[R_l - Z_0]F(t - l/u_0) = [R_l + Z_0]G(t + l/u_0), \tag{5.67}$$

with

$$F(t) + G(t) = 2V(t). \tag{5.68}$$

These are two functional equations for the two unknown functions $F(t)$ and $G(t)$. Once we solve for $F(t)$ and $G(t)$ individually, we can get the complete voltage and current distributions along the line, by addition and subtraction of (5.61) and (5.62), as

$$V(z, t) = \tfrac{1}{2}F(t - z/u_0) + \tfrac{1}{2}G(t + z/u_0), \tag{5.69}$$

$$Z_0 I(z, t) = \tfrac{1}{2}F(t - z/u_0) - \tfrac{1}{2}G(t + z/u_0). \tag{5.70}$$

Everything now depends on solving for the two unknown functions $F(t)$ and $G(t)$ from (5.67) and (5.68). This is not particularly difficult, although functional relations such as these may be less familiar than, say, differential equations. Before carrying the solution to the end, however, we want to interpret the results we have already. Such interpretation has special clarity when we restrict ourselves, temporarily, to a rather special case of the circuit we are analyzing. Afterward, we can return to the general circuit.

INTERPRETATION: MATCHED LINE

The simplest phenomenon associated with the transmission line occurs in the special case that we choose our load resistor R_l to be exactly equal to Z_0:

$$R_l = Z_0 = \sqrt{L/C}. \tag{5.71}$$

This choice is by no means a necessary one, but it does simplify not only the equations to be solved, but also their interpretation, and the phenomena to be observed. We therefore treat this special case first.

Recall that the inductance and capacitance per unit length, L and C, are properties of the transmission line structure, determined by the geometry and material constitution of the medium around the wires. Therefore, Z_0 is itself a property of the structure. The special case we are examining now corresponds to choosing a load resistor R_l that exactly matches the value of Z_0. If we attach a load resistor of exactly Z_0 ohms to the line, we say we have *matched* the line.

> **Example 5.3**
>
> *What load impedance will match a pair of parallel wires separated by 12 mm and each of diameter 2 mm, in air?* ◇
>
> The ratio of separation to radius is $s/a = 12$ and, in the previous examples, we have already calculated the inductance and capacitance per unit length as
>
> $$L = [\mu_0/\pi]\ln(12) = 0.994 \ \mu\text{H/m}, \quad C = \pi\varepsilon_0/\ln(12) = 11.2 \ \text{pF/m}.$$
>
> Hence, we find $Z_0 = \sqrt{L/C} = 298$ ohms as the matching load resistor. Once again, the result is an approximation valid for large ratios of s to a.

If we do match the line, the equations become quite simple to solve. The first of the two functional equations for $F(t)$ and $G(t)$ is

$$[R_l - Z_0]F(t - l/u_0) = (R_l + Z_0)G(t + l/u_0) \tag{5.72}$$

and the matching resistor makes the factor $[R_l - Z_0]$ become zero. This forces the function $G(t + l/u_0)$ to be zero, for all time t. That is, the function $G(t)$ is just zero, for all arguments t. The second of the two functional equations,

$$F(t) + G(t) = 2V(t), \tag{5.73}$$

then dictates that, for this matched case,

$$F(t) = 2V(t). \tag{5.74}$$

The function $F(t)$ is just twice the source voltage applied to the matched circuit.

From the expressions for the voltage and current distributions along the line in terms of the $F(t)$ and $G(t)$ functions,

$$V(z, t) = \tfrac{1}{2}F(t - z/u_0) + \tfrac{1}{2}G(t + z/u_0), \tag{5.75}$$

$$Z_0 I(z, t) = \tfrac{1}{2}F(t - z/u_0) - \tfrac{1}{2}G(t + z/u_0), \tag{5.76}$$

but now with $G = 0$, we get that

$$V(z, t) = \tfrac{1}{2}F(t - z/u_0) = V(t - z/u_0) \tag{5.77}$$

and

$$I(z, t) = \tfrac{1}{2}F(t - z/u_0)/Z_0 = V(t - z/u_0)/Z_0. \tag{5.78}$$

In this matched case, we have therefore that

$$V(z, t) = Z_0 I(z, t) \tag{5.79}$$

and, in particular (at $z = 0$), that

$$V(t) = Z_0 I(t). \tag{5.80}$$

Matching the line has evidently caused the voltage and current to be proportional at all values of z, the ratio being the impedance Z_0, not only at the load (where we forced it to be so) but all along the line and at the generator too.

We also now know, for this matched case, the voltage $V(z, t)$ at any position z, at any time t, in terms of the given generator voltage $V(t)$, as found in (5.77):

$$V(z, t) = V(t - z/u_0). \tag{5.81}$$

This equation says that the generator voltage $V(0, t) = V(t)$ becomes $V(z, t) = V(t - z/u_0)$ at any $z > 0$. In particular, the voltage across the (matched) load resistor $R_l = Z_0$ is

$$V(l, t) = V(t - l/u_0) = V(t - T), \tag{5.82}$$

which is the same as the voltage $V(t)$ provided by the generator, but *delayed* by the time interval

$$T = l/u_0 = l\sqrt{LC}. \tag{5.83}$$

At long last, we see action at a distance (the distance l from generator to matched load) after a delay (the time interval T). For this case, there is in fact nothing but this delay. Whatever signal $V(t)$ the voltage source may generate, be it a pulse, or an oscillation, or an amplitude-modulated sinusoidal carrier, or a pulse train representing a sequence of bits, or any signal imaginable, exactly that generated signal will be delivered to the matched load after the delay T. The transmission line with matched load is a pure delay element. Figure 5-8 shows an arbitrary pulse of some duration t_0

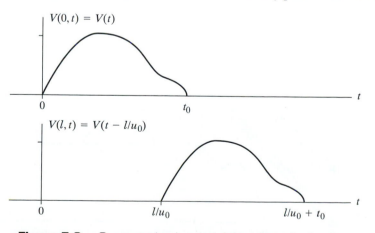

Figure 5-8   Generated pulse and delayed received pulse.

as seen at the generator (at $z = 0$) and its delayed version, as seen at the end of the line (at $z = l$).

The delay T is proportional to the length l of the line, with a proportionality constant $\sqrt{LC}$ that is a property of the transmission line. In fact, we have found that the voltage across the line at any location z develops in time exactly as does the generator voltage, except that it is delayed by the time interval z/u_0. This delay is proportional to the distance of the observation point from the source. The farther we are from the generator, the longer we have to wait to receive the signal generated by the source. That is, it takes time $t = \sqrt{LC}z$ to arrive at distance z from the source, so that the signal progresses along the line, at the fixed speed $u_0 = 1/\sqrt{LC}$. The generator voltage is transmitted along the transmission line at the characteristic speed of propagation u_0, which is a property of the line. Figure 5-9 suggests how an oscilloscope trace of a pulse would appear as measured at the front, the end, and some intermediate point along the transmission line. The straight line joining the several times of onset of the delayed pulses is intended to point out that the delays are proportional to the distance from the generator.

For the parallel-wire transmission line we have examined, the values of the inductance and capacitance per unit length were found to be

$$L = (\mu/\pi)\ln(s/a), \qquad C = (\pi\varepsilon)/\ln(s/a), \qquad (5.84)$$

so that the geometry of the structure (the ratio s/a of wire separation to wire radius) cancels in the product of L and C. Consequently, the characteristic speed is here determined by only the constitutive parameters μ and ε of the space around the wires, where the electromagnetic fields are present (not by the parameters of the wires; there are no fields in the perfectly conducting wires):

$$LC = \mu\varepsilon. \qquad (5.85)$$

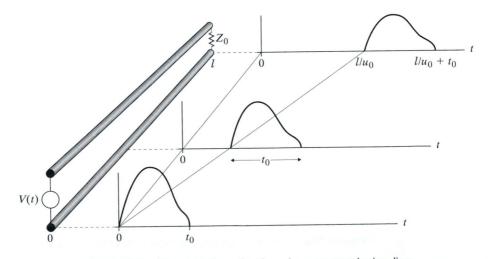

Figure 5-9   Progression of pulse along transmission line.

The speed is

$$u_0 = 1/\sqrt{LC} = 1/\sqrt{\mu\varepsilon} \qquad (5.86)$$

and if the medium around the wires is a vacuum or, for practical purposes also just air (which has very nearly the same permeability and permittivity as a vacuum), then the speed at which the signal is transmitted along the parallel-wire transmission line is

$$
\begin{aligned}
u_0 &= 1/\sqrt{\mu_0\varepsilon_0} \\
&= 1/\sqrt{(1.2566\ \mu\text{H/m})(8.8542\ \text{pF/m})} \\
&= (2.998)10^8\ \text{m/s} = c. \qquad (5.87)
\end{aligned}
$$

This is the vacuum speed of light; the signal developed at the generator is transmitted at the speed of light.

Incidentally, the precise value of the speed of light in a vacuum is

$$c = 299{,}792{,}458\ \text{m/s}. \qquad (5.88)$$

This universal constant has been set to this standard value by definition; in conjunction with an independent definition of the second, it defines precisely the length of a meter. Because the free-space permeability μ_0 has been defined as $(4\pi)10^{-7}$ H/m, the defined value of c also determines a precise value of the vacuum permittivity $\varepsilon_0 = 1/\mu_0 c^2$.

The preceding results apply when the load resistor has the particular value

$$R_l = Z_0 = \sqrt{L/C}. \qquad (5.89)$$

This is called the *characteristic impedance* of the transmission line. The line is matched when the load resistor is chosen to have this value. Under this matched condition, the ratio of voltage to current at any point along the transmission line is just this characteristic impedance. For the case of the parallel-wire line, the values of the inductance and capacitance parameters

$$L = (\mu/\pi)\ln(s/a), \qquad C = (\pi\varepsilon)/\ln(s/a), \qquad (5.90)$$

give as the characteristic impedance

$$Z_0 = (1/\pi)\sqrt{\mu/\varepsilon}\ln(s/a). \qquad (5.91)$$

This depends on the constitution of the medium surrounding the wires and also on the geometry of the structure. For a vacuum, the middle factor has the value

$$\sqrt{\mu_0/\varepsilon_0} = 376.73\ \text{ohms} = \eta_0 \qquad (5.92)$$

and is referred to as the *intrinsic impedance of free space*.

INTERPRETATION: WAVES

We have found the solution for the voltage along the parallel-wire transmission line when it is excited at one end by a voltage source that generates a signal $V(t)$ and is matched at the other end by a resistor exactly equal to the characteristic impedance Z_0. The result is that the voltage at location z at time t is

$$V(z, t) = V(t - z/c) \qquad (5.93)$$

if the medium around the wires is a vacuum (or air). We have already interpreted this as a delayed version of the generator signal $V(t)$, as seen as a function of time at a particular location z, with a delay z/c that is proportional to the distance from the source. We should also interpret this result as a function of z, for fixed t, to complement our understanding of it as a function of t, at fixed z.

The function $V(z, t)$ of position z and time t can be visualized as a function of t at a fixed location z and thought of as an oscilloscope trace of a time signal measured at that fixed location z along the line. However, when we think of this same voltage distribution as a function of z at a fixed time t, we are imagining that we can, in a sense, take an instantaneous "snapshot" of the voltage, this voltage somehow made visible and spread out before us along the length of the line, at a fixed instant of time. For a given signal $V(t)$ generated in time at $z = 0$, how will the snapshot look?

The answer, of course, is given by a plot of $V(z, t) = V(t - z/c)$ against z, at a specific instant t. This plot differs from a straightforward plot of the given $V(t)$ against time t in three respects. First, the variable z in z/c is rescaled by the factor c, as compared to the argument t in $V(t)$. Second, the plot is shifted by the addition of the fixed constant t in the argument of the function, so that, for example, the zero-argument point occurs not at the origin of the spatial plot but rather at the point $z = ct$ (the value of z at which the argument $t - z/c$ becomes zero). Third, the coordinate z has a minus sign in front of it, which results in a display of the signal shape V that appears reversed along the abscissa. Figure 5-10 shows a pair of pulses, a strong one followed by a weaker one, generated in time by the voltage source at $z = 0$ and also a snapshot of the spatial distribution of the voltage at one instant.

The plot features all three aspects of the change of abscissa from t to z. The widths of the pulses are scaled by the factor c; for example, the taller pulse has width t_1 in time but ct_1 in space. The shift is also evident in that, while the pulses begin at the time origin when plotted against t, they appear away from the origin in the snapshot. To get zero argument for the function $V(t)$, we just set $t = 0$ and notice the leading edge of the taller pulse at that time; to get zero argument for the function $V(t - z/c)$, we set $z = ct$ and this delimits the taller pulse in space. To get argument T, we set $t = T$ in $V(t)$ and find the end of the weaker pulse at that instant; to get argument T for the function $V(t - z/c)$, we must set $z = ct - cT$ and that becomes the location of the weaker pulse in space. Most strikingly, the pair of pulses appears

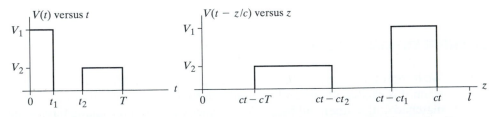

Figure 5-10   Views of voltage signal in time and of its spatial distribution.

reversed in the spatial plot, as compared to their time development. The taller pulse comes earlier in the time plot; it appears as the farther pulse in the spatial plot.

Another aspect of the spatial plot has great importance. We have thought of it as a snapshot at one instant of time, but suppose we could see a sequence of such snapshots as time elapses, as in a movie strip. The leading edge of the taller pulse is at location ct at time t. As time goes on, this location is at an ever greater distance from the generator. In fact, the rear of each pulse, or any of the points of the pulses, also occurs at points in space that move in time, all at the same speed c. Such motion is called *wave motion*. The pair of pulses moves along the z-axis, as a wave. Note that the earlier pulse, the stronger one, has had more time to travel along the line than has the weaker pulse, so that it has reached farther along z than has the weaker pulse; this is another way to understand the reversal of the signal when we change from a time plot to a spatial one.

Mathematically, any function of time and space in which the time coordinate and the space coordinate appear combined as the difference between time and (rescaled) space, or as the difference between space and (rescaled) time, is a *wave*. Any such function appears to move along the direction of increasing values of the distance coordinate, as time elapses, at a rate determined by the rescaling factor. Thus, both

$$f(t - z/u) \quad \text{and} \quad h(z - ut) \tag{5.94}$$

represent waves; each travels along z at the speed u. The former one is described by *waveform* $f(t)$ in time, as it evolves at $z = 0$; the latter one is expressed by the *waveshape* $h(z)$ in space, as seen at $t = 0$.

Suppose we wish to focus on some specific feature of either the waveform or the waveshape, such as its onset, or its peak, or a zero crossing. As time elapses, we must run along the z-axis at speed u, if we wish to keep up with any such feature. By running along z at that speed, we maintain the argument of $f(t - z/u)$, or of $h(z - ut)$, at the value that corresponds to the desired feature of the wave.

Example 5.4

Figure 5-11 shows a particular pulse waveform $f(\xi)$ plotted against its argument ξ. Besides the onset of the pulse at $\xi = 0$, the special arguments ξ_1, ξ_2, ξ_3 label the crest, an inflection point, and the end of the pulse, respectively. Show, in a plot of the wave $f(t - z/u)$ against z, where these special points of the waveform will appear in space at time t. ◇

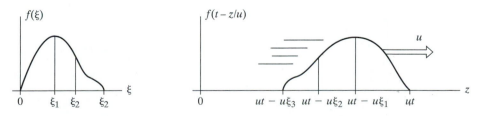

Figure 5-11 Waveform $f(\xi)$ versus ξ and wave $f(t - z/u)$ versus z.

The figure shows the snapshot of the waveshape. The onset of the pulse appears at $z = ut$; the other three specified features are located at the points $z = ut - u\xi_1, ut - u\xi_2, ut - u\xi_3$; all these locations move to the right as the time t elapses. Every point of the waveform moves at the speed u, so that the entire waveform moves at that speed; this motion is what makes $f(t - z/u)$ a wave. The arrow attached to the $f(t - z/u)$ plot is intended to suggest this wave motion: The pulse waveform, reversed and rescaled, is seen to slide along the z-axis, at speed u.

What we have found is that any signal generated at one end of a matched transmission line travels along the line, at the characteristic speed associated with the line. It spreads out along the line, appearing reversed in a spatial snapshot as compared to its temporal evolution. The wave moves along the line and is ultimately delivered to the matched load, exactly as generated (same strength, same duration, same waveform) but delayed by the time it takes to travel the length of the line at the speed of propagation of the wave. The transmission line thereby conveys, or transmits, any signal generated by the voltage source at $z = 0$ to the matched load at $z = l$, without distortion, with only the transmission delay l/u.

Although time may be thought of as unlimited in both the past and the future, the spatial extent of the transmission line is limited to the range $z = 0$ to $z = l$, so that the distance z in the wave $f(t - z/c)$ should be restricted to that range. The full waveform, reversed and rescaled, appears in space only after emission by the generator and before absorption by the matched load.

Figure 5-12 depicts the situation at a time τ when the second of two pulses transmitted along the line has not yet been fully absorbed by the load. The slanted portions of the plot show the voltage as a function of time t, at the source and at the load; the horizontal plot presents a snapshot of the voltage as a function of position z, at the instant τ. That time τ is later than $l/c + t_2$ (the arrival of the second, weaker pulse) but earlier than $l/c + T$ (the end of the second pulse, as seen at the load). The first,

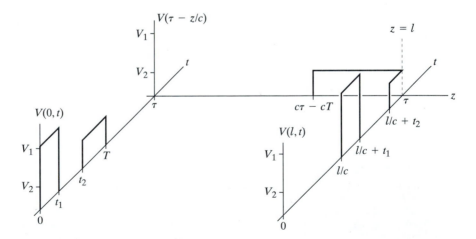

Figure 5-12   After emission of two pulses, before second one is absorbed.

stronger pulse has already been absorbed, between times l/c and $l/c + t_1$; the second one reached the load at time $l/c + t_2$ but only the initial portion of this second pulse has been absorbed when the time has reached τ. The trailing part of this second pulse is still spread in space, from location $c(\tau - T)$ to location l, and is moving to the load at the speed c.

MISMATCHED LINE

The transmission line equations in integral form that we examined had built into them the boundary conditions at the ends of the line. It is instructive to revert to the equations in differential form, which we obtained just before we developed the integral version. We can then find the general solution to the transmission line system, one that does not require a matched load.

The transmission line equations in integral form are

$$V(z, t) = V(0, t) - \partial\Phi(z, t)/\partial t, \qquad \Phi(z, t) = \int_0^z LI(z', t)\, dz', \qquad (5.95)$$

$$I(z, t) = I(0, t) - \partial Q(z, t)/\partial t, \qquad Q(z, t) = \int_0^z CV(z', t)\, dz'. \qquad (5.96)$$

As an alternative to applying the Maxwell equations to an infinitesimally short segment of transmission line, as we did earlier, we can extract the equations in differential form by simply differentiating the integral version, with respect to the distance z along the line. We get at once

$$\partial V(z, t)/\partial z = -L\, \partial I(z, t)/\partial t, \qquad (5.97)$$

$$\partial I(z, t)/\partial z = -C\, \partial V(z, t)/\partial t. \qquad (5.98)$$

These differential equations are to be supplemented with boundary conditions that describe what source and load are attached to the ends of the line. At this point, however, we seek the general solution to this pair of partial differential equations, without regard for the boundary conditions to be applied afterward to particularize that general solution.

We already have a clue as to the form of the solution, from our results for the case of a matched line. Based on that special case, let us try to solve the equations with functions of the form

$$V(z, t) = f(t - z/w), \qquad I(z, t) = g(t - z/w). \qquad (5.99)$$

We substitute these trial functions into the two partial differential equations, to find out what the two functions f and g may be and what the constant w should be. Let $s = t - z/w$. Then the various partial derivatives are

$$\partial V/\partial z = [df/ds]\, \partial s/\partial z = f'(s)[-1/w], \quad \partial V/\partial t = [df/ds]\, \partial s/\partial t = f'(s);$$

$$\partial I/\partial z = [dg/ds]\, \partial s/\partial z = g'(s)[-1/w], \quad \partial I/\partial t = [dg/ds]\, \partial s/\partial t = g'(s). \quad (5.100)$$

The differential equations therefore require of these trial functions that

$$f'(s)[-1/w] = -Lg'(s) \quad \text{and} \quad g'(s)[-1/w] = -Cf'(s). \quad (5.101)$$

This implies that

$$f'(s) = wLg'(s) = wLwCf'(s) = w^2LCf'(s), \quad (5.102)$$

which requires either that $f'(s)$ be zero, which means that $f(s)$ is a constant (this possibility is actually a special case of the following one), or else that

$$w^2LC = 1 \quad \text{or} \quad w = \pm 1/\sqrt{LC} = \pm u_0, \quad (5.103)$$

with *any* function $f(s)$ whatsoever. The corresponding $g(s)$ function is then found from $f'(s) = wLg'(s)$ to be related to whatever $f(s)$ is chosen by

$$f(s) = \pm\sqrt{L/C}g(s) = \pm Z_0 g(s), \quad (5.104)$$

except for an additive constant that can be absorbed in the final result, which is given in the following equations.

There are therefore two independent solutions for the voltage distribution, corresponding to the two signs of the square root. The general solution is a combination of both, using consistent choices of the signs:

$$V(z, t) = f_1(t - z/u_0) + f_2(t + z/u_0), \quad (5.105)$$

$$Z_0 I(z, t) = f_1(t - z/u_0) - f_2(t + z/u_0), \quad (5.106)$$

where $f_1(t)$ and $f_2(t)$ may be any functions (of a single variable) whatever. These two undetermined functions are to be found by applying the boundary conditions at the two ends of the transmission line.

To interpret this result, we note that we already know the meaning of the f_1 part of the general solution. It is a wave, of waveform $f_1(t)$, traveling along the positive z-axis, at speed $u_0 = 1/\sqrt{LC}$. The f_2 portion of the solution differs from this first one only in having a positive sign in front of z (or u_0), rather than a negative sign. By the same reasoning that identifies f_1 as a wave traveling along z, we recognize f_2 as a wave also, of waveform $f_2(t)$, but traveling along the negative z-axis, at the same speed u_0 (or, which amounts to the same thing, a wave along z traveling with negative speed $-u_0$).

This interpretation of f_2 is clear if we consider trying to keep up with some feature of the waveform $f_2(t)$, say its crest, as time elapses. To maintain a constant argument $t + z/u_0$ as time goes on, we must go to smaller values of z, at the rate u_0, so that this is a wave that travels along $-z$.

The general solution to the transmission line equations is a superposition of two arbitrary waves traveling in opposite directions along the line, at the same speed u_0. The voltage is the sum of a forward-traveling and a backward-traveling wave; the characteristic impedance Z_0 times the current is the difference between the same two traveling waves.

One transmission line problem differs from another only in the waveforms of the two oppositely traveling waves. These are to be found from the boundary conditions at the ends of the line.

Example 5.5

Short-circuited transmission line

A parallel-wire transmission line is short-circuited at its end at $z = l$ and a constant (battery) voltage V_0 is applied at the front end at $z = 0$ through a switch that is closed at time $t = 0$ and remains closed thereafter; see Figure 5-13. The voltage and current were both zero before the switch was closed. The parameters of the line are given, so that the characteristic impedance $Z_0 = \sqrt{L/C}$ and the propagation speed $u_0 = 1/\sqrt{LC} = c$ are known. Find the voltage and current along the line for all time after the switch is closed. ◈

First, we should dismiss the answer that is furnished by ordinary circuit theory, based on Kirchhoff's laws, which would assert that an infinite current flows in the wires as soon as the switch is closed. This cannot be correct, as any current that does begin to flow must generate a magnetic field and the time-varying magnetic flux then affects the voltage across the wires. Furthermore, the charge that accompanies the voltage generates a time-varying electric flux that affects the mmf and therefore the current in the wires.

The proper result is the superposition of oppositely traveling waves that we have predicted from Maxwell's equations, so the answer will be

$$V(z, t) = f(t - z/c) + g(t + z/c), \tag{5.107}$$

$$Z_0 I(z, t) = f(t - z/c) - g(t + z/c), \tag{5.108}$$

where $f(t)$ will be the forward-traveling waveform and $g(t)$ will be its backward-traveling counterpart. These two waveforms are to be found from the conditions imposed at the two ends. The boundary conditions are imposed by the short circuit at $z = l$ and by the applied voltage V_0 after $t = 0$:

$$V(l, t) = 0 \qquad \text{for all } t; \tag{5.109}$$

$$V(0, t) = 0 \qquad \text{for } t < 0, \qquad V(0, t) = V_0 \qquad \text{for } t > 0. \tag{5.110}$$

The first condition requires

$$f(t - l/c) + g(t + l/c) = 0 \qquad \text{for all } t, \tag{5.111}$$

in which the arguments of f and g differ by the constant $2l/c$. This tells us that, for any argument s,

$$g(s) = -f(s - 2l/c). \tag{5.112}$$

The second condition gives us

$$f(t) + g(t) = V_0 \qquad \text{for } t > 0. \tag{5.113}$$

Figure 5-13   Short-circuited transmission line.

Combining the two requirements on $f(t)$ and $g(t)$ yields for $t > 0$

$$f(t) = V_0 - g(t) \quad \text{or} \quad f(t) = V_0 + f(t - 2l/c). \tag{5.114}$$

This states that the waveform $f(t)$ is a function that, at any $t > 0$, equals V_0 plus an earlier value of f, earlier by $2l/c$ seconds. Since $f(t) = 0$ for $t < 0$, this suffices to give us $f(t)$, and therefore also $g(t)$, for all t, as follows.

Figure 5-14 illustrates the construction of the waveform $f(t)$ for all t, from its specification

$$f(t) = 0 \quad \text{for } t < 0, \quad f(t) = V_0 + f(t - 2l/c) \quad \text{for } t > 0. \tag{5.115}$$

We start on the negative time axis, along which $f(t)$ is just zero. Once we cross over to positive time, the function $f(t)$ becomes V_0 plus its earlier value. But, until t reaches $2l/c$, the earlier time is negative, so that the earlier voltage value is just zero and $f(t)$ in

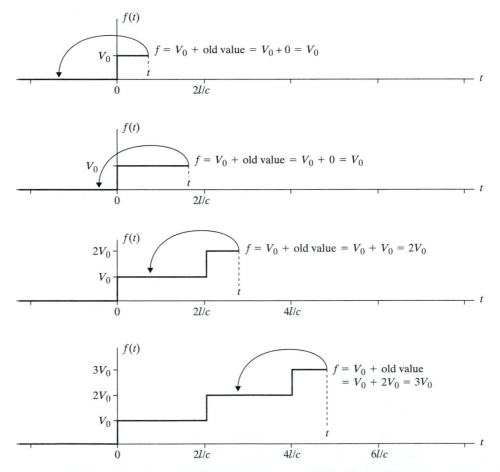

Figure 5-14   Stepwise construction of the $f(t)$ waveform.

the first $2l/c$ seconds after $t = 0$ is just V_0. The function $f(t)$ therefore begins as a step function of height equal to the battery voltage.

Once time t gets past $2l/c$ by just a bit, however, the value of $f(t)$ becomes V_0 plus an earlier value that corresponds to a small positive argument, so that the earlier value was already V_0, giving $2V_0$ for the value of $f(t)$ at such t. This second step upward persists until time t crosses the $4l/c$ point, whereupon $f(t)$ becomes V_0 plus an earlier value that was $2V_0$, giving yet another step to the $3V_0$ level.

This process of finding the $f(t)$ waveform by construction can be carried on indefinitely and yields a staircase function for $t > 0$, with steps of height V_0 and duration $2l/c$ seconds, as shown in Figure 5-15. The $g(t)$ waveform can then be formed at once from $g(t) = -f(t - 2l/c)$, by just delaying the $f(t)$ waveform by one step width and inverting it; this is also shown in Figure 5-15. We can readily confirm by inspection that $f(t) + g(t)$ is a single step function of height V_0, as required by (5.113).

Now that we have the two waveforms $f(t)$ and $g(t)$ for all t, we need only delay the first and advance the second one by z/c to arrive at the two constituent waves $f(t - z/c)$ and $g(t + z/c)$. These are to be added to give the voltage $V(z, t)$ for all z (from $z = 0$ to $z = l$) and all t; they are to be subtracted (and divided by Z_0) to give the current $I(z, t)$ as well.

In Figure 5-16, we have plotted snapshots of both these waves at the specific instant $t = 9l/4c$, against all z. For the f waveform, the negative sign in the argument requires reversal as well as rescaling; for the g waveform, there is only rescaling to get the spatial plot. Of course only the range of z from $z = 0$ to $z = l$ is relevant; this is the region between the vertical lines in Figure 5-16. The rest of the plot becomes informative when we imagine time elapsing, so that the forward waveform slides to the right and the backward wave slides to the left. The portions of the two waveforms that appear between the vertical lines are the ones to add and subtract to form the voltage and current distributions along the line.

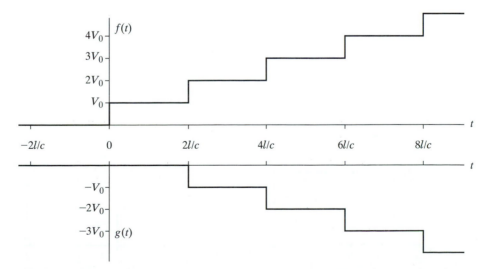

Figure 5-15   The $f(t)$ and $g(t)$ waveforms for the shorted transmission line.

For the specific time $t = 9l/4c$, adding the two waves between the vertical lines gives the voltage distribution $V(z, 9l/4c)$ and subtracting them yields Z_0 times the current distribution $I(z, 9l/4c)$; these are plotted in Figure 5-17. Near the generator, the current is $3V_0/Z_0$ but beyond $l/4$, the current is only $2V_0/Z_0$. The discontinuity occurs at location $ct - 2l$, which moves to the right as time elapses; this is suggested by the arrow attached to the step of current. Note that the voltage is indeed V_0 at $z = 0$ and zero at $z = l$, as it must be for all $t > 0$.

While this construction of the voltage and current distributions at a *specific* time illustrates the superposition of oppositely traveling waves that solves this, or any, transmission line problem, it offers little insight into how these distributions evolved into what we found. For that question, that of the transient problem, it is useful to follow the development of the voltage and current in a sequence of snapshots, as an aid to interpreting the solution.

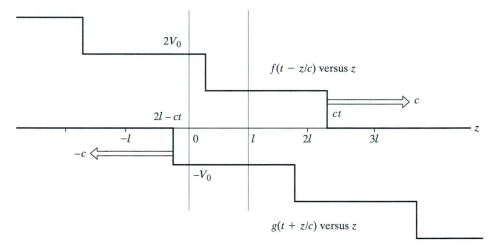

Figure 5-16   Snapshot of forward- and backward-traveling waves at $t = 9l/4c$.

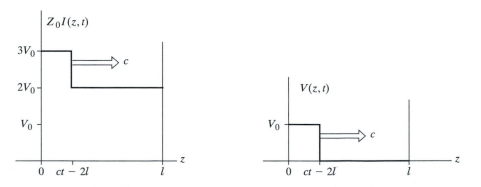

Figure 5-17   Current and voltage distributions along the shorted line, at $t = 9l/4c$.

Because the steps in the staircase waveforms $f(t)$ and $g(t)$ are of width $2l/c$ and the waves move toward each other, the step discontinuities will encounter each other every l/c seconds. We will therefore examine snapshots taken at instants within the first few time intervals of duration l/c.

The delay z/c of the forward-traveling wave $f(t - z/c)$ is limited to the range zero to l/c; the advance z/c of the backward-traveling wave $g(t + z/c)$ is in the same range. In the *first* time interval, $0 < t < l/c$, the argument of the f-waveform will range from $-l/c$ to l/c; the argument of the g-waveform will range from 0 to $2l/c$. In this time interval, therefore, the g-waveform remains zero and there is only the f-waveform; even the f-contribution is nonzero only for $z < ct$. The voltage, and also Z_0 times the current, is therefore a wave that travels from the battery to the short circuit, at speed c. This wave charges up the line to the amount CV_0 per unit length. At time t, the charging process has reached from the battery to the point $z = ct < l$. Figure 5-18 shows the voltage at time $3l/4c$; the current and the charge density have the identical appearance at this time.

At the end of this first time interval, the forward wave has reached the short circuit, leaving constant voltage $V(z, t) = V_0$ behind it. But this is not compatible with the presence of the short circuit at $z = l$, and the equal and opposite charges at the end of the line combine at the shorted end, discharging the line there, to zero volts. This is accomplished by the backward-traveling wave $g(t + z/c)$, superimposed on the forward one. In the *second* time interval $l/c < t < 2l/c$, the forward wave $f(t - z/c)$ has an argument between zero and $2l/c$, so that its value remains at V_0. In this same time interval, the backward wave $g(t + z/c)$ has an argument between l/c and $3l/c$. For locations z large enough for this argument to exceed $2l/c$, the g-wave has the value $-V_0$; nearer to the battery, the g-wave is still zero. The superposition is therefore such that the sum is zero for $z > 2l - ct$, while the difference is $2V_0$ there; for $z < 2l - ct$, the sum and difference are still V_0. This is shown in Figure 5-19, which is drawn for $t = 5l/4c$. The g-wave discharges the line as it travels from the short to the battery. This reduces the voltage to zero in its wake, but the discharge current is in the same direction as the charging current, so that the current near the short is doubled behind the backward wave.

When the discharging wave reaches the battery, at time $2l/c$, it has left behind it a line that has been completely discharged to zero volts. This is incompatible with the presence of the battery at $z = 0$ and a charging wave moves forward, again, from the battery to the shorted end, in the form of the next step of the f-wave. In the interval $2l/c < t < 3l/c$, this recharges the line to the voltage V_0, while adding yet another layer of current, as indicated in Figure 5-20, which is drawn as a snapshot taken at time $t = 11l/4c$.

The subsequent behavior should be clear. After the end of the third time interval, another discharging wave travels back to the battery, adding still more current. The waves that sweep back and forth along the line alternately charge and discharge it; each

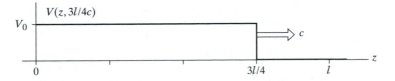

Figure 5-18   The source battery emits a wave that charges the line.

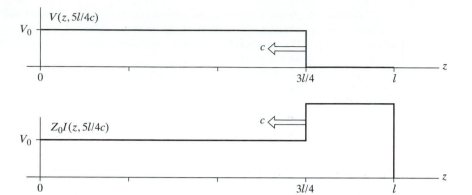

Figure 5-19   The backward wave discharges the line but adds to the current.

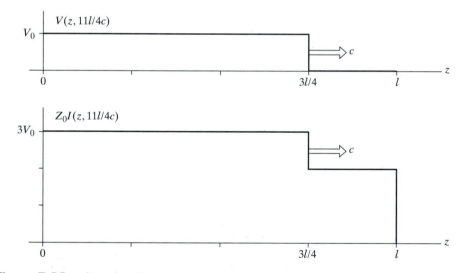

Figure 5-20   A recharging wave restores the voltage and adds a layer of current.

charging and discharging sweep, however, adds more current, in steps of V_0/Z_0 amperes. The current builds up indefinitely, in steps.

The same superpositions of forward and backward waves can be plotted as functions of time, at a particular observation point along the line, as the voltage and current that might be displayed on an oscilloscope with a probe at that point. The time development of the voltage and current at the point a distance $3l/4$ along the line is plotted in Figure 5-21.

Why the voltage appears as a periodic sequence of pulses and the current as a rising series of short and long steps can be understood in terms of the traveling waves that superimpose to form the voltage and current. After the switch is closed at $t = 0$, the observer at $z = 3l/4$ sees no change in the former zero voltage and current until the charg-

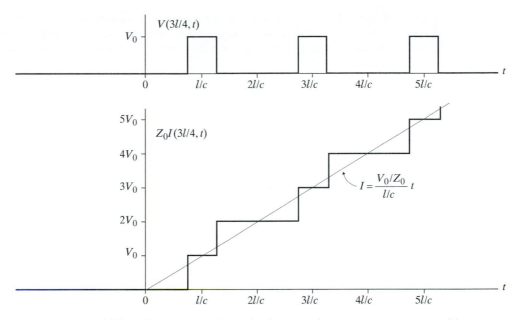

Figure 5-21   Evolution in time of voltage and current seen at $z = 3l/4$.

ing wave reaches the remote observation point, at time $t = 3l/4c$. The voltage then jumps to V_0 and the current to V_0/Z_0; these values are then maintained while the wave continues charging the remaining quarter of the line. There is also no further change until the discharge wave travels back toward the battery and reaches the observation point, at time $t = 5l/4c$. At that time, the voltage drops back to zero but the discharge current adds to the previous charging current, so that the current steps up to $2V_0/Z_0$. It remains at that level while the discharge wave travels the remaining three quarters of the line to the battery, and continues to maintain itself until the next charging wave covers the same distance back to the observation point, reaching it at time $t = 11l/4c$. The voltage is then restored and still more current is added at that time. This process continues indefinitely; the current at any point between the battery and the short increases in steps, each of V_0/Z_0 amperes, and each such step takes, on average, a time l/c.

On average, there is a linear increase in the current, which is indicated in Figure 5-21 by the straight line that passes through the center point of each step. The slope of this line is the step size, V_0/Z_0, divided by the average time per step, l/c. Such a linear ramp in response to a constant applied voltage is the behavior of an ideal, lumped inductor, such as L_0 in $dI/dt = V_0/L_0$. In fact, the ramp that averages out the steps in the rise of the current in the shorted line has a slope

$$\Delta I/\Delta t = (V_0/Z_0)/(l/c) = [V_0/\sqrt{L/C}]/[\sqrt{LC}l] = V_0/Ll. \qquad (5.116)$$

This corresponds to the rate of increase of current in an inductor of lumped inductance Ll; since L is the inductance per unit length of the line of length l, the average rate of rise is consistent with the line's acting as an inductor. In fact, the shorted line has the physical appearance of a one-turn coil. However, the actual current is a series of small steps, rather than a ramp, because of the distributed capacitance of the line. The battery does

> not really see a short circuit; it sees a distributed inductance and capacitance. The short circuit at the end of the line makes the structure primarily a one-turn inductor, modified by the capacitance between the wires. The current does rise indefinitely (until the wires melt!).

INCIDENT AND REFLECTED WAVES

The general solution to the transmission line equations is a superposition of two waves, one traveling forward and the other backward, both at the same characteristic speed of the line. The voltage distribution is the sum of the two waves; the characteristic impedance times the current distribution is the difference of the same two waves.

In the special case of a matched line, for which the load resistance exactly equals the characteristic impedance, there is only the forward-traveling wave. This is emitted by the generator and, after the propagation delay, is absorbed by the matched load. No backward-traveling wave is generated in the matched line. For any other load, the voltage and current at any point along the line is a superposition of two oppositely traveling waves.

As we discovered in the case of the short-circuited line, the transient development of the voltage and current distributions begins as it would if the line were matched. That is, the source emits a forward-traveling wave and that is all there is along the line, with the voltage and current distributions related by the characteristic impedance Z_0 of the line. This condition persists until the forward-traveling wave reaches the load, whereupon the incompatibility between the ratio of voltage to current in the forward wave and the ratio required by the mismatched load has to be resolved.

The resolution is achieved by the appearance at the load of the backward-traveling wave, with just the right amplitude and waveform to make the ratio of the sum of the two waves (the voltage) to their difference (Z_0 times the current) conform precisely to the requirements of the load. The backward-traveling wave is generated at the load, in response to the arrival of the forward wave there. We can think of the load as receiving the incoming wave and spitting back a wave of its own, to enforce the relationship between total voltage and total current that the load demands.

In these terms, the forward-traveling wave, from generator to load, is called the *incident* wave and the backward-traveling one is the *reflected* wave, bounced back by the mismatched load toward the generator. The resultant voltage distribution is the sum of the incident and reflected waves; the characteristic impedance times the current distribution is the difference of the same incident and reflected waves.

Because the reflected wave is generated at and by the load, it is often convenient to reexpress both the incident and the reflected waves in terms of coordinates appropriate to the location of the load, to make it easier to determine the waveform of the reflected wave. Instead of expressing the incident wave as $f(t - z/c)$ and the reflected wave as $g(t + z/c)$, which makes the waveforms $f(t)$ and $g(t)$ represent the signals seen at $z = 0$ (at the generator), we often prefer to write these two functions of z and t as incident wave: $f(t + [l - z]/c)$,

reflected wave: $g(t - [l - z]/c)$.

This makes the waveforms $f(t)$ and $g(t)$ be the signals observable at $z = l$ (at the load), instead of at the generator. We have simply traded the coordinate z, which measures distance from the generator, for the alternative coordinate $l - z$ that measures distance from the load.

Don't allow the signs in the arguments of the two functions to confuse you as to which way each one travels: With the minus sign between t and $l - z$, we have a wave that travels in the direction of increasing values of the coordinate $l - z$, which means in the direction from the load to the generator; this is the reflected wave. With the plus sign between t and $l - z$, we have a wave that travels in the direction of decreasing values of the coordinate $l - z$, which means in the direction from the generator to the load; this is the incident wave. Whenever there is any doubt as to the direction of travel, examine which way one would have to run to keep up with a fixed argument of the function; for $f(t + [l - z]/c)$, as time t goes on, we have to increase z (reduce $[l - z]$) to keep up with the wave, so this is the forward-traveling wave.

The advantage of this change of coordinate from z to $l - z$ is that it simplifies the description of the incident and reflected waves as seen at the load. At the load (i.e., at $z = l$), the voltage and current are, in this coordinate system,

$$V(l, t) = f(t) + g(t), \tag{5.117}$$

$$Z_0 I(l, t) = f(t) - g(t). \tag{5.118}$$

If the load is a pure resistance R not matched to the line, then the load requires

$$V(l, t) = RI(l, t). \tag{5.119}$$

Dividing the two previous equations gives

$$R/Z_0 = [f(t) + g(t)]/[f(t) - g(t)], \tag{5.120}$$

which inverts to

$$g(t) = \frac{R - Z_0}{R + Z_0} f(t). \tag{5.121}$$

This states that, for the case of a purely resistive mismatched load, the reflected waveform (as seen at the load) is just proportional to the incident waveform (as seen at the load), with *reflection coefficient*

$$\Gamma = \frac{R - Z_0}{R + Z_0}. \tag{5.122}$$

This gives the reflected waveform $g(t) = \Gamma f(t)$ in terms of the incident one, $f(t)$, at the load.

Elsewhere than at the load, the transmission line equations require the waveforms $f(t)$ and $g(t)$ to become the oppositely traveling waves

$$f(t + [l - z]/c) \quad \text{and} \quad g(t - [l - z]/c).$$

The voltage and current distributions are

$$V(z, t) = f(t + [l - z]/c) + g(t - [l - z]/c), \tag{5.123}$$

$$Z_0 I(z, t) = f(t + [l - z]/c) - g(t - [l - z]/c). \tag{5.124}$$

Note the following special cases of resistive loads.

a. If R is selected to be Z_0, then $\Gamma = 0$ and there is no reflected wave. The line is matched; all takes place along the line (from $z = 0$ to $z = l$) as if the line were infinitely long (meaning that the reflected wave one might otherwise expect never makes its appearance).

b. If $R = 0$ (a shorted line), then $\Gamma = -1$. The reflected wave at the load exactly cancels the incident one there, so that the voltage, which is the sum of the two, is maintained at zero at the short. But the current, which is the difference of the two waves, is doubled at the short. Elsewhere than at the short, however, the reflected wave does not cancel the incident one, because they travel in opposite directions: With $\Gamma = -1$, the voltage and current along the line are given by

$$V(z, t) = f(t + [l - z]/c) - f(t - [l - z]/c), \tag{5.125}$$

$$Z_0 I(z, t) = f(t + [l - z]/c) + f(t - [l - z]/c). \tag{5.126}$$

and the incident waveform, which is $f(t)$ only at the load, is still to be found, from the generator voltage or current at $z = 0$.

c. If $R = \infty$ (an open line), then $\Gamma = +1$. The reflected wave at the load exactly equals the incident one there, so that the current, which is the difference of the two, is maintained at zero at the open circuit. But the voltage, which is the sum of the two waves, is doubled at the open end. Elsewhere than at the open load, however, the reflected wave does not equal the incident one, since they travel in opposite directions: With $\Gamma = +1$, the voltage and current along the line are given by

$$V(z, t) = f(t + [l - z]/c) + f(t - [l - z]/c), \tag{5.127}$$

$$Z_0 I(z, t) = f(t + [l - z]/c) - f(t - [l - z]/c). \tag{5.128}$$

and the incident waveform, which is $f(t)$ only at the load, remains to be found, from the generator voltage or current at $z = 0$.

Example 5.6

One way to produce a narrow pulse (duration of the order of nanoseconds) is to charge an open-circuited transmission line (characteristic impedance Z_0, wave speed c, length l; switch S open, as in Figure 5-22), slowly, from a dc voltage source V_0, through a high resistance R_0. After the line is fully charged to V_0, the source is disconnected (switch S_0 opened) and then output switch S is closed suddenly to connect a matched load resistor $R = Z_0$ to the line. The intent is that the discharge current produce a narrow pulse across the load resistor.

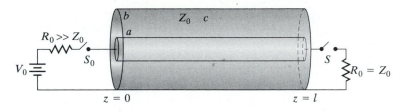

Figure 5-22   Apparatus for generating a narrow pulse.

Figure 5-22 depicts a coaxial line as the transmission line. This is a variety of two-conductor line in which one cylindrical wire is surrounded by a coaxial cylindrical shell that not only carries the return current but also shields the field region between the two conductors. The details of the properties of a coaxial line will be considered later; for present purposes, it is merely a transmission line with the given characteristic impedance Z_0 and wave speed c.

Consider the charging process (S open; S_0 closed at time $t = 0$; coaxial line initially uncharged). The one-way transit time along the coaxial line is $T = l/c$.

(a) Verify that the incident and reflected waves $f(t)$ and $g(t)$ at $z = 0$ are a sequence of unequal steps, such that in the time interval $2nT < t < 2(n + 1)T$, $f(t) = \frac{1}{2}V_0[1 - \Gamma^{n+1}]$, $g(t) = \frac{1}{2}V_0[1 - \Gamma^n]$. What is Γ?

(b) If $V_0 = 1000$ volts, $R_0 = 20Z_0$, and $T = 5$ nsec, how long must the charging phase continue in order to attain or exceed 999 volts at $z = 0$? «»

We are here using $f(t - z/c)$ and $g(t + z/c)$ as the forward and backward waves, so that $f(t)$ and $g(t)$ are the waves seen at the source. The sum of the two is the voltage; the difference is Z_0 times the current.

(a) At the open circuit at $z = l$, with $T = l/c$, we need

$$f(t - T) = g(t + T) \tag{5.129}$$

for all t to get zero current. At $z = 0$ we have $V_0 - R_0 I(0, t) = V(0, t)$ or

$$V_0 - (R_0/Z_0)[f(t) - g(t)] = f(t) + g(t). \tag{5.130}$$

Combining these two equations for f and g, we get a single equation for f:

$$f(t) = U + \Gamma f(t - 2T), \tag{5.131}$$

where

$$U = V_0[Z_0/(R_0 + Z_0)] \tag{5.132}$$

and

$$\Gamma = (R_0 - Z_0)/(R_0 + Z_0). \tag{5.133}$$

We can construct $f(t)$ from the functional equation in the same way we did for the shorted line. Since there was no $f(t)$ before $t = 0$, this waveform begins as a step of height U. Just after time $2T$, $f(t)$ becomes $U + \Gamma U$. Continuing the process of constructing $f(t)$, we find steps of unequal size. The first few of these steps attain the levels $U, U + \Gamma U, U + \Gamma(U + \Gamma U)$, etc. Let f_n and g_n be the levels of $f(t)$ and $g(t)$ in the time interval $2nT < t < 2(n + 1)T$. Then

$$f_n = U[1 + \Gamma + \Gamma^2 + \Gamma^3 + \cdots + \Gamma^n] = U[(1 - \Gamma^{n+1})/(1 - \Gamma)]$$

$$g_n = U[1 + \Gamma + \Gamma^2 + \Gamma^3 + \cdots + \Gamma^{n-1}] = U[(1 - \Gamma^n)/(1 - \Gamma)].$$

But $U/(1 - \Gamma) = V_0[Z_0/(R_0 + Z_0)]/[2Z_0/(R_0 + Z_0)] = \frac{1}{2}V_0$ as asserted.

(b) The input voltage is $V(0, t) = f(t) + g(t)$, so that in the nth time interval,

$$V(0, t) = \frac{1}{2}V_0[1 - \Gamma^{n+1} + 1 - \Gamma^n] = V_0[1 - \frac{1}{2}(1 + \Gamma)\Gamma^n].$$

We need n such that $\frac{1}{2}(1 + \Gamma)\Gamma^n < 0.001$. But $\Gamma = 19/21$, so that we solve $(21/19)^n \geq 20000/21$ to get $n \geq 68.5$ and $t = 2nT = 69(2T) = 690$ nsec.

Consider now the discharge process in the same pulse-forming apparatus.

Example 5.7

Assume the charging process has been completed in the apparatus of the previous example. Consider the discharging phase (S_0 open; line initially charged uniformly to V_0; switch S closed at time $t = 0$), beginning at a new time $t = 0$. Note that the boundary conditions include the open circuit at $z = 0$ and the matched load at $z = l$, while the initial conditions are that, at $t = 0$, the voltage along the line is V_0 and the current is zero.

(a) *Determine the duration T_1 and pulse strength V_1 of the output pulse, in relation to the parameters of the coaxial line and of the dc source.*

(b) *If $V_0 = 1000$ volts and $T = 5$ nsec, what are the duration T_1 and height V_1 of the output pulse?* ◈

(a) Using $f(t - z/c)$ and $g(t + z/c)$ as the incident and reflected waves, we impose $f(t) - g(t) = 0$ at the open circuit for all t and $g(t + T) = 0$ for $t > 0$ at the load. The initial conditions are $f(-z/c) + g(z/c) = V_0$ for the voltage condition and $f(-z/c) = g(z/c)$ for the current, for $0 < z/c < T$.

These equations yield $f(t) = g(t) = V_0/2$ for $-T < t < T$ and $f(t) = g(t) = 0$ for all other t. The output pulse is $V(l, t) = f(t - T) + g(t + T) = f(t - T) = V_0/2$ for $0 < t < 2T$ and zero thereafter. Thus, $V_1 = V_0/2, T_1 = 2T = 2l/c$.

(b) $V_1 = 500$ volts, $T_1 = 10$ nsec.

What this apparatus allows us to do is to charge the line slowly and to discharge it quickly.

The source that excites the transmission line need not be a voltage or current source. In the next example, it is a charged capacitor.

Example 5.8

A capacitor C_0 is initially charged to a voltage V_0 and is connected through an initially open switch to a transmission line of characteristic impedance Z_0, propagation speed c, length l, and matched load impedance. The switch is closed at time $t = 0$. See Figure 5-23.

(a) *How does the matching load impedance depend on the value of C_0?*

(b) *Obtain the voltage $V(t)$ across the matched load impedance, for all t.*

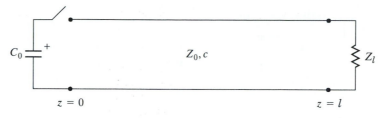

Figure 5-23 Capacitor connected to transmission line.

(c) *Find every instant of time (after t = 0) at which the voltage across the load attains $\frac{1}{2}V_0$.* «»

(a) The load impedance that matches the line is just Z_0, regardless of the source, and so is independent of C_0.

(b) The input impedance of the *matched* line is Z_0, so that the capacitor discharges through an equivalent resistance Z_0, with time constant Z_0C_0. The input voltage is therefore $V(0, t) = V_0 e^{-t/Z_0C_0}$ for $t > 0$ and this is delayed by l/c before appearing across the load. The load voltage $V(t) = V(l,t)$ is therefore $V(t) = 0$ for $t < l/c$ and $V(t) = V_0 e^{-(t-l/c)/Z_0C_0}$ for $t > l/c$.

(c) $V(t)$ attains $V_0/2$ twice; once in rising suddenly from zero to V_0 at time l/c and again when $e^{-(t-l/c)/Z_0C_0} = 0.5$ during the discharge phase. This is at time $t = l/c + Z_0C_0\ln2$.

The source that excites the transmission line need not be a constant one. In the next example, it is a pulse generator.

Example 5.9

A parallel-wire transmission line of length l, characteristic impedance Z_0, and wave speed c is bent into a circle and the two ends joined together, as in Figure 5-24. A voltage source $V_0(t)$ is connected at the point where the ends meet; it produces a single, narrow pulse of unit height at time t = 0. Assume that bending the line into a circle does not affect the characteristics and operation of the transmission line. Obtain the current $I_0(t)$ drawn from the source. «»

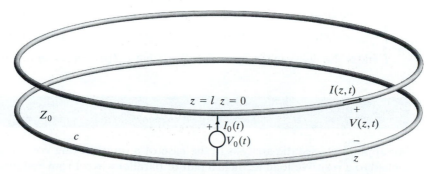

Figure 5-24 Parallel-wire transmission line with ends joined.

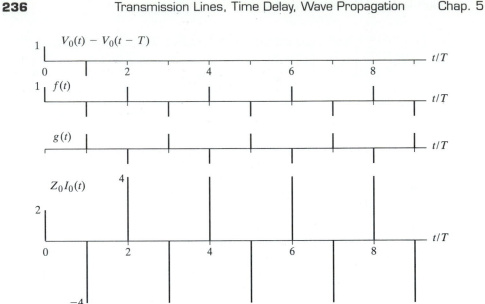

Figure 5-25   Pulse train on transmission line joined at its ends.

Using $f(t - z/c)$ and $g(t + z/c)$ as the incident and reflected waves, the conditions are that $f(t) + g(t) = V_0(t)$ because the source is at $z = 0$ and that $f(t - T) + g(t + T) = V_0(t)$ because the source is also at $z = l = cT$.

From the second of these two equations we get

$$g(t) = V_0(t - T) - f(t - 2T) \qquad (5.134)$$

and then, from the first,

$$f(t) = V_0(t) - V_0(t - T) + f(t - 2T). \qquad (5.135)$$

Since $V_0(t) - V_0(t - T)$ is here a pair of equal and opposite narrow pulses spaced T apart, we readily construct $f(t)$ as in Figure 5-25. Then we get the reflected wave as $g(t) = V_0(t) - f(t)$. The current supplied by the source is the difference between the current that enters the line at $z = 0$ and the current that arrives at $z = l$. It is therefore given by

$$Z_0 I_0(t) = f(t) - g(t) - f(t - T) + g(t + T) = 2f(t) - 2f(t - T), \qquad (5.136)$$

and is also shown in the figure; $Z_0 I(0) = 2$ but $Z_0 I(nT) = \pm 4$ for $n > 0$.

SUMMARY

We applied quasistatic analysis to the case of a time-varying voltage generator attached to a resistive load through a pair of parallel wires. From assumed charge and current distributions, we evaluated and superimposed the electric and magnetic fields generated by each of the wires and integrated these to relate the voltage and flux distributions to those charges and currents. This yielded a pair of distributed parameters of the parallel wires, their inductance L and capacitance C, per unit length.

Applying Maxwell's equations, we found the transmission line equations in differential form as

$$\frac{\partial V(z,t)}{\partial z} = -L \frac{\partial I(z,t)}{\partial t}, \tag{5.137}$$

$$\frac{\partial I(z,t)}{\partial z} = -C \frac{\partial V(z,t)}{\partial t}, \tag{5.138}$$

and in integral form as

$$V(z,t) = V(0,t) - \partial/\partial t \int_0^z LI(z',t)dz', \tag{5.139}$$

$$I(z,t) = I(0,t) - \partial/\partial t \int_0^z CV(z',t)dz'. \tag{5.140}$$

The latter are in a form ready-made for successive approximation, to extract the voltage and current distributions as infinite series of correction terms. Proceeding systematically, we were able to construct the entire infinite series and then to recognize them as Taylor series, which we could sum exactly.

The solution is simplest for the case of a matched line, whose load exactly matches the characteristic impedance $Z_0 = \sqrt{L/C}$ of the line. In that special case, we found that the matched transmission line conveys whatever signal is generated by the source to the load, without distortion, with only a delay. This represents action at a distance, with a delay. The delay corresponds to the time it takes to travel along the line at the characteristic speed $u_0 = 1/\sqrt{LC}$. The ideal, matched transmission line is simply a delay element.

The mathematical form of the solution for the matched line is that of a wave, like $f(t - z/u)$. This is any function whatever of a single argument, where that argument is formed from the difference between the time and a rescaled distance coordinate. Wave motion occurs because time does not stand still; it marches on, necessarily changing the wave's argument unless we move along the spatial coordinate at the characteristic speed that will maintain the argument fixed even as time elapses. To keep the argument $t - z/u$ fixed at any value ξ, we need to vary z as time elapses, at a rate such that $d\xi = 0 = dt - dz/u$, or at the speed $dz/dt = u$. If we move along z at that speed, we see a fixed value of the waveform, $f(\xi)$, despite the passage of time. Every point of the waveform therefore moves along z, at the constant speed u; that process is wave motion.

The general solution to the transmission line equations for a mismatched load was found to be a superposition of two waves, traveling in opposite directions along the line, both at the same characteristic speed. The voltage distribution $V(z,t)$ along the line is the sum of these two waves; the characteristic impedance times the current distribution $I(z,t)$ is the difference of the same two waves. The two waveforms are arbitrary functions; they are determined by the two conditions imposed by whatever may be attached to the two ends of the line, as generator and as load.

We examined at length the example of a short-circuited line to which a battery is suddenly applied. By enforcing the boundary conditions at the two ends of the line, we were able to construct the waveforms of both the incident and the reflected waves. The incident wave is emitted by the generator, travels along the line, eventually reaches the end, and encounters the short circuit. At that point, it bounces back, in the guise of the reflected wave, and upon reaching the source, bounces yet again as a later portion of the incident wave. The result is a wave that sweeps back and forth along the line, charging it in response to the applied battery voltage, and discharging it through the short circuit. The voltage at any location between the ends is a periodic sequence of pulses; the current rises, unsteadily, as a series of steps.

That Kirchhoff's laws, which predict infinite current immediately in this example, are useful at all stems from the brevity of the delays that they neglect. We found for the case of parallel wires that the speed of wave propagation is that of light in the medium around the wires. This makes the delays only nanoseconds, for people-sized structures. With waves bouncing back and forth this rapidly, the rise to large currents is perceived as nearly immediate. Nevertheless, for long structures or short time scales, the delays and the wave motion are crucial aspects of the operation of transmission lines to convey signals from one point to another.

We expressed the relation between the incident and reflected waves in terms of a reflection coefficient determined by the mismatch between the load and the characteristic impedance, for a resistive load. If the load is not purely resistive, the determination of the incident and reflected waves can become rather complicated and the transient formulation we have used becomes awkward. It may call for solution of differential equations, dictated by the relation between voltage and current imposed by the load circuitry. In such cases, it is expedient to recast the problem in the frequency domain, so that we deal with the sinusoidal steady state. That is what we turn our attention to next.

PROBLEMS

Electric field lines of parallel wires

5.1 We claim that the field lines of the electric field around two parallel wires, located at $\pm\hat{\mathbf{y}}(s/2)$, are given by the family of curves

$$r(\xi) = (s/2)[\hat{\mathbf{x}}\sin\theta + \hat{\mathbf{y}}\sinh\xi]/[\cosh\xi + \cos\theta],$$

in which the parameter θ distinguishes one field line from another, and that these curves are circles.

(a) Verify that $\mathbf{r}(\xi)$ is a field line, by confirming that $d\mathbf{r}/d\xi$ is parallel to the field $\hat{\mathbf{e}}_1/\rho_1 - \hat{\mathbf{e}}_2/\rho_2$ where $\hat{\mathbf{e}}$ is the unit vector away from either wire and ρ is the radial distance from the wire.

(b) Verify that $\mathbf{r}(\xi)$ traces out a circle as ξ varies, with the center of the circle at $\mathbf{c} = -(s/2)\hat{\mathbf{x}}\cot\theta$. What is its radius?

Magnetic field lines of parallel wires

5.2 We claim that the field lines of the magnetic field around two parallel wires, located at $\pm\hat{\mathbf{y}}(s/2)$ are given by the family of curves

$$\mathbf{r}(\theta) = (s/2)[\hat{\mathbf{x}}\sin\theta + \hat{\mathbf{y}}\sinh\xi]/[\cosh\xi + \cos\theta],$$

in which the parameter ξ distinguishes one field line from another, and that these curves are circles.

(a) Verify that $\mathbf{r}(\theta)$ is a field line, by confirming that $d\mathbf{r}/d\theta$ is parallel to the field $\hat{\mathbf{h}}_1/\rho_1 - \hat{\mathbf{h}}_2/\rho_2$, where $\hat{\mathbf{h}}$ is the azimuthal unit vector around either wire and ρ is the radial distance from the wire.

(b) Verify that $\mathbf{r}(\theta)$ traces out a circle as θ varies, with the center of the circle at $\mathbf{c} = (s/2)\hat{\mathbf{y}}\coth\xi$. What is its radius?

Capacitance of thick parallel wires

5.3 The field pattern around a pair of parallel thick wires can be obtained from that of infinitely thin ones, by positioning the thin line charges eccentrically with respect to the axes of the thick wires. The requirements on the field between the thick wires are that they satisfy Maxwell's equations, which the Coulomb field of the thin wires certainly do, and that the electric field be perpendicular to the conducting wire surfaces, which we claim can be achieved by proper positioning of the equivalent thin wires, as in Figure P5-1. The equivalent line charges carry the same charges per unit length $\pm q$ as the distributed ones on the thick wires.

The coordinate transformation

$$\mathbf{r}(\xi, \theta) = b[\hat{\mathbf{x}}\sin\theta + \hat{\mathbf{y}}\sinh\xi]/[\cosh\xi + \cos\theta]$$

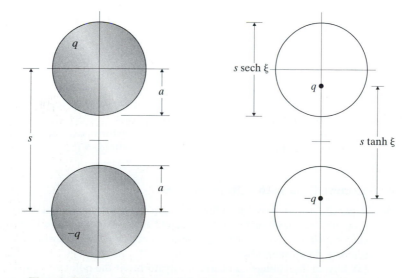

Figure P5-1   Thick parallel wires and equivalent line charges.

applied in the last two problems to a pair of parallel thin wires separated by $2b$ was shown to furnish field lines $\partial\mathbf{r}(\xi, \theta_0)/\partial\xi$ parallel to the electric field

$$\mathbf{E} = (q/2\pi\varepsilon_0)[\hat{\mathbf{e}}_1/\rho_1 - \hat{\mathbf{e}}_2/\rho_2]$$

of the thin wires. It was also shown that the curves $\partial\mathbf{r}(\xi_0, \theta)/\partial\theta$ are circles eccentric with the line charges at $y = \pm b$. These circles could be fitted to the cross section of the thick wires by proper choice of b and ξ_0. There remains to verify that the field lines $\partial\mathbf{r}/\partial\xi$ are perpendicular to the circles $\partial\mathbf{r}/\partial\theta$ defined by the appropriate ξ's to fit the thick wires, as suggested in Figure P5-1. The thick wires have radius a and their axes are separated by distance s.

(a) Confirm that the curves $\partial\mathbf{r}/\partial\xi$ and $\partial\mathbf{r}/\partial\theta$ are orthogonal where they meet.
(b) Obtain the exact capacitance per unit length $C = q/V$ by obtaining the emf V from the upper thick wire to the lower one.

Transmission line solution by Laplace transform

5.4   For readers sufficiently familiar with Laplace transform techniques, there is an easy alternative to the Taylor series summation in the text. For the matched line of length l, using the abbreviations $T = \sqrt{LCl}$ and $Z_0 = \sqrt{L/C} = R_l$, the infinite-order differential equation involving $V(t)$ and $I(t)$ appears as

$$V(l, t) = V - [T]d(Z_0 I)/dt + [T^2/2!]\,d^2V/dt^2 - [T^3/3!]d^3(Z_0 I)/dt^3 + \cdots$$
$$= R_l I(l, t) = Z_0 I - [T]\,dV/dt + [T^2/2!]d^2(Z_0 I)/dt^2 - [T^3/3!]\,d^3 V/dt^3 + \cdots.$$

Upon Laplace transformation of the equation, with zero initial conditions before time $t = 0$ for both $V(t)$ and $I(t)$ and with the Laplace transforms of $V(t)$ and $I(t)$ denoted $V'(s)$ and $I'(s)$, the equation becomes

$$V'(s) - [sT]Z_0 I'(s) + [s^2 T^2/2!]V'(s) - [s^3 T^3/3!]Z_0 I'(s) + \cdots =$$
$$Z_0 I'(s) - [sT]V'(s) + [s^2 T^2/2!]Z_0 I'(s) - [s^3 T^3/3!]V'(s) + \cdots$$

or

$$[1 + sT + s^2 T^2/2! + s^3 T^3/3! + \cdots]V'(s) = [1 + sT + s^2 T^2/2! + s^3 T^3/3! + \cdots]Z_0 I'(s).$$

This shows at once that $V'(s) = Z_0 I'(s)$.

(a) Obtain the Laplace transform $V'(l, s)$ of the load voltage $V(l, t)$, in terms of the transformed source voltage $V'(s)$.
(b) From known properties of the Laplace transform, express the load voltage $V(l, t)$ in the time domain in terms of the source voltage $V(t)$.

Open-circuited line with voltage source

5.5   A parallel-wire transmission line of length l, characteristic impedance Z_0, wave speed c is open circuited at the far end $z = l$ and has a constant voltage V_0 applied to it at $z = 0$ via a switch. The line is uncharged before time $t = 0$ and the switch is closed at $t = 0$ and remains closed thereafter.

(a) Sketch the voltage and current waveforms at the point $z = (4/5)l$.
(b) What are the time-averaged values of the voltage and current at that point?

Short-circuited line with current source

5.6   A parallel-wire transmission line of length l, characteristic impedance Z_0, wave speed c is short circuited at the far end $z = l$ and has a constant current I_0 applied to it at $z = 0$ via a switch. The line is uncharged before time $t = 0$ and the switch is closed at $t = 0$ and remains closed thereafter.

    (a) At any location z strictly between the source and the short, the voltage becomes a periodic sequence of positive and negative pulses. Give the pulsewidth, height, and time at its center for the first positive pulse and for the first negative pulse.

    (b) At any location z strictly between the source and the short, the current becomes a periodic sequence of pulses. The pulse shape will be a narrower flat-top pulse on top of a broader one. Give the pulsewidth, height, and time at its center for the first pulse, for both its broad and its narrower portions.

    (c) What are the time-averaged values of the voltage and current at z?

Exceeding rated voltage on an open-circuited line with current source

5.7   A parallel-wire transmission line of length l, characteristic impedance Z_0, wave speed c is open circuited at the far end $z = l$ and has a constant current I_0 applied to it at $z = 0$ via a switch. The line is uncharged before time $t = 0$ and the switch is closed at $t = 0$ and remains closed thereafter. The line is rated for a maximum allowable voltage of V_{max}. At which end of the line and at what time will the voltage first exceed the rated value?

Bounce diagram for propagation of voltage discontinuity

5.8   The countertraveling wave solution of the transmission line problem applies to discontinuous waveforms as well as to all others. When discontinuity is the only feature of the waveform applied to the line, it may be convenient to track the progress of the discontinuities in the incident and reflected waves by use of a bounce diagram.

    The bounce diagram is merely a space versus time plot with zigzag lines of slope $\pm c$ (or $\pm u_0$) representing the location of the discontinuity as the waves bounce back and forth between the mismatched ends of the line. The size of the discontinuity after each bounce is the reflection coefficient times the previous size of the discontinuity. For a constant source, the level attained after each discontinuity persists until the next discontinuity arrives. This makes it easy to trace the waveform at any location between the ends of the line.

    Consider the transmission line (Z_0, c) in Figure P5-2, loaded by resistance R_2 at the end $z = l = cT$. The reflection coefficient there is $\Gamma_2 = (R_2 - Z_0)/(R_2 + Z_0)$. The constant voltage V_0 from a battery of internal resistance R_1 is applied to the line at $z = 0$ through a switch closed at time $t = 0$.

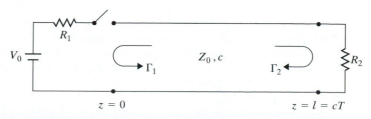

$z = 0$ $z = l = cT$

Figure P5-2   Propagation of discontinuity along mismatched line.

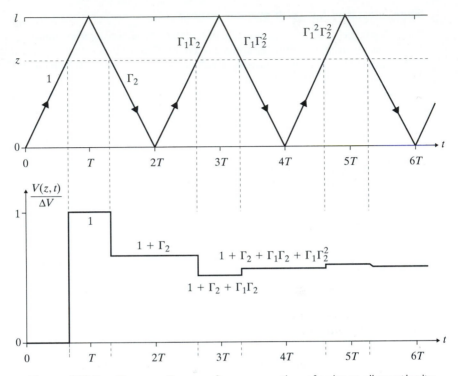

Figure P5-3   Bounce diagram for propagation of voltage discontinuity.

(a) Verify that the reflection coefficient at $z = 0$ is $\Gamma_1 = (R_1 - Z_0)/(R_1 + Z_0)$.
(b) Verify that the initial discontinuity applied at $z = 0$ is $\Delta V = V_0[Z_0/(R_1 + Z_0)]$.
    Figure P5-3 labels the successive parts of the zigzag line with the relative amplitudes of the discontinuities delivered by the incident and reflected waves. It also constructs the resultant voltage waveform at an intermediate point z along the line; the figure is drawn for the case $\Gamma_1 = 1/2, \Gamma_2 = -1/3$.
(c) What is the steady-state level of the voltage at z, in terms of $\Delta V, \Gamma_1, \Gamma_2$?
(d) What is the steady-state level of the voltage at z, in terms of V_0, R_1, R_2, Z_0? How does the characteristic impedance Z_0 affect the steady-state level?

Bounce diagram for propagation of current discontinuity

5.9   For the same conditions as in the previous problem, the bounce diagram can be adapted to construct the current waveform, by recognizing that the reflection coefficients for the current are $-\Gamma_1$ and $-\Gamma_2$, instead of Γ_1, Γ_2. This results in the modified bounce diagram and current waveform construction shown in Figure P5-4.
(a) What is the steady-state level of the current at z, in terms of $\Delta V, \Gamma_1, \Gamma_2$?
(b) What is the steady-state level of the current at z, in terms of V_0, R_1, R_2, Z_0? How does the characteristic impedance Z_0 affect the steady-state level?
(c) For the specific case of $\Gamma_1 = 1/2, \Gamma_2 = -1/3$, at what time does the current at $z = (2/3)l$ attain, and thereafter remain at, a level within 10% of the steady-state level?

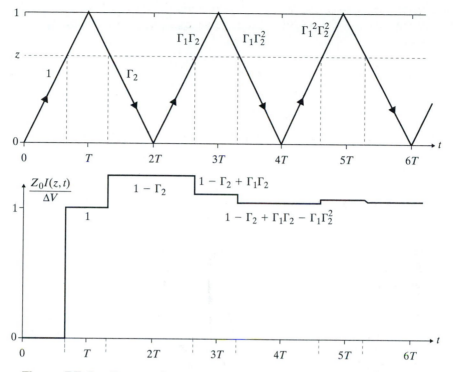

Figure P5-4 Bounce diagram for propagation of current discontinuity.

Parameters of wave propagation

5.10 An ideal transmission line with characteristic impedance Z_0 and wave speed c has a source at $z = 0$ and a load at $z = l$. The current along the line is known to be expressible in terms of two functions $I_1(\xi)$ and $I_2(\xi)$ and two positive constants α, β as

$$I(z, t) = I_1(\alpha t + \beta z) + I_2(\alpha t - \beta z).$$

(a) Which of the two terms represents a wave that travels from source to load?
(b) Find the ratio α/β.
(c) Write the corresponding expression for voltage $V(z, t)$, in terms of the functions I_1, I_2.

Properties of a current wave

5.11 A certain transmission line carries a current wave $I(z, t) = I_0 \,\mathrm{sech}\,(\alpha z + \beta t)$ between source and load, where α, β, I_0 are positive constants.
(a) In what direction does this wave propagate?
(b) At what speed does this wave propagate?
(c) On the basis of the preceding information, compare the load impedance at the end of the transmission line to the characteristic impedance of the line.

Solution of indeterminate circuit

5.12 Two ideal batteries of unequal voltages V_1 and V_2 and zero internal impedance are connected via switches to a pair of parallel wires (characteristic impedance Z_0, wave speed

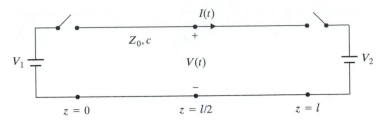

Figure P5-5   Classically indeterminate circuit.

c, length l, initially uncharged), as in Figure P5-5. The switches are closed simultaneously at time $t = 0$. Consider the point midway between the ends of the line and times that are multiples of $T = l/c$.

(a) Find the voltage $V(l/2, nT)$.

(b) Find the current $I(l/2, nT)$.

Transmission line with source between the ends

5.13 A two-wire transmission line (wave speed c, characteristic impedance Z_0) is loaded at both ends as shown in Figure P5-6 and is excited at its center point by a voltage source that generates a ramp pulse for time T, also shown.

(a) Write the general expressions for the voltage $v(z, t)$ and current $i(z, t)$, on both sides of the voltage source, without applying any boundary conditions.

(b) Sketch the voltage distribution $v(z, t_0)$ versus z, for all z from $z = 0$ to $z = l$, seen at a time t_0 such that $T < t_0 < l/2c$.

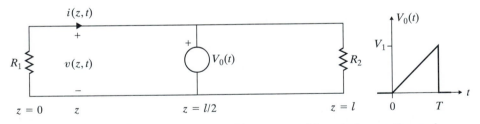

Figure P5-6   Transmission line with source midway between the ends.

Attempted discharge of uniformly charged line

5.14 A lossless, uniform line (wave speed c, characteristic impedance Z_0) of length cT is open circuited at both ends and is initially charged uniformly to voltage V_0. At time $t = 0$, the end at $z = cT$ is suddenly short circuited, in an attempt to discharge the line; see Figure P5-7.

(a) Sketch the voltage $V(0, t)$ at the open-circuited end, $z = 0$, for all time.

(b) Sketch the current $I(cT, t)$ at the short-circuited end, $z = cT$, for all time.

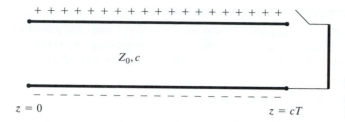

Figure P5-7   Initially uniformly charged transmission line is to be discharged.

Attempted discharge of oppositely charged lines

5.15 Two identical, lossless, uniform lines (wave speed c, characteristic impedance Z_0, length cT) are each open circuited at both ends and are initially charged uniformly but oppositely to voltages $\pm V_0$. At time $t = 0$, the two are suddenly joined, in an attempt to discharge the lines; see Figure P5-8.
   (a) Sketch the voltage $V(0, t)$ at the open-circuited end, $z = 0$, for all time.
   (b) Sketch the current $I(cT, t)$ at the junction, $z = cT$, for all time.

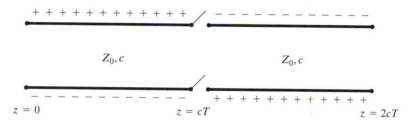

Figure P5-8   Initially oppositely charged transmission lines are to be discharged.

Parallel-wire line with ends joined

5.16 A parallel-wire transmission line of length l, characteristic impedance Z_0, and wave speed c is bent into a circle and the two ends joined together, as in Figure P5-9. A voltage source $V_0(t)$ is connected in series at the point where the ends meet; it produces a single, narrow pulse of unit height at time $t = 0$. Assume that bending the line into a circle does not affect the characteristics and operation of the transmission line. Obtain the current $I_0(t)$ drawn from the source.

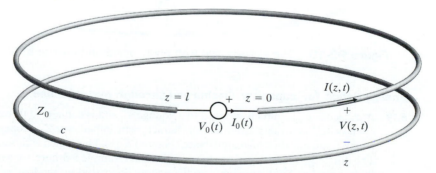

Figure P5-9   Parallel-wire transmission line with ends joined by source in series.

Resonant behavior of parallel-wire line with ends joined

5.17 For the parallel-wire transmission line bent into a circle and the two ends joined together, as in Example 5.9, consider a voltage source $V_0(t)$ that produces a train of narrow pulses of period $2T = 2l/c$ and unit height. Obtain the current $I_0(t)$ drawn from the source.

Matching of parallel-wire line with ends joined

5.18 For the parallel-wire transmission line bent into a circle and the two ends joined together, as in Example 5.9 (characteristic impedance Z_0, wave speed c, length $l = cT$), consider a voltage source $V_0(t)$ with an internal resistance R_0, as in Figure P5-10. The symmetry and the fact that the end of the line is also its beginning dictate that the forward- and the backward-traveling waves on the two sides of the junction point must be related, as shown in the figure.

     (a) Confirm that the forward-traveling wave at the junction point, $f(t)$, satisfies the functional equation $f(t) = AV_0(t) + Bf(t - T)$. What are the values of the two constants A, B?

     (b) For what special value of R_0 is the line "matched," in that no reflected wave is generated by R_0, leaving only $f(t) = CV_0(t)$? What is the constant C? Note that matching here does *not* mean that waves travel only away from the junction point; instead, it means that the wave that is incident onto the junction at $z = 0$ (which is also $z = l$) is entirely absorbed by R_0 and not reemitted.

     (c) Write the voltage $V(z, t)$ and scaled current $Z_0 I(z, t)$ at any point z along the line, under the matched condition, in terms of the source voltage $V_0(t)$.

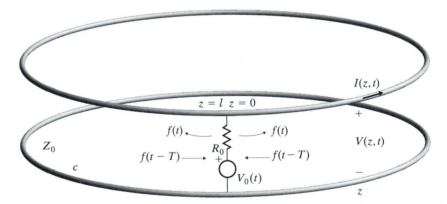

Figure P5-10   Parallel-wire line with ends joined and source to be matched.

Energy deficit for capacitor discharged through matched line

5.19 A capacitor C_0 is initially charged to a voltage V_0 and is connected through an initially open switch to a transmission line of characteristic impedance Z_0, propagation speed c, length l, and matched load impedance (Figure P5-11). The switch is closed at time $t = 0$. The discharge of the capacitor was examined in Example 5.8; here, we want to compare the energy lost by the capacitor to the energy dissipated by the load resistor and discover that they are not the same.

Recall that the capacitor energy is $\frac{1}{2}C_0V^2$, where $V = V(0, t)$ is the capacitor voltage at observation time t, and that the energy dissipated in the load resistor is the integral of V^2/R up to time t, where $V = V(l, t)$ is the voltage across the load resistor $R = Z_0$. For definiteness, let the length of the transmission line be $l = cZ_0C_0/4$ and set the observation time at $t = Z_0C_0$. Find the ratio of the energy dissipated in the load resistor by this time t to the energy drained from the capacitor by that time.

Note: The energy deficit will be explained and accounted for in Chapters 6 and 8.

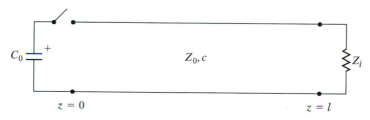

Figure P5-11   Capacitor connected to transmission line.

Impulse response of transmission line system

5.20   A transmission line of characteristic impedance Z_0 and wave speed c is loaded at $z = l$ by a resistance $R_2 = 2Z_0$ and is fed at $z = 0$ by a generator of internal impedance $R_1 = Z_0$, as in Figure P5-12. If the generator voltage $x(t)$ is the input and the voltage at $\frac{1}{2}l$ is the output $y(t)$, find the impulse response $h(t)$ of the system [the response to $\delta(t)$, a delta function input].

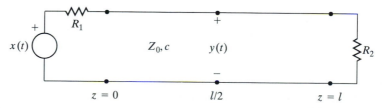

Figure P5-12   Transmission line system configuration for impulse response.

CHAPTER 6

Steady-State Wave Transmission and Plane Waves

We have found for the case of the parallel-wire transmission line that the speed of wave propagation along the line is $u_0 = 1/\sqrt{LC} = 1/\sqrt{\mu\varepsilon}$, which is the speed of light, $c = 1/\sqrt{\mu_0\varepsilon_0}$, if the medium around the wires is a vacuum. If the medium is a dielectric, with parameters μ, ε instead, the speed of propagation is correspondingly slower and the transmission delay longer. Our analysis showed that, for a matched line, there is no distortion of the shape, duration, or amplitude of any signal whatsoever transmitted along the line, only the delay that comes from propagation at a finite speed.

However, that conclusion is not realistic if the medium is not a vacuum. For any material medium, those parts of a signal that change rapidly and the portions that vary slowly are found to travel at different speeds along a transmission line. This results in distortion of the signal, as various parts of the signal arrive at different times. The medium behaves as if different values of μ and ε apply to different portions of the signal being transmitted.

This is indeed the case, and is a consequence of the fact that any material medium responds dynamically, not instantaneously, to applied fields. This means, for example, that we cannot simply assume that $\mathbf{D}(t) = \varepsilon\mathbf{E}(t)$ regardless of the time variation of $\mathbf{E}(t)$. Instead, the past history of $\mathbf{E}(\tau)$, for all $\tau \leq t$, contributes to the value of $\mathbf{D}(t)$ at the present time t, through a convolution of the electric field with the medium's impulse response. But we need not be frightened by this; Fourier analysis provides an easy way out, as follows.

SINUSOIDAL STEADY STATE

We know we can deal conveniently with dynamic linear systems of this sort by recasting the calculations from the time domain to the frequency domain, by Fourier analysis. This means expressing our signals as superpositions of sinusoidal oscillations of different frequencies and dealing with the system one frequency at a time. We refer to such individual oscillations as the sinusoidal steady state. The dynamic response is then embodied in that the proportionality between excitation and response becomes a function of frequency. This means that we can retain simple relations like **D** = ε**E** when the fields oscillate at a single frequency, but that the permittivity ε is a function of this frequency, not a constant.

Consideration of the sinusoidal steady state also allows us to deal easily with dynamic load impedances, which do not respond instantaneously like resistive loads, but include elements like capacitive or inductive reactances.

Finally, the sinusoidal steady state allows effective analysis of narrow-band signals, such as the amplitude-modulated sinusoidal carriers used in radio, microwave, and lightwave transmission.

COMPLEX EXPONENTIALS

The feature of the sinusoidal steady state that makes analysis so simple is that linear operations applied to a sinusoid of any single frequency result not in any other type of function but in a sinusoid again, and of exactly the same frequency, but with a new amplitude and phase. Such linear operations include not only amplification by a constant factor but also differentiations, integrations, delays, and combinations of these. Sinusoids, such as $f(t) = A \cos(\omega t + \varphi)$, survive linear operations unscathed, except for changes in their amplitude A and phase φ.

This feature is even more striking for exponential signals, like $g(t) = Ae^{st}$, which survive any linear operation unscathed, with only a change of amplitude A. Exponentials are *eigenfunctions* of linear, time-invariant systems, in that the response of such systems to any exponential is the same exponential function of time, except for a constant multiplicative factor called the *eigenvalue* (or transfer function). That sinusoids behave similarly (but change their phase as well as their amplitude) is because they can be expressed as exponentials, if we use imaginary exponents. The use of complex exponentials to represent sinusoidal oscillations allows changes in both the amplitude and phase of the sinusoid to be expressed by only a change of amplitude, at the cost of dealing with complex numbers as amplitude factors, rather than only real ones.

The exponential with pure imaginary exponent is actually a complex combination of two sinusoids (Euler's formula; see the Appendix in Chapter 10):

$$e^{j\omega t} = \cos \omega t + j \sin \omega t, \tag{6.1}$$

so that using the complex exponential entails carrying two real, sinusoidal signals throughout our calculations, simultaneously. A linear operation can result in multiplication of the complex exponential by a complex constant factor $A = ae^{j\theta}$ (with a and θ real), which is then interpreted in the time domain as acting separately on the two sinusoids: $e^{j\omega t} = \cos \omega t + j \sin \omega t$ becomes

$$Ae^{j\omega t} = ae^{j(\omega t + \theta)} = a \cos(\omega t + \theta) + ja \sin(\omega t + \theta). \tag{6.2}$$

Throughout a calculation involving only linear operations, we can always determine what the effect on the real sinusoid, either $\cos \omega t$ or $\sin \omega t$, has been by taking the real or the imaginary part of our complex expression. The need to deal with complex arithmetic is a small price to pay for the convenience of having all our linear operations expressed by simple multiplicative factors.

The only other cost of using complex exponentials rather than sinusoids is that we need to distinguish negative frequencies from positive ones. While using negative frequencies in the real expressions $\cos \omega t$ or $\sin \omega t$ would give us nothing new, the complex exponentials $e^{j\omega t}$ and $e^{-j\omega t}$ are independent signals, in that the latter is not expressible as some (complex) factor times the former. We will need both positive and negative frequencies in our superpositions of complex exponentials to form other signals. In fact, the real sinusoids are themselves superpositions of just two complex exponentials, the ones with the positive and negative versions of the actual frequency ω of oscillation:

$$\cos \omega t = \left[e^{j\omega t} + e^{-j\omega t} \right]/2, \qquad \sin \omega t = \left[e^{j\omega t} - e^{-j\omega t} \right]/2j. \tag{6.3}$$

Again, the need to include negative as well as positive frequencies in the spectrum is a tiny price to pay for the convenience of complex exponentials.

We will consider the steady state produced by sources of sinusoidal signals at frequency ω, also called harmonic signals, such as the voltage $v_0(t)$ with amplitude V_1 and phase θ_1

$$v_0(t) = V_1 \cos(\omega t + \theta_1) = \text{Re}\left[V_0 e^{j\omega t} \right], \tag{6.4}$$

which we give in terms of the complex amplitude

$$V_0 = V_1 e^{j\theta_1}, \tag{6.5}$$

in which V_1 and θ_1 are real but V_0 is complex. We express the signal as the complex voltage $V_0 e^{j\omega t}$, which includes $V_1 \cos(\omega t + \theta_1) + jV_1 \sin(\omega t + \theta_1)$ instead of only its real part. In effect, we are associating with our desired signal $v_0(t) = V_1 \cos(\omega t + \theta_1)$ another sinusoid at the same frequency, namely $v_{00}(t) = V_1 \sin(\omega t + \theta_1)$, to accompany the original signal $v_0(t)$. Using complex arithmetic, it is actually easier to deal with the pair of signals $v_0(t)$ and $v_{00}(t)$ than with just the first of the two.

Our electromagnetic fields will also be sinusoids at the same frequency ω; we express them by complex vector amplitudes. These comprise the amplitudes and phases of each vector component, as in

$$\mathbf{E} = \mathbf{E}(\omega) = E_1 e^{j\varphi_1}\hat{\mathbf{x}} + E_2 e^{j\varphi_2}\hat{\mathbf{y}} + E_3 e^{j\varphi_3}\hat{\mathbf{z}}. \tag{6.6}$$

We use a frequency-dependent permittivity, permeability, and conductivity to represent the dynamics of the medium's response to fields, as in

$$\mathbf{D}(\omega) = \varepsilon(\omega)\,\mathbf{E}(\omega), \qquad \mathbf{B}(\omega) = \mu(\omega)\,\mathbf{H}(\omega), \qquad \mathbf{J}(\omega) = \sigma(\omega)\,\mathbf{E}(\omega), \tag{6.7}$$

at frequency ω.

To interpret any result given in the frequency domain, we can revert to the time domain by attaching to the complex amplitude the harmonic factor $e^{j\omega t}$ and then taking the real part.

Example 6.1

A certain electric field in the sinusoidal steady state at frequency ω is expressed as $\mathbf{E} = (3 + j4)\hat{\mathbf{x}} + 12e^{j\pi/4}\hat{\mathbf{y}} - j\hat{\mathbf{z}}$. *Interpret this by giving the three components of the actual, physical electric field, in the time domain.* ◈

We form $\mathbf{E}e^{j\omega t}$ and take the real part. Recalling that

$$e^{j\omega t} = \cos \omega t + j \sin \omega t,$$

we get

$$E_x(t) = \mathrm{Re}\big[(3 + j4)e^{j\omega t}\big] = 3 \cos \omega t - 4 \sin \omega t;$$

$$E_y(t) = \mathrm{Re}\big[12e^{j\pi/4}e^{j\omega t}\big] = 12 \cos (\omega t + \pi/4);$$

$$E_z(t) = \mathrm{Re}\big[-je^{j\omega t}\big] = \sin \omega t.$$

Of course, the properties of trigonometric functions allow these to be rewritten in several ways, including $E_x(t) = 5 \cos (\omega t + 0.9273)$, where 5 is $\sqrt{3^2 + 4^2}$ and 0.9273 is $\tan^{-1}(4/3)$, and $E_y(t) = 8.4853[\cos \omega t - \sin \omega t]$ by use of $12 \cos \pi/4 = 12 \sin \pi/4 = 8.4853$.

SINUSOIDAL WAVES

Since a sinusoidal waveform is merely a special case of more general shapes, the result that the voltage and current distributions along a transmission line are comprised of waves that travel forward and backward is still valid in the sinusoidal steady state. But now we can add the stipulation that these two waveforms are themselves sinusoidal, at the same frequency, and will be represented by complex exponentials, through their complex amplitudes.

We should be aware, too, that the steady state implies that the forward and backward waves have been traveling back and forth forever, so that all the bouncing of waves at the two ends of the line has already taken place and the system has settled into its steady-state response. The only task remaining is to determine the two complex amplitudes to be assigned to the two traveling waves.

The general result is that the forward-traveling wave, seen as $f(t)$ at $z = 0$, becomes $f(t - z/u_0)$ elsewhere and that the backward-traveling wave, seen as $g(t)$ at $z = 0$, becomes $g(t + z/u_0)$ everywhere else. For the sinusoidal steady state, these statements become

$$f(t) = \Phi_0 \cos (\omega t + \varphi_0) = \mathrm{Re}\big[F_0 e^{j\omega t}\big] \quad \text{with} \quad F_0 = \Phi_0 e^{j\varphi_0}, \qquad (6.8)$$

so that

$$f(t - z/u_0) = \mathrm{Re}\big[F_0 e^{j\omega(t - z/u_0)}\big] = \mathrm{Re}\big[F_0 e^{-j\omega z/u_0} e^{j\omega t}\big]$$

$$= \mathrm{Re}\big[F(z)e^{j\omega t}\big] \quad \text{with} \quad F(z) = F_0 e^{-j\omega z/u_0}, \qquad (6.9)$$

which says that F_0 at $z = 0$ becomes $F(z) = F_0 e^{-j\omega z/u_0}$ elsewhere. Also,

$$g(t) = \Psi_0 \cos (\omega t + \psi_0) = \mathrm{Re}\big[G_0 e^{j\omega t}\big] \quad \text{with} \quad G_0 = \Psi_0 e^{j\psi_0}, \qquad (6.10)$$

so that

$$g(t + z/u_0) = \text{Re}\left[G_0 e^{j\omega(t+z/u_0)}\right] = \text{Re}\left[G_0 e^{j\omega z/u_0} e^{j\omega t}\right]$$

$$= \text{Re}\left[G(z)e^{j\omega t}\right] \quad \text{with} \quad G(z) = G_0 e^{j\omega z/u_0}, \quad (6.11)$$

which means that G_0 at $z = 0$ becomes $G(z) = G_0 e^{j\omega z/u_0}$ elsewhere. Writing, as an abbreviation, the constant

$$\beta = \omega/u_0 \quad (6.12)$$

allows us to state simply that the forward wave has complex amplitude

$$F(z) = F_0 e^{-j\beta z} \quad (6.13)$$

and that the backward wave has the amplitude

$$G(z) = G_0 e^{j\beta z}. \quad (6.14)$$

Correspondingly, the complex amplitudes of the voltage and current distributions along the line are given by the sum and difference of these complex numbers:

$$V(z) = F_0 e^{-j\beta z} + G_0 e^{j\beta z}, \quad (6.15)$$

$$Z_0 I(z) = F_0 e^{-j\beta z} - G_0 e^{j\beta z}. \quad (6.16)$$

The interpretation is, again, that the time-domain voltage is

$$v(z, t) = \text{Re}\left[V(z)e^{j\omega t}\right]$$

$$= \text{Re}\left[F_0 e^{j(\omega t - \beta z)} + G_0 e^{j(\omega t + \beta z)}\right], \quad (6.17)$$

and similarly for the current. This is a superposition of two oppositely traveling complex-exponential (sinusoidal) waves.

The sinusoidal, forward-traveling wave $e^{j(\omega t - \beta z)}$ is periodic, in both time and distance: It duplicates itself at time intervals $T = 2\pi/\omega$ apart and also at distance intervals $\lambda = 2\pi/\beta$ apart; this is because $e^{j\theta} = \cos\theta + j\sin\theta$ is periodic in θ, with period 2π. We need to become conversant with the words used to describe this sort of wave, as in the following list.

$$\omega = \text{frequency (radians/sec)}$$

$$\beta = \text{phase constant (radians/m)}$$

$$T = \text{period (sec)}$$

$$\lambda = \text{wavelength (m)}$$

$$\Phi(z, t) = \omega t - \beta z = \text{phase (radians)}.$$

The phase $\Phi(z, t) = \omega t - \beta z$ is a function of both z and t; this variable can be made to remain constant, even as time elapses, if observation position z is made to vary in time in such manner that

$$d\Phi = 0 = \omega \, dt - \beta \, dz, \quad \text{or} \quad dz/dt = \omega/\beta = u_0. \quad (6.18)$$

The speed ω/β at which one needs to move along the line to keep the phase fixed at some value is called the phase velocity, denoted v_p. It is the same as the characteristic velocity of the line. We see that the following relations hold:

$$v_p = u_0 = \omega/\beta = \lambda/T. \tag{6.19}$$

For the parallel-wire line, with inductance parameter $L = (\mu/2\pi)\ln(b/a)$ and capacitance parameter $C = 2\pi\varepsilon/\ln(b/a)$, this is

$$v_p = 1/\sqrt{LC} = 1/\sqrt{\mu\varepsilon} \tag{6.20}$$

and if the medium around the wires is a vacuum (air is almost the same), this is

$$v_p = 1/\sqrt{\mu_0\varepsilon_0} = c \quad \text{and} \quad \lambda = cT. \tag{6.21}$$

The wavelength and the period are related by the speed of light.

The operating frequency is most commonly cited not as ω in radians/sec but rather as f in cycles per second (hertz). The relationship is

$$\omega = 2\pi f, \tag{6.22}$$

because there are 2π radians in a cycle; correspondingly,

$$f = 1/T \quad \text{and} \quad f\lambda = c. \tag{6.23}$$

To cite yet another custom, if the medium of propagation is a vacuum, so that the phase velocity is the speed of light c, then we often denote the phase constant β by the letter k and the free-space wavelength by λ_0:

$$k = \omega/c = 2\pi/\lambda_0. \tag{6.24}$$

Finally, let us learn to recognize the nature of a pure imaginary exponential function of any distance coordinate x, such as $Ae^{-j\beta x}$, which may arise in the frequency domain. To interpret any such factor, we revert to the time domain by first attaching the $e^{j\omega t}$ time factor and then taking the real part. Assuming the complex coefficient A has magnitude $|A|$ and phase α, we get

$$f(x, t) = \text{Re}\left[Ae^{-j\beta x}e^{j\omega t}\right] = \text{Re}\left[|A|e^{j(\omega t - \beta x + \alpha)}\right]$$
$$= |A|\cos(\omega t - \beta x + \alpha) \tag{6.25}$$

and we recognize this as a sinusoidal wave, with amplitude $|A|$, initial phase α at $x = 0$, frequency ω, phase constant β, propagating along the direction of increasing values of x, at the speed ω/β. Henceforth, we will instantly recognize an exponential like $e^{-j\beta x}$ as a sinusoidal wave along x, at speed ω/β, without having to go through the details of reverting to the time domain.

TRANSMISSION LINES IN THE STEADY STATE

Because the reflected wave is determined from the incident one by the nature of the load at $z = l$, it is often convenient to use the alternative distance coordinate $l - z$, which represents the distance to the load, rather than the coordinate z that measures distance from the source at $z = 0$. The *incident* wave, which travels toward decreasing values of $l - z$ at speed $u_0 = \omega/\beta$, will appear in the frequency domain as a factor $e^{+j\beta(l-z)}$. The *reflected* wave, which progresses toward increasing values of the coordinate $l - z$, will be recognized by the factor $e^{-j\beta(l-z)}$. The voltage and current distributions are therefore expressed in the frequency domain as the complex amplitudes

$$V(z) = Fe^{+j\beta(l-z)} + Ge^{-j\beta(l-z)}, \tag{6.26}$$

$$Z_0 I(z) = Fe^{+j\beta(l-z)} - Ge^{-j\beta(l-z)}, \tag{6.27}$$

at any location z along the line. The complex voltage amplitude is the sum of the two wave amplitudes; Z_0 times the current amplitude is the difference of the same two wave amplitudes.

The quantities to be found from the boundary conditions at the two ends of the transmission line have been reduced to just the two complex constants, F and G. This is in contrast to the need to determine two full waveforms $f(t)$ and $g(t)$ for all time t when we face a transient problem, rather than the sinusoidal steady state.

We can now handle reactive loads as well as resistive ones with ease. Suppose the load at the end $z = l$ of the transmission line is expressed as a complex impedance Z_l. This will impose one condition on the two unknown constants F, G. We simply require that the complex voltage at the load be the load impedance Z_l times the complex current there. This calls for

$$V(l) = F + G \qquad \text{and} \qquad Z_0 I(l) = F - G \tag{6.28}$$

to have the ratio

$$\frac{V(l)}{I(l)} = Z_l = \frac{F + G}{(F - G)/Z_0}. \tag{6.29}$$

Inverting this relation, we get the ratio of G to F as $G/F = \Gamma_l$, where

$$\Gamma_l = \frac{Z_l - Z_0}{Z_l + Z_0}. \tag{6.30}$$

This ratio of the complex amplitude of the reflected wave to that of the incident wave, both as seen at the load, is the *voltage reflection coefficient at the load*. The name is appropriate because Γ_l gives the amount of reflected wave per unit of incident one in the sum that comprises the complex voltage amplitude:

$$V(l) = F[1 + \Gamma_l]. \tag{6.31}$$

For the current, we need the difference between the same two waves, so that the current reflection coefficient at the load is the negative of the one for the voltage. Rewriting the equations for the two distributions in terms of this ratio, we get

$$V(z) = F[e^{+j\beta(l-z)} + \Gamma_l e^{-j\beta(l-z)}], \tag{6.32}$$

$$Z_0 I(z) = F[e^{+j\beta(l-z)} - \Gamma_l e^{-j\beta(l-z)}]. \tag{6.33}$$

The remaining unknown F, the complex amplitude of the incident wave, can be determined from whatever information is given about the source at $z = 0$. For example, if the complex source voltage amplitude is given, as $V(0) = V_0$, we can solve for the unknown constant F and the complete distributions of voltage and current are given by

$$V(z) = V_0 \frac{e^{+j\beta(l-z)} + \Gamma_l e^{-j\beta(l-z)}}{e^{+j\beta l} + \Gamma_l e^{-j\beta l}}, \tag{6.34}$$

$$I(z) = \frac{V_0}{Z_0} \frac{e^{+j\beta(l-z)} - \Gamma_l e^{-j\beta(l-z)}}{e^{+j\beta l} + \Gamma_l e^{-j\beta l}}. \tag{6.35}$$

Even without using the specification of the source, we can determine the *impedance distribution along the line*, meaning the ratio of the complex voltage at any point to the complex current at that same point. This is

$$Z(z) = \frac{V(z)}{I(z)} = Z_0 \frac{e^{+j\beta(l-z)} + \Gamma_l e^{-j\beta(l-z)}}{e^{+j\beta(l-z)} - \Gamma_l e^{-j\beta(l-z)}}. \tag{6.36}$$

At the load, at $z = l$, this must equal the load impedance Z_l, which it does, because the ratio $[1 + \Gamma_l]/[1 - \Gamma_l]$ has been made to equal Z_l/Z_0. When evaluated at the source, at $z = 0$, this gives the *input impedance* of the loaded transmission line and tells us what load is being presented to the source. At any point z along the line, the impedance $Z(z)$ expresses the ratio of voltage to current that results from the superposition of the incident and reflected waves; *if there were only one wave, the impedance would be only Z_0, for all z.*

The important expression for the impedance along the line can be recast in terms of the load impedance Z_l directly. We simply replace the reflection coefficient by the ratio of the difference and sum of Z_l and Z_0 and then combine the exponentials with equal and opposite exponents into their trigonometric equivalents:

$$Z(z) = Z_0 \frac{e^{+j\beta(l-z)} + [Z_l - Z_0]e^{-j\beta(l-z)}/[Z_l + Z_0]}{e^{+j\beta(l-z)} - [Z_l - Z_0]e^{-j\beta(l-z)}/[Z_l + Z_0]}$$

$$= Z_0 \frac{Z_l[e^{j\beta(l-z)} + e^{-j\beta(l-z)}] + Z_0[e^{j\beta(l-z)} - e^{-j\beta(l-z)}]}{Z_0[e^{j\beta(l-z)} + e^{-j\beta(l-z)}] + Z_l[e^{j\beta(l-z)} - e^{-j\beta(l-z)}]}$$

$$= Z_0 \frac{Z_l 2\cos\beta(l-z) + Z_0 2j\sin\beta(l-z)}{Z_0 2\cos\beta(l-z) + Z_l 2j\sin\beta(l-z)}$$

$$= Z(z) = Z_0 \frac{Z_l + jZ_0 \tan\beta(l-z)}{Z_0 + jZ_l \tan\beta(l-z)}. \tag{6.37}$$

This comprehensive formula for the impedance at any point of a transmission line terminated with an arbitrary complex impedance Z_l is often expressed as giving the input impedance of a section of line of length l:

$$Z_{\text{in}} = Z_0 \frac{Z_l + jZ_0 \tan\beta l}{Z_0 + jZ_l \tan\beta l}. \tag{6.38}$$

Let us examine some salient properties of the powerful formula (6.37) for the impedance distribution $Z(z)$ along any transmission line.

1. The formula does give the required result at the load at $z = l$—namely, that $Z(l) = Z_0[Z_l/Z_0] = Z_l$. It also gives this same correct result for the input impedance of a line in the limit of zero length, $l = 0$.

2. If the line is matched, so that $Z_l = Z_0$, then the formula yields $Z(z) = Z_0$ for all z.

3. The formula gives us the input impedance for a short-circuited $(Z_l = 0)$ transmission line of length l:

$$Z_{in} = jZ_0 \tan \beta l \qquad \text{(shorted line).} \qquad (6.39)$$

Note that this impedance can vary over an infinite range of reactances, both positive (inductive) and negative (capacitive), depending on the length of the line l or on the operating frequency ω in $\beta = \omega/u_0$.

4. The formula gives us the input impedance for an open-circuited $(Z_l = \infty)$ transmission line of length l. To handle the infinite terms in the formula, we simply neglect Z_0 compared to the infinite load impedance; the infinities cancel and

$$Z_{in} = -jZ_0 \cot \beta l \qquad \text{(open line).} \qquad (6.40)$$

This impedance is also reactive and can range over all reactances as the length of the line or the operating frequency is varied.

5. We can also find the input impedance of a transmission line whose length to an arbitrary load impedance is exactly one-quarter wavelength. Since $\beta\lambda = 2\pi$, we get for this case $\beta l = \pi/2$ and $\tan \beta l = \infty$ and, upon again neglecting finite quantities compared to this infinite value of $\tan \beta l$, we find

$$Z_{in} = Z_0^2/Z_l \qquad \text{(quarter wavelength).} \qquad (6.41)$$

This means that the load impedance gets *inverted*, in the sense that a high resistance appears low at a quarter wavelength; a short circuit seems open; an open circuit acts shorted; an inductor behaves as a capacitance; a capacitor turns into an inductor. Only a matched load remains as it is.

6. Let us also verify and interpret the limit of short-length transmission lines, first for the shorted case. In the limit of $l \to 0$, we get $\tan \beta l \to \beta l$ and, upon recalling for the parallel-wire case that $Z_0 = \sqrt{L/C}$ and $\beta = \omega\sqrt{LC}$, the input impedance becomes for the shorted line,

$$Z_{in} = jZ_0\beta l = j\sqrt{L/C}\omega\sqrt{LC}l$$
$$= j\omega Ll \qquad \text{(short shorted line).} \qquad (6.42)$$

The short-length shorted parallel-wire line looks like a (one-turn) inductor, of total inductance Ll.

7. Next, for the open line, we get from $\cot \beta l \to 1/\beta l$,

$$Z_{in} = -jZ_0/\beta l = -j\sqrt{L/C}/\omega\sqrt{LC}l$$
$$= 1/j\omega Cl \qquad \text{(short open line).} \qquad (6.43)$$

The short-length open parallel-wire line appears as a capacitor, of total capacitance Cl, as we might expect from its structure.

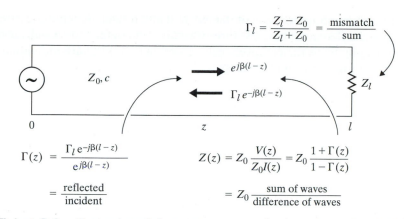

Figure 6-1   Illustration of three-step process for impedance distribution.

For all its usefulness, the formula for the impedance distribution along the line may appear formidable, opaque, or difficult to assimilate. For purposes of either calculation or interpretation, we can dispense entirely with the formula and its derivation. Instead, we can rely on three simple observations that lead to the same results, as in Figure 6-1. We have already observed that the *reflection coefficient at the load* is given by

$$\Gamma_l = \frac{Z_l - Z_0}{Z_l + Z_0}. \tag{6.44}$$

This can be understood and remembered easily, as it involves the mismatch between the load impedance and the characteristic impedance of the line. It is the ratio of the mismatch (difference, $Z_l - Z_0$) to the sum $(Z_l + Z_0)$ of these two impedances. The difference must be in the numerator, to guarantee that we get zero reflection when the line is matched. We can also appreciate that the numerator must be $Z_l - Z_0$ and not its negative, because the voltage reflection coefficient must surely be -1 when the line is shorted (so that setting $Z_l = 0$ in $[Z_l - Z_0]/[Z_l + Z_0]$ gives -1), to guarantee that the incident wave is canceled by the reflected one at the short.

Next, we need the ratio of the reflected and incident waves at any location z along the line. At the load, for unit incident-wave amplitude, the reflected wave has amplitude Γ_l. Elsewhere, the unit-amplitude incident wave is $e^{j\beta(l-z)}$ and the reflected wave becomes $\Gamma_l e^{-j\beta(l-z)}$. Consequently, we have for the ratio of the reflected to the incident wave amplitude at any position the *reflection coefficient distribution*

$$\Gamma(z) = \Gamma_l e^{-j2\beta(l-z)}. \tag{6.45}$$

Finally, the *impedance at any location* is the ratio of the voltage to the current there and these are the sum and difference of the incident and reflected waves there:

$$Z(z) = Z_0 \frac{1 + \Gamma(z)}{1 - \Gamma(z)}. \tag{6.46}$$

The numerator is the sum of the incident and reflected amplitudes; the denominator is their difference. These last three formulas lead to the same result, arithmetically and conceptually, as the comprehensive formula (6.37) but are individually easier to assimilate, interpret, and remember.

Example 6.2

A 300-ohm parallel-wire transmission line is one meter long and is loaded with a circuit whose lumped equivalent impedance is $(300 + j540)$ ohms when the line is operated at 500 MHz. What is the input impedance of this line? See Figure 6-2. ◇

The mismatch between the load and characteristic impedances is expressed by the ratio

$$Z_l/Z_0 = (300 + j540)/300 = 1 + j1.8 \qquad (6.47)$$

and the reflection coefficient at the load is

$$\Gamma_l = \frac{(1 + j1.8) - 1}{(1 + j1.8) + 1} = \frac{j1.8}{2 + j1.8}$$

$$= \frac{j1.8(2 - j1.8)}{2^2 + 1.8^2} = \frac{3.24 + j3.6}{7.24}$$

$$= 0.44751 + j0.49724 = 0.66896e^{j0.83798}. \qquad (6.48)$$

The reflection coefficient at the input end is therefore $\Gamma(0) = \Gamma_l e^{-j2\beta l}$ and we need $2\beta l$. This is, in radians,

$$2\beta l = 2\omega l/c = 4\pi f l/c = 4\pi(5 \cdot 10^8)1/(2.9979 \cdot 10^8) = 20.95845, \qquad (6.49)$$

so that

$$\Gamma(0) = 0.66896e^{-j20.12047} = 0.19762 - j0.63911. \qquad (6.50)$$

The input impedance is therefore

$$Z(0) = Z_0 \frac{1 + \Gamma(0)}{1 - \Gamma(0)}$$

$$= 300 \frac{1.19762 - j0.63911}{0.80238 + j0.63911} = 300 \frac{1.35748e^{-j0.49020}}{1.02581e^{j0.67261}}$$

$$= 300(1.32333e^{-j1.16282})$$

$$= 397.00e^{-j1.16282} = (157.5 - j364.4) \quad \text{ohms}. \qquad (6.51)$$

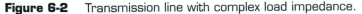

0 l

Figure 6-2 Transmission line with complex load impedance.

Note that the input impedance has a capacitive reactance in this case, despite the inductive reactance of the load.

Before leaving this example, we may wish to compare the preceding calculation with what is needed if we use the formula for input impedance. This is

$$Z_{in} = Z_0 \frac{(Z_l/Z_0) + j \tan \beta l}{1 + j(Z_l/Z_0)\tan \beta l}. \tag{6.52}$$

We have already calculated the mismatch ratio $Z_l/Z_0 = 1 + j1.8$ and βl, so we now need only $\tan \beta l = \tan(10.4792) = 1.76142$. We find

$$Z_{in} = 300 \frac{1 + j1.8 + j1.76142}{1 + j(1 + j1.8)1.76142}$$

$$= 300 \frac{1 + j3.56142}{-2.17055 + j1.76142}$$

$$= 300 \frac{3.69915e^{j1.29706}}{2.79533e^{j2.45987}}$$

$$= 300(1.32333e^{-j1.16282})$$

$$= (157.5 - j364.4) \qquad \text{ohms.} \tag{6.53}$$

This answer is, of course, the same one we obtained by the three-step process.

We should be aware of the need for maintaining a high degree of precision during these calculations, because of the occurrence of subtractions of comparable numbers, which leads to loss of accuracy, and also because of the evaluation of imaginary exponentials, or trigonometric functions. The latter often need high precision, in that additive multiples of 2π (or of π in the case of the tangent function) have no effect on the answer. For example, in calculating $2\beta l$ previously, we used

$$2\beta l = 2\omega l/c = 4\pi f l/c = 4\pi (5 \cdot 10^8)1/(2.9979 \cdot 10^8) = 20.95845,$$

but we might have been wiser to retain 2π as a factor, as in

$$2\beta l = 2\omega l/c = 4\pi f l/c = 2\pi(10^9)1/(2.9979 \cdot 10^8) = 2\pi(3.335641),$$

which entitles us to remove integers from the factor of 2π, so that the same result could have been obtained from exponential or trigonometric functions of $2\pi(0.335641)$ = 2.108895 radians as we got from 20.95845 radians. For long transmission lines, this approach helps maintain the needed precision.

Example 6.3

A 300-ohm twin-lead transmission line of length 2.027 meters is connected from an antenna to a 200-ohm FM radio receiver that can be tuned to any carrier frequency between 88 and 108 MHz; see Figure 6-3. Calculate the input impedance of the loaded line at each of the two extreme frequencies. (Assume the constitutive parameters of the material surrounding the parallel wires of the twin-lead line differ negligibly from those of a vacuum.) ◈

We have $Z_l = 200 \,\Omega$ and $Z_0 = 300 \,\Omega$, while $\beta l = 2\pi f l/c$ is either $\beta_1 l = 3.7385$ at $f = 88$ MHz or else $\beta_2 l = 4.5881$ at $f = 108$ MHz. From these we have

$$\tan \beta_1 l = 0.67958 \qquad \text{and} \qquad \tan \beta_2 l = 8.0069.$$

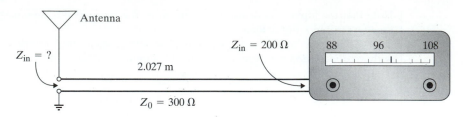

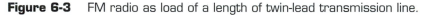

Figure 6-3   FM radio as load of a length of twin-lead transmission line.

If we use the formula for input impedance directly, we invoke

$$Z_{in} = Z_0[Z_l/Z_0 + j \tan \beta l]/[1 + j(Z_l/Z_0) \tan \beta l]$$

and find

$$Z_1 = 300[2/3 + j0.67958]/[1 + j(2/3)0.67958]$$
$$= (243 + j94) \, \Omega$$

as the input impedance at 88 MHz, or else

$$Z_2 = 300[2/3 + j8.0069]/[1 + j(2/3)8.0069]$$
$$= (442 + j45) \, \Omega$$

as the input impedance at 108 MHz.

The next example suggests a possible means for measuring the characteristic impedance of a transmission line, provided we have independent means for measuring input impedances.

Example 6.4

We measure the complex input impedance of a certain transmission line to be Z_1 when it is short-circuited and Z_2 when it is open-circuited instead.

   (a) *From just Z_1 and Z_2, determine the characteristic impedance Z_0 of the line.*

   (b) *Are there any frequencies at which this measurement technique fails?* «»

   (a) Since $Z_1 = jZ_0 \tan \beta l$ and $Z_2 = -jZ_0 \cot \beta l$, we have $Z_1 Z_2 = Z_0^2$ and we find Z_0 as $\sqrt{Z_1 Z_2}$.

   (b) At all frequencies for which $\beta l = \omega l/u_0$ is a multiple of $\pi/2$, the product $Z_1 Z_2$ is $\infty \cdot 0$, which is indeterminate.

Let's compare the voltages at the input and output of a transmission line, to establish a ratio we might loosely call the voltage "gain" of the line.

Example 6.5

A uniform transmission line (length l, characteristic impedance Z_0, wave speed c, perfect conductors and insulators) has an ideal voltage source (i.e., no internal impedance) $V_{in} e^{j\omega t}$ attached at $z = 0$ and a resistive load R at $z = l$. Let G be the ratio $|V_{out}/V_{in}|$ of the magnitude of the voltage $V_{out} e^{j\omega t}$ across the load resistor to that of the input voltage. See Figure 6-4.

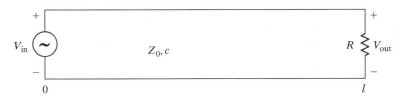

Figure 6-4 Can a transmission line have a gain greater than unity?

(a) *What is the maximum value of G attainable by varying the length l of the transmission line?*

(b) *Can the voltage across the load ever have a magnitude greater than that of the source (i.e., is a "gain" G > 1 possible)? If so, under what circumstances?* ◇

The voltage at any position is $V(z) = V_1[e^{j\beta(l-z)} + \Gamma_l e^{-j\beta(l-z)}]$, where $\Gamma_l = [R - Z_0]/[R + Z_0]$. Hence, $V_{in} = V_1[e^{j\beta l} + \Gamma_l e^{-j\beta l}]$ and $V_{out} = V_1[1 + \Gamma_l]$, which gives $V_{out}/V_{in} = [1 + \Gamma_l]/[e^{j\beta l} + \Gamma_l e^{-j\beta l}] = R/[R \cos \beta l + j Z_0 \sin \beta l]$. The magnitude of this ratio gives us $G = R/\sqrt{R^2 \cos^2 \beta l + Z_0^2 \sin^2 \beta l}$.

(a) The quantity $R^2 \cos^2 \beta l + Z_0^2 \sin^2 \beta l$ is a weighted average of R^2 and Z_0^2, varying between R^2 and Z_0^2 as βl changes, so that G varies between 1 and R/Z_0. Hence, $G_{max} = 1$ or R/Z_0, whichever is larger.

(b) $G \geq 1$ if $R > Z_0$ and peaks at R/Z_0 whenever $\cos \beta l = 0$.

Of course, the transmission line alone does not qualify as an amplifier; that the "gain" may here be greater than unity merely reflects reactive effects. The reactances are those that are distributed along the line, which we might naively consider to be merely resistive circuitry. Circuits with lumped reactive elements naturally exhibit high values of such "gain" near their resonant frequencies.

Transmission lines such as the parallel-wire line afford opportunities to attach lumped circuit elements, or perhaps voltage or current sources, at points between the ends, as well as at the ends. These can serve to modify the impedance presented to the source at the front end and may affect the mismatch of the load.

The method of analysis is still to write a superposition of forward and backward waves, with complex coefficients to be determined from the boundary conditions. What is new is that we must realize that the superposition of countertraveling waves applies only to segments of transmission line that are uniform, without discontinuities. This means that we must find *separate* superpositions of such waves *on both sides* of any discontinuity along the line, including the discontinuity presented by any lumped element attached at an intermediate point of the line. This is illustrated in the next example.

Example 6.6

A transmission line is loaded with a mismatched resistance R_0 at the end l and has a lumped capacitor C_0 at location l_0, as in Figure 6-5. What are the voltage and current distributions all along the line? What is the input impedance presented to the voltage source? Obtain numerical results for a 300-ohm line operated at 100 MHz, for the case that $R_0 = 50$ ohms and $C_0 = 5$ pF, for $l = 2$ m and $l_0 = 1.5$ m. ◇

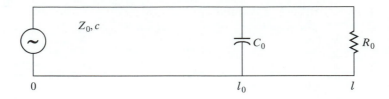

Figure 6-5   Transmission line with lumped element at intermediate point.

The voltage and current distributions are separate versions of the superposition of incident and reflected waves to the left and to the right of the capacitor. Referring the distance coordinate to the location of the capacitor, we can write for the first section of the line

$$V(z) = V_1 e^{j\beta(l_0 - z)} + V_2 e^{-j\beta(l_0 - z)} \qquad \text{for } 0 < z < l_0; \tag{6.54}$$

$$Z_0 I(z) = V_1 e^{j\beta(l_0 - z)} - V_2 e^{-j\beta(l_0 - z)} \qquad \text{for } 0 < z < l_0; \tag{6.55}$$

with the unknown complex amplitudes V_1, V_2. But for the second section we need two additional unknowns V_3, V_4:

$$V(z) = V_3 e^{-j\beta(z - l_0)} + V_4 e^{j\beta(z - l_0)} \qquad \text{for } l_0 < z < l; \tag{6.56}$$

$$Z_0 I(z) = V_3 e^{-j\beta(z - l_0)} - V_4 e^{j\beta(z - l_0)} \qquad \text{for } l_0 < z < l. \tag{6.57}$$

Take careful note of the signs in the exponents. We have chosen to write the distance coordinates as positive quantities (i.e., $l_0 - z$ in front of the capacitor but $z - l_0$ beyond it). In either case, the forward (incident) wave has the same sign in the expressions for both V and $Z_0 I$ and the backward (reflected) wave has a different sign in the two expressions.

We have four unknown constants, V_1, V_2, V_3, V_4, and we have four conditions to determine them. We know the source voltage at $z = 0$ and the load resistance R_0 at $z = l$; we also know that, at $z = l_0$, we have continuity of the voltage but discontinuity of the current, by the amount that is diverted to the capacitor ($j\omega C_0$ times the voltage there):

$$V_1 + V_2 = V_3 + V_4, \tag{6.58}$$

$$V_1 - V_2 = V_3 - V_4 + j\omega Z_0 C_0 (V_1 + V_2). \tag{6.59}$$

Adding to these the conditions at the two ends of the entire line, we get four equations for four unknowns and can solve for all the unknowns.

We can also use the usual methods based on the reflection coefficient, but always keeping the two segments of the line separate. The reflection coefficient at the load is, for the given mismatch in impedance,

$$\Gamma_2(l) = (R_0 - Z_0)/(R_0 + Z_0) = (50 - 300)/(50 + 300) = -5/7, \tag{6.60}$$

where the subscript refers to the second segment. It may be helpful to recall that this is the ratio of the reflected wave at the load, $V_4 e^{j\beta(l - l_0)}$, to the incident one there, $V_3 e^{-j\beta(l - l_0)}$. This reflection coefficient gets transferred to the location of the capacitor as

$$\Gamma_2(l_0) = \Gamma_2(l) e^{-j2\beta(l - l_0)}, \tag{6.61}$$

where it represents the ratio V_4/V_3. The phase $2\beta(l - l_0)$ is

$$4\pi f(l - l_0)/c = 4\pi 10^8(0.5)/(2.9979 \cdot 10^8) = 2.0958 \text{ rad} \qquad (6.62)$$

and the reflection coefficient at the capacitor is

$$\Gamma_2(l_0) = -(5/7)e^{-j2.0958} = 0.35804 + j0.61807. \qquad (6.63)$$

The input impedance for the second segment, as seen at the capacitor, is therefore

$$Z(l_0) = Z_0[1 + \Gamma_2(l_0)]/[1 - \Gamma_2(l_0)]$$

$$= Z_0(1.35804 + j0.61807)/(0.64196 - j0.61807)$$

$$= Z_0 1.67435 e^{j1.19354} = Z_0(0.616774 + j1.55661) \qquad (6.64)$$

and this appears in parallel with the impedance of the capacitor at that location. The capacitor has impedance $1/j\omega C_0$ or

$$Z_0/j\omega Z_0 C_0 = Z_0(1/j2\pi f Z_0 C_0) = Z_0(-j1.061033) \qquad (6.65)$$

so that the parallel combination of the impedance of the second segment and this capacitor gives for the load impedance at l_0 for the first segment

$$Z_1(l_0) = Z(l_0)(1/j\omega C_0)/[Z(l_0) + 1/j\omega C_0]$$

$$= Z_0(1.65161 - j0.65442)/(0.61677 + j0.49558)$$

$$= Z_0 2.24536 e^{-j1.05412} = Z_0(1.10919 - j1.95226). \qquad (6.66)$$

This is now the load impedance for the first segment, which has length l_0. The corresponding reflection coefficient is

$$\Gamma_1(l_0) = [Z_1(l_0) - Z_0]/[Z_1(l_0) + Z_0]$$

$$= 0.68034 e^{-j0.76815} \qquad (6.67)$$

and this gets transferred to the source as

$$\Gamma_1(0) = \Gamma_1(l_0)e^{-j2\beta l_0} = 0.68034 e^{-j0.77250}. \qquad (6.68)$$

This corresponds to the input impedance

$$Z_{\text{in}} = Z_0[1 + \Gamma_1(0)]/[1 - \Gamma_1(0)] = Z_0 2.23397 e^{-j1.05604}$$

$$= (329.95 - j583.35) \quad \text{ohms.} \qquad (6.69)$$

For comparison, note that if the capacitor were absent, the phase βl would be 4.19169 radians, the reflection coefficient would be $0.36072 + j0.61651$ at the generator, and the input impedance would be

$$Z_{\text{in}} = (186.29 + j468.97) \quad \text{ohms} \quad \text{(no capacitor).} \qquad (6.70)$$

Note particularly that the addition of the capacitor has effected a drastic change in the input impedance of the line.

POWER TRANSFER

If we look upon the generator as the supplier of power or energy and the load as the receiver or absorber of that energy, the transmission line becomes the channel along which the power is conveyed from source to load. We then should examine how efficiently the energy gets transferred when the voltage and current along the line are comprised of countertraveling waves. We know we can avoid the reflected wave by matching the line. But do we really improve the power transfer when we match the line?

To explore these issues, we must look at power and energy transfer along the transmission line. Power and energy are quantities related quadratically, not linearly, to the voltage and current distributions. Consequently, we are safe only if we perform the calculations in the time domain; if we venture to obtain information about power and energy in the frequency domain, we must perform calculations that reflect what is appropriate to do in the time domain.

If the generator provides voltage $v(0, t)$ and current $i(0, t)$ at $z = 0$, then it supplies power $p(0, t) = v(0, t)i(0, t)$ to the transmission line. If the load has voltage $v(l, t)$ across it and current $i(l, t)$ through it, then it is absorbing power $p(l, t) = v(l, t)i(l, t)$. At any location z along the transmission line, the voltage is $v(z, t)$ across the line and the current along it is $i(z, t)$, so that the power being supplied to the region beyond z is $p(z, t) = v(z, t)i(z, t)$.

In the simplest case, that of a matched line, if the generator produces signal $V_0(t)$, then the voltage distribution is $v(z, t) = V_0(t - z/c)$ and the current is $i(z, t) = V_0(t - z/c)/Z_0$ if we assume propagation at the vacuum speed of light. Then the power supplied by the source is $p_0(t) = V_0^2(t)/Z_0$ and the power at location z is

$$p(z, t) = V_0^2(t - z/c)/Z_0 = p_0(t - z/c), \tag{6.71}$$

which is seen to propagate as a wave, at the same speed. In particular, the power delivered to the matched load at $z = l$ is $p(l, t) = p_0(t - l/c)$, which is the same as the power supplied by the source, but delayed by the time l/c of wave propagation from source to load.

In the steady state, the voltage at some point may be

$$v(t) = V_0 \cos(\omega t + \varphi_0) = \mathrm{Re}\{V e^{j\omega t}\}, \tag{6.72}$$

where the complex amplitude V is $V_0 e^{j\varphi_0}$. If the complex impedance at that point is $Z = |Z|e^{j\psi}$, then the current there is

$$i(t) = \mathrm{Re}\{(V/Z)e^{j\omega t}\} = \mathrm{Re}\{(|V|/|Z|)e^{j(\omega t + \varphi_0 - \psi)}\}$$
$$= (|V|/|Z|)\cos(\omega t + \varphi_0 - \psi). \tag{6.73}$$

It follows that the power at that point is given by

$$p(t) = v(t)i(t) = (|V|^2/|Z|)\cos(\omega t + \varphi_0)\cos(\omega t + \varphi_0 - \psi). \tag{6.74}$$

This is, of course, a function of time. As a product of two sinusoids at frequency ω, it too is an oscillation, but at twice this frequency and generally not centered about zero power level. To see this quantitatively, we use a trigonometric identity to rewrite the product of two cosines, as in

$$p(t) = (|V|^2/2|Z|)[\cos\psi + \cos(2\omega t + 2\varphi_0 - \psi)]. \tag{6.75}$$

This exhibits the time function $p(t)$ as a superposition of a constant and a rapidly oscillating sinusoid. For most purposes, the rapid, double-frequency, zero-centered oscillation elicits little or no response from the circuitry, apparatus, or instrument that receives this power. The time-averaged power, however, is most often the relevant quantity. We see that the time average of $p(t)$, over any whole number of its oscillations, is the constant part of this superposition:

$$p_0 = \langle p(t) \rangle = \tfrac{1}{2}(|V|^2/|Z|)\cos\psi. \tag{6.76}$$

As a time average, this is of course a constant. We obtained this constant by appending the factor $e^{j\omega t}$ to the complex voltage and current amplitudes in the frequency domain, taking the real parts to convert them to the time domain, multiplying them, and then averaging the product. Starting with the frequency-domain description by complex amplitudes V and $I = V/Z$, we can arrive at this constant more directly by rewriting it as

$$\begin{aligned}
p_0 &= \tfrac{1}{2}(|V|^2/|Z|)\cos\psi = \mathrm{Re}\{\tfrac{1}{2}(|V|^2/|Z|)e^{j\psi}\} \\
&= \mathrm{Re}\{\tfrac{1}{2}|V|^2/(|Z|e^{-j\psi})\} = \mathrm{Re}\{\tfrac{1}{2}VV^*/Z^*\} = \mathrm{Re}\{\tfrac{1}{2}VI^*\}.
\end{aligned} \tag{6.77}$$

We can therefore get the time-averaged power at a point of a circuit where the complex voltage and current amplitudes are known to be V and I as

$$p_0 = \langle p(t) \rangle = \mathrm{Re}\left[\tfrac{1}{2}VI^*\right]. \tag{6.78}$$

That is, we can simply multiply the complex voltage amplitude by the complex conjugate of the current amplitude and then take one-half the real part of the complex product to arrive at the time-averaged power—a welcome short cut.

Let us look at this for the case of a point along a transmission line that carries incident and reflected waves. The complex voltage and current are given in terms of the incident and reflected voltage amplitudes at the load, V_1, V_2, by

$$V(z) = V_1 e^{j\beta(l-z)} + V_2 e^{-j\beta(l-z)} = V_1 e^{j\beta(l-z)}[1 + \Gamma(z)], \tag{6.79}$$

$$Z_0 I(z) = V_1 e^{j\beta(l-z)} - V_2 e^{-j\beta(l-z)} = V_1 e^{j\beta(l-z)}[1 - \Gamma(z)]. \tag{6.80}$$

The time-averaged power at the point z of the line is therefore (for real Z_0)

$$\begin{aligned}
p_0 &= \mathrm{Re}\left[\tfrac{1}{2}V(z)I^*(z)\right] = \mathrm{Re}\{(|V_1|^2/2Z_0)[1 + \Gamma(z)][1 - \Gamma^*(z)]\} \\
&= \mathrm{Re}\{(|V_1|^2/2Z_0)[1 + \Gamma(z) - \Gamma^*(z) - |\Gamma(z)|^2]\} \\
&= (|V_1|^2/2Z_0)[1 - |\Gamma|^2],
\end{aligned} \tag{6.81}$$

because $\Gamma(z) - \Gamma^*(z)$ is purely imaginary. Since $|\Gamma(z)|$ is independent of z, the time-averaged power is not a function of z, which expresses conservation of power along the ideal, lossless transmission line.

Note that the dependence of the average power p_0 on the reflection coefficient Γ is not obvious in this expression, because the complex amplitude V_1 of the steady-state incident voltage depends on the reflection coefficient. A more transparent version of the formula can be expressed in terms of the voltage or the current at any particular point z_0 at which either one is known. Since, from (6.79) and (6.80),

$$|V_1|^2 = \frac{|V(z_0)|^2}{|1 + \Gamma(z_0)|^2} = \frac{Z_0^2|I(z_0)|^2}{|1 - \Gamma(z_0)|^2}, \tag{6.82}$$

we have the alternative expressions

$$p_0 = \frac{|V(z_0)|^2}{2Z_0} \frac{1 - |\Gamma|^2}{|1 + \Gamma(z_0)|^2} = \frac{Z_0|I(z_0)|^2}{2} \frac{1 - |\Gamma|^2}{|1 - \Gamma(z_0)|^2}. \tag{6.83}$$

Example 6.7

An ideal voltage generator of 10 volts rms is connected at $z = 0$ to a 300-Ω transmission line of length $l = 34.286 \lambda$. The line is terminated by load impedance $Z_l = (3 + j2)Z_0$. Find the time-averaged power delivered to the load. ◈

We are given the voltage amplitude at $z = 0$, so we need the reflection coefficient there. The reflection coefficient at the load is

$$\Gamma_l = (2 + j2)/(4 + j2) = (3 + j)/5 = 0.63246\, e^{j0.32175} \tag{6.84}$$

and at the input to the line it is

$$\begin{aligned}
\Gamma(0) &= 0.63246 e^{j0.32175} e^{-j2\beta l} \\
&= 0.63246 e^{j0.32175} e^{-j4\pi(34.286)} \\
&= 0.63246 e^{j0.32175} e^{-j4\pi(0.286)} \\
&= 0.63246 e^{-j3.27223} = -0.62707 + j0.08239. \tag{6.85}
\end{aligned}$$

We removed an integer multiple of 4π in the complex exponential, to help maintain the precision of the result. Noting that 10 volts rms corresponds to an amplitude of $V(0) = 10\sqrt{2}$ volts, the power at the input to the line is therefore

$$\begin{aligned}
p_0 &= (10^2/300)[1 - (0.63246)^2]/|0.37293 + j0.08239|^2 \\
&= (1/3)[1 - 0.4]/0.14586 = 1.371 \text{ watt.} \tag{6.86}
\end{aligned}$$

This is the average power supplied by the generator and, since the line is lossless, it is also the time-averaged power delivered to and dissipated by the load.

Several aspects of this last example may be baffling—and therefore instructive when explained. The mismatched load in the example receives 1.371 watt from the source. Had this load been attached directly to the source, it would have drawn only $\text{Re}\{|V(0)|^2/2Z_l\} = \text{Re}\{10^2/[300(3 + j2)]\} = 0.0769$ watt from the generator. Has the transmission line somehow acted as a power amplifier? Furthermore, had the line been matched, the reflection coefficient would have been zero all along the line and the load would have absorbed only $|V(0)|^2/2Z_0 = 10^2/300 = 0.333$ watt instead of 1.371 watt. How can it be that more power can be transferred from source to load in the mismatched case than in the matched one?

That we draw more power with the transmission line interposed between source and load than without it is due to the impedance-transforming property of a transmission line. The input impedance of the mismatched line differs from the load impedance; it is the real part of the input admittance that determines the power drawn from the voltage source and absorbed by the load. The seemingly amplified power transfer is merely a manifestation of the constructive interference of the incident and reflected waves along the line.

As to the apparent achievement of a more efficient power transfer with a mismatched load than with a matched one, this arises from confusion over the use of the term *matching* in two different contexts. We must now clarify this terminology.

If we are interested in the transfer of power from source to load, and perhaps in designing the transmission line circuit to optimize it, then we should be more realistic about the generator. Real sources cannot deliver unlimited amounts of current; they have an internal impedance. Once we include the generator's internal impedance in the circuit attached to the transmission line, we can discuss the steady-state power transfer along the line. If the source did have zero internal impedance, it could, in principle, deliver unlimited power.

Figure 6-6 shows a generator of complex internal impedance Z_g and complex amplitude V_g attached to a transmission line of length l and characteristic impedance Z_0; the line is loaded by complex impedance Z_l. If we replace the circuitry beyond the input end at $z = 0$ with the equivalent input impedance of the loaded line, Z_{in}, we see a voltage divider and obtain the voltage at $z = 0$ as

$$V(0) = V_g Z_{in}/[Z_g + Z_{in}]. \tag{6.87}$$

The average power transferred to the input end of the line is therefore

$$p(0) = \tfrac{1}{2}\mathrm{Re}\big[V(0)I^*(0)\big] = \tfrac{1}{2}\mathrm{Re}\left\{V_g \frac{Z_{in}}{(Z_g + Z_{in})} \frac{V_g^*}{(Z_g + Z_{in})^*}\right\}$$

$$= \frac{\tfrac{1}{2}|V_g|^2 R_{in}}{|Z_g + Z_{in}|^2}, \tag{6.88}$$

where $R_{in} = \mathrm{Re}\, Z_{in}$ is the resistive part of the input impedance.

This is the usual circuit-theoretic expression for steady-state average power transfer to a circuit of input impedance Z_{in} when fed by a source of complex amplitude V_g and internal impedance Z_g. As is readily verified and well known, we can optimize this power transfer for a given internal impedance Z_g by selecting the input impedance Z_{in} of the attached circuit to be the complex conjugate of the internal impedance of the source:

$$Z_{in} = Z_g^* \qquad \text{(matching for maximum power transfer)}. \tag{6.89}$$

The optimum power transfer is then $p_{opt} = |V_g|^2/8R_g$, where R_g is the resistive part of the generator's internal impedance. This selection of the load impedance to be attached directly to the nonideal source is known as matching the load to the source,

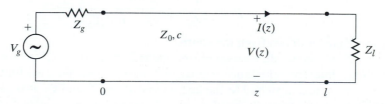

Figure 6-6   Generator with internal impedance attached to loaded line.

in circuits parlance. Despite the use of the same terminology, this matching to achieve optimum power transfer is not the same as the matching of a transmission line's load impedance Z_l to its characteristic impedance Z_0 to preclude reflections from the load back to the generator:

$$Z_l = Z_0 \qquad \text{(matching for elimination of reflection).} \tag{6.90}$$

The benefits of matching in the transmission-line sense are those that accrue from having only unidirectional waves from generator to load, without reflected waves that may damage the generator, or echoes that may distort the signal that modulates the sinusoidal carrier wave, or long-duration transients. The benefits of matching for maximized power transfer are optimized power reception for a given level of generator power, or minimized generator power for a given level of required receiver power.

If we want the benefits of both maximized power transfer and reflectionless transmission, then we should attempt both types of matching. We would then seek a transmission line whose characteristic impedance matches the generator's internal impedance (if resistive) and also choose a load impedance that matches those other two impedances.

Before illustrating these considerations, let us obtain a formula for the power transfer along a transmission line, from a source with internal impedance to a mismatched load. We need only combine (6.83) at $z_0 = 0$; that is,

$$p_0 = \frac{|V(0)|^2}{2Z_0} \frac{1 - |\Gamma(0)|^2}{|1 + \Gamma(0)|^2} \tag{6.91}$$

with (6.87), which expresses $V(0)$ in terms of the generator voltage V_g:

$$V(0) = V_g \frac{Z_{in}}{Z_g + Z_{in}} = \frac{[1 + \Gamma(0)]/[1 - \Gamma(0)]}{Z_g/Z_0 + [1 + \Gamma(0)]/[1 - \Gamma(0)]} \tag{6.92}$$

We have here also replaced the input impedance by its expression in terms of the reflection coefficient at the input to the line, $\Gamma(0)$. We can define the reflection coefficient at the source in terms of the mismatch between the generator impedance Z_g and the characteristic impedance Z_0 of the transmission line:

$$\Gamma_g = \frac{Z_g - Z_0}{Z_g + Z_0} \qquad \text{or} \qquad Z_g/Z_0 = \frac{1 + \Gamma_g}{1 - \Gamma_g}. \tag{6.93}$$

Substituting into the previous equation and simplifying gives the formula

$$p_0 = \frac{|V_g|^2}{8Z_0} \frac{\{1 - |\Gamma(0)|^2\}|1 - \Gamma_g|^2}{|1 - \Gamma(0)\Gamma_g|^2} \tag{6.94}$$

for the power drawn from the source.

Note that we do recover our previous formula (6.91) if the source is ideal; that is, if the internal impedance is zero, making $\Gamma_g = -1$. We also find that the special case of a source matched to the line, so that $Z_g = Z_0$ and $\Gamma_g = 0$, gives

$$p_0 = \{|V_g|^2/8Z_0\}\{1 - |\Gamma(0)|^2\} \qquad \text{(matched source).} \tag{6.95}$$

Further, the case of a matched load $(\Gamma(0) = 0)$ but mismatched source yields

$$p_0 = \{|V_g|^2/8Z_0\}|1 - \Gamma_g|^2 \qquad \text{(matched load)}. \qquad (6.96)$$

Finally, if both the load impedance Z_l and the source impedance Z_g are matched to the line, the optimum power transfer $p_0 = |V_g|^2/8Z_0$ is achieved.

Example 6.8

A voltage generator of 10 volts rms and internal impedance $Z_g = 50\ \Omega$ is connected at $z = 0$ to a 300-Ω transmission line of length $l = 34.286\ \lambda$. The line is terminated by impedance $Z_l = (3 + j2)Z_0$. Find the time-averaged power delivered to the load. ◈

This is the same configuration as in the previous example, except that the source now has an internal impedance. We have already calculated the reflection coefficient at the line input, $\Gamma(0) = -0.62707 + j0.08239$, as in (6.85). Combining this with $\Gamma_g = -5/7$ in the formula (6.94) gives us $p_0 = 0.4767$ watt as the power drawn from the source and transferred to the load.

For comparison, consider the case of a load impedance matched, in the sense of optimized power transfer, to the source but connected to it by means of a transmission line that is mismatched to both the source and load.

Example 6.9

A voltage generator of 10 volts rms and internal impedance $Z_g = 50\ \Omega$ is connected at $z = 0$ to a 300-Ω transmission line of length $l = 34.286\ \lambda$. The line is terminated by impedance $Z_l = 50\ \Omega$, to match the source but not the line. Find the time-averaged power delivered to the load. ◈

This time, the reflection coefficient at the load and at the source are both the same, $\Gamma_l = \Gamma_g = -5/7$, but we need to recalculate the reflection coefficient at the input end:

$$\Gamma(0) = (-5/7)e^{-j2\beta l}$$

$$= 0.714286e^{j\pi}e^{-j4\pi(34.286)} = 0.714286e^{j\pi}e^{-j4\pi(0.286)}$$

$$= 0.714286e^{-j0.45239} = 0.64243 - j0.31223.$$

We then calculate

$$\frac{\{1 - |\Gamma(0)|^2\}|1 - \Gamma_g|^2}{|1 - \Gamma(0)\Gamma_g|^2} = \frac{\{24/49\}[12/7]^2}{|1.45888 - j0.22302|^2} = 0.66086$$

and since $|V_g|^2/8Z_0 = 1/12$ watt, we get only $p_0 = 55.07$ mW as the power delivered to the load through the mismatched transmission line.

By contrast, had we used a 50-Ω transmission line to join the 50-Ω source to the 50-Ω load, we would have transferred $|V_g|^2/8Z_0 = 500$ mW from source to load in the then fully matched circuit.

PLANE WAVES

The waves we have been bouncing back and forth between the generator and load of the parallel-wire transmission line have field patterns that are intimately tied to the supporting structure and are guided by the wires. We want now to take a closer look at the fields, approximately midway between the two wires and particularly for the case of wires that are far apart. We will thereby be examining electromagnetic fields that subsist and propagate far away from their supporting or guiding structure.

Early in the last chapter, we obtained the fields around a pair of parallel wires. The electric field was

$$\mathbf{E} = \frac{q(z, t)}{2\pi\varepsilon_0}\left(\frac{\hat{\mathbf{e}}_1}{\rho_1} - \frac{\hat{\mathbf{e}}_2}{\rho_2}\right), \tag{6.97}$$

where the $\hat{\mathbf{e}}$'s are unit vectors directed away from each wire at the observation point and the ρ's are the perpendicular distances from the wires; $q(z, t)$ is the charge per unit length on one wire. The magnetic field was

$$\mathbf{H} = \frac{I(z, t)}{2\pi}\left(\frac{\hat{\mathbf{h}}_1}{\rho_1} - \frac{\hat{\mathbf{h}}_2}{\rho_2}\right), \tag{6.98}$$

where the $\hat{\mathbf{h}}$'s are unit vectors directed azimuthally around the wires at the observation point; $I(z, t)$ is the current in one wire. Since $\hat{\mathbf{h}} = \hat{\mathbf{z}} \times \hat{\mathbf{e}}$, we have $Z\mathbf{H} = \hat{\mathbf{z}} \times \mathbf{E}$ at every point, where Z is the ratio $q(z, t)/\varepsilon_0 I(z, t)$. We already know this ratio; for a unidirectional wave along the line, it is merely a constant:

$$Z = q(z, t)/\varepsilon_0 I(z, t) = CV(z, t)/\varepsilon_0 I(z, t)$$
$$= CZ_0/\varepsilon_0 = C\sqrt{L/C}/\varepsilon_0 = \sqrt{LC}/\varepsilon_0 = \sqrt{\mu_0/\varepsilon_0} = \eta_0. \tag{6.99}$$

We want to examine these fields in the vicinity of the point midway between the wires, at any axial location z, especially when the wires are very far apart. To facilitate this, let's put the z-axis midway between the wires and look at the field $\mathbf{E}(x, y, z, t)$ at points near that axis, so that $|x| \ll s$ and $|y| \ll s$, and expand this function in a series of powers of x/s and y/s; s is the separation between the wires. We write $\hat{\mathbf{e}}/\rho$ as $\rho\hat{\mathbf{e}}/\rho^2$, which is

$$\rho\hat{\mathbf{e}}/\rho^2 = \left[\mathbf{r} \mp (s/2)\hat{\mathbf{x}}\right]/|\mathbf{r} \mp (s/2)\hat{\mathbf{x}}|^2$$
$$= \{\hat{\mathbf{x}}[x \mp (s/2)] + \hat{\mathbf{y}}y\}/\{x^2 + y^2 + s^2/4 \mp sx\} \tag{6.100}$$

if $\mathbf{r} = \hat{\mathbf{x}}x + \hat{\mathbf{y}}y$ represents the transverse position vector to the observation point and the wires are at $x = \pm(s/2)$. The difference $\hat{\mathbf{e}}_1/\rho_1 - \hat{\mathbf{e}}_2/\rho_2$ becomes

$$\hat{\mathbf{e}}_1/\rho_1 - \hat{\mathbf{e}}_2/\rho_2 = \frac{\hat{\mathbf{x}}[x + (s/2)] + \hat{\mathbf{y}}y}{x^2 + y^2 + s^2/4 + sx} - \frac{\hat{\mathbf{x}}[x - (s/2)] + \hat{\mathbf{y}}y}{x^2 + y^2 + s^2/4 - sx}$$
$$= \frac{-[\hat{\mathbf{x}}x + \hat{\mathbf{y}}y]2sx + \hat{\mathbf{x}}s[x^2 + y^2 + s^2/4]}{(x^2 + y^2 + s^2/4)^2 - s^2x^2}$$

$$= \frac{\hat{\mathbf{x}}[s^3/4 + s(y^2 - x^2)] + \hat{\mathbf{y}}[2sxy]}{s^4/16 + (s^2/2)(y^2 - x^2) + (x^2 + y^2)^2}$$

$$= \frac{(4/s)\{\hat{\mathbf{x}}[1 + (4/s^2)(y^2 - x^2)] + \hat{\mathbf{y}}[8xy/s^2]\}}{1 + (8/s^2)(y^2 - x^2) + (16/s^4)(x^2 + y^2)^2}. \tag{6.101}$$

As the separation s between the wires becomes large, we can expand this expression in powers of $1/s$ and retain only the two lowest powers. This gives

$$\hat{\mathbf{e}}_1/\rho_1 - \hat{\mathbf{e}}_2/\rho_2 \approx (4/s)\{\hat{\mathbf{x}}[1 - (4/s^2)(y^2 - x^2)] + \hat{\mathbf{y}}[8xy/s^2]\}. \tag{6.102}$$

The field on the axis, midway between the wires, is therefore

$$\mathbf{E}(0, 0, z, t) = (4/s)\hat{\mathbf{x}}q(z, t)/2\pi\varepsilon_0 = \hat{\mathbf{x}}E_0(z, t) \tag{6.103}$$

and, finally, the field in the vicinity of the axis is, approximately,

$$\mathbf{E}(x, y, z, t) \approx E_0(z, t)\{\hat{\mathbf{x}}[1 - (4/s^2)(y^2 - x^2)] + \hat{\mathbf{y}}[8xy/s^2]\}. \tag{6.104}$$

The magnetic field for a unidirectional wave is correspondingly

$$\mathbf{H}(x, y, z, t) = \hat{\mathbf{z}} \times \mathbf{E}/\eta_0$$

$$\approx E_0(z, t)\{\hat{\mathbf{y}}[1 - (4/s^2)(y^2 - x^2)] - \hat{\mathbf{x}}[8xy/s^2]\}/\eta_0. \tag{6.105}$$

In the limit of infinite separation between the wires, $s \to \infty$, the dependence on x and y disappears and these fields become uniform in x and y:

$$\mathbf{E}(x, y, z, t) = E_0(z, t)\hat{\mathbf{x}}, \qquad \mathbf{H}(x, y, z, t) = E_0(z, t)\hat{\mathbf{y}}/\eta_0. \tag{6.106}$$

Figure 6-7 shows electric and magnetic field lines near the midpoint between two parallel wires in a transverse plane, for large but finite separation of the wires, and also for infinite separation.

But what is the source of these fields? For finite separation of the wires, it is the charge per unit length on one of them, $q(z, t) = (\pi\varepsilon_0/2)sE_0(z, t)$, which increases

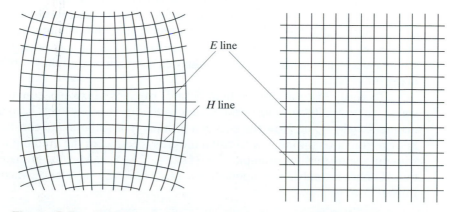

Figure 6-7   Field lines far from parallel wires, and infinitely far from them.

with the wire separation s. This just means that to maintain the field strength E_0 midway between the wires as they are increasingly separated requires more and more charge on the wires, as we may expect. For infinite separation, the charge per unit length would have to become infinite to retain a finite field strength infinitely far from the sources.

In the limit of infinite separation of the source charges in the wires, there is no source in any portion of finite space. Another way to look at this is that, in this limit, the uniform (in the transverse xy-plane) fields

$$\mathbf{E}(x, y, z, t) = E_0(z, t)\hat{\mathbf{x}}, \qquad \mathbf{H}(x, y, z, t) = E_0(z, t)\hat{\mathbf{y}}/\eta_0$$

are sourceless fields, in the sense of the normal modes that we examined in Chapter 4. These field patterns satisfy Maxwell's equations without any sources to generate them. Their strength E_0 represents an undetermined amplitude; they are homogeneous solutions of Maxwell's equations. Such fields can be superimposed, forming fundamental building blocks for field patterns generated by actual sources.

Although these sourceless fields are uniform in the xy-plane, they do depend on z and t. The function $E_0(z, t)$ is proportional to the charge density $q(z, t)$, which we know constitutes a wave $q(z - ct)$ when the wires have finite separation. We may wonder whether the wave properties survive the process of achieving infinite separation of the wires or, effectively, in the absence of the wires. We can confirm that they do by simply verifying that the uniform propagating fields satisfy the source-free Maxwell equations.

In a vacuum, away from any sources, electromagnetic fields may exist, provided they satisfy

$$\nabla \times \mathbf{E} = -\mu_0 \partial \mathbf{H}/\partial t, \qquad \nabla \times \mathbf{H} = \varepsilon_0 \partial \mathbf{E}/\partial t \qquad (6.107)$$

at each point and at each instant. These are a pair of vector partial differential equations. Generally, their solution comprises three-dimensional fields, but here we are interested in verifying that the electric field $E_0(z, t)\hat{\mathbf{x}}$ and its companion magnetic field, which both vary in only one dimension, can satisfy the equations.

It must be emphasized that the $\mathbf{E}$ and $\mathbf{H}$ fields we are examining are severely constrained, in that they vary in only one direction, z, and only in their vector magnitudes, as they never tilt away from their orientations along $\hat{\mathbf{x}}$ and $\hat{\mathbf{y}}$. These constraints are by no means required by Maxwell's equations, and they surely eliminate many other solutions.

To verify that the electric field $E_0(z, t)$ satisfies Maxwell's equations if it is a wave $f(z - ct)$, its curl must be equated to the appropriate time derivative of the magnetic field E_0/η_0. Now the curl of a product of a scalar and a constant vector is known from a vector identity to equal the cross product of the gradient of the scalar with the vector; thus, for example, $\nabla \times (\hat{\mathbf{x}} E_0) = \nabla E_0 \times \hat{\mathbf{x}}$. Also, the gradient of a function of only z has only a z-component: $\nabla E_0(z, t) = \hat{\mathbf{z}}(\partial E_0/\partial z)$. Hence, Maxwell's equations reduce to

$$(\partial E_0/\partial z)\hat{\mathbf{z}} \times \hat{\mathbf{x}} = -\mu_0(\partial[E_0/\eta_0]/\partial t)\hat{\mathbf{y}} \qquad (6.108)$$

and

$$(\partial[E_0/\eta_0]/\partial z)\hat{\mathbf{z}} \times \hat{\mathbf{y}} = \varepsilon_0(\partial E_0/\partial t)\hat{\mathbf{x}}. \tag{6.109}$$

Since $\hat{\mathbf{z}} \times \hat{\mathbf{x}} = \hat{\mathbf{y}}$ and $\hat{\mathbf{z}} \times \hat{\mathbf{y}} = -\hat{\mathbf{x}}$, the equations require of $E_0(z, t)$ that

$$\partial E_0/\partial z = -(\mu_0/\eta_0)\partial E_0/\partial t \tag{6.110}$$

and

$$\partial E_0/\partial z = -(\varepsilon_0\eta_0)\partial E_0/\partial t. \tag{6.111}$$

Because $\eta_0 = \sqrt{\mu_0/\varepsilon_0}$, we have

$$\mu_0/\eta_0 = \varepsilon_0\eta_0 = \sqrt{\mu_0\varepsilon_0} = 1/c, \tag{6.112}$$

so that there is only one requirement on $E_0(z, t)$, namely

$$\partial E_0/\partial z = -(1/c)\partial E_0/\partial t. \tag{6.113}$$

The spatial derivative is to be proportional to the time derivative. This is indeed satisfied by any function of the combined variable $z - ct$:

$$E_0(z, t) = f(z - ct), \tag{6.114}$$

which is a wave, of any shape $f(z)$, traveling along the z-axis at the speed of light, as was to be verified. We have thereby confirmed that the transversely uniform field pattern propagating along z, obtained in the limit of infinite separation of the parallel wires, satisfies Maxwell's equations without sources and is therefore a sourceless normal mode of empty space.

WAVE DESIGNATORS

The electromagnetic field configuration comprising a wave that propagates in a fixed direction (the $+z$ direction), that does not vary at all (is uniform) in the plane perpendicular to that direction, and whose field vectors have fixed directions in space can now be described in full detail. For any waveform $F(t)$, such fields are given by

$$\mathbf{E}(z, t) = \hat{\mathbf{e}}F(t - z/c), \quad \mathbf{H}(z, t) = \hat{\mathbf{h}}F(t - z/c)/\eta_0, \quad \hat{\mathbf{e}} \times \hat{\mathbf{h}} = \hat{\mathbf{z}}. \tag{6.115}$$

The unit vectors $\hat{\mathbf{e}}$, $\hat{\mathbf{h}}$ are perpendicular to each other and to the direction of propagation, $\hat{\mathbf{z}}$. $\mathbf{E}$, $\mathbf{H}$, and $\hat{\mathbf{z}}$ form a right-handed orthogonal triad of vectors, at every point and instant.

The electric and magnetic fields vary in space (along z) and in time, as a wave motion. In time, the wave progresses along z, at the speed of light, c. In fact, light is itself an electromagnetic wave, but of a more complex type, with certain fluctuations and coherence properties.

At any point in space, the electric and magnetic fields vary in time in unison, or *in phase*. At any instant, the waveshapes of the two fields correspond exactly. The waveshape (versus z) is in fact the same as the waveform (versus t), but displayed backward and rescaled by the factor c. At any point and instant, the magnitudes of the electric and magnetic field strengths are related by a constant factor, $\eta_0 = 376.7$ ohms, the *impedance of free space*.

Because the simple wave we have described differs from a host of waves of other, more complicated types, a number of specific terms are in use to designate various features that may not be shared by more general forms.

At any instant, the locations in space at which the electric and the magnetic fields of this wave each have some fixed value are those points where the phase, or the quantity $(t - z/c)$, is some fixed constant. This collection of points forms a plane, and this plane is perpendicular to the z-axis. As time elapses, any such plane of fixed field values moves along z, perpendicular to itself, at the speed of light. This type of wave is referred to as a *plane wave*.

The field vectors $\mathbf{E}, \mathbf{H}$ of this wave are everywhere and always perpendicular to the direction of propagation. Accordingly, this is called a *transverse electromagnetic wave*, or a TEM wave.

The field vectors here do not tilt; their directions remain constant, each along an unchanging line directed along $\hat{\mathbf{e}}$ and $\hat{\mathbf{h}}$, at every point and for all time. The behavior of the field vectors is referred to as the wave's *polarization* and this wave is said to be *linearly polarized*.

The wave we consider here has no spatial variation of the fields, other than its propagation along z. That is, the only variation in space is along z, as a wave motion. The wave has the same appearance for all values of x, y; it is called a *uniform* wave. In particular, it is not a "beam"; a beam has only a finite transverse extent and hence is inherently not independent of the transverse coordinates x, y.

The wave is a "uniform, linearly polarized, transverse electromagnetic plane wave," of arbitrary waveform, in a vacuum.

Although Maxwell's equations are satisfied by this TEM plane wave for *any* waveform whatsoever, the wave equation does prescribe an important relationship between the time scale T of the waveform that evolves at any one point and the distance scale L of the waveshape along the direction of propagation, as seen at any one instant: The scales are related by $L = cT$, where c is the vacuum speed of light.

Caution: The uniform plane wave is a permissible solution to the Maxwell equations alone, without sources and without boundary conditions. It is physically quite unrealistic, in that the fields extend undiminished all the way out to infinity in any transverse plane. It is, however, feasible to set up a wave that is a close facsimile of this uniform one, varying along the z-direction and almost undiminished over vast distances perpendicular to z, vast on the scale L of variation along z. For example, radiation from the sun reaches the earth essentially as a plane wave and substantially uniform, by virtue of the immense distances involved.

SINUSOIDAL PLANE WAVES

Let the waveform $f(t)$ of a plane wave at $z = 0$ be a complex exponential, which represents a pair of steady-state sinusoids at frequency ω: $f(t) = e^{j\omega t}$. Elsewhere than at the $z = 0$ plane, the general vacuum result says that the emitted signal $f(t)$ appears delayed, as $f(t - z/c)$. For the complex exponential, this is $e^{j\omega(t-z/c)}$ as a special case of the general result. We rewrite this as $e^{j(\omega t - kz)}$, where $k = \omega/c$. A number of designators are in common use to refer to the various features of such complex signals.

The plane wave is now complex and takes the form

$$\mathbf{E} = \hat{\mathbf{e}}E_0 e^{j(\omega t - kz)}, \quad \mathbf{H} = \hat{\mathbf{h}}(E_0/\eta_0)e^{j(\omega t - kz)}, \quad \hat{\mathbf{e}} \times \hat{\mathbf{h}} = \hat{\mathbf{z}}. \tag{6.116}$$

The various parts are identified as follows.

The coefficient of the complex exponential, E_0, is a complex number, the *complex amplitude* of the electric field. As such, it has a real and an imaginary part, or else a magnitude and a phase: $E_0 = |E_0|e^{j\varphi_0}$. Consequently, upon combining the exponentials, the complex electric field can also be expressed as $\mathbf{E} = \hat{\mathbf{e}}|E_0|e^{j(\omega t - kz + \varphi_0)}$ and its real part is $\hat{\mathbf{e}}|E_0|\cos(\omega t - kz + \varphi_0)$, which is a physically meaningful plane wave's real electric field. It is sinusoidal at frequency ω; the sinusoid has peak amplitude $|E_0|$ and initial phase φ_0 at $z = 0$. The electric field vector is directed along the real unit vector $\hat{\mathbf{e}}$.

The real quantity $\Phi = \omega t - kz + \varphi_0$ is the *phase* of the wave.

The time coefficient, ω, is the *angular frequency* of the sinusoidal signal, measured in radians per second.

Since the function $e^{j\xi}$ is periodic in ξ, with period 2π, the signal $e^{j\omega t}$, or also $e^{j(\omega t - kz)}$, is periodic in time, with period $\tau = 2\pi/\omega$. The reciprocal of the period is the cyclical frequency, $f = 1/\tau = \omega/2\pi$, the number of full cycles of sinusoidal oscillation per second, measured in hertz.

The distance coefficient, k, is the *phase constant* of the sinusoidal waveshape, measured in radians per meter. Since the function $e^{j\xi}$ is periodic in ξ, with period 2π, the waveshape e^{-jkz}, or also $e^{j(\omega t - kz)}$, is periodic in space, with period $\lambda = 2\pi/k$, called the *wavelength*.

As with the time and distance scales of any waveform propagating in a vacuum, the period and the wavelength are related by the speed of light. The expressions

$$\lambda = c\tau \quad \text{or} \quad \omega = kc \quad \text{or} \quad f\lambda = c \tag{6.117}$$

are equivalent statements of how the time scale of the waveform, τ, and the distance scale of the waveshape, λ, are related.

The fields maintain a fixed value wherever and whenever the phase $\Phi(z, t)$ remains fixed. As time elapses, $\Phi(z, t)$ stays fixed only if the z-coordinate is allowed to increase at the appropriate rate dz/dt to keep Φ constant, or to make $d\Phi = 0$. Since $d\Phi = \omega \, dt - k \, dz$, the rate of progression along z that would be necessary to maintain a fixed phase (and hence fixed fields) is the *phase velocity* $v_p = dz/dt$ that achieves $d\Phi = 0$, namely

$$v_p = \omega/k. \tag{6.118}$$

The phase velocity is hence the *ratio of the coefficients of t and of z in the phase*. It is also the product of the cyclical frequency f and the wavelength λ, since $\omega = 2\pi f$ and

$1/k = \lambda/2\pi$. For the plane wave propagating in a vacuum, the phase velocity is the speed of light: $v_p = c$.

To this point, we have designated the single direction of propagation as the z-direction. As soon as waves traveling in different directions are to be considered, it becomes appropriate to define a vector that points in the direction of propagation. This may be a unit vector $\hat{\mathbf{n}}$, whereupon the distance coordinate along the direction of wave motion is no longer just z but now becomes $\hat{\mathbf{n}} \cdot \mathbf{r}$, where $\mathbf{r} = (x, y, z)$ is the position vector from the origin to any point in space. The phase for a wave that propagates along the direction of $\hat{\mathbf{n}}$ is then $\Phi(\mathbf{r}, t) = \omega t - k\hat{\mathbf{n}} \cdot \mathbf{r}$. It is convenient to combine the phase constant k with the unit vector $\hat{\mathbf{n}}$ to form the *wave vector* $\mathbf{k} = k\hat{\mathbf{n}}$. This points in the direction of phase progression and has the magnitude $|\mathbf{k}| = k = \omega/v_p = 2\pi/\lambda$. The *wavefront* is the surface (in our case, planar) of some fixed phase Φ_0, identified analytically by the constraint $\omega t - \mathbf{k} \cdot \mathbf{r} = \Phi_0$. This is the equation of a *plane* perpendicular to the direction of $\mathbf{k}$ or $\hat{\mathbf{n}}$, and this plane's distance d from the origin is given by $\hat{\mathbf{n}} \cdot \mathbf{r}$. This is just $(\omega t - \Phi_0)/k$, and this distance increases in time at the phase velocity ω/k. This wavefront is depicted in Figure 6-8, for one instant of time. The $\mathbf{E}$ and $\mathbf{H}$ fields are in this plane, perpendicular to the wave vector $\mathbf{k}$.

To summarize, the complex plane wave $\mathbf{E}e^{j(\omega t - \mathbf{k} \cdot \mathbf{r})}$ is distinguished and identified by its frequency ω, its wave vector $\mathbf{k}$, and its complex amplitude $\mathbf{E}$. Its real part, and also its imaginary part, is a sinusoidal plane wave that propagates along $\mathbf{k}$, at the speed $\omega/|\mathbf{k}|$. When we express the field in the time domain as $\mathrm{Re}[\mathbf{E}e^{j\omega t}]$, the factor $\mathbf{E}$ is a complex constant vector. If we need to avoid confusion between the time- and frequency-domain descriptions of the fields, we will use script type to denote the real physical field in the time domain:

$$\mathcal{E} = \mathcal{E}(\mathbf{r}, t) = \mathrm{Re}[\mathbf{E}e^{-j\mathbf{k} \cdot \mathbf{r}} e^{j\omega t}]$$

$$= (\mathrm{Re}\,\mathbf{E})\cos(\omega t - \mathbf{k} \cdot \mathbf{r}) - (\mathrm{Im}\,\mathbf{E})\sin(\omega t - \mathbf{k} \cdot \mathbf{r}). \tag{6.119}$$

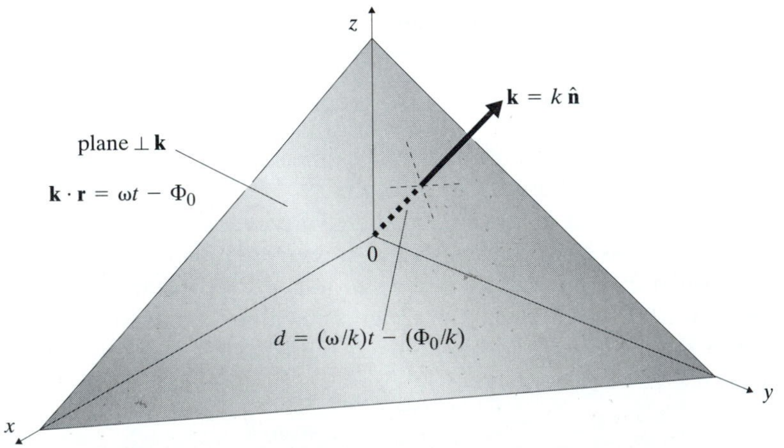

Figure 6-8   Planar wavefront propagating along $\mathbf{k}$.

Example 6.10

(a) *Write a TEM plane wave, $\mathcal{E}(x, y, z, t)$, $\mathcal{H}(x, y, z, t)$, with polarizations along $\hat{\mathbf{e}}$ and $\hat{\mathbf{h}}$, of waveform $F(t)$, but traveling along the direction of the vector $\mathbf{b} = 4\hat{\mathbf{x}} - 3\hat{\mathbf{y}} + 12\hat{\mathbf{z}}$ instead of along $\hat{\mathbf{z}}$.*

(b) *If the polarization $\hat{\mathbf{e}}$ is in the xy-plane, with positive x- and y-components, find $\hat{\mathbf{e}}$ and $\hat{\mathbf{h}}$.* «»

(a) Since the unit vector in the direction of propagation is $\mathbf{b}/|\mathbf{b}|$, we have $\mathcal{E}(x, y, z, t) = \hat{\mathbf{e}}G(x, y, z, t)$ and $\mathcal{H}(x, y, z, t) = \hat{\mathbf{h}}G(x, y, z, t)/\eta_0$, with $G(x, y, z, t) = F(t - \mathbf{b} \cdot \mathbf{r}/|\mathbf{b}|c) = F(t - [4x - 3y + 12z]/13c)$.

(b) Since $\hat{\mathbf{e}}$, $\hat{\mathbf{h}}$, and $\mathbf{b}$ form an orthogonal triad, $\hat{\mathbf{e}} \cdot \mathbf{b} = 0$, which gives

$$\hat{\mathbf{e}} = (3\hat{\mathbf{x}} + 4\hat{\mathbf{y}})/5 \quad \text{and} \quad \hat{\mathbf{h}} = \mathbf{b} \times \hat{\mathbf{e}}/|\mathbf{b}| = (-48\hat{\mathbf{x}} + 36\hat{\mathbf{y}} + 25\hat{\mathbf{z}})/65.$$

Example 6.11

The electric field of a sinusoidal plane wave is linearly polarized along the x-axis and travels along the y-axis, in vacuum. At time $t = 0$, the field is zero at the point $(3, 0, 2)$ and attains its peak value of $3\,V/m$ at the point $(5, 8, -3)$. All position coordinates are in meters.

(a) *Express this wave's electric field $\mathcal{E}$ as a real, time-domain field.*

(b) *Express it as a complex, frequency-domain field $\mathbf{E}$.*

(c) *What is the lowest possible frequency, f, of this wave?* «»

(a) Since the initial field is zero at $y = 0$, its form must be $\hat{\mathbf{x}}E_0 \sin(ky - \omega t)$. The specifications call for $E_0 = 3$ V/m, and since the peak occurs at phase $\pi/2$, we must have $ky_0 = \pi/2 + l2\pi$, where l is an integer, at $y_0 = 8$ m. Also, $\omega = kc$, so we have all the parameters, k, ω, E_0, within an undetermined integer l.

(b) By comparing $\hat{\mathbf{x}}E_0 \sin(ky - \omega t)$ with $\mathrm{Re}[\mathbf{E}e^{j\omega t}]$, we identify the complex amplitude as $\mathbf{E} = \hat{\mathbf{x}}jE_0 e^{-jky}$.

(c) The lowest possible frequency is the one that makes k the smallest it can be. This happens for $l = 0$, giving $k = \pi/16$ m^{-1}, so that $\omega = kc = \pi c/16$ and $f = \omega/2\pi = 9.37$ MHz.

POLARIZATION

Compared to the simple plane wave $\mathcal{E}(\mathbf{r}, t) = \hat{\mathbf{e}}E_0 f(t - z/c)$ in the time domain that we studied first, the sinusoidal plane wave $\mathbf{E} = \mathbf{E}_0 e^{-j\mathbf{k} \cdot \mathbf{r}}$ in the complex frequency domain, which is $\mathcal{E}(\mathbf{r}, t) = \mathrm{Re}[\mathbf{E}e^{j\omega t}] = \mathrm{Re}[\mathbf{E}_0 e^{-j\mathbf{k} \cdot \mathbf{r}}e^{j\omega t}]$ in the time domain, is more general in some respects but more specialized in others. The former could have any waveform $f(t)$ at all, but the latter is only sinusoidal at frequency ω. On the other hand, the former was allowed to vary only along z, but the latter varies along any direction, given by the vector $\mathbf{k}$. The former propagates at the vacuum speed of light, c, regardless of the waveform $f(t)$, but the latter's phase velocity is $v_p = 1/\sqrt{\mu(\omega)\varepsilon(\omega)}$, which varies with frequency for dielectric materials other than a vacuum. Most important, the former was required to have field vectors that remain fixed in direction, along $\hat{\mathbf{e}}$ and $\hat{\mathbf{h}}$, but this restriction need not be imposed when we solve Maxwell's equations in the frequency domain. In fact, the wave vectors $\mathcal{E}, \mathcal{H}$ can tilt in space as well as vary in strength; they can reorient themselves as time progresses, but in a particular manner consistent with their being harmonic (sinusoidal) in time.

The behavior of the field vectors at some fixed observation point as time elapses is termed the *polarization* of the wave. The wave field vectors previously considered were of a type that vary only in magnitude but don't tilt in their direction; that behavior was termed *linear polarization*. For the more general sinusoidal plane waves, how the field vectors vary in strength and how they tilt or twist in direction are related.

Just as the time variation of the position vector $\mathbf{r} = \mathbf{r}(t)$ in a mechanics problem describes an orbit or trajectory of, say, a particle moving through space, so does any vector function of time, like $\mathcal{E}(t)$, describe some curve in the space in which that vector may be "plotted," as in Figure 6-9. Thus, *any vector function of one scalar variable plots as a curve* in the space of that vector; that curve is parametrized by the scalar argument of the function. If we could visualize the vector $\mathcal{E}(\mathbf{r}_0, t) = \mathcal{E}(t)$ at a fixed observation point $\mathbf{r}_0$ as it twists and turns and waxes and wanes in time, what curve would the tip of the $\mathcal{E}(t)$ vector trace out?

Analytically, the answer is readily deduced by converting from the frequency domain to the time domain. If $\mathbf{E} = \mathbf{E}_1 + j\mathbf{E}_2$ is the given complex constant vector, then the time variation of the actual field is given by

$$\mathcal{E}(t) = \text{Re}\{\mathbf{E}e^{j\omega t}\} = \mathbf{E}_1 \cos \omega t - \mathbf{E}_2 \sin \omega t, \tag{6.120}$$

where $\mathbf{E}_1$ and $\mathbf{E}_2$ are the real and imaginary parts of $\mathbf{E}$, hence real vectors. This gives the curve traced by $\mathcal{E}(t)$ in parametric form and can readily be plotted, by evaluating $\mathcal{E}$ at many instants of time. The curve lies entirely in the plane formed by the real and imaginary parts of $\mathbf{E}$, and this plane is perpendicular to the wave vector $\mathbf{k}$. The parametric form of the equation for this curve may not be so familiar to us, however, as compared to an alternative form, one that relates the components of the $\mathcal{E}$ vector to each other without reference to the time parameter. Conversion of the parametric form to a nonparametric equation requires only the elimination of the time t.

This elimination is easily achieved by solving for the trigonometric functions of time in (6.120), in terms of the field components. When the cosine and sine functions

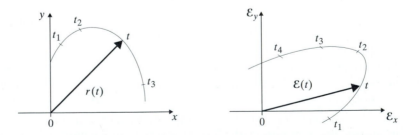

Figure 6-9   Orbit $\mathbf{r}(t)$ of a particle; curve traced by $\mathcal{E}(t)$ vector.

of time are squared and added, the sum is unity and time has been eliminated from the equation. Since the real and imaginary parts of $\mathbf{E}$ lie in the plane perpendicular to $\mathbf{k}$, they have only two orthogonal components in that plane; let these be called the x- and y-components. It follows that (6.120) can be expressed as the matrix equation

$$\begin{bmatrix} \mathcal{E}_x(t) \\ \mathcal{E}_y(t) \end{bmatrix} = \begin{bmatrix} E_{1x} & E_{2x} \\ E_{1y} & E_{2y} \end{bmatrix} \begin{bmatrix} \cos\omega t \\ -\sin\omega t \end{bmatrix} \tag{6.121}$$

or, more simply, as

$$\mathbf{F} = \mathbf{M}\mathbf{\Omega}, \tag{6.122}$$

where $\mathbf{F}$ comprises the components of the variable $\mathcal{E}(t)$, $\mathbf{M}$ the components of the real vectors $\mathbf{E}_1$ and $\mathbf{E}_2$ as its columns, and $\mathbf{\Omega}$ the trigonometric functions. Solving for the latter, we get

$$\mathbf{\Omega} = \mathbf{M}^{-1}\mathbf{F}, \tag{6.123}$$

where the two entries in $\mathbf{\Omega}$ are $\cos\omega t$ and $-\sin\omega t$, so that multiplication by the transpose of the matrix yields the sum of their squares, which is unity:

$$\mathbf{\Omega}'\mathbf{\Omega} = 1 = \mathbf{F}'(\mathbf{M}^{-1})'(\mathbf{M}^{-1})\mathbf{F} = \mathbf{F}'(\mathbf{M}\mathbf{M}')^{-1}\mathbf{F}, \tag{6.124}$$

where primes indicate transposes of the matrices and we have used the fact that the inverse of a product of matrices is the product of their inverses, but with the factors in reverse order.

The relation $\mathbf{F}'\mathbf{G}\mathbf{F} = 1$ is a quadratic form in the components of $\mathbf{F}$, or of the vector variable $\mathcal{E}$, or in $X = \mathcal{E}_x$ and $Y = \mathcal{E}_y$, its rectangular components in the plane. Fully expanded, this is a quadratic form like $AX^2 + 2BXY + CY^2 = 1$, where A, B, B, C are the four entries in the symmetric real matrix $\mathbf{G} = (\mathbf{M}\mathbf{M}')^{-1}$. This quadratic form represents a conic section in the XY-plane; it can be a hyperbola, parabola, or ellipse, depending primarily (assuming either A or C is positive, which they both are) on the sign of the discriminant $AC - B^2$. But this discriminant is just the determinant of the matrix $\mathbf{G}$, which is the reciprocal of the determinant of the product $\mathbf{M}\mathbf{M}'$; this, in turn, equals the product of the determinants of the two factors and these, being transposes of each other, have equal determinants. Hence, the discriminant of $\mathbf{G}$ is here $1/(\det \mathbf{M})^2$, which is invariably positive, so that the curve must be an ellipse, or a degenerate form of an ellipse (a circle or a straight line) in special cases. We conclude that the curve traced out by the field is an ellipse; we say the general harmonic plane wave is *elliptically polarized*.

This designation is descriptive of the behavior of the field vector as time goes on. Giving the field in the frequency domain by specifying a complex constant vector $\mathbf{E}$ implies that the time-domain variation of the actual field $\mathcal{E}(t)$ must be sinusoidal in time, but this requires only that *each vector component oscillate sinusoidally*; the entire vector is able to change its direction as well as vary in strength, but these two variations are now seen to be constrained to make the tip of the $\mathcal{E}(t)$ vector trace out an ellipse in the plane perpendicular to the direction $\mathbf{k}$ of propagation.

Figure 6-10 shows the ellipse formed by the motion of the tip of the $\mathcal{E}(t)$ vector in time, as a resultant of the summation of the two vectors $\mathbf{E}_1 \cos\omega t$ and $-\mathbf{E}_2 \sin\omega t$,

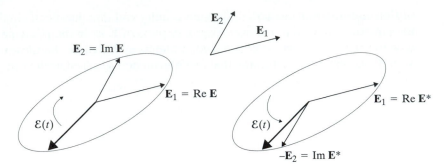

Figure 6-10   Ellipses traced out by $\mathcal{E}(t)$, based on complex vectors $\mathbf{E}$ and $\mathbf{E}^*$

where $\mathbf{E}_1$ and $\mathbf{E}_2$ are the real and imaginary parts of the complex constant vector $\mathbf{E}$; these are fixed and are also shown in the drawing. The figure also indicates that the same ellipse can be traced out, but in the opposite direction, by the $\mathcal{E}(t)$ vector that corresponds to the complex conjugate $\mathbf{E}^*$ of the vector $\mathbf{E}$ used for the first ellipse.

Which way the field vector rotates around the ellipse is a matter about which there is much confusion, but there need be no big mystery about it. Since $\mathcal{E}(t) = \mathbf{E}_1 \cos \omega t - \mathbf{E}_2 \sin \omega t$, the rotating $\mathcal{E}(t)$ vector is along $\mathbf{E}_1$ when $\omega t = 0$ (and then not again until $\omega t = 2\pi$), but it lies along $\mathbf{E}_2$ when $\omega t = -\pi/2$, earlier than when it reaches $\mathbf{E}_1$. Thus, the sense of the rotation is from Im $\mathbf{E}$ toward Re $\mathbf{E}$. Whether this is clockwise or counterclockwise rotation depends on the configuration of the $\mathbf{E}_1$ and $\mathbf{E}_2$ vectors (i.e., on which way the imaginary part $\mathbf{E}_2$ would have to turn to reach the real part $\mathbf{E}_1$, through the smallest angle). If it's that simple, why is there any confusion? It is the designations used to distinguish the two senses of rotation that may be perplexing.

Since clockwise as seen from the front becomes counterclockwise when viewed from the rear, an unambiguous designation requires specification of whether the rotating vector is being visualized from the direction in which the wave is either approaching or receding from the observer; the sense is reported in terms of the "handedness" of the rotation. By the engineers' standards, with the thumb pointing in the direction of advance of the wave, the polarization is either right-handed or left-handed, according to the hand whose fingers conform to the direction of rotation. In Figure 6-10, if the waves are approaching the reader, then the one on the left is termed *left-hand elliptically polarized (LHEP), wave approaching* while the other is *right-hand ... (RHEP)*.

Unfortunately and incredibly, there is disagreement about this designation. Physicists, especially those versed in optics, traditionally refer to the two cases with the opposite designations (but without specifying "wave approaching"; they prefer to have the thumb point to where the wave is coming from!). There is a further source of confusion in the physicists' preference for the complex exponential to be written $e^{-i\omega t}$, compared to the engineers' $e^{j\omega t}$. The use of i versus j is inconsequential, but the sign change reverses the roles of the real and imaginary parts of **E** in determining the sense of rotation. Still another ambiguity arises from the two signs in the full expression for the phase, $\omega t \pm kz$, which pertains to the direction of advance of the wave. The combination of all these ambiguities effectively frustrates attempts to memorize a rule as to which sense of rotation is designated right or left. It seems safest to compare the orientation of the $\mathcal{E}(t)$ field at two easily distinguished times, say at $\omega t = 0$ and $\omega t = \pi/2$, when the complex exponential is purely real and purely imaginary, respectively, making it especially simple to get $\mathcal{E}(t)$ from **E**. One should also be specific about the direction of advance of the wave.

Example 6.12

Identify the sense of rotation of each of the following elliptically polarized plane waves, using the engineering standard [RHEP or LHEP handedness, with wave approaching and $e^{j(\omega t - kz)}$ dependence]. E_0 is a real scale factor.

(a) *Given orthogonal components:* $\quad \mathbf{E} = E_0\{\hat{\mathbf{x}}[5 - j3] + \hat{\mathbf{y}}[-2 + j]\}$

(b) *Given real and imaginary parts:* $\quad \mathbf{E} = E_0\{[2\hat{\mathbf{x}} + 3\hat{\mathbf{y}}] + j[5\hat{\mathbf{x}} - 6\hat{\mathbf{y}}]\}$

(c) *Given components as magnitude and phase:*

$$\mathbf{E} = E_0\{\hat{\mathbf{x}}5e^{j\pi/8} - \hat{\mathbf{y}}3e^{-j\pi/6}\} \qquad \langle\!\rangle$$

In every case, the sense of rotation is from the direction of $\mathbf{E}_2 = \text{Im}\,\mathbf{E}$ to that of $\mathbf{E}_1 = \text{Re}\,\mathbf{E}$, through the smaller of the two angles between them. Refer to Figure 6-11.

(a) $\mathbf{E}_1 = 5\hat{\mathbf{x}} - 2\hat{\mathbf{y}}$ and $\mathbf{E}_2 = -3\hat{\mathbf{x}} + \hat{\mathbf{y}}$. Here, $\mathbf{E}_2$ is in the second quadrant, at an angle $\tan^{-1}(1/3)$ to the negative x-axis, while $\mathbf{E}_1$ is in the fourth quadrant, at angle $\tan^{-1}(2/5)$ to the x-axis. The smaller of the two angles that would bring $\mathbf{E}_2$ to $\mathbf{E}_1$ requires rotation through the third quadrant, so that the wave is RHEP.

(b) $\mathbf{E}_1 = 2\hat{\mathbf{x}} + 3\hat{\mathbf{y}}$ and $\mathbf{E}_2 = 5\hat{\mathbf{x}} - 6\hat{\mathbf{y}}$. Here, rotation from $\mathbf{E}_2$ to $\mathbf{E}_1$ clearly gives an RHEP wave.

(c) $\mathbf{E}_1 = 4.62\hat{\mathbf{x}} - 2.60\hat{\mathbf{y}}$ and $\mathbf{E}_2 = 1.91\hat{\mathbf{x}} + 1.50\hat{\mathbf{y}}$: Polarization is LHEP.

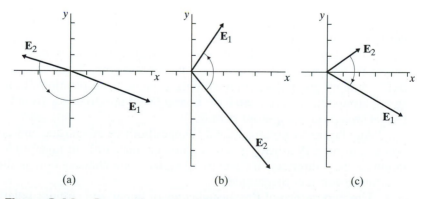

(a)   (b)   (c)

Figure 6-11   Sense of rotation from $\mathbf{E}_2$ to $\mathbf{E}_1$ gives handedness of wave.

LINEAR AND CIRCULAR POLARIZATION

There are two important special cases of the behavior of the field in time, one in which the ellipse collapses to a straight line, the other in which it opens out to a circle.

The case of linear polarization occurs when the real and imaginary parts of the **E** vector are parallel, so that $\mathbf{E} = (E_1 + jE_2)\hat{\mathbf{e}}$. The derivation of the equation for the curve traced by $\mathcal{E}(t)$ then fails, because the matrix M in (6.122) is singular when $\mathbf{E}_1$ and $\mathbf{E}_2$ are parallel, so that there is no inverse matrix. The behavior in time is then given by $\mathcal{E}(t) = (E_1 \cos \omega t - E_2 \sin \omega t)\hat{\mathbf{e}}$, which remains aligned with the fixed, real, unit vector $\hat{\mathbf{e}}$ (or opposite to it) but does not rotate or tilt in any way; this is linear polarization.

The other special case is that in which the ellipse reduces to a circle. This happens when the quadratic form $AX^2 + 2BXY + CY^2 = 1$ satisfies two conditions: It has no cross term, so that $B = 0$, and the ellipse's semiaxes are equal, which happens if $A = C$. In terms of the equivalent quadratic form $\mathbf{F'GF} = 1$, the matrix $\mathbf{G} = (\mathbf{MM'})^{-1}$ is required to be proportional to the unit matrix (the constant of proportionality is then interpretable as the inverse square of the radius of the resultant circle). This condition is satisfied when $\mathbf{M'}$ (the transpose of matrix $\mathbf{M}$) is proportional to $\mathbf{M}^{-1}$ (the inverse of matrix $\mathbf{M}$). In that case, the product $\mathbf{MM'}$ is the same as $\mathbf{M'M}$ and writing out the condition that $\mathbf{M'M} \propto \mathbf{I}$ shows that the real and imaginary parts of $\mathbf{E}$ are orthogonal vectors $(\mathbf{E}_1 \cdot \mathbf{E}_2 = 0)$ and are of equal magnitude $(E_1^2 = E_2^2)$. The complex vector $\mathbf{E}$ then has the form $\mathbf{E} = (\hat{\mathbf{e}}_1 + j\hat{\mathbf{e}}_2)E$, where E is real and the real unit vectors $\hat{\mathbf{e}}_1$ and $\hat{\mathbf{e}}_2$ are perpendicular to each other and to the direction of propagation of the wave. The time-domain field is then the vector $\mathcal{E}(t) = \hat{\mathbf{e}}_1 E \cos \omega t - \hat{\mathbf{e}}_2 E \sin \omega t$ and it has the following properties. Its magnitude is just E and remains constant in time; its direction is at an angle ωt to that of vector $\hat{\mathbf{e}}_1$; this angle increases uniformly in time, so that the real physical electric field vector turns, its tip tracing out a circle as it sweeps around in the direction from $\hat{\mathbf{e}}_2$ toward $\hat{\mathbf{e}}_1$. This is circular polarization.

For example, the complex constant $\mathbf{E} = (\hat{\mathbf{x}} - j\hat{\mathbf{y}})E$ represents a right-hand circularly polarized (RHCP) plane wave if the wave travels along $\hat{\mathbf{z}}$, the complex exponential is $e^{j(\omega t - kz)}$, and the engineers' standards are adopted. That the sense of rotation is right-handed is clear from the fact that the rotation is from the imaginary part, here $-\hat{\mathbf{y}}$, to the real part, here $\hat{\mathbf{x}}$, as time elapses.

Note that although all polarizations involve periodic motion of the real field vector $\mathcal{E}(t)$, with period $2\pi/\omega$, circular polarization entails a uniform rotation of the field, while elliptic polarization exhibits a nonuniform rotation: It takes as long to sweep around the ellipse from $\mathbf{E}_2$ to $\mathbf{E}_1$ as it then takes to sweep from $\mathbf{E}_1$ to $-\mathbf{E}_2$; both partial sweeps take a quarter period.

An elliptically polarized field changes both its magnitude and direction, periodically in time. A linearly polarized one changes only in magnitude while maintaining a fixed direction; a circularly polarized one changes only in direction while retaining a constant magnitude.

The importance of the special cases of linear and circular polarization is enhanced by the fact that a wave of any polarization, generally elliptic, can be expressed as a superposition of two simpler partial waves. The two partial waves can be linearly polarized, at right angles to each other; alternatively, they can be circularly polarized, with opposite senses.

The first assertion is confirmed simply by recognizing that an arbitrary complex vector $\mathbf{E}$ is always expressible as $\mathbf{E} = \hat{\mathbf{x}}E_x + \hat{\mathbf{y}}E_y$, where the first term is evidently linearly polarized along $\hat{\mathbf{x}}$ and the second is linearly polarized along $\hat{\mathbf{y}}$. Note that E_x and E_y are complex numbers, in general.

The second assertion, that a superposition of oppositely sensed circularly polarized waves can comprise any elliptically polarized wave, is verified by noting that *any* complex vector $\mathbf{E}$ can be expressed as a combination of the circularly polarized vectors $\hat{\mathbf{x}} \pm j\hat{\mathbf{y}}$. It is in fact an identity that $\mathbf{E} = R(\hat{\mathbf{x}} - j\hat{\mathbf{y}}) + L(\hat{\mathbf{x}} + j\hat{\mathbf{y}})$, where $R = (E_x + jE_y)/2$ and $L = (E_x - jE_y)/2$. Note again that E_x and E_y are complex numbers, so that, despite appearances, R and L are not necessarily complex conjugate numbers. The magnitudes of the complex numbers R, L give the radii of the circles traced by the two circularly polarized field vectors.

Example 6.13

Find A, B, C, D in the following complex amplitude coefficients of $e^{j(\omega t - kz)}$ if
   (a) $\mathbf{E} = E_0\{[3\hat{\mathbf{x}} + 4\hat{\mathbf{y}}] + j[-5\hat{\mathbf{x}} + A\hat{\mathbf{y}}]\}$ *is a linearly polarized wave;*
   (b) $\mathbf{E} = E_0\{\hat{\mathbf{x}}[3 + j4] + \hat{\mathbf{y}}[-5 + jB]\}$ *is a linearly polarized wave;*
   (c) $\mathbf{E} = E_0\{[3\hat{\mathbf{x}} + C\hat{\mathbf{y}}] + j[4\hat{\mathbf{x}} + D\hat{\mathbf{y}}]\}$ *is circularly polarized (RHCP).*
In each case, E_0 is real. ◇

   (a) The real and imaginary parts $[3\hat{\mathbf{x}} + 4\hat{\mathbf{y}}]$ and $[-5\hat{\mathbf{x}} + A\hat{\mathbf{y}}]$ must be proportional, so $A = (-5/3)4 = -20/3$.
   (b) The real and imaginary parts $[3\hat{\mathbf{x}} - 5\hat{\mathbf{y}}]$ and $[4\hat{\mathbf{x}} + B\hat{\mathbf{y}}]$ must be proportional, so $B = (4/3)(-5) = -20/3$.
   (c) For an RHCP wave, the real and imaginary parts $\mathbf{E}_1$ and $\mathbf{E}_2$ must be related by $\hat{\mathbf{z}} \times \mathbf{E}_2 = \mathbf{E}_1$. Here, we need $\hat{\mathbf{z}} \times [4\hat{\mathbf{x}} + D\hat{\mathbf{y}}] = [3\hat{\mathbf{x}} + C\hat{\mathbf{y}}]$, so that $C = 4$ and $D = -3$.

Example 6.14

The combination of two circularly polarized waves, $\mathbf{E}e^{j(\omega t - kz)}$, with

$$\mathbf{E} = E_0\{(3 + j2)[\hat{\mathbf{x}} + j\hat{\mathbf{y}}] + (1 - j3)[\hat{\mathbf{x}} - j\hat{\mathbf{y}}]\}$$

and with E_0 real is an elliptically polarized wave. Is it RHEP or LHEP? ◇

   The real and imaginary parts are $\mathbf{E}_1 = 4\hat{\mathbf{x}} - 5\hat{\mathbf{y}}$ and $\mathbf{E}_2 = -\hat{\mathbf{x}} + 2\hat{\mathbf{y}}$, and rotation from $\mathbf{E}_2$ to $\mathbf{E}_1$ through the smaller angle runs from the second to the fourth quadrant of the plane, through the first quadrant. Hence the wave is LHEP.

OPTICALLY ACTIVE MEDIA

There are media in which the speed of wave propagation depends on the wave's polarization. In particular, left and right circularly polarized waves may travel at different speeds in certain media, notably in a magnetized plasma such as the ionosphere, or in certain *optically active* crystals, or even in a sugar solution. Without pausing to consider the underlying mechanism responsible for the difference in wave speeds, let's look at its consequences in the next example.

Example 6.15

Let the two wave speeds for the two senses of rotation of circularly polarized waves in an optically active medium be such that the phase constant $k = \omega/v$ is $k_L = k_0 + \kappa$ for the LHCP wave and $k_R = k_0 - \kappa$ for the RHCP wave. Suppose that, at $z = 0$, a wave is linearly polarized, along x, and propagates along z in the active medium. What will its polarization be at some arbitrary distance z? ◈

If κ were zero, so that all polarizations propagated at the same speed, making $k = k_0$, then the original linearly polarized field $\mathbf{E}(0) = E_0\hat{\mathbf{x}}$ at $z = 0$ would become $\mathbf{E}(z) = E_0 e^{-jk_0 z}\hat{\mathbf{x}}$ at distance z, which is still linearly polarized, still along $\hat{\mathbf{x}}$ (since the real and imaginary parts of $\mathbf{E}$ are both along $\hat{\mathbf{x}}$). Instead, we know that *circularly* polarized waves have different phase constants in this medium and we need to determine what happens to a *linearly* polarized field. But any polarization, including a linear one, can be expressed as a superposition of two oppositely sensed circular polarizations, for each of which we know the phase constant. We rewrite the original field at $z = 0$ as

$$\mathbf{E}(0) = E_0\hat{\mathbf{x}} = \tfrac{1}{2}E_0(\hat{\mathbf{x}} + j\hat{\mathbf{y}}) + \tfrac{1}{2}E_0(\hat{\mathbf{x}} - j\hat{\mathbf{y}}) \tag{6.125}$$

and assign each circularly polarized partial wave its own propagation factor:

$$\mathbf{E}(z) = \tfrac{1}{2}E_0(\hat{\mathbf{x}} + j\hat{\mathbf{y}})e^{-j(k_0+\kappa)z} + \tfrac{1}{2}E_0(\hat{\mathbf{x}} - j\hat{\mathbf{y}})e^{-j(k_0-\kappa)z}. \tag{6.126}$$

Now, to identify the polarization at distance z, we may collect terms so as to compare the real and imaginary parts more easily.

$$\begin{aligned} E(z) &= \tfrac{1}{2}E_0 e^{-jk_0 z}\big[(\hat{\mathbf{x}} + j\hat{\mathbf{y}})e^{-j\kappa z} + (\hat{\mathbf{x}} - j\hat{\mathbf{y}})e^{+j\kappa z}\big] \\ &= E_0 e^{-jk_0 z}\big[\hat{\mathbf{x}}(e^{j\kappa z} + e^{-j\kappa z})/2 + \hat{\mathbf{y}}(e^{j\kappa z} - e^{-j\kappa z})/2j\big] \\ &= E_0 e^{-jk_0 z}[\hat{\mathbf{x}}\cos\kappa z + \hat{\mathbf{y}}\sin\kappa z] = E_0 e^{-jk_0 z}\hat{\mathbf{e}}, \end{aligned} \tag{6.127}$$

instead of $E_0\hat{\mathbf{x}}$ at $z = 0$. The real and imaginary parts are both along the real vector $\hat{\mathbf{e}}$, so that this $\mathbf{E}(z)$ is still linearly polarized, but now along the unit vector $\hat{\mathbf{e}} = (\hat{\mathbf{x}}\cos\kappa z + \hat{\mathbf{y}}\sin\kappa z)$ instead of along $\hat{\mathbf{x}}$. The new direction of linear polarization makes angle κz with the original direction. Figure 6-12 shows the superposition of the two circularly polarized waves and the resultant rotated linear polarization.

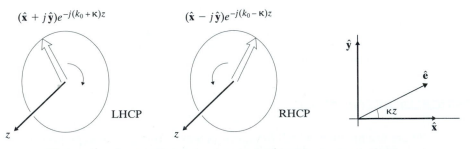

Figure 6-12   Rotation of polarization in optically active medium.

The result is that the wave in this medium remains linearly polarized everywhere, but the orientation of its polarization tilts through an angle κz over an interval of z meters, where κ is half the difference between the phase constants of the two constituent circularly polarized partial waves. The orientation of the field vector tilts progressively from point to point along the direction of propagation, at the rate of κ radians per meter. An instantaneous "snapshot" of the electric field would show a twist in its orientation, as suggested in Figure 6-13. This, however, is only its spatial behavior; at any fixed point, the temporal behavior is merely a sinusoidal oscillation along a fixed direction in space, which defines a linear polarization.

Certain sugars in solution are chemically equivalent but distinguishable from each other by the sense of rotation they impart to the polarization of a light wave that traverses them (dextrose or levulose).

Faraday rotation refers to a similar progressive tilt of the optical wave polarization in a crystal in which the phase velocities of the circularly polarized components of the light are different and dependent on an applied magnetic field. In the latter case, there is a preferred direction in space (that of the applied field) and, upon reflection from a mirror, the wave polarization twists even more. In the case of sugar solutions, there is no such global asymmetry and a reflected wave's polarization unwinds back to its original orientation upon its return.

Optical fibers can also exhibit wave rotatory effects when their phase velocities differ for different polarizations, either by design or by imperfect fabrication that affects the cylindrical symmetry. Such effects can be used to impart modulation on a carrier signal.

The polarization of light from natural or artificial sources exhibits randomness and requires a statistical description. *Unpolarized* light refers to a superposition of many waves, of all possible polarizations, equally distributed so that no particular polarization predominates. Sunlight or thermal radiation emanating from many, independent sources has this character. If, however, the statistical distribution of all the possible polarizations is not uniform in the superposition of waves from many sources, then the light is termed *partially polarized*.

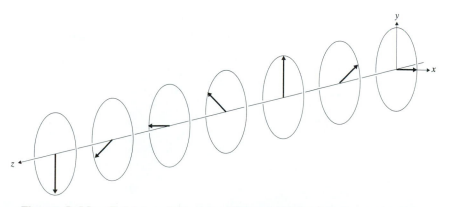

Figure 6-13   Twist in orientation of linear polarization, at one instant.

INCIDENT AND REFLECTED PLANE WAVES

We have considered plane waves at frequency ω propagating in a vacuum, for which the wave speed is $\omega/k = c = 1/\sqrt{\mu_0 \varepsilon_0}$ and the wave impedance is $E/H = \eta_0 = \sqrt{\mu_0/\varepsilon_0}$. In other dielectric media, the same sort of wave can propagate, but with a different wave speed $\omega/\beta = 1/\sqrt{\mu\varepsilon}$ and with another wave impedance $E/H = \eta = \sqrt{\mu/\varepsilon}$.

Suppose we have a plane wave propagating in a dielectric medium along some direction we call z. If the medium fills all space, this wave can travel forever in it. But suppose the wave encounters a new medium, beyond the plane $z = z_0$. That is, we have two media in the universe: For $z < z_0$, the dielectric has constitutive parameters μ_1, ε_1 and for $z > z_0$, the medium is given by μ_2, ε_2. Can the wave in medium 1, at frequency ω, with phase constant $\beta_1 = \omega\sqrt{\mu_1\varepsilon_1}$, and with polarization $\mathbf{E}_0$ just cross the boundary plane at $z = z_0$ and continue onward into medium 2? If it could, it would have to change its wave speed, from $1/\sqrt{\mu_1\varepsilon_1}$ to $1/\sqrt{\mu_2\varepsilon_2}$, and its wave impedance would have to become $\eta_2 = \sqrt{\mu_2/\varepsilon_2}$ instead of $\eta_1 = \sqrt{\mu_1/\varepsilon_1}$. Is this acceptable to Maxwell?

The plane waves in each of the two media are acceptable, but we must also ensure that the fields on the boundary plane $z = z_0$ satisfy the boundary conditions. We recall that these require that the tangential electric and magnetic fields be continuous across the boundary plane.

First, notice that both the $\mathbf{E}$ field and the $\mathbf{H}$ field are entirely tangential to the boundary plane, because the wave is encountering the plane head-on (we say the wave in medium 1 is *normally incident* onto the plane) and the wave fields, being perpendicular to the direction z of propagation, are necessarily tangential to the plane; see Figure 6-14, in which the plane faces direction $\hat{\mathbf{n}} = \hat{\mathbf{z}}$.

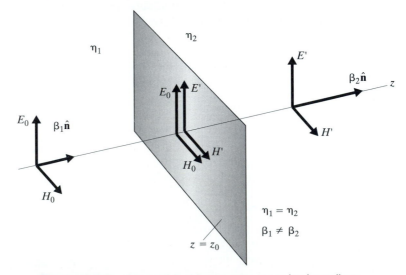

Figure 6-14   Normal incidence onto matched medium.

Second, let's say that the incident wave, with polarization $\mathbf{E}_0$, crosses the plane and becomes the new wave, with polarization $\mathbf{E}'$. Since $\mathbf{E}_0$ and $\mathbf{E}'$ are tangential to the plane, they must be equal, to maintain continuity. Also, since continuity must be maintained for all time, the frequency of oscillation ω must be the same on both sides of the plane; otherwise, the fields would get out of step and not continue to be equal.

But third, $\mathbf{H}_0$ has magnitude $H_0 = E_0/\eta_1$ while $\mathbf{H}'$ has magnitude $H' = E'/\eta_2$ and these fields too must be the same, by continuity. The directions of the vectors $\mathbf{H}_0$ and $\mathbf{H}'$ are the same, since both are perpendicular to z and to their already equal electric partners $\mathbf{E}_0$ and $\mathbf{E}'$, but are their magnitudes the same? The magnitudes H_0 and H' are equal only if $\eta_1 = \eta_2$, which means that both media have the same wave impedance, or that the wave impedance of medium 2 is *matched* to that of medium 1. We conclude that a plane wave normally incident onto the interface plane between two media can cross over and simply continue into the second medium only if the wave impedances of the two media are matched. The wave then merely changes its speed to accommodate the new medium: $\mathbf{E}_0 e^{j(\omega t - \beta_1 z)}$ becomes $\mathbf{E}_0 e^{j(\omega t - \beta_2 z)}$ in medium 2. There are other changes as well; if $\eta_1 = \eta_2$ but $\beta_1 \neq \beta_2$, then μ and ε are different in the two media, so that although $\mathbf{E}$ and $\mathbf{H}$ are continuous, $\mathbf{D}$ and $\mathbf{B}$ are discontinuous across the interface plane, which these tangential fields are allowed to be.

What does it take for the media to be matched? The condition $\eta_1 = \eta_2$ is trivially satisfied if the two media are actually identical, but then there is really no boundary plane and $\beta_1 = \beta_2$ as well. If the media are dissimilar, we need $\mu_1/\varepsilon_1 = \mu_2/\varepsilon_2$ or, more simply, $\mu_2/\mu_1 = \varepsilon_2/\varepsilon_1$; that is, the changes in permeability and in permittivity must be in the same ratio. This is not readily achieved, however, as materials with permeabilities significantly different from μ_0 are not usually transparent to light, or to the plane waves we are considering. The more practical and important question is therefore, What happens when the two media are mismatched?

From our experience with waves guided by transmission lines and encountering mismatched obstacles, we should expect the answer to be that a reflected wave is generated at the interface plane. This is in fact the case, and we want to develop the details of that process.

That a backward-traveling wave can appear in medium 1 should be clear too. For the incident wave, the field is $\mathbf{E}_0 e^{-j\mathbf{k}\cdot\mathbf{r}}$ with $\mathbf{k} = \beta_1\hat{\mathbf{z}}$; for the reflected wave, it is $\mathbf{E}_1 e^{-j\mathbf{k}'\cdot\mathbf{r}}$ with $\mathbf{k}' = -\beta_1\hat{\mathbf{z}}$. The wave vector $\mathbf{k}$ is reversed, to $\mathbf{k}' = -\mathbf{k}$; its magnitude β_1 is the same, since the wave speed is still that of medium 1 and the frequency is unchanged, as already noted. Note carefully, however, that the direction of the magnetic field is that of $\mathbf{k} \times \mathbf{E}_0$ for the incident wave but along $\mathbf{k}' \times \mathbf{E}_1 = -\mathbf{k} \times \mathbf{E}_1$ for the reflected one. This means that if the polarizations $\mathbf{E}_0$ and $\mathbf{E}_1$ are both along $\hat{\mathbf{x}}$, say, then $\mathbf{H}_0$ is along $\hat{\mathbf{y}}$ but $\mathbf{H}_1$ is along $-\hat{\mathbf{y}}$. We then have for the superposition of both the incident and reflected waves in medium 1

$$E_x = E_0 e^{-j\beta_1 z} + E_1 e^{j\beta_1 z}, \tag{6.128}$$

$$\eta_1 H_y = E_0 e^{-j\beta_1 z} - E_1 e^{j\beta_1 z}. \tag{6.129}$$

The electric field is the sum of two oppositely traveling plane waves; the wave impedance η_1 of the medium times the magnetic field is the difference of the same two waves.

If this sounds familiar, it should. Uniform plane waves traveling forward and backward along a fixed direction in a homogeneous medium behave just as do voltage and current waves along a uniform transmission line. The electric field plays the role of the voltage and the magnetic field that of the current. The medium's wave impedance (η_0 for a vacuum) takes the place of the transmission line's characteristic impedance Z_0. To complete the analogy, we look into what entities can play the roles of source and load for the plane waves.

SOURCE OF PLANE WAVES

Since the plane waves themselves were defined as source free, we can expect to locate a source for such waves only at a spot where they cease to be the continuous functions they are supposed to be. That is, we need to find or create a plane of discontinuity of the fields. The boundary conditions do not permit a discontinuity in tangential electric field, but they do allow one in the tangential magnetic field, provided there is a current sheet of appropriate strength at the discontinuity plane. We therefore conceive of our source for the plane waves as an infinitely extended current sheet. Let's place it along the plane $z = 0$ and let it oscillate at frequency ω, as $\mathbf{K}e^{j\omega t}$. This surface current sheet acts as an antenna, generating and emitting the plane wave in the z-direction, at the frequency ω and with the polarization determined by the orientation of $\mathbf{K}$.

Because the electric field must be the same on both sides of the source plane, we must actually be getting two plane waves, one emitted on the $z > 0$ side and another radiated on the other side; if we were emitting on only one side, there would be a discontinuity in the tangential electric field. Only the magnetic field can be discontinuous; the amount of the discontinuity equals the strength of the current sheet and the direction of the current is perpendicular to the jump in magnetic field, with its sense dictated by the right-hand rule. Figure 6-15 shows the configuration of the source and emitted waves.

Note that the right-hand rule puts the direction of the surface current *opposite* to that of the electric field of both emitted waves. The discontinuity in magnetic field is $H - (-H) = 2H$, so that is the strength of the surface current. Put another way, a current sheet of strength K in a homogeneous medium emits equal and opposite plane waves with magnetic field strengths $\pm\frac{1}{2}K$.

That the source generates just one wave on each side may seem obvious to some or mysterious to others. The question is, Why isn't there a superposition of a forward- and a backward-traveling wave on each side of the source plane? That there is only one wave on each side is a nontrivial statement of a separate physical principle, called the *radiation condition*. This states that any source can only radiate fields away from itself; it cannot "suck in" fields from a sourceless environment. If waves are arriving at the source, they must have been generated elsewhere, by some other source or by some discontinuity that reflects the wave emitted by the source; they could not be the direct creation of the source itself.

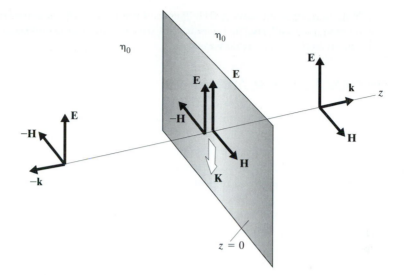

Figure 6-15   Current sheet as source of plane waves.

Note also that the plane-wave source we are considering is as impractical as is the plane wave itself, as both are infinitely extended in any transverse plane. Source planes that are large compared to the wavelength of the radiation do, however, emit reasonable facsimiles of plane waves. We should consider our ideal plane waves and the current-sheet plane-wave generator as limiting cases of large-cross-section beams and of large-scale planar antennas.

Example 6.16

What surface current sheet $\mathbf{K}e^{j\omega t}$ at $z = 0$ can generate the circularly polarized wave $\mathbf{E}e^{j(\omega t - kz)}$ with $\mathbf{E} = E_0[\hat{\mathbf{x}} - j\hat{\mathbf{y}}]$ in the vacuum space $z > 0$? What does this source generate on the other side? ◈

As before, the complex electric field amplitude will be the same on both sides at $z = 0$ but the waves travel in opposite directions:

$$\mathbf{E}e^{j(\omega t - kz)} \quad \text{for } z > 0 \quad \text{and} \quad \mathbf{E}e^{j(\omega t + kz)} \quad \text{for } z < 0.$$

The magnetic field amplitude is $(\hat{\mathbf{z}} \times \mathbf{E}/\eta_0)e^{j(\omega t - kz)}$ for $z > 0$ but it is $(-\hat{\mathbf{z}} \times \mathbf{E}/\eta_0)$ $e^{j(\omega t + kz)}$ for $z < 0$. The current sheet at $z = 0$ is $\hat{\mathbf{z}} \times \Delta\mathbf{H}$, which is just $\mathbf{K}e^{j\omega t}$ $= (-2\mathbf{E}/\eta_0)e^{j\omega t} = (-2E_0/\eta_0)[\hat{\mathbf{x}} - j\hat{\mathbf{y}}]e^{j\omega t}$.

The interpretation of this complex expression is obtained in the time domain, by taking the real part; this is $\mathcal{K}(t) = (-2E_0/\eta_0)[\hat{\mathbf{x}}\cos\omega t + \hat{\mathbf{y}}\sin\omega t]$ and shows the current density vector to be rotating, at frequency ω, in the sense from $\hat{\mathbf{x}}$ to $\hat{\mathbf{y}}$, at every point of the plane, simultaneously.

On the other side of the source plane, the electric field amplitude is the same as in the front, but propagates in the opposite direction:

$$\mathbf{E}e^{j(\omega t + kz)} = E_0[\hat{\mathbf{x}} - j\hat{\mathbf{y}}]e^{j(\omega t + kz)}.$$

Note that, while the wave is RHCP in the front $(z > 0)$, the reversed direction of propagation makes it LHCP in the rear $(z < 0)$. The rotating-current sheet emits oppositely sensed circularly polarized waves.

LOAD IMPEDANCES FOR PLANE WAVES

As to the equivalent for plane waves of a load at the end of a transmission line, we have already encountered the special case of a dissimilar medium that is matched to the sourced one $(\eta_1 = \eta_2)$, and this second medium acts as a matched load on the first medium. We now investigate the consequences of normal incidence onto the interface plane between sourced region 1 and unsourced medium 2, with the media mismatched $(\eta_1 \neq \eta_2)$.

The matched-media solution is now no longer feasible. The incident wave's impedance ratio $E/H = \eta_1$ is appropriate for the first region, but is wrong (does not satisfy Maxwell's equations) in the second medium, which demands $E/H = \eta_2$ instead. The incident wave cannot simply cross the boundary plane and continue into the second medium, leaving the wave unaffected in the first region. We need to satisfy Maxwell's equations within the two regions, consistent with the source of the incident wave, *and* to satisfy the boundary conditions at the interface. The latter requirement involves matching the tangential **E** and **H** fields on the two sides of the boundary plane; this matching of the fields must be achieved at every point of the interface, at every instant of time.

The fields on both sides of the plane must oscillate at the frequency ω of the source; otherwise, even if we succeeded in matching the total tangential fields across the boundary at one instant, they would not remain matched to each other at subsequent times, as they fail to oscillate in unison. Furthermore, because the incident wave is directed normally to the interface, any value of the fields is attained simultaneously at every point of the boundary. Matching at one boundary point will therefore achieve matching at every point of the interface. Whatever fields appear on the two sides will need to attain their values simultaneously at every point of the plane. These fields can therefore also be normally directed waves.

Maxwell's equations allow solutions in the form of a superposition of *two* oppositely sensed traveling waves, both with the same wave speed but with oppositely directed wave vectors:

$$E_x = E_0 e^{-j\beta_1 z} + E_1 e^{j\beta_1 z}, \qquad \eta_1 H_y = E_0 e^{-j\beta_1 z} - E_1 e^{j\beta_1 z}.$$

However, since region 2 is unsourced, it cannot carry a wave that travels *toward* the interface. If it did, there would have to be some source for it somewhere, even if only at infinity, but none has been specified in that region. The boundary plane can *emit* or *scatter* fields, in the form of waves, into both regions, *away from* the interface; it cannot, however, draw them in from a source-free region, to conform to the radiation condition.

What we are left to deal with is a steady state with two waves, oppositely sensed, in region 1, plus one wave in region 2. One of the two waves in the sourced region is the incident wave; the *scattered* fields are the *reflected* wave, in the sourced region, and the *transmitted* wave, in the unsourced one. All the waves oscillate at the original frequency of the source, in order for the fields to match across the plane for all time (this statement would have to be modified if the interface plane were moving, instead of stationary). In order that the scattered waves, as well as the incident one, satisfy Maxwell's equations within the two regions, they must have propagation constants $\beta_1 = \omega \sqrt{\mu_1 \varepsilon_1}$ in region 1 and $\beta_2 = \omega \sqrt{\mu_2 \varepsilon_2}$ in region 2. The time- and space dependence of the three waves must therefore be as follows:

$$\text{incident:} \qquad e^{j(\omega t - \beta_1 z)} \qquad \text{in 1,} \tag{6.130}$$

$$\text{reflected:} \qquad e^{j(\omega t + \beta_1 z)} \qquad \text{in 1,} \tag{6.131}$$

$$\text{transmitted:} \qquad e^{j(\omega t - \beta_2 z)} \qquad \text{in 2.} \tag{6.132}$$

Each of these waves has its own strength and polarization, given by its electric field vector $\mathbf{E}$, which is perpendicular to the z-axis (or, more generally, to $\hat{\mathbf{n}}$) and to the corresponding magnetic field $\mathbf{H}$. As indicated in Figure 6-16, the incident wave has a given field amplitude $\mathbf{E}_0$ and a corresponding, known field $\mathbf{H}_0$ related to it by the wave impedance of the sourced region, $\mathbf{H}_0 = \hat{\mathbf{n}} \times \mathbf{E}_0 / \eta_1$. The reflected wave has some unknown field amplitude $\mathbf{E}_1$, together with a magnetic field $\mathbf{H}_1$ related to it by

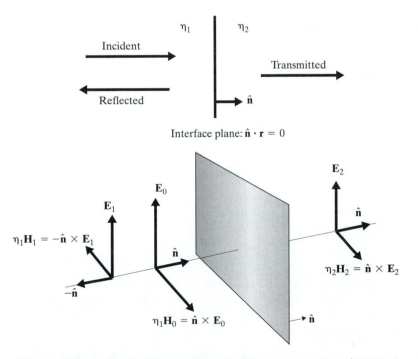

Figure 6-16   Field patterns of the incident and scattered waves.

the same wave impedance η_1 but directed so that the wave vector is along $-\hat{\mathbf{n}}$, instead of $\hat{\mathbf{n}}$. The transmitted wave has an unknown field $\mathbf{E}_2$; its magnetic field is oriented along $\hat{\mathbf{n}} \times \mathbf{E}_2$ but with a magnitude related to that of $\mathbf{E}_2$ by the wave impedance η_2 instead of η_1. There are hence two (vector) unknowns, $\mathbf{E}_1$ and $\mathbf{E}_2$, to be found in terms of the given $\mathbf{E}_0$ of the specified incident wave. There are two conditions to be imposed on these two unknowns: that the total $\mathbf{E}$ field and the total $\mathbf{H}$ field (which are tangential to the interface) both be continuous across the boundary plane. With two conditions for two unknowns, a solution should be forthcoming.

The total electric field in region 1 is given by

$$\mathbf{E} = \mathbf{E}_0 e^{-j\beta_1 z} + \mathbf{E}_1 e^{+j\beta_1 z}; \tag{6.133}$$

more generally, z is replaced by $\hat{\mathbf{n}} \cdot \mathbf{r}$. The field in region 2 is just

$$\mathbf{E} = \mathbf{E}_2 e^{-j\beta_2 z}. \tag{6.134}$$

The corresponding magnetic fields in the two regions are

$$\mathbf{H} = \left[\hat{\mathbf{n}} \times \mathbf{E}_0 / \eta_1\right] e^{-j\beta_1 z} + \left[-\hat{\mathbf{n}} \times \mathbf{E}_1 / \eta_1\right] e^{+j\beta_1 z} \tag{6.135}$$

in medium 1 and

$$\mathbf{H} = \left[\hat{\mathbf{n}} \times \mathbf{E}_2 / \eta_2\right] e^{-j\beta_2 z} \tag{6.136}$$

in medium 2.

The requirements of continuity of the tangential electric and magnetic fields at all points on the interface $z = 0$ (or $\hat{\mathbf{n}} \cdot \mathbf{r} = 0$) are expressed by

$$\mathbf{E}_0 + \mathbf{E}_1 = \mathbf{E}_2, \tag{6.137}$$

$$\left[\hat{\mathbf{n}} \times \mathbf{E}_0 / \eta_1\right] + \left[-\hat{\mathbf{n}} \times \mathbf{E}_1 / \eta_1\right] = \left[\hat{\mathbf{n}} \times \mathbf{E}_2 / \eta_2\right]. \tag{6.138}$$

Forming the cross product of the first of these with $\hat{\mathbf{n}}$ yields the pair

$$\hat{\mathbf{n}} \times \mathbf{E}_0 + \hat{\mathbf{n}} \times \mathbf{E}_1 = \hat{\mathbf{n}} \times \mathbf{E}_2, \tag{6.139}$$

$$\hat{\mathbf{n}} \times \mathbf{E}_0 - \hat{\mathbf{n}} \times \mathbf{E}_1 = \left[\eta_1 / \eta_2\right]\hat{\mathbf{n}} \times \mathbf{E}_2, \tag{6.140}$$

and adding these last two versions of the two conditions eliminates the unknown vector $\mathbf{E}_1$:

$$2\hat{\mathbf{n}} \times \mathbf{E}_0 = \left[1 + (\eta_1 / \eta_2)\right]\hat{\mathbf{n}} \times \mathbf{E}_2, \tag{6.141}$$

so that

$$\hat{\mathbf{n}} \times \mathbf{E}_2 = \left[2\eta_2 / (\eta_2 + \eta_1)\right]\hat{\mathbf{n}} \times \mathbf{E}_0 \tag{6.142}$$

and (6.139) gives $\hat{\mathbf{n}} \times \mathbf{E}_1$ as $\hat{\mathbf{n}} \times \mathbf{E}_2 - \hat{\mathbf{n}} \times \mathbf{E}_0$ or

$$\hat{\mathbf{n}} \times \mathbf{E}_1 = \left[(\eta_2 - \eta_1)/(\eta_1 + \eta_2)\right]\hat{\mathbf{n}} \times \mathbf{E}_0. \tag{6.143}$$

Finally, by virtue of the general vector identity

$$\mathbf{E} \equiv (\hat{\mathbf{n}} \times \mathbf{E}) \times \hat{\mathbf{n}} + (\hat{\mathbf{n}} \cdot \mathbf{E})\hat{\mathbf{n}} \tag{6.144}$$

and the fact that, for the waves in question, $\hat{\mathbf{n}} \cdot \mathbf{E} = 0$, we can extract the full vectors $\mathbf{E}$ from $\hat{\mathbf{n}} \times \mathbf{E}$ by crossing again with $\hat{\mathbf{n}}$ on both sides of the equations:

$$\mathbf{E}_1 = \frac{\eta_2 - \eta_1}{\eta_2 + \eta_1} \mathbf{E}_0, \tag{6.145}$$

$$\mathbf{E}_2 = \frac{2\eta_2}{\eta_2 + \eta_1} \mathbf{E}_0. \tag{6.146}$$

These are the electric field amplitudes of the scattered waves, in terms of the specified incident field. The total electric field in region 1 is given by

$$\mathbf{E} = \mathbf{E}_0 \{ e^{-j\beta_1 z} + \Gamma_e e^{+j\beta_1 z} \}, \tag{6.147}$$

where

$$\Gamma_e = \frac{\eta_2 - \eta_1}{\eta_2 + \eta_1} \tag{6.148}$$

acts as a reflection coefficient, a scalar factor that relates the complex vector amplitude of the reflected wave electric field in medium 1 to that of the incident wave. This reflection coefficient has a familiar form: the ratio of the mismatch between the wave impedances of the two media to their sum. We see that η_1 plays the role of the characteristic impedance of medium 1, while η_2 acts as the load impedance appended to medium 1 by the presence of the unbounded medium 2.

The field in region 2 is

$$\mathbf{E} = \mathbf{E}_0 T_e e^{-j\beta_2 z}, \tag{6.149}$$

where the *transmission coefficient*

$$T_e = \frac{2\eta_2}{\eta_2 + \eta_1} \tag{6.150}$$

relates the complex vector amplitude of the transmitted wave electric field in medium 2 to that of the incident wave in region 1.

The associated magnetic fields in the two regions are, from (6.135) and (6.136),

$$\mathbf{H} = \mathbf{H}_0 \{ e^{-j\beta_1 z} - \Gamma_e e^{+j\beta_1 z} \}, \tag{6.151}$$

in medium 1 and

$$\mathbf{H} = \mathbf{H}_0 T_h e^{-j\beta_2 z}, \tag{6.152}$$

in medium 2, where $\mathbf{H}_0 = \hat{\mathbf{n}} \times \mathbf{E}_0 / \eta_1$ and

$$T_h = \frac{2\eta_1}{\eta_2 + \eta_1} \tag{6.153}$$

is the transmission coefficient for the magnetic field. The reflection coefficient for the magnetic field can also be expressed as $\Gamma_h = -\Gamma_e$, to account for the opposite direction of the magnetic field for the reflected wave.

Note that $T_e = 1 + \Gamma_e$, since the transmitted field is the sum of the incident and reflected ones at the interface; that is the essence of the boundary conditions there. These results represent the solution to the problem of normal incidence onto the interface between mismatched media.

Example 6.17

For normal incidence onto the interface between two nonmagnetic, lossless dielectrics with $\varepsilon_1 = 1.5\varepsilon_0$ and $\varepsilon_2 = 2.5\varepsilon_0$, what are the reflection and transmission coefficients? ◈

For nonmagnetic media, $\mu_1 = \mu_2 = \mu_0$ and for perfect dielectrics the impedances are $\sqrt{\mu_0/\varepsilon}$. The impedance mismatch is therefore given by $\eta_2/\eta_1 = \sqrt{1.5/2.5} = 0.7746$. Hence, the reflection coefficient is found as

$$\Gamma_e = (0.7746 - 1)/(0.7746 + 1) = -0.1270$$

and the transmission coefficient is $T_e = 1 + \Gamma_e = 0.8730$, for the electric fields.

The negative sign of Γ_e is due to the fact that $\eta_2 < \eta_1$ here; it means that the reflected electric field at the interface is directed oppositely to the incident one. Correspondingly, the reflected magnetic field has the same direction as the incident one, as is confirmed by $\Gamma_h = -\Gamma_e = +0.1270$. Also, $T_h = 1 + \Gamma_h = 1.1270$ indicates that the transmitted magnetic field is stronger than the incident one, and in the same direction.

A simple mnemonic for the reflected and transmitted wave amplitudes is obtained by noting that the *transfer impedance* that relates the *incident* electric field to the *transmitted* magnetic field is

$$E_0/H_2 = \tfrac{1}{2}(\eta_1 + \eta_2), \tag{6.154}$$

which is merely the average of the two impedances and an easily remembered compromise between the two, giving the transmitted magnetic field generated by the incident electric field.

The next example illustrates one scheme for constructing a perfect absorber for normally incident plane waves. It makes use of a *resistive sheet*, a plane with a surface conductivity σ_s or resistivity r_s such that an electric field **E** in that plane is accompanied by a current sheet **K** proportional to it:

$$\mathbf{E} = r_s \mathbf{K} \qquad \text{or} \qquad \mathbf{K} = \sigma_s \mathbf{E}.$$

Example 6.18

A plane wave $\mathbf{E}_0 e^{-jkz}$ is incident onto a resistive sheet of surface conductivity σ_s. The sheet is coincident with the xy-plane and the media on the two sides of the sheet are both a vacuum. At a distance l beyond the sheet (i.e., at $z = l$), we place a perfectly conducting plane. See Figure 6-17.

   (a) *Design the surface conductivity σ_s and the distance l to make the combination of resistive sheet and conducting plane a perfect absorber for the incident wave (i.e., to generate no reflection back into the $z < 0$ region).*

   (b) *What are the fields in the region $0 < z < l$?*

   (c) *What is the surface current **K** on the resistive sheet?*

   (d) *What is the surface current $\mathbf{K}_0$ on the conducting plane?* ◈

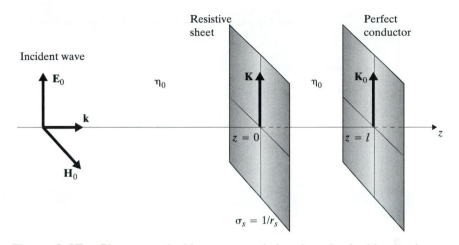

Figure 6-17   Plane wave incident onto resistive sheet backed by conductor.

If we succeed in eliminating reflection at $z = 0$, then there is only $E_0 e^{-jkz}$ with its $H_0 = (E_0/\eta_0)e^{-jkz}$ in the region $z < 0$. Between the resistive sheet and the conducting plane, we should have countertraveling waves $E_1 e^{-jkz} + E_2 e^{jkz}$. This superposition must be zero at the perfectly conducting plane and must also be continuous with the incident wave at the resistive sheet. We therefore have the following two conditions on the two unknowns E_1 and E_2:

$$E_1 e^{-jkl} + E_2 e^{jkl} = 0 \quad \text{and} \quad E_1 + E_2 = E_0.$$

(a) Solving for E_1 and E_2 in terms of the given E_0, we get

$$E_1 = E_0 e^{jkl}/[2j \sin kl] \quad \text{and} \quad E_2 = -E_0 e^{-jkl}/[2j \sin kl].$$

Beyond the resistive sheet, the magnetic field is given by

$$\eta_0 H = E_1 e^{-jkz} - E_2 e^{jkz} = \{E_0 e^{jk(l-z)} + E_0 e^{-jk(l-z)}\}/[2j \sin kl]$$

$$= E_0 \cos k(l - z)/j \sin kl,$$

which becomes $\eta_0 H = -jE_0 \cot kl$ at $z = 0$. In front of the sheet, the field is given by $\eta_0 H = E_0$. The discontinuity equals the surface current, so that

$$\eta_0 K = E_0[1 + j \cot kl].$$

But $K = \sigma_s E$, so this is to be $\eta_0 \sigma_s E_0$, giving as the condition for no reflection

$$\eta_0 \sigma_s = 1 + j \cot kl.$$

As the left side is real, we need both $\eta_0 \sigma_s = 1$ and $\cot kl = 0$. The design calls for $r_s = \eta_0$ and $l = \lambda_0/4$. A resistive sheet of resistivity equal to the impedance of free space, placed a quarter wavelength in front of a perfectly conducting plane, forms a perfect absorber for the incident plane wave.

(b) Since $kl = \pi/2$, the fields are

$$E = E_1 e^{-jkz} + E_2 e^{jkz} = \{E_0 e^{jk(l-z)} - E_0 e^{-jk(l-z)}\}/[2j \sin kl]$$

$$= E_0 \sin k(l - z)/\sin kl = E_0 \sin (\pi/2 - kz)/\sin \pi/2$$

$$= E_0 \cos kz,$$

$$\eta_0 H = E_0 \cos k(l - z)/j \sin kl = E_0 \cos (\pi/2 - kz)/j \sin \pi/2 = -jE_0 \sin kz.$$

(c) $\eta_0 K = E_0[1 + j \cot \pi/2] = E_0$ or $K = E_0/\eta_0$.

(d) The current sheet at the conducting plane equals the tangential magnetic field there, which is given by $\eta_0 H = -jE_0 \sin kz$ at $z = l$, giving $\mathbf{K}_0 = -j\mathbf{E}_0/\eta_0$.

The combination of the resistive sheet of resistivity η_0 and the conducting plane $\lambda_0/4$ away provides a matched load for the incident wave, only a quarter wavelength thick.

SUMMARY

Because real media respond to electromagnetic fields in a dynamic way, time-domain analysis with fixed constitutive parameters ε, μ is not realistic. Instead, we turn to frequency-domain analysis, which focuses on sinusoidal signals in the steady state, one frequency at a time. Fourier analysis allows any realistic signal to be decomposed into its sinusoidal components, so we retain generality. We actually use complex exponentials rather than real sinusoids, so that all linear operations reduce to multiplicative factors and the exponential time functions drop out of our equations. The price of this convenience is twofold: that we need complex arithmetic and that we must distinguish positive and negative frequencies. That modest investment pays off in ease of analysis.

Accordingly, we examined the propagation of sinusoidal waves along transmission lines. The incident and reflected waves then appear as complex exponential functions of distance along the line:

$$V(z) = V_1 e^{+j\beta(l-z)} + V_2 e^{-j\beta(l-z)}, \tag{6.155}$$

$$Z_0 I(z) = V_1 e^{+j\beta(l-z)} - V_2 e^{-j\beta(l-z)}. \tag{6.156}$$

These are the coefficients of the $e^{j\omega t}$ factor; the frequency ω is a given quantity, determined by the source, and the factor β is ω/u_0. The equations apply to a uniform, lossless line with characteristic speed u_0 and characteristic impedance Z_0. *Uniform* means that, between the two ends of the segment of transmission line, there are no changes in geometry or constitution and no discontinuities such as might be imposed by lumped elements attached to the line. *Lossless* means that the line is made of materials that are perfect conductors and insulators, which do not convert electrical energy into heat, drawing off some of the power.

The solution to any problem involving such transmission lines entails merely finding the two complex constants, V_1 and V_2, in the preceding two equations. The conditions that determine these two constants are imposed by the source at one end and the load at the other.

We learned the vocabulary associated with sinusoidal waves, including the frequency (ω and $f = \omega/2\pi$), the phase constant (β or $k = \omega/c$), the period, and the wavelength. We recognize any pure imaginary exponential function of distance, like $e^{-j\beta x}$, to be a wave that travels along the direction of increasing distance x, at speed ω/β.

We found that most questions about the response of a transmission line with a given source and load can be answered by combining three fundamental equations. The first gives the complex reflection coefficient at the location of a given load impedance, in terms of the mismatch between the load and the characteristic impedance:

$$\Gamma_l = \frac{Z_l - Z_0}{Z_l + Z_0}. \tag{6.157}$$

The second one gives the reflection coefficient at any location, in terms of that at the load, as the ratio of a reflected wave to an incident one:

$$\Gamma(z) = \Gamma_l \exp\left[-j2\beta(l - z)\right]. \tag{6.158}$$

The third gives the impedance at any location, as the ratio of the sum of the incident and reflected waves to the difference of the same two waves:

$$Z(z) = Z_0 \frac{1 + \Gamma(z)}{1 - \Gamma(z)}. \tag{6.159}$$

These equations give us the impedance distribution at every point of the line, so that we can determine circuit variables like input impedance for any segment of the transmission line. We noted that when a lumped element appears at a point intermediate between the source and load at the two ends, we must treat the segments on each side of that element separately.

We discussed power transfer along a lossless transmission line. Since this involves a nonlinear (multiplicative or quadratic) operation on the voltage or current distribution, we either calculate the power in the time domain, as the product of the voltage and current time functions, or else take a short cut to the crucial time-averaged power directly in the frequency domain, by finding half the real part of the product of the voltage and current complex amplitudes, one of them complex conjugated. We stressed the important distinction between matching for optimum power transfer and matching to eliminate reflections from a mismatched load.

We explored transmission line behavior of fields far removed from their supporting and guiding wire structure, so far removed, in fact, that these fields are entirely source free, or normal modes of empty space. They are uniform, TEM sinusoidal plane waves, of the form $\mathbf{E}e^{j(\omega t - kz)}$, with $\mathbf{E}$ a constant vector. We found that the electric field has the wave behavior of the voltage of a physical transmission line and the accompanying magnetic field $\mathbf{H}e^{j(\omega t - kz)}$ acts as its current. The electric and magnetic field vectors are perpendicular to each other and to the direction of propagation. Their magnitudes are related by the wave impedance $\eta = \sqrt{\mu/\varepsilon}$, a property of the medium that plays the role of the characteristic impedance Z_0 of a physical transmission line.

There are a number of differences between what we encounter in dealing with a physical transmission line and the behavior of the plane waves in a homogeneous medium or vacuum. One is that, in the absence of any structure at all, the direction of propagation of the plane wave may be any at all and we replace the distance coordinate z with the more general one $\hat{\mathbf{n}} \cdot \mathbf{r}$ along the direction of an arbitrary unit vector $\hat{\mathbf{n}}$; it is then convenient to define the wave vector $\mathbf{k} = (\omega/c)\hat{\mathbf{n}}$ (for a vacuum, the wave speed is c) and write the plane wave as $\mathbf{E}e^{j(\omega t - \mathbf{k} \cdot \mathbf{r})}$.

Another major difference is that, for a plane wave, the quantity that propagates is not merely a scalar (V or I) but a vector ($\mathbf{E}$ or $\mathbf{H}$). This entails an orientation of the vector as well as its magnitude and raises the question of how the vector's direction may change in time as the wave propagates. The behavior of the field vector in time is described as the wave's *polarization*. Because we specialize plane waves to the sinusoidal steady state, the behaviors of the vector's magnitude and of its direction are related, and we found that the most general polarization is elliptical, with the field vector tracing out an ellipse in time. We explored elliptic polarization and its special cases of linear and circular polarization. The most general elliptic polarization can be expressed as a superposition of either two linear ones or two oppositely sensed circular ones.

Yet another difference arises from the unbounded nature of the plane waves. They are infinitely extended in the transverse directions; consequently, the sources and loads we consider are also unbounded planes or planar interfaces between media. We also learned the *radiation condition*, which dictates that only outbound waves can be emitted by any source.

Plane waves provide important building blocks for expressing more complicated field structures as superpositions of TEM waves. Just as the $e^{j\omega t}$ time factor allows a decomposition of any realistic signal into its Fourier spectrum, so does the e^{-jkz} space factor permit a similar Fourier analysis of any spatial field structure.

Despite the differences, the remarkable result is the similarity in wave behavior of plane waves and of waves guided by a physical transmission line. The most striking similarity is perhaps the formula for the reflection coefficient that arises when either wave encounters a discontinuity. The discontinuity may be some lumped impedance or line termination in the case of a transmission line or a transverse planar interface between media or simply a planar conductor or resistive sheet in the case of plane waves. We have

$$\Gamma_l = \frac{Z_l - Z_0}{Z_l + Z_0} \tag{6.160}$$

for the reflection coefficient at the load of a transmission line; we get

$$\Gamma_e = \frac{\eta_2 - \eta_1}{\eta_2 + \eta_1} \tag{6.161}$$

for the reflection coefficient at an interface between dissimilar media.

We are now poised to look into matters that affect many design decisions, specifically those that involve the measurement, control, and matching of impedance in both transmission line structures and in media for the propagation of plane waves.

PROBLEMS

Input impedance of mismatched transmission line

6.1 Calculate the input impedance of 2.027 meters of 300-ohm twin-lead transmission line connected at its other end to a 200-ohm FM radio receiver tuned to a carrier frequency of 100 MHz. Assume the constitutive parameters of the material surrounding the parallel wires of the twin-lead line differ negligibly from those of a vacuum.

Length of line with known load and input impedances

6.2 The input impedance of a length of 75-ohm transmission line operated at 200 MHz and loaded by $(50 - j60)$ Ω is known to be $(50 + j60)$ Ω. What is the minimum possible length of the line, if its wave speed is $c/2$?

Length of line with real input impedance

6.3 The input impedance of a 50-ohm transmission line (wave speed c) operated at 100 MHz and loaded by $(30 - j50)$ Ω is to be made real.
(a) What is the minimum possible length of the line?
(b) What is the input impedance for that length?

Operating frequency for incident and reflected waves in quadrature

6.4 The reflected wave at the input of a 50-ohm transmission line (wave speed c) of length $l = 1.25$ m and loaded by $(75 - j75)$ Ω is to be in time quadrature (90° out of phase) with the incident wave.
(a) What is the lowest possible operating frequency?
(b) What is the input impedance at that frequency?

Characteristic impedance determined from load and line length

6.5 The reflected wave at the input of a certain transmission line is found to have no quadrature (90° out of phase) component with the incident wave when the line is loaded by $(30 - j40)$ Ω and the line length makes $\beta l = 0.5453$.
(a) Find the reflection coefficient Γ_0 at the input.
(b) Find the (real) characteristic impedance Z_0 of the line.

Measurement of characteristic impedance

6.6 We measure the complex input impedance of a certain transmission line to be Z_1 when it is short-circuited at some arbitrary distance from the source and Z_2 when it is short-circuited at a location a quarter wavelength closer to the source instead.
(a) From just Z_1 and Z_2, determine the characteristic impedance Z_0 of the line.
(b) Are there any frequencies at which this measurement technique fails?

Voltage "gain" of transmission line

6.7 A uniform, lossless transmission line (length l, characteristic impedance Z_0, wave speed c) has an ideal voltage source (i.e., no internal impedance) $V_{in}e^{j\omega t}$ attached at $z = 0$ and a reactive load jX at $z = l$. Let G be the "gain" of the line, the ratio $|V_{out}/V_{in}|$ of the amplitude of the voltage $V_{out}e^{j\omega t}$ across the load reactance to that of the input voltage. For what value of the reactance X will the gain G be infinite?

Voltage source between short and match

6.8   A parallel-wire transmission line with characteristic impedance Z_0 and phase constant β is shorted at the end $z = 0$ and matched at the end $z = l$. A voltage source $V_0 e^{j\omega t}$ is placed at $z = l_0$, as in Figure P6-1.
   (a) Obtain the voltage and current $V(z)$, $I(z)$ on both sides of the source.
   (b) What current I_0 does the source supply?

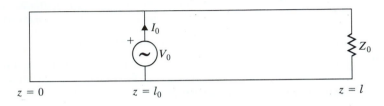

Figure P6-1   Voltage source positioned between short and matched load.

Current source between short and match

6.9   A parallel-wire transmission line with characteristic impedance Z_0 and phase constant β is shorted at the end $z = 0$ and matched at the end $z = l$. A current source $I_0 e^{j\omega t}$ is placed at $z = l_0$, as in Figure P6-2.
   (a) Obtain the voltage and current $V(z)$, $I(z)$ on both sides of the source.
   (b) What voltage V_0 does the source develop?

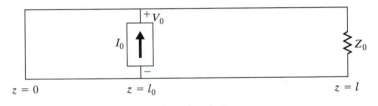

Figure P6-2   Current source positioned between short and matched load.

Voltage source midway between short and open circuit

6.10   A parallel-wire transmission line, with phase constant β, characteristic impedance Z_0, and length $2l$, is shorted at the end $z = 0$, open-circuited at the end $z = 2l$, and has a voltage source V_0 at frequency ω attached across it at the middle position $z = l$, as in Figure P6-3.
   (a) Obtain the voltage and current $V(z)$, $I(z)$ on both sides of the source.
   (b) What current I_0 does the source supply?

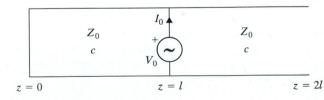

Figure P6-3   Voltage source midway between short and open circuit.

Two sources along a matched line

6.11 A parallel-wire transmission line is one wavelength long and is matched at both ends. Equal and opposite voltage sources are placed one third of the way from each end, as in Figure P6-4. Find the current midway along the line.

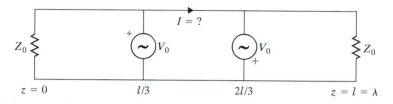

Figure P6-4 Equal and opposite sources on matched line.

Resonance on line with two synchronized sources

6.12 A parallel-wire transmission line of length l, characteristic impedance Z_0, and wave speed c has two identical steady-state voltage sources at frequency ω and of complex amplitude V_0 connected at the two ends and perfectly synchronized in phase; see Figure P6-5.
(a) Obtain the complex source current amplitude $I(0)$.
(b) What is the lowest frequency ω at which the line is resonant [i.e., infinite current $I(0)$ for finite voltage V_0]?

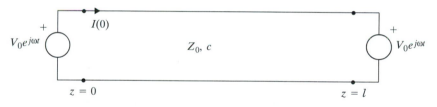

Figure P6-5 Line with two synchronized sources.

Resonance on parallel-wire line with ends joined

6.13 A parallel-wire transmission line of length l, characteristic impedance Z_0, and wave speed c is bent into a circle and the two ends joined together, as in Figure P6-6. A steady-state voltage source at frequency ω and of complex amplitude V_0 is connected at the point where the ends meet. Assume that bending the line into a circle does not affect the characteristics and operation of the transmission line.
(a) Obtain the complex current I_0 drawn from the source.
(b) What is the lowest frequency ω at which the line is resonant (i.e., infinite current I_0 for finite voltage V_0)?

Effect of transmission line on resonant frequency of *LC* load

6.14 The resonant frequency of a parallel combination of an inductance L and capacitance C is $\omega_0 = 1/\sqrt{LC}$. How is this resonant frequency affected if the combination is the load of a short transmission line (Z_0, c) of length $l \ll c/\omega_0$?

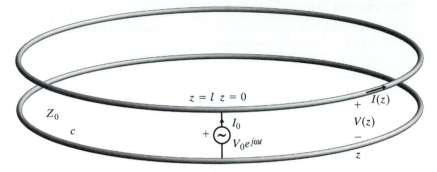

Figure P6-6   Parallel-wire transmission line with ends joined.

Maximum and minimum mismatch for given power reflection

6.15 If the power reflection coefficient $|\Gamma|^2$ for a transmission line with a mismatched load resistor is to be kept no greater than 5%, what is the maximum ratio of the load resistance to the characteristic impedance of the line? What is the minimum such ratio?

Power in terms of standing wave parameters

6.16 A lossless transmission line of characteristic impedance Z_0 has a mismatched load and is at least a half-wavelength long. We measure the standing wave and find the maximum and minimum magnitudes of the voltage along the line, V_{max} and V_{min}. Express the power along the line in terms of these parameters.

Power on matched and mismatched transmission lines

6.17 An ideal voltage source feeds one watt to a matched line (parameters Z_0, β, l), with $\beta l = 1.620$. If now the voltage source remains unchanged but the load resistor is changed to $R = (3/2)Z_0$, how much power is delivered to the load?

Power division between two resistors along a transmission line

6.18 An ideal voltage source feeds one watt to a matched line (parameters Z_0, β, l), with $\beta l = 1.620$. If now the voltage source remains unchanged but the resistor at $z = l$ is changed to $R_2 = (3/2)Z_0$ and a second resistor $R_1 = 3Z_0$ is attached at the point $z = (2/3)l$, how much power is delivered to each of the two resistors?

Power delivered to load resistor on parallel-wire line with ends joined

6.19 A parallel-wire transmission line of length l, characteristic impedance Z_0, and phase constant β is bent into a circle and the two ends joined together, as in Figure P6-7. A steady-state voltage source at frequency ω and of complex amplitude V_0 is connected to a resistor R_0 and the combination both feeds and loads the line at the point where the ends meet. Assume that bending the line into a circle does not affect the characteristics and operation of the transmission line.
(a) Obtain the power dissipated in the resistor.
(b) What choice of resistance R_0 will maximize the dissipated power?

Maximum power transfer from nonideal source via transmission line

6.20 A nonideal voltage source with internal impedance Z_g is to be connected to a load, whose impedance we can choose, through a transmission line of characteristic impedance

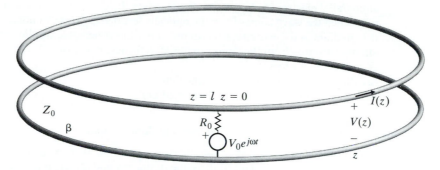

Figure P6-7   Parallel-wire transmission line with ends joined and load resistor.

Z_0, phase constant β, length l. We seek to choose the load impedance Z_l so as to maximize the power transfer from source to load.

(a) Verify that if the line length is zero (direct connection to load) or any even multiple of a quarter wavelength, the optimum load impedance is $Z_l = Z_g^*$.

(b) What is the optimum load impedance if the line length is an odd multiple of a quarter wavelength?

Wavelengths for common frequency ranges

6.21   Find the wavelengths (in a vacuum) for plane waves at the following frequencies.

    (a) 60 Hz (power lines)   **(b)** 1 kHz (audible vibration)

    (c) 1 MHz (am radio)   **(d)** 100 MHz (fm radio, TV)

    (e) 2.45 GHz (microwave oven, radar)   **(f)** $(4.8)10^{14}$ Hz (red light)

Polarizations of a family of waves

6.22   A family of nine plane waves is given by $\mathbf{E}_{mn} = E_0[j^m\hat{\mathbf{x}} + j^n\hat{\mathbf{y}}]e^{-jkz}$, where m and n range from 1 to 3. The time dependence is $e^{j\omega t}$, and E_0 is a complex constant scalar. Which of these waves are linearly polarized? Which are circularly polarized? For the linearly polarized ones, give the direction of the polarization; for the circularly polarized ones, give the sense of rotation.

Quarter-wave plate

6.23   A *birefringent* material is one that propagates waves with two different, orthogonal linear polarizations at different speeds. Certain crystals (for example, mica) have this property; they can be used to convert linearly polarized waves into circularly polarized ones. Let c/n_1 and c/n_2 be the wave speeds for waves polarized along the x- and y-axes, respectively.

    A *quarter-wave plate* made of such material has a thickness l such that the phase difference between waves initially polarized along $\hat{\mathbf{x}}$ and $\hat{\mathbf{y}}$ is $\pi/2$ after traversing the distance l.

(a) How thick (in free-space wavelengths) is a quarter-wave plate?

(b) If a wave is initially linearly polarized along $\hat{\mathbf{x}} + \hat{\mathbf{y}}$, what will its polarization be at the other end of this plate?

Field reversal in reflected wave

6.24   When a linearly polarized plane wave is normally incident onto the interface between two dielectric media, the two scattered waves (reflected and transmitted) either have antiparallel electric fields and parallel magnetic fields or vice versa. One of these applies when the wave comes from the optically less dense (lower ε) nonmagnetic dielectric medium and encounters the denser one; the other case applies when the light attempts to cross from the denser material to the less dense one. Which scattered fields are antiparallel when

(a) the wave encounters a denser medium;

(b) the wave encounters a less dense medium?

Cancellation of field scattered by two incident waves

6.25 Each of two mismatched semi-infinite dielectrics (impedances η_1 and η_2) has a source that generates a plane wave at the same frequency and directs it normally onto the interface plane, as shown in Figure P6-8.

(a) How should the two incident-wave amplitudes, E_1 and E_2, be related in order not to have any transmitted wave E_4 in medium 2? Give E_2/E_1.

(b) What is the reflected-wave amplitude E_3 in medium 1 under the conditions of part (a)? Give E_3/E_1.

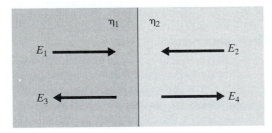

Figure P6-8   Two plane waves incident onto an interface from both sides.

Reversal of field scattered by two incident waves

6.26 Each of two mismatched semi-infinite dielectrics (impedances η_1 and η_2) has a source that generates a plane wave at the same frequency and directs it normally onto the interface plane, as shown in Figure P6–8.

(a) How should the two incident-wave amplitudes, E_1 and E_2, be related in order to have the transmitted wave E_4 in medium 2 be exactly $-E_1$? Give E_2/E_1.

(b) What is the reflected-wave amplitude E_3 in medium 1 under the conditions of part (a)? Give E_3/E_1.

Nulls of moving reflector

6.27 A plane wave encountering a perfectly conducting plane at $z = l$ at normal incidence results in a superposition of an incident and a reflected wave, $E(z, t) = E_0 e^{j(\omega t + k[l-z])} - E_0 e^{j(\omega t - k[l-z])}$, which is zero at the reflector. There are nulls also at locations z such that $k[l - z] = n\pi$ (n = positive integer), a half-wavelength apart (since $k = 2\pi/\lambda$). Suppose now that the reflecting plane is moving, uniformly at speed u, so that it is at $z = l_0 + ut$ at time t.

(a) Verify that the reflected wave's frequency ω' is not that of the incident wave. What is the Doppler shift $\Delta\omega = \omega' - \omega$?

(b) Where are the nulls of the superimposed incident and reflected waves?

Radar speed trap

6.28 A plane wave encountering a perfectly conducting plane at $z = l$ at normal incidence results in a superposition of an incident and a reflected wave, $E(z, t) = E_0 e^{j(\omega t + k[l-z])} - E_0 e^{j(\omega t - k[l-z])}$. Suppose that the reflecting plane is moving, uniformly at speed u, so that it is at $z = l_0 + ut$ at time t. The resultant Doppler-shifted reflected wave combines with the incident one to give $E(z, t) = E_0 e^{j(\omega t + k[l_0-z])} - E_0 e^{j(\omega' t - k'[l_0-z])}$. The shifted ω' and k' are determined by having $E(l_0 + ut, t)$ be zero. At $z = 0$, the traffic officer applies the combined field to a square-law detector, which effectively measures the mean-squared field or just $\text{Re}(\frac{1}{2} EE^*)$. Noting that $u \ll c$, obtain the output signal of the detector. Verify that measuring the frequency Ω of the detector output furnishes the speed of the reflecting target.

Eavesdropping with laser beams

6.29 A plane wave encounters an interface plane at normal incidence. The interface vibrates in conformity with an acoustic signal that strikes it. The interface may model an office window; a conversation inside the office would cause the sound waves to set the window into vibration. Pretend that the vibration is only one-dimensional [i.e., that the window acts as a diaphragm that moves only perpendicular to itself, following the sound signal $s(t)$]. The plane wave incident on the window from outside encounters it at $z = l_0 + s(t)$ and is reflected, with some reflection coefficient Γ. The eavesdropper at $z = 0$ applies the combined field to a square-law detector, which effectively measures the mean-squared field or just $\text{Re}(\frac{1}{2} EE^*)$. Verify that the output of the detector is a phase-modulated signal, so that the sound signal $s(t)$ may be extracted from this output by standard phase or frequency discriminator circuitry.

CHAPTER 7

Impedance Matching Techniques and Oblique Waves

STANDING WAVES

When a source of sinusoidal signal at some frequency is applied to a transmission line terminated in some mismatched load, there are generated an incident wave and a reflected wave that are each also sinusoidal, at the same frequency. Each of the two waves has its own magnitude and phase, and they travel along the line in opposite directions, at the same speed. The superposition of the two countertraveling sinusoids that forms the voltage distribution will be in phase at some locations, out of phase at others. The amplitude of the sinusoidal voltage oscillation (in time) will therefore be stronger at some locations and weaker at others. We say that the two waves interfere with each other; there is *constructive interference* wherever they add in phase and *destructive interference* wherever they combine out of phase, tending to cancel each other.

The locations of constructive and destructive interference generally do not remain fixed. Crests and troughs and nodes (points of complete cancellation to zero amplitude) all move along the line as time elapses, but not in a steady motion. For example, if we have a unit-amplitude incident wave and a reflection coefficient of magnitude Γ and phase φ, then the voltage distribution will be

$$V(z, t) = \cos(\omega t - \beta z) + \Gamma \cos(\omega t + \beta z + \varphi). \tag{7.1}$$

If we wish to keep up with a node of this superposition, we need to move along z with a motion $z(t)$ that maintains the voltage at zero for all time:

$$V(z(t), t) = 0 = \cos \omega t \cos \beta z + \sin \omega t \sin \beta z$$
$$+ \Gamma \cos(\omega t + \varphi) \cos \beta z - \Gamma \sin(\omega t + \varphi) \sin \beta z \tag{7.2}$$

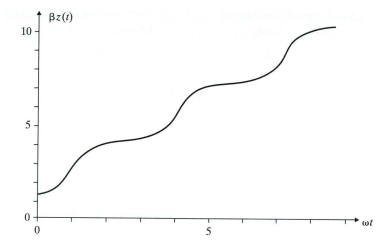

Figure 7-1   Motion of node of interfering incident and reflected waves.

We can solve for $z = z(t)$ from the ratio of $\sin \beta z$ to $\cos \beta z$, as in

$$\tan[\beta z(t)] = \frac{\Gamma \cos(\omega t + \varphi) + \cos \omega t}{\Gamma \sin(\omega t + \varphi) - \sin \omega t}. \tag{7.3}$$

This motion is clearly not uniform if $0 < \Gamma < 1$. Figure 7-1 shows the motion of a node for the case of $\Gamma = 0.5$ and $\varphi = \pi/3$.

One exception to our assertion of nonuniform motion of points on the superposition of incident and reflected waves is, of course, the case of zero reflection. There is then only the incident wave, and every point of that wave moves steadily at the speed ω/β.

Another exception occurs if the magnitude of the reflection coefficient is one. In that case, the voltage distribution we just considered becomes

$$V(z, t) = \cos(\omega t - \beta z) + \cos(\omega t + \beta z + \varphi), \tag{7.4}$$

which reduces, by trigonometric identities, to

$$V(z, t) = 2 \cos(\omega t + \varphi/2) \cos(\beta z + \varphi/2). \tag{7.5}$$

This distribution does not move at all; for example, the node that exists at the point z such that $\beta z = (\pi - \varphi)/2$ remains there for all time. This voltage distribution is a superposition of countertraveling waves of equal amplitudes. These waves interfere with each other, combining into a distribution that stands still, and the superposition is called a *standing wave*.

Note that a standing wave is not a wave at all; its distribution does not progress along z as time goes on. Instead, it is a sinusoidal distribution in space of the amplitude of a sinusoidal oscillation in time; it is comprised of oppositely traveling waves of equal amplitudes and equal speeds.

The current distribution in a standing wave is out of phase with the voltage that accompanies it, in both space and time:

$$Z_0 I(z,t) = \cos(\omega t - \beta z) - \cos(\omega t + \beta z + \varphi)$$

$$= 2\sin(\omega t + \varphi/2)\sin(\beta z + \varphi/2). \qquad (7.6)$$

The current is therefore also a sinusoidal oscillation, with an amplitude that varies sinusoidally along the line. Note that the current is a maximum where the voltage is a minimum (in magnitude), and vice versa.

We encountered standing waves in the case of sinusoidal excitation of a shorted line, for which the complex reflection coefficient Γ_l is -1, and for the open line, for which $\Gamma_l = +1$. In both cases, the magnitude of the reflection coefficient is $|\Gamma_l| = 1$. The same is true for a transmission line with any purely reactive load, either inductive or capacitive, because if $Z_l = jX$, then

$$|\Gamma_l| = \frac{|jX - Z_0|}{|jX + Z_0|} = 1, \qquad (7.7)$$

assuming the characteristic impedance Z_0 is purely resistive. We recognize pure standing waves in the frequency domain by a complex reflection coefficient of magnitude unity.

The nonuniform motion such as that of a node in case $0 < \Gamma < 1$ is not normally observable, as high frequencies preclude the usual instruments from following the fine-scale details of the motion. What is readily observable, however, is the spatial envelope of the oscillating signal, with the time variation averaged out by either rectification or squaring and averaging. The resultant spatial variation is the *interference pattern* of the countertraveling waves; it is referred to as a *standing wave* pattern. It is "standing" because no motion remains after the time averaging; it is not a "wave" at all and is so designated only because it exhibits an oscillatory distribution in space. This standing wave or interference pattern is easily measured and highly informative, as we shall now see.

IMPEDANCE MEASUREMENT AND STANDING WAVE RATIO

At high frequencies, direct measurements of impedance along a transmission line, or of the impedance of some load attached to the line, are difficult to make. The quantities that we can readily measure include the amplitude of the sinusoidal voltage at any accessible point along the line, as well as the distances along the line of these accessible points. If we can make a range of positions along the line accessible to voltage amplitude measurements, along an interval at least one-half wavelength long, then we can deduce the impedance at the load and anywhere else from these measurements, as explained in this section.

In the case of a transmission line contained within a shielded structure, such as a coaxial line or a waveguide, accessibility is commonly provided by inserting a *slotted line* at a convenient place between source and load. This is a section of line in

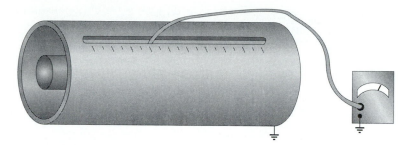

Figure 7-2 Structure of slotted line for coaxial transmission line.

which a narrow slot has been cut in the axial direction to allow a probe to penetrate to the field region inside the shield. The probe should penetrate just deeply enough to pick up a signal proportional to the electric field inside, without excessively disturbing the field pattern. The probe is connected to a voltmeter that provides a reading proportional to the root-mean-square (rms) voltage at the location of the probe. The slot allows the probe to be moved along the length of the line, so that the rms voltage can be measured as a function of distance along the line. Because the rms voltage is proportional to the magnitude of the complex voltage, we can get a measurement of $|V(z)|$ versus z, over a range of z at least one-half wavelength long. A ruler scribed along the slot allows easy measurement of the distance from probe to load. Figure 7-2 shows the essential structure of a slotted line for a coaxial transmission line.

To see how we can extract load impedance measurements from the readings provided by the slotted line, we need to express the magnitude of the complex voltage $V(z)$ along the line as a function of distance z. This voltage is

$$V(z) = V_0\left[e^{j\beta(l-z)} + \Gamma_l e^{-j\beta(l-z)}\right] \tag{7.8}$$

for a line terminated at $z = l$ by a complex load impedance Z_l that imposes a complex reflection coefficient $\Gamma_l = |\Gamma_l|\exp j\varphi_l$. The factor V_0 is the complex amplitude of the incident wave. We rewrite the voltage distribution as

$$V(z) = V_0\, e^{j\beta(l-z)}\left[1 + |\Gamma_l|e^{j\{\varphi_l - 2\beta(l-z)\}}\right] \tag{7.9}$$

and designate the variable phase angle ψ as

$$\psi(z) = \varphi_l - 2\beta(l - z). \tag{7.10}$$

The magnitude is therefore

$$|V(z)| = |V_0|\left|1 + |\Gamma_l|e^{j\psi(z)}\right| \tag{7.11}$$

and this is what is measured as a function of distance to load, $l - z$.

How this varies with z can be understood with the aid of Figure 7-3, which shows the complex quantity $w(z) = 1 + |\Gamma_l|e^{j\psi(z)}$ plotted in the complex plane. As z varies, only the angle $\psi(z)$ changes, so that the locus traced by $w(z)$ is a circle, centered at the point $1 + j0$ and with radius $|\Gamma_l|$. The magnitude $|w(z)|$ that we require

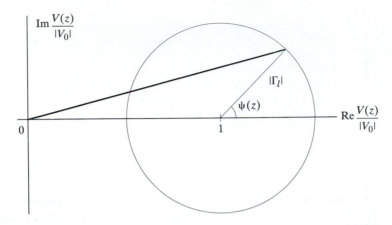

Figure 7-3   Crank diagram of magnitude of voltage distribution.

is the length of the line from the origin to the point on the circle. As we move the probe, the point cranks around the circle and the magnitude varies, from a minimum of $1 - |\Gamma_l|$ when ψ becomes an odd multiple of π to a maximum of $1 + |\Gamma_l|$ when ψ is an even multiple of π. The plot is known as a *crank diagram*.

Algebraically, the voltmeter reading as the observation point moves along the line is given by

$$|V(z)| = |V_0||1 + |\Gamma_l|e^{j\psi(z)}| = |V_0|[w(z)w^*(z)]^{1/2}$$
$$= |V_0|\{1 + 2|\Gamma_l|\cos\psi(z) + |\Gamma_l|^2\}^{1/2} \qquad (7.12)$$

and this is plotted in Figure 7-4, for the case of $\Gamma_l = 0.6\exp[j\pi/3]$. What we see as the probe is moved is an interference pattern formed by the superposition of the

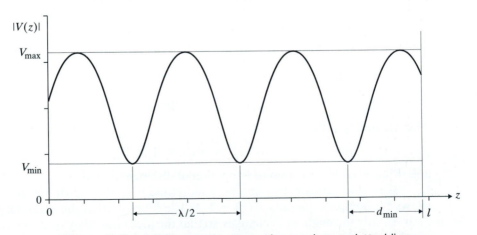

Figure 7-4   Interference pattern observed on a slotted line.

incident and reflected waves. If there were no reflected wave, $|\Gamma_l| = 0$, the magnitude of the voltage would remain constant; on the crank diagram, the circle would shrink to zero radius. For a pure standing wave, for which $|\Gamma_l| = 1$ and the incident and reflected waves are of equal strength, the minima in the interference pattern would reach zero; on the crank diagram, the circle would pass through the origin.

The general case of $0 < |\Gamma_l| < 1$ is often described as a *partial standing wave*; it is only the envelope of the oscillation that is standing in this case. Note in the interference pattern that the distortion of the sinusoid introduced by the square root operation makes the minima sharper than the maxima; consequently, the minima can be located more precisely than can the maxima. Note also that successive minima are spaced a half wavelength apart.

Observations of the interference pattern suffice to determine the complex reflection coefficient Γ_l for the line and, from this, the complex load impedance. It is a particularly simple measurement to find the maximum and minimum levels of $|V(z)|$ as the probe moves along the slot; these levels are labeled V_{max} and V_{min} on Figure 7-4. It is also easy to locate a particular minimum of the interference pattern and measure its distance d_{min} from the load. These readings provide all the information needed to determine the load impedance.

Since $V_{max} = |V_0|[1 + |\Gamma_l|]$ and $V_{min} = |V_0|[1 - |\Gamma_l|]$, we can eliminate the unknown amplitude $|V_0|$ of the incident wave by taking the ratio of the maximum to the minimum of the interference pattern. This ratio has been named the *standing wave ratio* or SWR, more specifically the *voltage standing wave ratio* (VSWR). Our measurement gives us the VSWR as

$$\text{VSWR} = V_{max}/V_{min} \tag{7.13}$$

and the theory tells us that this number is

$$\text{VSWR} = \frac{1 + |\Gamma_l|}{1 - |\Gamma_l|}. \tag{7.14}$$

Note that the reflection coefficient for current is the negative of the one for voltage, so that this ratio is the same for current and for voltage and hence is often referred to as just the SWR. However, the locations of the maxima and minima are not the same for current and for voltage, and these locations enter into the determination of the full reflection coefficient (not just its magnitude) from the SWR. For such purposes, we therefore need to know whether the given SWR was measured from the voltage or the current. Problem 7.5 illustrates the difference this makes.

Note also that the VSWR is a number between 1, for a matched line, and ∞, for the case of a pure standing wave. We invert the formula to obtain the *magnitude* of the complex reflection coefficient from our easy measurement of the VSWR.

$$|\Gamma_l| = \frac{\text{VSWR} - 1}{\text{VSWR} + 1}. \tag{7.15}$$

We still need the phase of the reflection coefficient.

We measure the distance d_{min} from the location of a convenient minimum of the interference pattern to the load, and we either know the wavelength λ along the line (from the operating frequency and the constitution of the line) or else we also measure the distance between successive minima to get $\lambda/2$. This measurement gives us $\psi(z)$ at the location of the minimum of the pattern. But the value of $\psi(z)$ at a minimum must be π, or any odd multiple of π. We therefore have

$$\psi_{min} = m\pi = \varphi_l - 2\beta(l - z) = \varphi_l - 4\pi d_{min}/\lambda, \tag{7.16}$$

where m is an odd integer. Inverting this formula, we find the unknown phase φ_l of the complex reflection coefficient at the load as

$$\varphi_l = \pi[m + 4d_{min}/\lambda]. \tag{7.17}$$

Since additive multiples of 2π are insignificant to the value of the phase, we can choose the odd integer m (possibly negative) to reduce the phase φ_l to the primary interval $-\pi$ to π. This, together with the magnitude of Γ_l we got from the VSWR, gives the full complex reflection coefficient at the load.

Finally, from the complex reflection coefficient at the load, Γ_l, and the known characteristic impedance of the transmission line, Z_0, we can obtain the complex load impedance, Z_l, as

$$Z_l = Z_0 \frac{1 + \Gamma_l}{1 - \Gamma_l}. \tag{7.18}$$

From this, we can of course get the impedance anywhere along the line, including the impedance presented to the generator. Thus, the slotted line measurements give all the information we may need about the voltage, current, and impedance distributions along the line.

Example 7.1

Suppose slotted line measurements along a 300-ohm transmission line give voltage maxima of 6.0 V and minima of 2.5 V and that there are successive minima at distances 1.20 m and 1.75 m from an unknown load. Find the complex impedance of the load. ◇

The maximum and minimum voltage give the VSWR as $6.0/2.5 = 2.4$, so that the magnitude of the reflection coefficient is

$$|\Gamma| = (2.4 - 1)/(2.4 + 1) = 0.4118. \tag{7.19}$$

The locations of successive minima tell us that $(1.75 - 1.20) = 0.55$ m is half of a wavelength. For the minimum at 1.20 m from the load, the quantity

$$4d_{min}/\lambda = 2(1.20\ \text{m})/(0.55\ \text{m}) = 4.3636 \tag{7.20}$$

tells us that the phase of the reflection coefficient is, for m an odd integer,

$$\varphi_l = \pi[m + 4.3636]$$
$$= \pi[-5 + 4.3636] = -1.9992\ \text{radians,} \tag{7.21}$$

so that the reflection coefficient at the load is

$$\Gamma_l = 0.4118e^{-j1.9992} = -0.17105 - j0.37455. \tag{7.22}$$

This gives for the normalized load impedance

$$\frac{Z_l}{Z_0} = \frac{1 - 0.17105 - j0.37455}{1 + 0.17105 + j0.37455}$$

$$= \frac{0.82895 - j0.37455}{1.17105 + j0.37455} = \frac{0.90964e^{-j0.42438}}{1.22949e^{j0.30956}}$$

$$= 0.73985e^{-j0.73395} = 0.54936 - j0.49555 \tag{7.23}$$

and we get the load impedance, with $Z_0 = 300$ ohm, to be

$$Z_l = (164.81 - j148.67) \text{ ohm.} \tag{7.24}$$

This impedance has resistive and capacitive parts.

As an aid to appreciating the nature of wave interference in various transmission line circuits, let us calculate the VSWR in the sections of line we encountered in some examples in the last chapter.

Example 7.2

A 300-ohm parallel-wire transmission line is one meter long and is loaded with a circuit whose lumped equivalent impedance is $(300 + j540)$ ohms when the line is operated at 500 MHz. What is the VSWR along this line? See Figure 7-5. ◈

All we need is the magnitude of the reflection coefficient along the line. The mismatch between the load and characteristic impedances is

$$Z_l/Z_0 = (300 + j540)/300 = 1 + j1.8 \tag{7.25}$$

and the reflection coefficient at the load is

$$\Gamma_l = \left[(1 + j1.8) - 1\right]/\left[(1 + j1.8) + 1\right] = 0.66896e^{j0.83798}, \tag{7.26}$$

so that $|\Gamma| = 0.66896$ and

$$\text{VSWR} = (1 + 0.66896)/(1 - 0.66896) = 5.042. \tag{7.27}$$

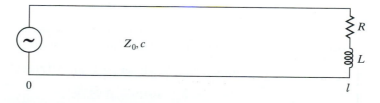

Figure 7-5 Transmission line with complex load impedance.

Example 7.3

A transmission line is loaded with a mismatched resistance R_0 at the end l and has a lumped capacitor C_0 at location l_0, as in Figure 7-6. Find the VSWR in each section of the 300-Ω line when $R_0 = 50\ \Omega$, $l = 2\ m$, $C_0 = 5\ pF$, and $l_0 = 1.5\ m$, for a 100-MHz source. ◈

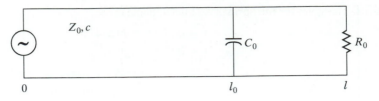

Figure 7-6   Transmission line with capacitor at intermediate point.

The superposition of incident and reflected waves is different in each section of the line, so that the magnitude of the reflection coefficient differs in the two sections. Each uniform section of transmission line therefore has its own VSWR. The mismatch at the end of the line gives

$$\Gamma_l = (50 - 300)/(50 + 300) = -5/7 \tag{7.28}$$

at the load, so that in the far section

$$|\Gamma_2| = 5/7 \quad \text{and} \quad \text{VSWR}_2 = 6.000. \tag{7.29}$$

For the nearer section, we need the parallel combination of the input impedance of the second section and the capacitor. As found in the last chapter, $\Gamma_2(l_0) = \Gamma_2(l)e^{-j2\beta(l-l_0)}$, the phase $2\beta\,(l - l_0)$ is $4\pi f(l - l_0)/c = 2.0958$ radians, and the reflection coefficient just beyond the capacitor is

$$\Gamma_2(l_0) = -(5/7)e^{-j2.0958} = 0.35804 + j0.61807. \tag{7.30}$$

The input impedance for the second segment, as seen at the capacitor, is

$$Z(l_0) = Z_0\big[1 + \Gamma_2(l_0)\big]/\big[1 - \Gamma_2(l_0)\big] = Z_0(0.616774 + j1.55661)$$

and this appears in parallel with the impedance of the capacitor at that location. The capacitor's impedance is

$$1/j\omega C_0 = Z_0/j\omega Z_0 C_0 = Z_0(1/j2\pi f Z_0 C_0) = Z_0(-j1.061033),$$

so that the parallel combination of the impedance of the second segment and this capacitor gives for the load impedance at l_0 for the first segment

$$Z_1(l_0) = \big[Z(l_0)\big]\big[1/j\omega C_0\big]/\big[Z(l_0) + 1/j\omega C_0\big]$$
$$= Z_0(1.10919 - j1.95226). \tag{7.31}$$

This is now the load impedance for the first segment; the corresponding reflection coefficient is

$$\Gamma_1(l_0) = \big[Z_1(l_0) - Z_0\big]/\big[Z_1(l_0) + Z_0\big] = 0.68034e^{-j0.76815}, \tag{7.32}$$

so that, finally,

$$\text{VSWR}_1 = (1 + 0.68034)/(1 - 0.68034) = 5.257. \tag{7.33}$$

IMPEDANCE MATCHING WITH A TUNING STUB

We have noted that a significant change in the input impedance of a transmission line can be achieved by attaching a reactive lumped element at some intermediate point along the line. This raises the question of whether an otherwise mismatched line could be matched by the addition of such elements. If so, then the reflection from the mismatched load could be eliminated, without altering the load itself.

We can take this notion one step farther, replacing the reactive lumped element by a segment of another transmission line, one that is itself terminated in, say, a short circuit. We already know that a suitable length of such shorted line can present any reactive impedance at its input end. The aim is to choose the length of this additional segment of shorted line so as to make the parallel combination of this segment with the original mismatched line appear to have its reactive part canceled, leaving only a resistive part. If we can also choose the location at which to attach this shorted line segment so that the remaining resistive part of the combined impedance is adjusted to be the characteristic impedance of the line, the combination will present a matched load to the source.

We are thus led to the configuration in Figure 7-7, in which a length s of shorted line, acting as a *tuning stub*, has been attached across the original line at a distance d from the mismatched load. The length s and the position d are to be chosen to *tune out* the mismatched impedance, leaving a matched line. Let us see how this tuning stub can be designed.

From the connection point onward, both lengths of transmission line are uniform segments, each of characteristic impedance Z_0. The shorted one is of length s; the mismatched one has length d. From the formula for impedance distribution, we know that the shorted stub has an input impedance

$$Z_s = jZ_0 \tan \beta s \qquad (7.34)$$

and also that the input impedance for the segment of mismatched line beyond the stub's connection point is

$$Z_d = Z_0 \frac{Z_l + jZ_0 \tan \beta d}{Z_0 + jZ_l \tan \beta d}. \qquad (7.35)$$

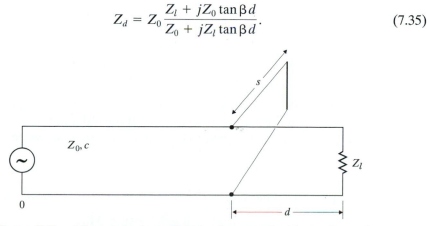

Figure 7-7 Mismatched parallel-wire line matched by tuning stub.

These two input impedances appear in parallel at the connection point. This combined impedance is then the load impedance for the line from the source to the connection point, and we intend to make this combined impedance equal the characteristic impedance. This will match the line, as far as the segment from source to connection point is concerned; there will then be no reflected wave returning from the connection point to the source.

To make the parallel combination of impedances match Z_0 entails making the reactive part cancel to zero and making the resistive part equal Z_0. For these two conditions, we have available as design parameters the two lengths, s and d. With two equations for two unknowns, we may be able to obtain a solution.

It is expedient to express the matching conditions in terms of the reflection coefficients of the two segments at their connection point. For the shorted stub, the reflection coefficient, -1, at the short becomes $\Gamma_s = -e^{-j2\beta s}$ at its input. For the mismatched main line, the reflection coefficient $\Gamma_l = (Z_l - Z_0)/(Z_l + Z_0)$ at the load becomes $\Gamma_d = \Gamma_l e^{-j2\beta d}$ at the connection point. The impedances of the two segments are

$$Z_s = Z_0[1 + \Gamma_s]/[1 - \Gamma_s] \quad \text{and} \quad Z_d = Z_0[1 + \Gamma_d]/[1 - \Gamma_d] \qquad (7.36)$$

and their parallel combination is to be set equal to Z_0. More conveniently, the parallel admittances, $1/Z$, are to add to the combined admittance $1/Z_0$:

$$\frac{1}{Z_0}\frac{1 - \Gamma_s}{1 + \Gamma_s} + \frac{1}{Z_0}\frac{1 - \Gamma_d}{1 + \Gamma_d} = \frac{1}{Z_0} \qquad (7.37)$$

or

$$[1 - \Gamma_s]/[1 + \Gamma_s] + [1 - \Gamma_d]/[1 + \Gamma_d] = 1. \qquad (7.38)$$

This simplifies to

$$\Gamma_s = [1 - \Gamma_d]/[1 + 3\Gamma_d]. \qquad (7.39)$$

At this point, we recall that we know the magnitudes of the two reflection coefficients, Γ_s and Γ_d, but not their phases:

$$\Gamma_s = -e^{-j2\beta s}; \quad \Gamma_d = \Gamma_l e^{-j2\beta d}; \qquad (7.40)$$

$$\Gamma_l = (Z_l - Z_0)/(Z_l + Z_0) = |\Gamma_l|e^{j\varphi_l}. \qquad (7.41)$$

The reflection coefficient at the original load, Γ_l, is known, in both magnitude and phase; the unknowns are s and d. We can get the unknown phase of Γ_d at once, by requiring the right side of (7.39) to have unit magnitude. We write

$$\Gamma_d = |\Gamma_l|e^{j\psi}, \quad \text{where} \quad \psi = \varphi_l - 2\beta d , \qquad (7.42)$$

and require the numerator and denominator in (7.39) to have the same magnitude:

$$|1 - |\Gamma_l|e^{j\psi}|^2 = |1 + 3|\Gamma_l|e^{j\psi}|^2. \qquad (7.43)$$

Upon expanding the squared magnitudes as products of the quantities and their complex conjugates, this reduces to

$$\cos\psi = -|\Gamma_l|, \tag{7.44}$$

which gives us the position d of the connection point, from $2\beta d = \varphi_l - \psi$. This also determines Γ_d for use in (7.39), to give us $2\beta s$ as well, which provides the length s of the shorted stub. Let us look at a numerical example.

Example 7.4

Find the position and length of the shorted tuning stub that will match a transmission line terminated by impedance $Z_l = (3 + j2)Z_0$. ◇◇

We need the reflection coefficient at the mismatched load, in both magnitude and phase, namely

$$\Gamma_l = (Z_l - Z_0)/(Z_l + Z_0) = |\Gamma_l|e^{j\varphi_l}. \tag{7.45}$$

This is

$$\Gamma_l = (3 + j2 - 1)/(3 + j2 + 1) = (2 + j2)/(4 + j2)$$
$$= (3 + j)/5 = \left[\sqrt{10}/5\right]e^{j\,\tan^{-1}(1/3)}$$
$$= 0.63246e^{j0.32175} \tag{7.46}$$

and we need ψ such that $\cos\psi = -0.63246$; there are two such values of ψ in any range of 2π radians and we can write

$$\psi = \pm2.25552, \tag{7.47}$$

to which we can add multiples of 2π. This gives

$$2\beta d = \varphi_l - \psi = 0.32175 \mp 2.25552 + m2\pi \tag{7.48}$$

with m any integer; the smallest suitable distance to the load is given by

$$2\beta d = 2.57727. \tag{7.49}$$

We can express the distance d in terms of the free-space wavelength $\lambda = 2\pi/\beta$ as

$$d/\lambda = 0.205092. \tag{7.50}$$

However, we have not been given an operating frequency or wavelength that would have allowed us to translate this into an actual distance in, say, meters.

To find the length of the stub, we express the reflection coefficient of the segment of line beyond the connection point, Γ_d, as

$$\Gamma_d = |\Gamma_l|e^{j\psi} = 0.63246e^{-j2.25552} = -0.40000 - j0.48990, \tag{7.51}$$

which gives

$$[1 - \Gamma_d]/[1 + 3\Gamma_d] = \Gamma_s = -e^{-j2\beta s} = e^{j(\pi - 2\beta s)}$$
$$= [1.4000 + j0.48990]/[-0.20000 - j1.46970]$$
$$= e^{j2.04266}. \tag{7.52}$$

This yields, with n any integer,

$$\pi - 2\beta s = 2.04266 + n2\pi \tag{7.53}$$

or, for the shortest suitable stub length,

$$2\beta s = \pi - 2.04266 = 1.09893. \tag{7.54}$$

In terms of wavelength, this is

$$s/\lambda = 0.08745. \tag{7.55}$$

Use of the alternative solution for ψ would give, instead of these shortest distances, $d/\lambda = 0.34612$ and $s/\lambda = 0.41255$. We can also add multiples of a half wavelength to the distances d and s to arrive at other solutions that may be more convenient in certain cases.

We have constructed the solution to this matching problem so as to guarantee no reflection in the section of transmission line between the source and the stub. But what of the section beyond it? That portion of the line is still mismatched and there is both an incident and a reflected wave there. It may be helpful to our understanding of the action of the tuning stub to exhibit the voltage, current, and impedance distributions in that section of the line.

Having successfully matched the part of the line before the stub, we know that the voltage and current entering the connection point are expressible as V_0 and V_0/Z_0. Beyond the stub's location at $z = l - d$, where l is the length of the original mismatched line, the voltage and current are superpositions of waves:

$$V(z) = V_1 e^{-j\beta(z-l+d)} + V_2 e^{j\beta(z-l+d)}, \tag{7.56}$$

$$Z_0 I(z) = V_1 e^{-j\beta(z-l+d)} - V_2 e^{j\beta(z-l+d)}. \tag{7.57}$$

The amplitudes V_1 and V_2 can be determined by imposing the conditions of continuity of the voltage at the connection point

$$V_1 + V_2 = V_0 \tag{7.58}$$

and of discontinuity of the current, by the amount that is diverted to the shorted stub—namely, $V_0/(jZ_0 \tan \beta s)$:

$$V_1 - V_2 = V_0 - V_0/(j \tan \beta s). \tag{7.59}$$

The results are

$$V_1 = V_0[1 - 1/j2 \tan \beta s], \qquad V_2 = V_0/j2 \tan \beta s. \tag{7.60}$$

The impedance distribution $Z(z)$ beyond the connection point is therefore, upon writing ξ for $z - l + d$,

$$\frac{Z(z)}{Z_0} = \frac{V_1 e^{-j\beta\xi} + V_2 e^{j\beta\xi}}{V_1 e^{-j\beta\xi} - V_2 e^{j\beta\xi}} = \frac{(1 - j2 \tan \beta s)e^{-j\beta\xi} - e^{j\beta\xi}}{(1 - j2 \tan \beta s)e^{-j\beta\xi} + e^{j\beta\xi}}. \tag{7.61}$$

This varies from $(-j \tan \beta s)/(1 - j \tan \beta s)$ at the connection point $(\xi = 0)$ to

$$\frac{(1 - j2 \tan \beta s)e^{-j\beta d} - e^{j\beta d}}{(1 - j2 \tan \beta s)e^{-j\beta d} + e^{j\beta d}} \qquad \text{at the mismatched load } (\xi = d).$$

For the preceding numerical example, we found $\beta s = 0.54947$ and $\beta d = 1.28863$, so that the impedance ratio does end up as

$$\frac{(1 - j2[0.61237])e^{-j1.28863} - e^{j1.28863}}{(1 - j2[0.61237])e^{-j1.28863} + e^{j1.28863}} = 3 + j2,$$

as required.

Figure 7-8 traces the locus of the impedance, in the complex plane of $Z(z)/Z_0$ as position z varies from the connection point at $z = l - d$ to the mismatched load at $z = l$. Note that at the stub, the normalized impedance is $[3 - j\sqrt{24}]/11$, this being the impedance that, combined in parallel with the shorted stub's impedance $j\sqrt{6}/4$, matches the line. At the load, the normalized impedance becomes $(3 + j2)$, as specified.

Incidentally, if the locus of the impedance distribution $Z(z)$ in the complex plane appears to be an arc of a circle, that is because it is. And this is not a coincidence; details appear in Problem 7.6.

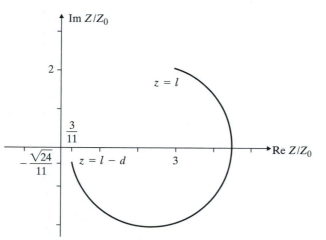

Figure 7-8 Locus of impedance distribution beyond the tuning stub.

Example 7.5

Find the VSWR in all three portions of the line loaded by $Z_l = (3 + j2)Z_0$ but matched by the shorted tuning stub, as in the previous example. ◇

Since we have already calculated the reflection coefficient in the various sections of line in the course of the tuning stub calculations, we need only make use of their magnitudes to obtain the VSWR in each section. The formula is

$$\text{VSWR} = (1 + |\Gamma|)/(1 - |\Gamma|). \tag{7.62}$$

We had a mismatch $Z_l/Z_0 = 3 + j2$ beyond the stub's connection point, which gave a reflection coefficient of magnitude $|\Gamma| = 0.63246$ there. From this,

$$\text{VSWR} = (1 + 0.63246)/(1 - 0.63246) = 4.442 \tag{7.63}$$

beyond the stub. In front of the stub, we created a matched line, so that we have VSWR $= 1$ there. Along the stub, we have a shorted line with $\Gamma_l = -1$, which has unit magnitude, so that VSWR $= \infty$ in the shorted stub, indicative of a pure standing wave.

SMITH CHART

Before electronic calculators and computers became commonplace tools, the calculations illustrated in the previous section for transmission line circuit design were quite tedious. The Smith chart was developed to facilitate these calculations and to provide a graphical interpretation of the relationships among impedances, admittances, and reflection coefficient distributions along the line. Though now of lesser importance for carrying out the actual calculations, the chart still serves well for graphical display and analysis.

The Smith chart is a rendering of the unit circle in the complex plane of the reflection coefficient Γ, but with points inside and on that circle labeled by their corresponding complex impedance values $Z = R + jX$, normalized to the characteristic impedance Z_0. This is done by plotting and labeling lines of fixed resistance R for varying reactance, and of fixed reactance X for varying resistance, in the Γ-plane. These lines of constant R and of constant X turn out to be circles and arcs of circles, respectively. A simplified version of the Smith chart, with only a few loci of constant-R and constant-X circles plotted, is shown in Figure 7-9.

The center of the chart, marked with a tiny circle, is the origin of the Γ-plane ($\Gamma = 0$), corresponding to a matched impedance. The outermost circle is the unit circle ($|\Gamma| = 1$), so that the rightmost point of the chart is $\Gamma = 1(|Z| = \infty)$, the leftmost one is $\Gamma = -1(Z = 0)$, the uppermost one is $\Gamma = j$, and the lowest one is $\Gamma = -j$. The curves on the chart are labeled not by Γ but by

$$Z/Z_0 = r + jx = \frac{1 + \Gamma}{1 - \Gamma}. \tag{7.64}$$

This enables us to plot the electrical state at any location along the transmission line in terms of the impedance there; for example, if the impedance at some point of the transmission line (perhaps at the load) is known, say, $Z = (3 + j2)Z_0$, that point appears on the chart at the intersection of the circle labeled $r = 3$ and of the arc labeled $x = 2$. If we move to another point along the transmission line, by a displacement Δz, the magnitude of Γ remains fixed while its phase changes in proportion to the displacement, by $2\beta\Delta z$ radians. On the chart, therefore, the displacement along the transmission line is represented by motion *along a circle centered at the origin* and with the radius determined by the location of the initial point. After tra-

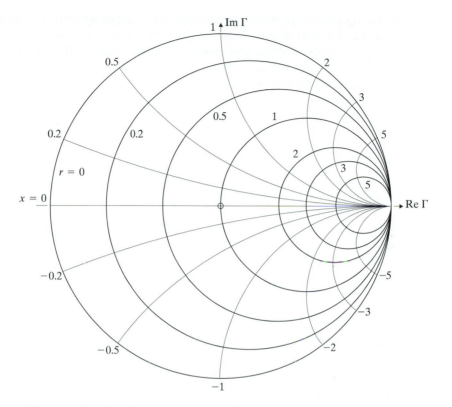

Figure 7-9   Smith chart: circles of fixed R and of fixed X in Γ-plane.

versing the appropriate angular measure of arc along that circle, we can read the new impedance value at the new location, directly off the chart.

Some of the properties of the Smith Chart that are helpful in its interpretation follow.

a. Only the inside (and edge) of the unit circle in the complex Γ-plane is relevant, since $0 \le |\Gamma| \le 1$; this condition corresponds to the requirement that the resistive part of any physically realizable impedance be nonnegative.

b. Lines of constant resistance are circles, centered on the real axis of the Γ-plane; all of these are tangent to the point $\Gamma = 1$. The radius of the circular locus labeled r is $1/(r + 1)$; its center is at $\Gamma = r/(r + 1)$. The zero-resistance locus is the unit circle itself.

c. Lines of constant reactance are also circles, centered on the vertical line through the point $\Gamma = 1$ in the Γ-plane; all of these are tangent to the point $\Gamma = 1$. The radius of the circular locus labeled x is $1/|x|$; its center is outside the unit circle, at $\Gamma = 1 + j/x$. For zero reactance, however, the circle has infinite radius and reduces to the straight line that forms the real axis in the Γ-plane. Within the unit circle, the constant-reactance loci are only arcs of circles.

d. Two points on the chart that are related to each other by reflection through the origin (i.e., such that Γ at one of the points is the negative of the value at the other) have reciprocal complex impedance values: If the first point (at some complex Γ) corresponds to $r + jx$, then the reflected point (at $-\Gamma$) corresponds to normalized impedance $1/(r + jx)$.

e. Motion along the transmission line shows up on the chart as movement along a circle concentric with the unit circle itself (i.e., centered at the origin). The angular progression $\Delta\varphi$ along this circle is proportional to the distance Δz traveled along the transmission line, as

$$\Delta\varphi = 4\pi\,\Delta z/\lambda, \tag{7.65}$$

where λ is the wavelength. Motion along the line toward the load corresponds to counterclockwise traversal of the circle; motion toward the generator is traced by clockwise movement around the circle. Displacement by one-half wavelength brings us fully around the circle. Movement by a quarter wavelength along the line brings us to the diametrically opposite point of the circle; hence, by (d), to the reciprocal of the normalized impedance.

f. A short circuit is represented by $r = 0$, $x = 0$, which is at $\Gamma = -1$, the leftmost point of the chart. An open circuit is at $\Gamma = 1$, the rightmost point. A matched line is depicted by the center point of the chart, at $\Gamma = 0$ ($r = 1$, $x = 0$). For a purely reactive load, we traverse the unit circle $|\Gamma| = 1$, the outermost circle.

g. We can read the VSWR corresponding to any load impedance by noting where the circle of travel intersects the positive real axis. The resistance value at that intersection point equals the VSWR.

h. The Smith chart serves equally well as a chart of normalized admittances labeling points inside the unit circle in the complex Γ-plane. We simply reinterpret the full circles as constant-g loci and the circular arcs as constant-b lines, where

$$Y = G + jB = (g + jb)/Z_0 = 1/Z = (1/Z_0)/(r + jx). \tag{7.66}$$

The loci and their labels are identical to the ones labeled as impedances.

You can demonstrate many of the above properties, as suggested in the problems.

Let us illustrate the use of the Smith chart in the design of the shorted tuning stub that matches an otherwise mismatched line, as considered analytically in Example 7.4.

Example 7.6

On a Smith chart, plot the circle of travel for the mismatched line loaded by $Z_l = (3 + j2)Z_0$ and show graphically how to obtain the position and length of a shorted tuning stub that can match the line. ◇

On the Smith chart, we draw the circle centered at the origin and passing through the point at the intersection of the $r = 3$ and $x = 2$ circles; that intersection point is shown in Figure 7-10 as the point labeled L on the bold circle of travel.

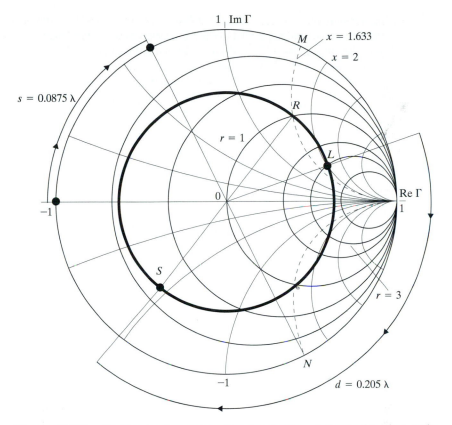

Figure 7-10   Smith chart construction for design of shorted tuning stub.

As we recede from the load toward the generator, we travel around the circle, clockwise, maintaining the magnitude of the reflection coefficient but varying the impedance, or admittance, as we go. Since the tuning stub in parallel with a section of the line will add only a reactive admittance and is intended to give a summed admittance of unity, we are searching for a location where the conductive part of the admittance is already unity. So we look for the places where the circle of travel intersects the circle labeled $g = 1$ (if the labels represent admittances) or $r = 1$ (if impedances). There are two such points on the circle of travel; one is labeled R in the figure. Since we are representing the impedances, we must be at the diametrically opposed point, labeled S, to get an impedance whose reciprocal is an admittance with a normalized conductance of unity. Having found a point S where this does happen, we can tell its distance d from the load by measuring the angle we have traveled around the circle to get from L to S. This angle is $4\pi d/\lambda$, and we measure $d/\lambda = 0.205$, as indicated.

There remains to find the length of the shorted tuning stub. Its purpose is to add a susceptance b that cancels the one found at point R. We read this susceptance as $b = 1.633$, and since we need its negative for cancellation, we follow the $b = -1.633$ arc until we reach the outer circle, at point N. On the outer circle (the unit circle itself), the impedances or admittances are purely imaginary. Diametrically opposite point N is the point where the reflection coefficient $\Gamma = -1$ at the short is transformed into that for the impedance needed in parallel to cancel the susceptance of the main line there. We measure the angle of travel from the short to that point as $s/\lambda = 0.0875$, as indicated, giving us the length of the shorted stub.

Because we are dealing here with combinations of impedances in parallel, the entire graphical calculation is simpler in terms of admittances. The chart is identical when labeled as complex admittances; the starting point, however, is at the point diametrically opposed to the $3 + j2$ point on the circle of travel, because the load was given as an impedance, not as an admittance. Also, the short at the end of the stub represents an infinite admittance and is located at the rightmost point of the admittance chart, not the leftmost one.

The reader should be forewarned that what we have referred to as the circle of travel on the Smith chart, for a lossless line, actually becomes a spiral in the more realistic case of a lossy transmission line, which is considered in the next chapter.

REFLECTION AT A MIRROR

There is a strong and useful analogy between wave propagation along a transmission line and the behavior of plane waves. Reflection at a mismatched load is one such analog. An important example is reflection from a short circuit at the end of a transmission line; the plane-wave analog is reflection at the interface between a dielectric and a perfect conductor.

The perfect conductor does not tolerate any tangential electric field along its surface, so the incident wave's electric field has to be canceled by the reflected wave. The electric-field reflection coefficient $\Gamma_e = (\eta_2 - \eta_1)/(\eta_2 + \eta_1)$ must therefore be $\Gamma_e = -1$, regardless of the impedance η_1 of the dielectric in front of the interface. This is consistent with assigning the perfect conductor a wave impedance $\eta_2 = 0$, which is itself consistent with there being zero electric field at the surface, since $\eta = E/H$ and E must be zero while H need not vanish at the surface.

The reflection coefficient -1 makes the reflected electric field equal and opposite to the incident one, canceling it at the surface of the conductor. The amplitude transmission coefficient $T_e = 2\eta_2/(\eta_2 + \eta_1) = 0$, confirming that there is no transmitted electric field. But for the magnetic field, the reflection coefficient is $\Gamma_h = -\Gamma_e = +1$, which means that the reflected magnetic field equals the incident one and results in a doubled magnetic field in front of the conducting wall. Correspondingly, the magnetic field transmission coefficient is $T_h = 2\eta_1/(\eta_1 + \eta_2) = 2$, regardless of the impedance of the dielectric in front of the interface. This means that there is a transmitted magnetic field just inside the perfect conductor, of magnitude twice that of the incident field, as required by continuity of the tangential magnetic field across the interface.

However, this doubled magnetic field that appears just inside the wall of the perfect conductor survives only on the surface of the conductor; it penetrates to no discernible depth within the perfect conductor. We perceive a discontinuity in the

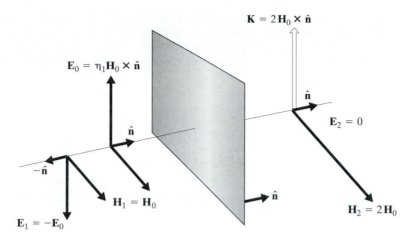

Figure 7-11 Reflection at a perfectly conducting mirror.

magnetic field, from a finite value $2H_0$ just inside the conducting wall to zero value at any nonzero depth within the perfect conductor. This discontinuity must be accompanied by a surface current K just inside the conductor's surface. The strength of this surface current equals the discontinuity, namely $2H_0$, and it is directed at right angles to the magnetic field at the surface, along the direction of the incident electric field E_0, as indicated in Figure 7-11.

For clarity, this figure is an exploded view, to show the separate field configurations of each wave. The fields shown are actually those at the reflecting surface; in particular, the current sheet **K** that corresponds to the discontinuity in magnetic field, from $2H_0$ to zero, exists only on the surface.

The perfectly conducting wall acts as a mirror. The incident electric field is reversed and the incident magnetic field is doubled at the wall. A current flows along the surface of the conducting wall, induced by the incident electric field; its magnitude is twice that of the incident magnetic field. The fields in the dielectric are, if $\hat{\mathbf{n}} = \hat{\mathbf{z}}$ and the conducting plane is at $z = 0$,

$$\mathbf{E} = \mathbf{E}_0 e^{-j\beta_1 z} - \mathbf{E}_0 e^{+j\beta_1 z} = -2j\mathbf{E}_0 \sin \beta_1 z, \tag{7.67}$$

$$\mathbf{H} = \mathbf{H}_0 e^{-j\beta_1 z} + \mathbf{H}_0 e^{+j\beta_1 z} = 2\mathbf{H}_0 \cos \beta_1 z. \tag{7.68}$$

This is a pure standing wave. Its time-domain version is

$$\mathcal{E}(\mathbf{r}, t) = 2\mathbf{E}_0 \sin \beta_1 z \sin \omega t, \qquad \mathcal{H}(\mathbf{r}, t) = 2\mathbf{H}_0 \cos \beta_1 z \cos \omega t, \tag{7.69}$$

if $\mathbf{E}_0$ is real. There is no propagation, only oscillation in a fixed, sinusoidal spatial pattern. The electric and magnetic fields are in time quadrature, 90° out of phase.

Example 7.7

If the wave incident normally onto a perfectly conducting plane is circularly polarized, say RHCP, what is the polarization of the reflected wave? ◈

Although we are dealing with a complex vector $\mathbf{E} = E_0[\hat{\mathbf{x}} - j\hat{\mathbf{y}}]$ it is still the case that the reflected wave's electric field is just $-\mathbf{E}$, which is $-E_0[\hat{\mathbf{x}} - j\hat{\mathbf{y}}]$. This is still circularly polarized, and the time-domain electric field spins in the same direction as does that of the incident wave. However, the reflected wave is traveling in the opposite direction, so that it is LHCP.

Example 7.8

A plane-wave source at $z = 0$ carries a surface current sheet $\mathbf{K}_s$. At $z = l$, we place a perfectly conducting reflecting plane. The medium is a vacuum.

(a) *Obtain the fields on both sides of the source plane, in terms of $\mathbf{K}_s$.*

(b) *What is the surface current $\mathbf{K}_r$ on the reflecting plane, in terms of $\mathbf{K}_s$?*

(c) *Construct an equivalent transmission line analog of this problem, valid for both sides of the source plane.* ◈

(a) The reflection coefficient $\Gamma_e = -1$ at $z = l$ makes the fields between the two planes

$$\mathbf{E} = \mathbf{E}_0 e^{jk(l-z)} - \mathbf{E}_0 e^{-jk(l-z)} = 2j\mathbf{E}_0 \sin k(l - z), \tag{7.70}$$

$$\mathbf{H} = \mathbf{H}_0 e^{jk(l-z)} + \mathbf{H}_0 e^{-jk(l-z)} = 2\mathbf{H}_0 \cos k(l - z), \tag{7.71}$$

in terms of the as yet undetermined field amplitude $\mathbf{E}_0$, with $\mathbf{H}_0 = \hat{\mathbf{z}} \times \mathbf{E}_0/\eta_0$.

At the source plane, the electric field is $2j\mathbf{E}_0 \sin kl$, and this must be continuous across the source. With an unbounded vacuum on the other side of the source plane, the fields there must be outgoing only, $\mathbf{E}'e^{jkz}$ and $\mathbf{H}'e^{jkz}$ with $\mathbf{H}' = -\hat{\mathbf{z}} \times \mathbf{E}'/\eta_0$. Continuity of the electric field at $z = 0$ makes $\mathbf{E}' = 2j\mathbf{E}_0 \sin kl$, so that $\mathbf{H}' = -2j\mathbf{H}_0 \sin kl$. We now have a discontinuity in the tangential magnetic field at the source plane, from $\mathbf{H} = 2\mathbf{H}_0 \cos kl$ on the positive side to $\mathbf{H}' = -2j\mathbf{H}_0 \sin kl$ on the other side. That discontinuity determines the source surface current:

$$\mathbf{K}_s = \hat{\mathbf{z}} \times (\mathbf{H} - \mathbf{H}') = 2\hat{\mathbf{z}} \times \mathbf{H}_0(\cos kl + j \sin kl) = 2\hat{\mathbf{z}} \times \mathbf{H}_0 e^{jkl}$$

$$= 2\hat{\mathbf{z}} \times (\hat{\mathbf{z}} \times \mathbf{E}_0/\eta_0)e^{jkl} = -2(\mathbf{E}_0/\eta_0)e^{jkl}. \tag{7.72}$$

Hence, $\mathbf{E}_0 = -\frac{1}{2}\eta_0\mathbf{K}_s e^{-jkl}$ and the fields between the planes (for $0 < z < l$) are

$$\mathbf{E} = -j\eta_0\mathbf{K}_s e^{-jkl} \sin k(l - z), \qquad \mathbf{H} = -\hat{\mathbf{z}} \times \mathbf{K}_s e^{-jkl} \cos k(l - z), \tag{7.73}$$

while those on the other side of the source plane ($z < 0$) are

$$\mathbf{E} = -j\eta_0\mathbf{K}_s \sin kl e^{-jk(l-z)} \qquad \mathbf{H} = j\hat{\mathbf{z}} \times \mathbf{K}_s \sin kl e^{-jk(l-z)}. \tag{7.74}$$

(b) The surface current at the reflector is $\mathbf{K}_r = \mathbf{H} \times \hat{\mathbf{z}}$, with $\mathbf{H}$ evaluated at that plane, at $z = l$. This is $[-\hat{\mathbf{z}} \times \mathbf{K}_s e^{-jkl}] \times \hat{\mathbf{z}}$, or $\mathbf{K}_r = -\mathbf{K}_s e^{-jkl}$.

(c) Using the analogy of electric field to voltage and magnetic field to current, we represent the plane-wave generator as a current source and the reflector as a short circuit. Since the space on the other side of the source plane is unbounded, there is only the outgoing wave there and no reflected one, so we represent that side of the circuit as a matched load Z_0. Since $E/H = \eta_0$ for a unidirectional wave, the characteristic impedance of the equivalent transmission line is $Z_0 = \eta_0$. The equivalent line is shown in Figure 7-12.

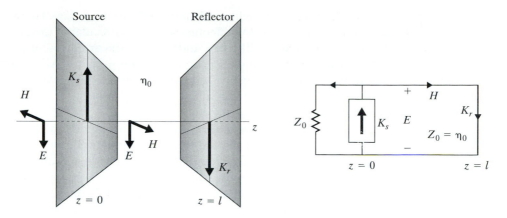

Figure 7-12 Plane-wave source and reflector; transmission line equivalent.

ANTIREFLECTION COATING

Many design problems involving the transmission of waves of any sort entail techniques for matching otherwise mismatched impedances. We have seen that normal incidence of a plane wave onto the interface between two media whose impedances are mismatched results in reflection as well as transmission. If our purpose is to transmit the electromagnetic signal or energy into the second, unsourced medium, then the reflection is an undesirable effect that reduces the efficiency of the process. We will now see that reflection from a mismatched medium can be overcome, by interposing a layer of a suitable other medium. This result has many applications, notably in the coating of optical lenses to ensure that as much light as possible can enter the lens itself. What the antireflection coating does is to cancel the reflected wave, as will be seen.

With a layer of finite thickness interposed between two semi-infinite media, the system consists of three media, as indicated in Figure 7-13. There is the semi-infinite, sourced medium of impedance η_1; there is the layer of impedance η_2 and thickness d; finally, there is the semi-infinite, unsourced medium, of impedance η_3, into

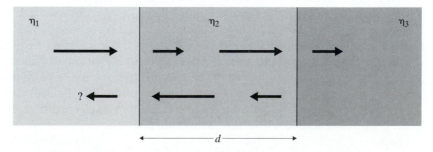

Figure 7-13 Antireflection coating between mismatched media.

which we wish to transmit all the light incident from the first medium. There are now two interfaces, which are parallel to each other and perpendicular to the direction of propagation. The impedances of the first and third media are given and the third medium is mismatched to the first one $(\eta_3 \neq \eta_1)$, or else there would be no need to interpose the antireflection coating. The task is to design the thickness d and the impedance η_2 of the layer to achieve zero reflection back into the first medium, and hence maximal transmission into the third one.

If we can arrange for the impedance η_1 of the sourced medium to be matched to the impedance that applies just beyond the interface between the first medium and the layer, then the reflection at that interface will be zero and our goal will be achieved. However, the impedance just inside the layer is *not* just the intrinsic impedance η_2 of the layer itself, because the layer is not infinitely thick. Rather, the impedance at the first interface is affected by the existence of the second one at a distance d beyond the first one. The actual impedance that is to be matched to that of the semi-infinite first medium, η_1, can be expected to be some combination of both η_2 of the layer and η_3 of the medium beyond it. We need to find that corrected impedance and then set it equal to η_1 in order to match it to the first medium and thereby reduce the reflection to zero.

The matching of impedances at an interface simply ensures that the tangential electric and magnetic fields be continuous across that interface, with fewer than two waves on each side, by making the ratio of the tangential E and H fields conform to that of the incident wave alone. When there is only one wave, that ratio is just the intrinsic impedance η (in this case, η_1) of the medium. When, however, there are two, oppositely traveling waves in a medium of intrinsic impedance η (here, η_2), the ratio Z of the tangential electric and magnetic fields becomes a function of the relative strengths of the two waves and varies from point to point along their direction of propagation.

When the incident wave encounters the interface with the layer, there should be both reflection and transmission; the latter results in a wave that travels forward within the intermediate layer. This wave then encounters the second interface, at which there should again be reflection and transmission. Of these, the reflected wave is a backward-traveling wave within the layer; upon reaching the first interface, it should itself result in transmission (into the original, sourced medium) and reflection (back into the layer). This last wave joins with the earlier one to form a modified, forward-going wave that then encounters the second interface, repeating the process yet again to modify the backward-traveling wave as well. In the steady state, there are two oppositely traveling waves within the layer and one overall forward-going wave in the third medium, besides the incident wave and the overall reflected one in the first medium; the last of these is what we are trying to eliminate.

The expedient way to get the expression for the impedance (ratio of tangential E and H fields) anywhere in the system is to work backward from the region furthest from the source. Here, this is medium 3 which, being unsourced, carries only one wave (directed away from the interface with the layer) and therefore has a wave impedance equal to that of the medium, η_3. In the layer, however, there are two, oppositely directed waves, so that the wave impedance is not just η_2 of the medium but is affected by the presence of the third medium beyond the interface, acting as an electrical "load" on the "circuit." To get the ratio of E to H, we need to know the relative strengths of the two waves within the layer. But we already know how the reflected wave is related to the forward-traveling one at the last interface: The field structure at the last interface is exactly the three-wave situation that we have studied, with an incident wave within the layer (here of unknown amplitude, however) and a reflected wave of amplitude Γ times the unknown incident-wave amplitude and a transmitted wave in the last medium, of amplitude T times the unknown incident-wave amplitude. The matching of tangential fields at the last interface is identical to that for a two-medium, single-interface, three-wave problem and yields once again

$$\Gamma = \Gamma_{32} = (\eta_3 - \eta_2)/(\eta_3 + \eta_2) \quad \text{and} \quad T = T_{32} = 2\eta_3/(\eta_3 + \eta_2). \quad (7.75)$$

The double-subscript notation here indicates that the reflection and transmission coefficients apply to the interface between medium 2 and medium 3. We can therefore write the fields within the layer and beyond it in terms of the still-unknown amplitude E_0 of the forward-traveling wave in the layer:

$$\text{In 2:} \quad E(z) = E_0 e^{-j\beta_2 z} + \Gamma_{32} E_0 e^{+j\beta_2 z}, \quad (7.76)$$

$$\eta_2 H(z) = E_0 e^{-j\beta_2 z} - \Gamma_{32} E_0 e^{+j\beta_2 z}; \quad (7.77)$$

$$\text{In 3:} \quad E(z) = \eta_3 H(z) = T_{32} E_0 e^{-j\beta_3 z}. \quad (7.78)$$

We have set the z-axis along the direction of propagation and put the origin $z = 0$ at the last interface, which then puts the first interface at $z = -d$. The unknown amplitude E_0 drops out from the ratio of $E(z)$ to $H(z)$; this ratio is the effective impedance in the layer and is a function of position, z:

$$Z_2(z) = \eta_2 \frac{\exp(-j\beta_2 z) + \Gamma_{32} \exp(+j\beta_2 z)}{\exp(-j\beta_2 z) - \Gamma_{32} \exp(+j\beta_2 z)}. \quad (7.79)$$

The impedance is complex. Note that, at the last interface, this impedance reduces properly to

$$Z_2(0) = \eta_2[(1 + \Gamma_{32})/(1 - \Gamma_{32})] = \eta_2[2\eta_3/2\eta_2] = \eta_3, \quad (7.80)$$

to match the impedance beyond the interface.

Moving now to the interface between media 1 and 2, the fields are

$$\text{In 2, for } z = -d: \quad E(-d) = E_0 e^{+j\beta_2 d} + \Gamma_{32} E_0 e^{-j\beta_2 d}, \quad (7.81)$$

$$\eta_2 H(-d) = E_0 e^{+j\beta_2 d} - \Gamma_{32} E_0 e^{-j\beta_2 d}, \quad (7.82)$$

and the impedance just beyond the first interface is

$$Z_2(-d) = \eta_2 \frac{\exp(+j\beta_2 d) + \Gamma_{32} \exp(-j\beta_2 d)}{\exp(+j\beta_2 d) - \Gamma_{32} \exp(-j\beta_2 d)}. \quad (7.83)$$

This corrected impedance, to be used instead of η_2 alone, exhibits the effect of loading by medium 3. The fields in the sourced medium are

$$\text{In 1, for } z < -d: \quad E(z) = E_1 e^{-j\beta_1(z+d)} + \Gamma_{21}E_1 e^{+j\beta_1(z+d)}, \quad (7.84)$$

$$\eta_1 H(z) = E_1 e^{-j\beta_1(z+d)} - \Gamma_{21}E_1 e^{+j\beta_1(z+d)}, \quad (7.85)$$

where E_1 is the given amplitude of the incident wave at the interface and Γ_{21} is the reflection coefficient that we are trying to force to zero. Imposing continuity of tangential fields at this interface results in the same expression for the reflection coefficient as for the two-medium case, except that we replace the unbounded-medium impedance η_2 with the loaded impedance $Z_2(-d)$ in $\Gamma = (\eta_2 - \eta_1)/(\eta_2 + \eta_1)$:

$$\Gamma_{21} = \frac{Z_2(-d) - \eta_1}{Z_2(-d) + \eta_1}. \quad (7.86)$$

Our purpose is to eliminate reflection in medium 1, by adjusting the effective impedance $Z_2(-d)$ to match the impedance of the sourced medium:

$$Z_2(-d) = \eta_1, \quad (7.87)$$

or

$$\eta_1/\eta_2 = \left[e^{j\beta_2 d} + \Gamma_{32}e^{-j\beta_2 d}\right]/\left[e^{j\beta_2 d} - \Gamma_{32}e^{-j\beta_2 d}\right] \quad (7.88)$$

or

$$\frac{\eta_1}{\eta_2} = \frac{(\eta_3 + \eta_2)e^{j\beta_2 d} + (\eta_3 - \eta_2)e^{-j\beta_2 d}}{(\eta_3 + \eta_2)e^{j\beta_2 d} - (\eta_3 - \eta_2)e^{-j\beta_2 d}}, \quad (7.89)$$

or

$$\frac{\eta_1}{\eta_2} = \frac{\eta_3 \cos \beta_2 d + j\eta_2 \sin \beta_2 d}{\eta_2 \cos \beta_2 d + j\eta_3 \sin \beta_2 d}. \quad (7.90)$$

This matching condition requires that a real number be equated to a complex one, giving us two conditions on the two design variables η_2 and $\beta_2 d$, one from the real part, another from the imaginary part. Cross multiplying,

$$\eta_1 \eta_2 \cos \beta_2 d + j\eta_1 \eta_3 \sin \beta_2 d = \eta_2 \eta_3 \cos \beta_2 d + j\eta_2^2 \sin \beta_2 d \quad (7.91)$$

and, since β_2 and all the η's are real, the matching conditions are

$$(\eta_1 - \eta_3)\eta_2 \cos \beta_2 d = 0 \quad (7.92)$$

and

$$(\eta_1 \eta_3 - \eta_2^2) \sin \beta_2 d = 0. \quad (7.93)$$

Since the first and last media are mismatched (or else the antireflection layer would be superfluous), $\eta_1 \neq \eta_3$; also, the possibility $\eta_2 = 0$ (a perfectly conducting layer) would make the coating opaque. Hence, the first of these two simultaneous matching conditions requires $\cos \beta_2 d = 0$ or

$$\beta_2 d = \pi/2 \quad (\text{or} \quad \beta_2 d = [n + \tfrac{1}{2}]\pi, \quad n = \text{integer}) \quad (7.94)$$

and the second condition then requires

$$\eta_2 = \sqrt{\eta_1\eta_3}. \tag{7.95}$$

These are indeed real quantities, so that the matching is feasible. As $\beta_2 = 2\pi/\lambda_2$, where λ_2 is the wavelength in the layer, the condition $\beta_2 d = \pi/2$ is the same as

$$d = \lambda_2/4, \tag{7.96}$$

(or, using $[n + \frac{1}{2}]\pi$, $d = n\lambda_2/2 + \lambda_2/4$, with n any integer).

The interpretation should be clear. The antireflection coating is to have an impedance η_2 that is the geometric mean of the two mismatched impedances of the outer media, and its thickness is to be a quarter wavelength, as measured in the layer material, or else an integral number of half wavelengths more than that. This sort of layer is referred to as a quarter-wave matching section; for light, the wavelengths are short and the required layer is very thin, a mere coating, made of the appropriate material. For nonmagnetic materials, the requirement is for $\varepsilon_2 = \sqrt{\varepsilon_1\varepsilon_3}$.

The antireflection coating works by making the wave reflected from the first interface and the wave reflected from the second one (and then transmitted back from the layer to the sourced medium) exactly cancel each other. Note that a coating so designed will be 100% effective in eliminating reflection only at the design frequency, and at other discrete frequencies at which the wavelength in the layer is $(n + \frac{1}{2})\lambda_2/2$, with n an integer, assuming the materials are not lossy. When operated at other frequencies, the reflection coefficient, given in (7.86), will not be zero, although it may remain quite small over a rather broad band of frequencies. A succession of intervening layers between mismatched media can be designed to extend or shape that bandwidth.

> **Example 7.9**
>
> *Design an antireflection coating for a medium made of glass $(\varepsilon/\varepsilon_0 = 2.5)$, to be effective for waves incident normally onto the surface from air, at the free-space wavelength $\lambda_0 = 0.6$ μm.* ◇
>
> We want to penetrate from air, with impedance $\eta_0 = \sqrt{\mu_0/\varepsilon_0}$, to the glass, with impedance $\eta = \sqrt{\mu_0/\varepsilon}$. The antireflection layer should have its impedance $\eta_1 = \sqrt{\mu_0/\varepsilon_1}$ equal to $\sqrt{\eta_0\eta}$ or $\varepsilon_1/\varepsilon_0 = \sqrt{\varepsilon/\varepsilon_0} = \sqrt{2.5} = 1.581$.
>
> The wavelength in the layer will be $\lambda_1 = u_1/f$, where f is the frequency of the light and $u_1 = 1/\sqrt{\mu_0\varepsilon_1} = c/\sqrt{\varepsilon_1/\varepsilon_0} = c/1.257$ is the speed of light in the layer. But $c/f = \lambda_0$, so that $\lambda_1 = \lambda_0/1.257$. The thickness of the layer should be $\lambda_1/4 = \lambda_0/5.030 = 0.119$ μm. The antireflection coating is a thin film.
>
> In practice, suitable materials with just the right permittivity may not always be available, and it is advantageous to resort to multiple layers instead of just one. The multiplicity provides more freedom to use available materials and can allow the antireflection film to be effective at many wavelengths.

The antireflection coating for matching otherwise mismatched optical media has an exact analog in the realm of transmission lines, as in the next example.

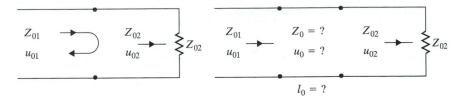

Figure 7-14   Quarter-wave matching section interposed between two lines.

Example 7.10

If the load on a transmission line of characteristic impedance Z_{01} and wave speed u_{01} is a length of matched line of characteristic impedance Z_{02} and wave speed u_{02}, with $Z_{01} \neq Z_{02}$, then line 1 is clearly mismatched. We wish to eliminate reflections from the end of line 1 by interposing a section of yet another line, of impedance Z_0, wave speed u_0, and length l_0 between the two original transmission lines, as in Figure 7-14. Determine Z_0 and $\psi_0 = \omega l_0/u_0$ so as to achieve the desired reflectionless transmission to the end of line 2, with the shortest possible matching section. ◈

The load on the interposed matching section is just Z_{02} and the impedance at its input is therefore $Z_0[Z_{02} + jZ_0 \tan \psi_0]/[Z_0 + jZ_{02} \tan \psi_0]$. Line 1 will be matched if we set this equal to Z_{01}. The resulting equation is

$$Z_0 Z_{02} + jZ_0^2 \tan \psi_0 = Z_0 Z_{01} + jZ_{01}Z_{02} \tan \psi_0 \tag{7.97}$$

or

$$Z_0(Z_{02} - Z_{01}) \cot \psi_0 = j(Z_{01}Z_{02} - Z_0^2). \tag{7.98}$$

Since the left side is real and the right side is imaginary, both are zero. This gives $Z_{01}Z_{02} = Z_0^2$ and, since $Z_{01} \neq Z_{02}$, also $\cot \psi_0 = 0$. The shortest matching section is obtained for $\psi_0 = \pi/2$. The matching section has as its characteristic impedance the geometric mean of those of the two original lines, $Z_0 = \sqrt{Z_{01}Z_{02}}$, and $\psi_0 = \omega l_0/u_0 = 2\pi l_0/\lambda = \pi/2$ makes $l_0 = \lambda/4$, a quarter-wavelength long.
    The section of transmission line interposed between the original two mismatched ones to eliminate reflections is called a *quarter-wave matching section*.

OBLIQUE INCIDENCE

For normal incidence of plane waves onto the interface between dissimilar media, it is relatively easy to satisfy the boundary conditions of continuity of tangential electric and magnetic fields because both fields are parallel to the boundary plane in their entirety. Continuity of tangential fields then becomes simply continuity of the total fields, which are related by the wave impedance of each medium. When the plane wave is incident at an angle to the normal, however, the electric and magnetic fields cannot both be tangential to the interface; at most one of them can be and usually neither one is. For such oblique incidence, as depicted in Figure 7-15, it is only the tangential components of the two fields that must be continuous, not the total fields, and

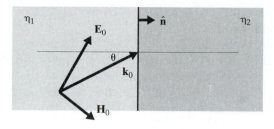

Figure 7-15 Oblique incidence onto interface between two media.

the impedance that relates these components must first be established before the reflection and transmission properties can be found.

A change in viewpoint about the obliquely incident wave, however, can make this more challenging problem appear as a mere extension of the previously derived results for normal incidence. It is possible to trade the feature of obliqueness for uniformity; that is, we can make the uniform, oblique wave appear to be normally incident (transverse) but nonuniform. The nonuniformity is a small price to pay for reduction of the problem to a previously solved one.

OBLIQUE WAVES AS NONUNIFORM TRANSVERSE WAVES

The incident wave is a uniform one, meaning that its amplitude is the same everywhere: In

$$\mathbf{E}(\mathbf{r}) = \mathbf{E}_0 \exp(-j\mathbf{k}_0 \cdot \mathbf{r}) \qquad \mathbf{H}(\mathbf{r}) = \mathbf{H}_0 \exp(-j\mathbf{k}_0 \cdot \mathbf{r}), \qquad (7.99)$$

the vector amplitudes $\mathbf{E}_0$ and $\mathbf{H}_0$ are constants. These two constant vectors are, in fact, related by the impedance η_1 of the sourced medium, as $\mathbf{E}_0 = \eta_1 \mathbf{H}_0$. The field amplitudes $\mathbf{E}_0$ and $\mathbf{H}_0$ are perpendicular to each other and to the wave vector $\mathbf{k}_0$, forming a right-handed orthogonal triad. The interface plane, given by $\hat{\mathbf{n}} \cdot \mathbf{r} = 0$, is perpendicular to unit vector $\hat{\mathbf{n}}$, which points into the unsourced medium 2. That the wave is obliquely incident means that the wave vector $\mathbf{k}_0$ is not parallel to $\hat{\mathbf{n}}$; instead, it makes an angle θ with the normal to the interface, with $0 < \theta < \pi/2$. The plane that contains both $\mathbf{k}_0$ and $\hat{\mathbf{n}}$ is termed the *plane of incidence* of the wave.

Now a uniform plane wave that propagates along $\mathbf{k}_0$ can be viewed as a nonuniform plane wave that travels along $\hat{\mathbf{n}}$. For example, for $\hat{\mathbf{n}} = \hat{\mathbf{z}}$,

$$\mathbf{E}(\mathbf{r}) = \mathbf{E}_0 \exp(-j\mathbf{k}_0 \cdot \mathbf{r}) = \mathbf{E}_0 \exp[-j(k_x x + k_y y + k_z z)]$$
$$= \mathbf{E}_0 \exp[-j(k_x x + k_y y)] \exp(-jk_z z) = \mathbf{E}_0'(x, y) \exp(-jk_z z). \qquad (7.100)$$

The last version has the appearance of a wave that propagates along z and hence is normally incident onto the interface, but with a complex amplitude that varies from point to point, instead of being constant. That is, the uniform plane wave along $\mathbf{k}_0$ appears as a nonuniform plane wave along z. The nonuniformity shows up in particular on the interface plane $z = 0$, on which the field is not constant but varies as $\mathbf{E}_0'(x, y) = \mathbf{E}_0 \exp[-j(k_x x + k_y y)]$. The boundary condition on the interface is that the tangential components of both $\mathbf{E}_0'(x, y)$, and the corresponding $\mathbf{H}_0'(x, y)$, must be continuous at *every* point of the xy-plane.

For true normal incidence, making the fields continuous at one point of the boundary plane automatically made it continuous at every point of the plane, since the fields were constant on the interface. For oblique incidence, we can *pretend* that the wave is still normally incident but is nonuniform; the variation on the interface requires us to ensure continuity at each point of the boundary plane.

Let's be more general than for an interface that coincides with the xy-plane. The wave vector $\mathbf{k}_0$ can always be decomposed into components perpendicular and parallel to the normal to the interface, $\hat{\mathbf{n}}$, whatever its direction:

$$\mathbf{k}_0 = \mathbf{p}_0 + \boldsymbol{\beta}_0 \quad \text{with} \quad \mathbf{p}_0 = \hat{\mathbf{n}} \times \mathbf{k}_0 \times \hat{\mathbf{n}} \quad \text{and} \quad \boldsymbol{\beta}_0 = (\hat{\mathbf{n}} \cdot \mathbf{k}_0)\,\hat{\mathbf{n}}, \quad (7.101)$$

by an identity of vector algebra. Hence, the incident uniform, but oblique, wave can always be expressed as a nonuniform but normally incident one:

$$\mathbf{E}(\mathbf{r}) = \mathbf{E}_0 \exp(-j\mathbf{k}_0 \cdot \mathbf{r}) = \mathbf{E}_0 \exp(-j\mathbf{p}_0 \cdot \mathbf{r})\exp(-j\boldsymbol{\beta}_0 \cdot \mathbf{r})$$

$$= \{\mathbf{E}_0 \exp(-j\mathbf{p}_0 \cdot \mathbf{r})\}\exp(-j\beta_0\hat{\mathbf{n}} \cdot \mathbf{r}). \quad (7.102)$$

To satisfy Maxwell's equations, $\mathbf{E}_0$ and $\mathbf{k}_0$ are restricted to satisfy

$$\mathbf{E}_0 \cdot \mathbf{k}_0 = 0 \qquad \text{and} \qquad \mathbf{k}_0 \cdot \mathbf{k}_0 = \omega^2/v_1^2, \quad (7.103)$$

where v_1 is the phase velocity, at frequency ω, for medium 1. The last equation can be rewritten as

$$\mathbf{p}_0 \cdot \mathbf{p}_0 + \beta_0^2 = \omega^2/v_1^2 \qquad \text{or} \qquad p_0^2 + \beta_0^2 = \omega^2/v_1^2. \quad (7.104)$$

The accompanying magnetic field is also expressible as a nonuniform plane wave along $\hat{\mathbf{n}}$:

$$\mathbf{H}(\mathbf{r}) = \mathbf{H}_0 \exp(-j\mathbf{k}_0 \cdot \mathbf{r}) = \{\mathbf{H}_0 \exp(-j\mathbf{p}_0 \cdot \mathbf{r})\}\exp(-j\beta_0\hat{\mathbf{n}} \cdot \mathbf{r}), \quad (7.105)$$

where $\omega\mu_1 \mathbf{H}_0 = \mathbf{k}_0 \times \mathbf{E}_0$, by the Faraday-Maxwell law.

Note that, since $\hat{\mathbf{n}}$ and $\mathbf{k}_0$ are given as part of the specification of the interface and of the oblique incidence, both $\mathbf{p}_0$ and $\boldsymbol{\beta}_0$ are known quantities. The source (infinitely remote from the interface) determines the frequency ω and the *direction* of the wave vector $\mathbf{k}_0$; the constitutive parameters of the sourced medium determine its phase velocity v_1 and hence the *magnitude* ω/v_1 of the wave vector.

SNELL'S LAWS

For the case of true normal incidence, the boundary conditions resulted in scattered waves, reflected and transmitted from the interface. For oblique incidence that appears as normal incidence of a nonuniform wave, it may be expected (and will be verified) that reflected and transmitted waves will again be generated at the interface, but now also nonuniform, to allow for continuity at each point of the boundary plane. The nonuniform scattered waves will also correspond to obliquely propagating uniform plane waves, but with new directions of propagation. The frequencies of the scattered waves must be identical to that of the incident one (assuming the interface is stationary; otherwise a Doppler shift arises). If the frequencies do not match, it becomes impossible to satisfy the boundary conditions at *all* times even if we succeed in satisfying them at any one time.

To verify these assertions, assume an obliquely reflected wave in medium 1 along some wave vector $\mathbf{k}_1$ (to be found),

$$\mathbf{E}_1(\mathbf{r}) = \mathbf{E}_1 \exp(-j\mathbf{k}_1 \cdot \mathbf{r}) = \{\mathbf{E}_1 \exp(-j\mathbf{p}_1 \cdot \mathbf{r})\} \exp(-j\boldsymbol{\beta}_1 \cdot \mathbf{r}), \qquad (7.106)$$

and an obliquely transmitted wave in medium 2 along some wave vector $\mathbf{k}_2$ (also to be determined),

$$\mathbf{E}_2(\mathbf{r}) = \mathbf{E}_2 \exp(-j\mathbf{k}_2 \cdot \mathbf{r}) = \{\mathbf{E}_2 \exp(-j\mathbf{p}_2 \cdot \mathbf{r})\} \exp(-j\boldsymbol{\beta}_2 \cdot \mathbf{r}), \qquad (7.107)$$

and confirm that these can be made to satisfy Maxwell's equations in the two media and also the boundary conditions at the interface. That $\mathbf{E}_1(\mathbf{r})$ is a reflected wave means that $\boldsymbol{\beta}_1$ is opposed in direction to $\boldsymbol{\beta}_0$ (i.e., $\boldsymbol{\beta}_1$ is along $-\hat{\mathbf{n}}$ to propagate back into the sourced medium, while $\boldsymbol{\beta}_0$ is along $\hat{\mathbf{n}}$) and that $\mathbf{E}_2(\mathbf{r})$ is a transmitted wave means that $\boldsymbol{\beta}_2$ has the same direction as $\boldsymbol{\beta}_0$ (i.e., $\boldsymbol{\beta}_2$ is along $+\hat{\mathbf{n}}$, to propagate onward in the second medium).

To satisfy Maxwell's equations within the two semi-infinite media, the scattered waves must conform to

$$\mathbf{E}_1 \cdot \mathbf{k}_1 = 0 \quad \text{and} \quad \omega\mu_1\mathbf{H}_1 = \mathbf{k}_1 \times \mathbf{E}_1 \quad \text{and} \quad \mathbf{k}_1 \cdot \mathbf{k}_1 = \omega^2/v_1^2, \qquad (7.108)$$

$$\mathbf{E}_2 \cdot \mathbf{k}_2 = 0 \quad \text{and} \quad \omega\mu_2\mathbf{H}_2 = \mathbf{k}_2 \times \mathbf{E}_2 \quad \text{and} \quad \mathbf{k}_2 \cdot \mathbf{k}_2 = \omega^2/v_2^2. \qquad (7.109)$$

The last relations in this pair are expressible in terms of the tangential and normal components of the wave vectors as

$$p_1^2 + \beta_1^2 = \omega^2/v_1^2, \qquad p_2^2 + \beta_2^2 = \omega^2/v_2^2, \qquad (7.110)$$

corresponding to the incident wave's $p_0^2 + \beta_0^2 = \omega^2/v_1^2$. Note that, so far, only the right sides of the equations (7.110) are known, but both p_0 and β_0 are also known.

At the interface plane $\hat{\mathbf{n}} \cdot \mathbf{r} = 0$ the total electric fields are

$$\mathbf{E}_0(\mathbf{r}) + \mathbf{E}_1(\mathbf{r}) = \mathbf{E}_0 \exp(-j\mathbf{p}_0 \cdot \mathbf{r}) + \mathbf{E}_1 \exp(-j\mathbf{p}_1 \cdot \mathbf{r}) \quad \text{in medium 1}, \qquad (7.111)$$

$$\mathbf{E}_2(\mathbf{r}) = \mathbf{E}_2 \exp(-j\mathbf{p}_2 \cdot \mathbf{r}) \qquad \qquad \text{in medium 2}, \qquad (7.112)$$

and these vary from point to point on the interface. The magnetic fields have exactly the same spatial dependences. The requirement is to ensure that the tangential components of both the electric and the magnetic fields be continuous across the interface, at every point of the plane. It is not sufficient to ensure this at just one point; continuity must hold at every point of the boundary plane.

The key result is the following one. The only way the tangential components of the fields can be matched across the plane at *every* point of the interface is for the spatial dependence along the plane to be *identical* for all three waves. This is not a sufficient condition (since the tangential components still need to be matched at some point) but it is surely a necessary one: If the spatial dependences were not identical, matching at one point would preclude matching at every other point. The exponential (actually sinusoidal) dependences will be identical only if

$$\mathbf{p}_0 = \mathbf{p}_1 = \mathbf{p}_2 \equiv \mathbf{p}. \tag{7.113}$$

The three wave vectors' tangential components are now all to be called simply $\mathbf{p}$. Note that $\mathbf{p}$ is known, being fully determined by the direction of the incident wave vector $\mathbf{k}_0$, in relation to the orientation of the interface, $\hat{\mathbf{n}}$: $\mathbf{p} = \hat{\mathbf{n}} \times \mathbf{k}_0 \times \hat{\mathbf{n}}$.

Since we know the magnitude of each of the three wave vectors and we now also know their tangential components, their normal components are immediately obtainable, so that the scattered wave vectors are now fully determined. That is, we know the directions of the obliquely propagating reflected and transmitted waves. Since $k_1^2 = k_0^2 = \omega^2/v_1^2$ and $\boldsymbol{\beta}_1$ and $\boldsymbol{\beta}_0$ are opposed, the reflected wave vector must be

$$\mathbf{k}_1 = \mathbf{p}_1 + \boldsymbol{\beta}_1 = \mathbf{p} - \boldsymbol{\beta}_0 = \mathbf{p} - \beta_0\hat{\mathbf{n}} \tag{7.114}$$

(compare $\mathbf{k}_0 = \mathbf{p} + \beta_0\hat{\mathbf{n}}$), while the transmitted wave vector is

$$\mathbf{k}_2 = \mathbf{p}_2 + \boldsymbol{\beta}_2 = \mathbf{p} + \beta_2\hat{\mathbf{n}} \quad \text{with} \quad \beta_2 = \sqrt{(\omega^2/v_2^2) - p^2}. \tag{7.115}$$

Figure 7-16 shows how the oblique incident and scattered wave vectors are related. The equation $\mathbf{p}_0 = \mathbf{p}_1 = \mathbf{p}_2 \equiv \mathbf{p}$ embodies the familiar Snell's laws of reflection and refraction, as follows.

The angles of incidence, of reflection, and of refraction are defined as the angles, measured in the plane of incidence (the plane of $\mathbf{k}_0$ and $\hat{\mathbf{n}}$), between $\mathbf{k}$ and $\boldsymbol{\beta}$ for each wave. Since, as shown in Figure 7-16, the tangential components of the $\mathbf{k}$'s are unique and $\boldsymbol{\beta}_1 = -\boldsymbol{\beta}_0$, the angles between $\mathbf{k}_0$ and $\boldsymbol{\beta}_0$ (or $\hat{\mathbf{n}}$) and between $\mathbf{k}_1$ and $\boldsymbol{\beta}_1$ (or $-\hat{\mathbf{n}}$) must be the same and we have the following law:

Snell's law of reflection: The angle of incidence (θ in Figure 7-16) equals the angle of reflection *and* the reflected wave vector $\mathbf{k}_1$ is in the plane of incidence (of $\mathbf{k}_0$ and $\hat{\mathbf{n}}$).

Furthermore, the tangential component $\mathbf{p}$ of any of the wave vectors is given in terms of the angle to the normal, θ, by $p = k \sin \theta$ and, since all $\mathbf{p}$'s are the same, we have

$$k_2 \sin \theta' = k_0 \sin \theta, \tag{7.116}$$

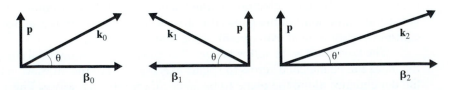

Figure 7-16   Wave vectors of incident, reflected, and transmitted waves.

to yield the angle θ' of the transmitted wave; the magnitudes of the wave vectors are given by $k_0 = \omega/v_1$ and $k_2 = \omega/v_2$. A dimensionless way of specifying the phase velocity in a medium is to give its *index of refraction n*, defined by

$$n = c/v_p. \tag{7.117}$$

For a dielectric, $n = \sqrt{\mu\varepsilon/\mu_0\varepsilon_0}$. Consequently, (7.116) can be expressed nondimensionally by multiplying by c/ω, leaving a companion law:

Snell's law of refraction: The angle of refraction (θ' in Figure 7-16) is related to the angle of incidence θ by

$$n_2 \sin\theta' = n_1 \sin\theta \tag{7.118}$$

and the refracted (transmitted) wave vector $\mathbf{k}_2$ is in the plane of incidence (of $\mathbf{k}_0$ and $\hat{\mathbf{n}}$).

Snell's law is most commonly stated as just (7.118), because that is what yields the angle of refraction, but it is important to realize that each of the two laws has two parts. All these laws follow from the fact that the incident and scattered waves share the tangential wave vector component $\mathbf{p}$.

Example 7.11

A beam of light travels horizontally in air and enters a glass prism whose cross section is an equilateral triangle. The prism rests horizontally on one of its sides, as in Figure 7-17. The glass has refractive index $n = 1.5$. We ignore all reflected beams, including those internal to the prism, and we assume the beam behaves as would a true, infinitely extended plane wave. At what angle θ to the horizontal does the beam emerge from the prism? ◈

Since the angles at which the faces of the prism meet are all $60°$, the angle of incidence to the normal of the entry face is $\theta_0 = 30°$. By Snell's law, the angle of refraction inside the prism is given by $n \sin\theta_1 = \sin\theta_0$, since air has unity refractive index, so that $\theta_1 = \sin^{-1}(1/2n) = 19.5°$. This is $30° - 19.5° = 10.5°$ below the horizontal, and the geometry makes the angle of incidence at the exit face $\theta_2 = 60° - \theta_1 = 40.5°$. Snell's

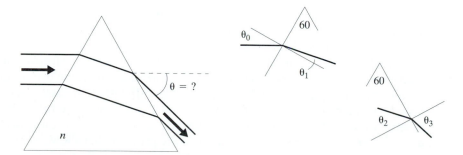

Figure 7-17 Refraction upon entering and exiting an equilateral prism.

law gives $n \sin \theta_2 = \sin \theta_3$ at that face, which yields $\theta_3 = 77.1°$ for the angle to the normal. Once again, the geometry makes the output angle below the horizontal $\theta = \theta_3 - 30° = 47.1°$.

TRANSVERSE ELECTRIC AND TRANSVERSE MAGNETIC POLARIZATIONS

The requirements of Snell's laws make it *possible* to satisfy continuity of tangential E and H fields at every point of the interface *if* we succeed in satisfying continuity at one point. To *achieve* continuity requires that the amplitudes of the scattered waves be properly related to that of the incident wave.

Returning to (7.111) and (7.112), at the interface plane $\hat{\mathbf{n}} \cdot \mathbf{r} = 0$, the total electric fields are

$$\mathbf{E}_0(\mathbf{r}) + \mathbf{E}_1(\mathbf{r}) = \left[\mathbf{E}_0 + \mathbf{E}_1\right]\exp(-j\mathbf{p} \cdot \mathbf{r}) \qquad \text{on side 1,} \qquad (7.119)$$

$$\mathbf{E}_2(\mathbf{r}) = \mathbf{E}_2 \exp(-j\mathbf{p} \cdot \mathbf{r}) \qquad \text{on side 2,} \qquad (7.120)$$

and the total magnetic fields are, correspondingly,

$$\mathbf{H}_0(\mathbf{r}) + \mathbf{H}_1(\mathbf{r}) = \left[\mathbf{H}_0 + \mathbf{H}_1\right]\exp(-j\mathbf{p} \cdot \mathbf{r}) \qquad \text{on side 1,} \qquad (7.121)$$

$$\mathbf{H}_2(\mathbf{r}) = \mathbf{H}_2 \exp(-j\mathbf{p} \cdot \mathbf{r}) \qquad \text{on side 2.} \qquad (7.122)$$

The magnetic field of each wave is perpendicular to both the wave vector and the electric field of that wave and its magnitude is related to that of the electric field by the impedance η of the medium. Since the $\exp(-j\mathbf{p} \cdot \mathbf{r})$ factor is common to the fields of all these waves, the requirement of continuity of tangential fields at every point of the plane is met by ensuring that the constant amplitudes be related by

$$\hat{\mathbf{n}} \times (\mathbf{E}_0 + \mathbf{E}_1) = \hat{\mathbf{n}} \times \mathbf{E}_2, \qquad (7.123)$$

$$\hat{\mathbf{n}} \times (\mathbf{H}_0 + \mathbf{H}_1) = \hat{\mathbf{n}} \times \mathbf{H}_2. \qquad (7.124)$$

The incident-wave field amplitudes $\mathbf{E}_0$ and $\mathbf{H}_0$ are given; for the scattered waves, the magnetic fields are related to the electric ones by the impedance of each medium, but there remain two vector unknowns, $\mathbf{E}_1$ and $\mathbf{E}_2$. Although (7.123 and 7.124) provide no information at all about the normal components of the two unknown vectors, there is actually a sufficient number of equations to determine the vectors, because it is also known that $\mathbf{E}_1$ and $\mathbf{E}_2$ are perpendicular to $\mathbf{k}_1$ and $\mathbf{k}_2$, respectively, which are fully known from Snell's laws.

Elaborate matrix methods or complicated coordinate transformations could be used to extract the unknown scattered wave polarizations from the two continuity conditions, but a simpler approach is available. The difficulty to be overcome is that, for oblique incidence, the natural coordinate systems for the wave (formed by the orthogonal vectors $\mathbf{E}, \mathbf{H}, \mathbf{k}$) and for the interface (formed by the orthogonal vectors $\boldsymbol{\beta}$, $\mathbf{p}, \boldsymbol{\beta} \times \mathbf{p}$) are not compatible. The feature of the problem that allows considerable simplification is that the unit vector $\hat{\mathbf{e}}$ along the direction of $\boldsymbol{\beta} \times \mathbf{p}$ is perpendicular to *both* $\hat{\mathbf{n}}$ and $\mathbf{k}$. The vector $\hat{\mathbf{e}}$ therefore lies *in the plane of the interface* and also *in the*

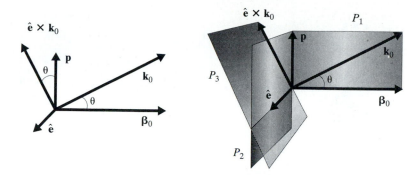

Figure 7-18 Planes of incidence P_1, of interface P_2, and of polarization P_3.

plane of polarization (the plane of the **E** and **H** fields, perpendicular to **k**) *of each of the three waves.* The vector $\hat{\mathbf{e}}$ therefore "belongs" to all the coordinate systems, of the interface and of each of the three participating waves.

As illustrated in Figure 7-18, for the incident wave, the field vectors lie in the plane P_3 of $\hat{\mathbf{e}}$ and of $\hat{\mathbf{e}} \times \mathbf{k}_0$, both of which are orthogonal to $\mathbf{k}_0$. For any of the waves, all of the vectors $\hat{\mathbf{n}}, \boldsymbol{\beta}, \mathbf{k}, \mathbf{p}$, and $\hat{\mathbf{e}} \times \mathbf{k}$ lie in the plane of incidence P_1; the unit vector $\hat{\mathbf{e}}$ is perpendicular to all of these and lies in both the interface plane P_2 and the polarization plane P_3.

We can take advantage of this special status of the vector $\hat{\mathbf{e}}$ by decomposing the electric and magnetic fields of each of the three waves into components along $\hat{\mathbf{e}}$ and along $\hat{\mathbf{e}} \times \mathbf{k}$. More than a mere decomposition into orthogonal components is involved here, because if we pair the component of **E** along $\hat{\mathbf{e}}$ with the component of **H** along $-\hat{\mathbf{e}} \times \mathbf{k}$ and separately pair the component of **H** along $\hat{\mathbf{e}}$ with the component of **E** along $\hat{\mathbf{e}} \times \mathbf{k}$, we arrive at two pairs of fields that each satisfy Maxwell's equations for a wave that propagates along **k** and has the impedance η of the medium. This follows from the relation $\eta\mathbf{H} = \mathbf{k} \times \mathbf{E}/k$ demanded by Maxwell's equations, for any of the waves. If the operation $\mathbf{k} \times \mathbf{E}/k$ is applied to the decomposition of the field into $\mathbf{E} = E_e\hat{\mathbf{e}} + E_m\hat{\mathbf{e}} \times \mathbf{k}/k$, there results $E_m\hat{\mathbf{e}} - E_e\hat{\mathbf{e}} \times \mathbf{k}/k$, which is exactly the decomposition of $\eta\mathbf{H}$ called for previously (i.e., $\eta H_e = E_e$ and $\eta H_m = E_m$). The result is that a wave propagating along **k** with *arbitrary* polarization can be decomposed into just *two partial waves*, each traveling along **k** but with special polarizations that orient *either* **E** *or* **H** entirely in the plane of the interface, along $\hat{\mathbf{e}}$.

This remarkable decomposition of any wave into two partial waves is illustrated in Figure 7-19. We designate the direction $\hat{\mathbf{n}}$ normal to the interface longitudinal and that along $\hat{\mathbf{e}}$ transverse. Then the partial wave with the electric field along $\hat{\mathbf{e}}$ is referred to as *transverse electric* (TE) and the partial wave with the magnetic field along $\hat{\mathbf{e}}$ is referred to as *transverse magnetic* (TM). The TE wave is also referred to as having "perpendicular" polarization (because its **E** is perpendicular to the plane of incidence), or "horizontal" polarization (appropriate if the interface is the ground), or "s" polarization (from the German word for perpendicular, *senkrecht*). The TM wave is also referred to as having "parallel" polarization (because its **E** is parallel to the plane

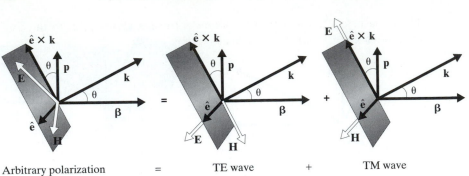

Figure 7-19   Arbitrary wave as superposition of TE and TM partial waves.

of incidence), or "vertical" polarization (appropriate if the interface is the ground), or "p" polarization (from the German word for parallel, *parallel*).

The simplification in the boundary conditions after decomposition into TE and TM partial waves should now be clear. For each partial wave, the field that is the transverse one must be made continuous across the interface, as in the case of normal incidence.

$$\mathbf{E}_0 + \mathbf{E}_1 = \mathbf{E}_2 \quad \text{for TE}; \qquad \mathbf{H}_0 + \mathbf{H}_1 = \mathbf{H}_2 \quad \text{for TM}. \tag{7.125}$$

For each partial wave, the field that does have a longitudinal component must have only its transverse component continuous across the interface. The field component tangential to the interface is

$$H \cos\theta \quad \text{for TE}; \qquad E\cos\theta \quad \text{for TM}. \tag{7.126}$$

The continuity conditions are the same as for normal incidence, for each partial wave. They involve the impedance mismatch, where *impedance* now refers to the ratio of tangential electric and magnetic field components. For the TE and TM partial waves, the tangential fields are the complete transverse one and a projection of the one that has a longitudinal component. The tangential components are therefore related by *equivalent impedances*

$$E/(H\cos\theta) = \eta\sec\theta \qquad \text{for TE}, \tag{7.127}$$

$$(E\cos\theta)/H = \eta\cos\theta \qquad \text{for TM}, \tag{7.128}$$

instead of just the impedance η of the medium in which the partial wave propagates.

The same continuity conditions, but with equivalent impedances, yield the same reflection and transmission coefficients, but using the new impedances

$$Z^{\text{TE}} = \eta\sec\theta, \qquad Z^{\text{TM}} = \eta\cos\theta. \tag{7.129}$$

Thus, the reflection and transmission coefficients that apply to the fields that are transverse in each partial wave are given by

$$\Gamma_e^{TE} = \frac{Z_2^{TE} - Z_1^{TE}}{Z_2^{TE} + Z_1^{TE}} = \frac{\eta_2 \sec \theta_2 - \eta_1 \sec \theta_1}{\eta_2 \sec \theta_2 + \eta_1 \sec \theta_1}, \qquad (7.130)$$

$$T_e^{TE} = \frac{2Z_2^{TE}}{Z_2^{TE} + Z_1^{TE}} = \frac{2\eta_2 \sec \theta_2}{\eta_2 \sec \theta_2 + \eta_1 \sec \theta_1}, \qquad (7.131)$$

for the electric field of the TE partial wave, and by

$$\Gamma_h^{TM} = \frac{Z_1^{TM} - Z_2^{TM}}{Z_1^{TM} + Z_2^{TM}} = \frac{\eta_1 \cos \theta_1 - \eta_2 \cos \theta_2}{\eta_1 \cos \theta_1 + \eta_2 \cos \theta_2}, \qquad (7.132)$$

$$T_h^{TM} = \frac{2Z_1^{TM}}{Z_1^{TM} + Z_2^{TM}} = \frac{2\eta_1 \cos \theta_1}{\eta_1 \cos \theta_1 + \eta_2 \cos \theta_2} \qquad (7.133)$$

for the magnetic field of the TM partial wave.

These are the *Fresnel formulas*, written for those fields of the partial waves that are entirely transverse. Their counterparts for those fields that have a longitudinal component as well $\left(\text{i.e., } \Gamma_h^{TE}, T_h^{TE}, \Gamma_e^{TM}, T_e^{TM}\right)$ could also be written in this way, but they would apply only to the transverse projection of the total field that is not wholly transverse; these would then need additional trigonometric factors to get them to apply to the entire E field of the TM wave and the entire H field of the TE wave. But this is wholly unnecessary, since the reflection and transmission coefficients for the transverse fields suffice to yield the other field directly, by relating the entire E and H fields of either partial wave by the medium's impedance, η.

Note that, in the Fresnel equations, θ_1 is given for the incident wave and θ_2 must be precalculated from Snell's law of refraction. The major result is that the expressions for reflection and transmission coefficients for the tangential fields (which are the complete fields for the transverse ones) are given by the same mismatch equations as for normal incidence; that is,

$$\Gamma_e = (Z_2 - Z_1)/(Z_2 + Z_1), \qquad T_e = 2Z_2/(Z_2 + Z_1), \qquad (7.134)$$

$$\Gamma_h = (Z_1 - Z_2)/(Z_1 + Z_2), \qquad T_h = 2Z_1/(Z_1 + Z_2), \qquad (7.135)$$

but using the appropriate impedances, $\eta \sec \theta$ for TE and $\eta \cos \theta$ for TM, and the coefficients that apply to the transverse fields of each partial wave, Γ_e, T_e for TE, Γ_h, T_h for TM. If the given incident wave is neither TE nor TM, its given polarization should first be decomposed into the two partial transverse waves, by finding the field components along $\hat{e}$ and along $\hat{e} \times k$; the reflected and transmitted partial waves can also, if desired, be added together, vectorially, to obtain the overall reflected and transmitted waves.

Note also that the impedance mismatch across the interface depends now not only on the two impedances η_1 and η_2 but also on the angle of incidence θ_1 and on the polarization (TE or TM). Two major consequences of this new dependence will be seen to be that, first, we can "tune" to a desired impedance, perhaps a matching one, by tilting the interface with respect to the incident wave, and second, that an encounter with an interface can be used to favor one polarization (TE or TM) over the other.

Example 7.12

A plane wave is incident onto the interface between two nonmagnetic dielectrics at angle $\theta_1 = 36.87°$ to the normal. The permittivities of the media are $\varepsilon_1 = 1.5\varepsilon_0$ for the sourced one, $\varepsilon_2 = 2.5\varepsilon_0$ for the other. Obtain the reflection and transmission coefficients for both the TE and the TM polarizations; compare the results with the normal-incidence case. ◈

The wave velocities in the two media are, since $\mu_1 = \mu_2 = \mu_0$ for nonmagnetic materials, $v_1 = c/\sqrt{1.5} = c/1.225$ and $v_2 = c/\sqrt{2.5} = c/1.581$, so that $n_1 = 1.225$ and $n_2 = 1.581$. Only their ratio is relevant here: $v_1/v_2 = \sqrt{\varepsilon_2/\varepsilon_1} = n_2/n_1 = 1.291 = \eta_1/\eta_2$. This impedance ratio is directly relevant to the case of normal incidence, but the effective impedances are different for the TE and TM components of the obliquely incident wave.

Snell's law yields the angle of refraction for both partial waves: $n_1 \sin \theta_1 = n_2 \sin \theta_2$, with $n_1 = 1.225$, $n_2 = 1.581$, and $\sin \theta_1 = 0.6000$ yields $\sin \theta_2 = 0.4649$ or $\theta_2 = 27.70°$. We will need $\sec \theta_2 = 1.1295$, and $\cos \theta_2 = 0.8854$, as well as $\sec \theta_1 = 1.2500$ and $\cos \theta_1 = 0.8000$.

The impedances for each partial wave are

$$Z_1^{\text{TE}} = \eta_1 \sec \theta_1 = (\eta_0/1.225) \sec \theta_1 = 1.0204 \, \eta_0, \tag{7.136}$$

$$Z_2^{\text{TE}} = \eta_2 \sec \theta_2 = (\eta_0/1.581) \sec \theta_2 = 0.7144 \, \eta_0, \tag{7.137}$$

$$Z_1^{\text{TM}} = \eta_1 \cos \theta_1 = (\eta_0/1.225) \cos \theta_1 = 0.6531 \, \eta_0, \tag{7.138}$$

$$Z_2^{\text{TM}} = \eta_2 \cos \theta_2 = (\eta_0/1.581) \cos \theta_2 = 0.5600 \, \eta_0, \tag{7.139}$$

so that the impedance mismatch for the given angle of incidence is

$$(Z_2/Z_1)^{\text{TE}} = 0.7001, \qquad (Z_2/Z_1)^{\text{TM}} = 0.8574 \tag{7.140}$$

for the two partial waves. This may be compared to the material impedance mismatch, which is relevant to normal incidence,

$$(Z_2/Z_1)^{\text{TEM}} = \eta_2/\eta_1 = 1.225/1.581 = 0.7748 \tag{7.141}$$

(the designation TEM, for transverse electromagnetic, is applied to the case that both the electric and magnetic fields are transverse, which occurs at normal incidence). Note that the TEM mismatch is intermediate between those of the TE and TM cases; in fact, it is the geometric mean of the two.

We obtain the reflection coefficients for the two partial waves as

$$\Gamma_e^{\text{TE}} = [(Z_2/Z_1)^{\text{TE}} - 1]/[(Z_2/Z_1)^{\text{TE}} + 1] = -0.1764, \tag{7.142}$$

$$\Gamma_h^{\text{TM}} = [1 - (Z_2/Z_1)^{\text{TM}}]/[1 + (Z_2/Z_1)^{\text{TM}}] = 0.07677 \tag{7.143}$$

and compare these with $\Gamma_e = -\Gamma_h = [(\eta_2/\eta_1) - 1]/[(\eta_2/\eta_1) + 1] = -0.1269$ for the TEM (normal incidence) case. These tell us that the reflected electric field of the TE partial wave is 18% in strength and opposed in direction, compared to the incident TE wave's electric field, while the reflected magnetic field of the TM partial wave is 8% in strength and with the same orientation, compared to the incident TM wave's magnetic field. Of course, the magnitude of the reflected magnetic field for the TE wave is obtainable from that of the reflected electric field as $H = E/\eta_1$ and the magnitude of the reflected electric field for the TM wave is related to that of the reflected magnetic field by $E = \eta_1 H$.

The transmission coefficients for the two partial waves are

$$T_e^{\text{TE}} = 2(Z_2/Z_1)^{\text{TE}}/[(Z_2/Z_1)^{\text{TE}} + 1] = 0.8236, \tag{7.144}$$

$$T_h^{\text{TM}} = 2/[1 + (Z_2/Z_1)^{\text{TM}}] = 1.0768; \tag{7.145}$$

we can compare these with $T_e = 2(\eta_2/\eta_1)/[(\eta_2/\eta_1) + 1] = 0.8731$ and with $T_h = 2/[1 + (\eta_2/\eta_1)] = 1.1269$ for the TEM (normal incidence) case. These tell us that the transmitted electric field of the TE partial wave is 82% in strength and in the same direction, compared to the incident TE wave's electric field, while the transmitted magnetic field of the TM partial wave is 108% in strength and with the same orientation, compared to the incident TM wave's magnetic field. Of course, the magnitude of the transmitted magnetic field for the TE wave is obtainable from that of the transmitted electric field as $H = E/\eta_2$ and the magnitude of the transmitted electric field for the TM wave is related to that of the transmitted magnetic field by $E = \eta_2 H$.

Knowing the reflected and transmitted wave vectors $\mathbf{k}_1$ and $\mathbf{k}_2$ from Snell's laws and having found the transverse field for both the reflected and the transmitted partial waves, the full fields of each partial wave can be constructed for the TE and the TM waves individually and, by superposition of the partial waves, for the complete reflected and transmitted waves.

The incident, reflected, and refracted partial waves are depicted for this example in Figure 7-20 for the TE component and in Figure 7-21 for the TM part. (In these figures, the field vectors are not shown to scale, but their orientations are consistent with the results calculated previously.)

BREWSTER'S ANGLE

Since the effective impedance on each side of the interface depends on the angle of incidence, it may be possible to match otherwise mismatched media by judiciously selecting some particular angle at which an obliquely incident wave encounters the

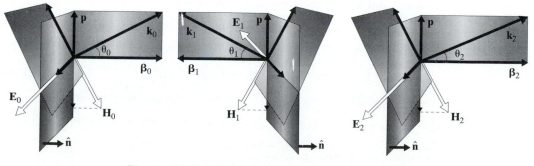

Figure 7-20 Incident and scattered TE waves.

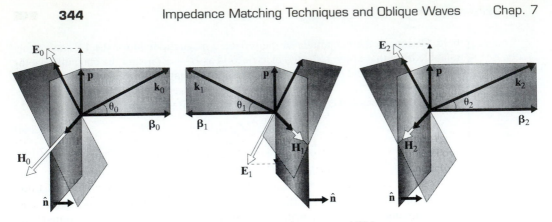

Figure 7-21 Incident and scattered TM waves.

boundary plane. If we succeed, then the refracted light should enter the sourceless medium without reflection and all of the incident signal and energy should be transferred across the interface.

To match impedances for a TE wave, we need $Z_1^{\text{TE}} = Z_2^{\text{TE}}$, or

$$\eta_1 \sec \theta_1 = \eta_2 \sec \theta_2 \qquad (7.146)$$

while, for a TM wave, matching occurs if $Z_1^{\text{TM}} = Z_2^{\text{TM}}$, or

$$\eta_1 \cos \theta_1 = \eta_2 \cos \theta_2. \qquad (7.147)$$

In either case, the angles θ_1, θ_2 are related by Snell's law,

$$n_1 \sin \theta_1 = n_2 \sin \theta_2. \qquad (7.148)$$

We are given the impedances η and indices of refraction n of the two mismatched media (for nonmagnetic materials, one of these suffices to determine the other) and we wish to choose an angle of incidence θ_1 that will make the reflection coefficient vanish, by matching the effective impedances for one or the other of the partial waves. To obtain θ_1, we need to eliminate θ_2 from the pair of equations (Snell's law and an impedance-matching condition) for each of the two partial waves. We can achieve this by squaring and adding the expressions for $\sin \theta_2$ and $\cos \theta_2$, as follows.

For TE: $\qquad \cos^2 \theta_2 = (\eta_2/\eta_1)^2 \cos^2 \theta_1,$ $\qquad (7.149)$

For TM: $\qquad \cos^2 \theta_2 = (\eta_1/\eta_2)^2 \cos^2 \theta_1,$ $\qquad (7.150)$

For both: $\qquad \sin^2 \theta_2 = (n_1/n_2)^2 \sin^2 \theta_1.$ $\qquad (7.151)$

Adding each pair and replacing $\cos^2 \theta_2 + \sin^2 \theta_2 = 1$ by $\cos^2 \theta_1 + \sin^2 \theta_1$ yields

For TE: $\quad 1 = (\eta_2/\eta_1)^2 \cos^2 \theta_1 + (n_1/n_2)^2 \sin^2 \theta_1 = \cos^2 \theta_1 + \sin^2 \theta_1,$ $\quad (7.152)$

For TM: $\quad 1 = (\eta_1/\eta_2)^2 \cos^2 \theta_1 + (n_1/n_2)^2 \sin^2 \theta_1 = \cos^2 \theta_1 + \sin^2 \theta_1.$ $\quad (7.153)$

Collecting the sine and cosine terms separately gives

For TE:
$$\tan^2\theta_1 = \frac{(\eta_2/\eta_1)^2 - 1}{1 - (n_1/n_2)^2} = \frac{(\mu_2/\mu_1)(\varepsilon_1/\varepsilon_2) - 1}{1 - (\mu_1/\mu_2)(\varepsilon_1/\varepsilon_2)}, \tag{7.154}$$

For TM:
$$\tan^2\theta_1 = \frac{(\eta_1/\eta_2)^2 - 1}{1 - (n_1/n_2)^2} = \frac{(\mu_1/\mu_2)(\varepsilon_2/\varepsilon_1) - 1}{1 - (\mu_1/\mu_2)(\varepsilon_1/\varepsilon_2)}. \tag{7.155}$$

Since, for the range of possible incidence angles $(0 < \theta_1 < \pi/2)$, $\tan\theta_1$ ranges from zero to infinity, it will be possible to find a matching angle θ_1, provided merely that the right sides of these equations are positive.

In most practical cases in which such matching is used, the two media are nonmagnetic $(\mu_1 = \mu_2 = \mu_0)$; significantly magnetic materials are not likely to be transparent. The matching conditions are then reduced to

For TE:
$$\tan^2\theta_1 = [(\varepsilon_1/\varepsilon_2) - 1]/[1 - (\varepsilon_1/\varepsilon_2)] = -1, \tag{7.156}$$

for which there can never (for any ε's) be a solution, and

For TM:
$$\tan^2\theta_1 = [(\varepsilon_2/\varepsilon_1) - 1]/[1 - (\varepsilon_1/\varepsilon_2)] = \varepsilon_2/\varepsilon_1, \tag{7.157}$$

for which there is always (for any ε's) a solution. Thus, nonmagnetic media can be matched and the reflection eliminated, but *only for the TM partial wave*, and then only at the specific angle of incidence θ_B, called *Brewster's angle*, given by $\tan\theta_B = \sqrt{\varepsilon_2/\varepsilon_1}$ or

$$\theta_B = \tan^{-1}(n_2/n_1). \tag{7.158}$$

For example, light can enter a plate of glass of refractive index 1.6, from air, without reflection if the light is polarized as a TM wave and is incident at an angle of 58° to the normal. If the light is polarized as a TE wave, there will be some reflection at this, and any other, angle of incidence.

If the light is arbitrarily polarized and incident at the Brewster angle, then only the TE partial wave is reflected. When Brewster's angle is used in this way to select one of the two polarizations (TE) out of the mixture of both in an arbitrary incident wave and thereby achieve a reflected wave that has pure TE polarization and no TM component at all, θ_B is called the *polarizing angle*. Such reflected light could, for example, be made to impinge onto still another plate of glass, this time at only TM polarization with respect to the new interface, whereupon that light is perfectly transmitted into the new plate.

We note that, for incidence at Brewster's angle θ_B, the angle of refraction θ_2, given by (7.148), is found from

$$\sin\theta_2 = (n_1/n_2)\sin\theta_B = \cot\theta_B \sin\theta_B = \cos\theta_B \tag{7.159}$$

to be

$$\theta_2 = \pi/2 - \theta_B. \tag{7.160}$$

This indicates not only that the refracted light propagates at the angle that is complementary to the incident Brewster's angle but also that

$$\tan\theta_2 = \cot\theta_B = n_1/n_2 = \tan\theta'_B, \tag{7.161}$$

which means that the output light travels in the second medium at the angle that would be Brewster's angle θ'_B for transmission from the *second* medium to the *first*. That is, an interface at Brewster's angle will transmit in *either* direction without reflection, for TM incidence. The output window of a laser is usually set at Brewster's angle and the light properly polarized to take advantage of this perfect transmission.

FAILURE OF SNELL'S LAW

There is the possibility of failure of Snell's law, in the sense that the equation for the angle of refraction, $n_1 \sin \theta_1 = n_2 \sin \theta_2$, might have no solution under certain circumstances—at least not without reinterpretation of the results. The circumstances that lead to such failure actually represent an important optical effect, one that is crucial to the operation of optical fibers.

We first ask which way the wave that impinges obliquely upon an interface will bend as it crosses the plane, toward or away from the normal to the boundary? That is, will the angle of refraction be smaller or larger than the angle of incidence? The answer is given by Snell's law, considered as an equation that yields the angle of refraction θ_2 as a function of a variable angle of incidence θ_1.

$$\theta_2 = \sin^{-1}\left[(n_1/n_2)\sin\theta_1\right]. \tag{7.162}$$

Whether θ_2 is smaller or larger than θ_1 depends on whether n_1 is smaller or larger than n_2; that is, on whether the wave slows down or speeds up upon crossing the interface, from phase velocity $v_1 = c/n_1$ to $v_2 = c/n_2$.

If the wave travels from a medium of faster speed (optically less dense) to one of slower speed (optically denser), $v_1 > v_2$ or $n_1 < n_2$, then the refracted wave bends toward the normal, $\theta_2 < \theta_1$. In crossing from a denser to a less dense medium, the wave bends away from the normal. This is illustrated in Figure 7-22, in which the semicircles represent the loci of the wave vectors as the angle of incidence varies.

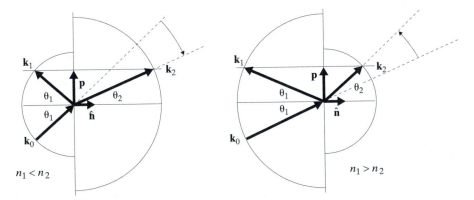

Figure 7-22   Refraction toward $(n_1 < n_2)$ or away from $(n_1 > n_2)$ the normal.

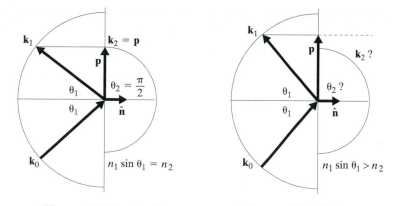

Figure 7-23   Limiting case, and failure, of Snell's law.

Their radii are $(\omega/c)n_1$ and $(\omega/c)n_2$ to the left and the right, respectively, of the interface. Since the tangential component $\mathbf{p}$ of each wave vector has to be the same, the shorter wave vector (smaller index n) needs to bend further away from the normal to accommodate its tangential component.

Now the tangential component of all the wave vectors, $\mathbf{p}$, is determined by the incident wave and the refracted wave has to accommodate to it. But what if the magnitude of the tangential component, p, exceeds that of the entire wave vector of the transmitted wave, $k_2 = (\omega/c)n_2$? How can the refracted wave vector $\mathbf{k}_2$ have a tangential component $\mathbf{p}$ larger than its own magnitude? This will be called for when $n_1 > n_2$ (transmission from a denser to a less dense medium) *and* the angle of incidence is too large for the $\mathbf{p}$ vector to fit into the smaller semicircle beyond the interface. This is illustrated in Figure 7-23, which shows first the limiting case, for which $\mathbf{k}_2$ becomes entirely tangential and equals $\mathbf{p}$, and then a case for which no refracted wave vector $\mathbf{k}_2$ can conform to the required tangential component $\mathbf{p}$.

Mathematically, this failure to determine the refracted wave corresponds to the case that not only is $n_1 > n_2$ but the incidence angle is large enough that even $n_1 \sin \theta_1 > n_2$, whereupon Snell's law requests an angle of refraction θ_2 such that $\sin \theta_2 > 1$. There is no solution in such cases and we need to determine what happens then.

REACTIVE IMPEDANCES

What happens when Snell's law fails to yield a solution for the angle of refraction, and hence a solution for the transmitted wave vector $\mathbf{k}_2$, is that Maxwell's equations nevertheless continue to hold, and so do the boundary conditions. The apparent incompatibility of the two is only a failure of our interpretation of the wave vector. The difficulty comes from our having decomposed the wave vector $\mathbf{k}_2$ into tangential and normal components at the interface, with the tangential component, prescribed by the boundary conditions to equal the tangential component of the incident $\mathbf{k}_0$, too large for $\mathbf{k}_2$, as

$$\mathbf{k}_2 = \mathbf{p} + \boldsymbol{\beta}_2, \qquad p > k_2 = (\omega/c)n_2. \qquad (7.163)$$

That is, the boundary conditions require $\mathbf{p}$ to be large, while Maxwell's equations still demand that

$$\mathbf{k}_2 \cdot \mathbf{k}_2 = p^2 + \beta_2^2 = (\omega/c)^2 n_2^2, \quad \text{despite} \quad p^2 > (\omega/c)^2 n_2^2. \qquad (7.164)$$

The only way to preserve both Maxwell's equations and the boundary conditions that they imply is to make β_2^2 negative:

$$\beta_2^2 = -\alpha^2 \qquad \text{with} \qquad \alpha^2 = p^2 - (\omega/c)^2 n_2^2. \qquad (7.165)$$

This reconciles the two requirements, but it leaves β_2 imaginary, which requires reinterpretation of the vector $\boldsymbol{\beta}_2 = \beta_2 \hat{\mathbf{n}} = \pm j\alpha\hat{\mathbf{n}}$. Note that α is known to us, from (7.165) with p determined by the incident wave, but we do not yet know how to choose the sign of the square root of α^2.

To interpret the result and choose the proper sign for $\sqrt{\alpha^2}$ we note that the transmitted wave takes the form

$$\mathbf{E}_2(\mathbf{r}) = \mathbf{E}_2 e^{-j\mathbf{k}_2 \cdot \mathbf{r}} = \mathbf{E}_2 e^{-j(\mathbf{p}+\boldsymbol{\beta}_2)\cdot\mathbf{r}} = \mathbf{E}_2 e^{-j\mathbf{p}\cdot\mathbf{r}} e^{-j\boldsymbol{\beta}_2\cdot\mathbf{r}}$$

$$= \mathbf{E}_2 \exp[-j\mathbf{p}\cdot\mathbf{r}] \exp[-j(\pm j\alpha\hat{\mathbf{n}})\cdot\mathbf{r}] = \mathbf{E}_2 \exp[-j\mathbf{p}\cdot\mathbf{r}] \exp[\pm\alpha\hat{\mathbf{n}}\cdot\mathbf{r}]$$

$$= \{\mathbf{E}_2 \exp[\pm\alpha\hat{\mathbf{n}}\cdot\mathbf{r}]\} \exp[-j\mathbf{p}\cdot\mathbf{r}]. \qquad (7.166)$$

This is seen to be a *nonuniform plane wave* that propagates along the vector $\mathbf{p}$ but with an amplitude that varies from point to point within medium 2. With the $e^{j\omega t}$ factor restored, the field is the propagating wave $e^{j(\omega t - \mathbf{p}\cdot\mathbf{r})}$ that progresses along the direction of $\mathbf{p}$, but with the varying amplitude $\mathbf{E}_2 \exp[\pm\alpha\hat{\mathbf{n}}\cdot\mathbf{r}]$. We see that the nonuniformity of the wave is here essential; it is not a mere rewriting of a uniform but obliquely propagating wave as a nonuniform but normally propagating one. We also note that medium 2 is the semi-infinite region $\hat{\mathbf{n}}\cdot\mathbf{r} > 0$, so that at an infinite distance from the interface, the factor $\exp[\pm\alpha\hat{\mathbf{n}}\cdot\mathbf{r}]$ either becomes infinitely strong if the upper sign is used, or else becomes zero if the lower one is adopted. The former choice is physically not admissible, and we are left with the exponentially attenuating nonuniform plane wave

$$\mathbf{E}_2(\mathbf{r}) = \mathbf{E}_2 e^{-\boldsymbol{\alpha}\cdot\mathbf{r}} e^{-j\mathbf{p}\cdot\mathbf{r}} \qquad (7.167)$$

as the electric field that is transmitted into the unsourced medium when Snell's law fails, where we define the attenuation vector $\boldsymbol{\alpha}$ as

$$\boldsymbol{\alpha} = \alpha\hat{\mathbf{n}} = \sqrt{p^2 - (\omega/c)^2 n_2^2}\,\hat{\mathbf{n}}. \qquad (7.168)$$

This field propagates parallel to the interface plane, along $\mathbf{p}$, but attenuates exponentially in the direction perpendicular to that plane; it does not propagate (neither normally nor obliquely) *into* the second medium, and neither the field nor the energy reaches to infinite distance from the plane, in the less dense, faster-phase-velocity medium 2. In medium 1, there are still the incident and reflected waves of the usual, obliquely propagating type

$$\mathbf{E}_0(\mathbf{r}) = \mathbf{E}_0 e^{-j\boldsymbol{\beta}_0 \cdot \mathbf{r}} e^{-j\mathbf{p} \cdot \mathbf{r}}, \qquad \mathbf{E}_1(\mathbf{r}) = \mathbf{E}_1 e^{-j\boldsymbol{\beta}_1 \cdot \mathbf{r}} e^{-j\mathbf{p} \cdot \mathbf{r}}, \qquad (7.169)$$

with $\boldsymbol{\beta}_1 = -\boldsymbol{\beta}_0$. The component of each wave vector normal to the interface, β if real or α if imaginary, satisfies the relations

$$p^2 + \beta_0^2 = p^2 + \beta_1^2 = (\omega/c)^2 n_1^2, \qquad p^2 - \alpha^2 = (\omega/c)^2 n_2^2. \qquad (7.170)$$

Figure 7-24 contrasts the configuration of wave vectors for the two cases, $p < (\omega/c)n_2$ and $p > (\omega/c)n_2$.

On the interface plane $\hat{\mathbf{n}} \cdot \mathbf{r} = 0$, all three waves vary as $e^{-j\mathbf{p}\cdot\mathbf{r}}$ with the same $\mathbf{p}$, so that the tangential electric and magnetic fields will be continuous at all points of the boundary plane if they are made to match at one of its points. To achieve this matching, the continuity conditions can again be expressed in terms of the mismatch of impedances on the two sides of the interface. However, when the exponential variation of the wave represents propagation on one side and attenuation on the other, the impedances on the two sides must be defined properly, in conformity with Maxwell's equations.

To extend the definitions of the wave impedances for the TE and TM polarizations from the cases that feature propagation normal to the interface (as well as tangentially), with $\mathbf{k} = \mathbf{p} + \boldsymbol{\beta}$, to the ones that demand attenuation instead, using $\mathbf{k} = \mathbf{p} - j\boldsymbol{\alpha}$, we must reconcile our impedance designations in the natural coordinate system of the interface,

$$\text{For TE waves:} \qquad \mathbf{E} = Z^{\text{TE}} \mathbf{H} \times \hat{\mathbf{n}} \qquad (7.171)$$

$$\text{For TM waves:} \qquad Z^{\text{TM}} \mathbf{H} = \hat{\mathbf{n}} \times \mathbf{E}, \qquad (7.172)$$

with the requirements of Maxwell's equations for a plane wave $e^{-j\mathbf{k}\cdot\mathbf{r}}$, with $\mathbf{k}$ complex, as $j\mathbf{k} = \boldsymbol{\alpha} + j\mathbf{p}$. The two Maxwell equations become

$$[\boldsymbol{\alpha} + j\mathbf{p}] \times \mathbf{E} = j\omega\mu\mathbf{H}, \qquad -[\boldsymbol{\alpha} + j\mathbf{p}] \times \mathbf{H} = j\omega\varepsilon\mathbf{E} \qquad (7.173)$$

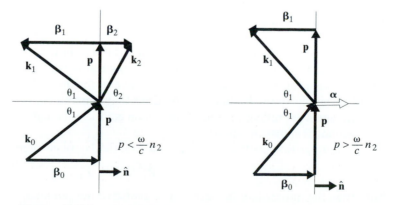

Figure 7-24   Propagation versus attenuation when $n_2 < n_1$.

and these apply to both the TE and TM partial waves. We cross these two equations with $\hat{\mathbf{n}}$ and note that $\hat{\mathbf{n}} \cdot [\boldsymbol{\alpha} + j\mathbf{p}] = \alpha$. We also note that $\hat{\mathbf{n}} \cdot \mathbf{E}^{\mathrm{TE}} = 0$ and $\hat{\mathbf{n}} \cdot \mathbf{H}^{\mathrm{TM}} = 0$ reduce the triple cross product to just α times the field. We therefore find for the TE waves

$$\{[\boldsymbol{\alpha} + j\mathbf{p}] \times \mathbf{E}\} \times \hat{\mathbf{n}} = j\omega\mu\mathbf{H} \times \hat{\mathbf{n}} = \alpha\mathbf{E} = j\omega\mu\mathbf{E}/Z^{\mathrm{TE}} \tag{7.174}$$

and for the TM waves

$$\hat{\mathbf{n}} \times \{-[\boldsymbol{\alpha} + j\mathbf{p}] \times \mathbf{H}\} = j\omega\varepsilon\hat{\mathbf{n}} \times \mathbf{E} = \alpha\mathbf{H} = j\omega\varepsilon Z^{\mathrm{TM}}\mathbf{H}. \tag{7.175}$$

The extended definitions of the TE and TM wave impedances when the waves attenuate along $\hat{\mathbf{n}}$ are therefore

$$Z^{\mathrm{TE}} = j\omega\mu/\alpha, \qquad Z^{\mathrm{TM}} = \alpha/j\omega\varepsilon. \tag{7.176}$$

To interpret these results, we note first that these wave impedances do reduce to our previous expressions when the transmitted wave does propagate; that is, when $\mathbf{k} = \mathbf{p} + \boldsymbol{\beta}$ instead of $\mathbf{k} = \mathbf{p} - j\boldsymbol{\alpha}$, we can replace α with $j\beta$ to recover the earlier versions

$$Z^{\mathrm{TE}} = j\omega\mu/j\beta = \omega\mu/[k\cos\theta] = \eta\sec\theta, \tag{7.177}$$

$$Z^{\mathrm{TM}} = j\beta/j\omega\varepsilon = [k\cos\theta]/\omega\varepsilon = \eta\cos\theta. \tag{7.178}$$

We note next that the modified impedances for the attenuating waves—namely, $Z^{\mathrm{TE}} = j\omega\mu/\alpha$ and $Z^{\mathrm{TM}} = \alpha/j\omega\varepsilon$—are purely *reactive* impedances, in that $Z = jX$, and that Z is inductive ($X > 0$) for the TE wave and capacitive ($X < 0$) for the TM polarization. This is in contrast to the purely resistive impedances of the obliquely propagating cases, $Z^{\mathrm{TE}} = \eta\sec\theta$, $Z^{\mathrm{TM}} = \eta\cos\theta$.

It is also interesting to note that, even in the case of the attenuated waves, the geometric mean of $Z^{\mathrm{TE}} = j\omega\mu/\alpha$ and $Z^{\mathrm{TM}} = \alpha/j\omega\varepsilon$ is just the impedance η of the medium, as was true also in the propagating case.

TOTAL INTERNAL REFLECTION

We have obtained reactive impedances for the TE and TM partial waves in the cases for which Snell's laws fail to yield a propagating refracted wave. These impedances still represent the ratio of tangential electric and magnetic fields at the interface; therefore, applying the boundary conditions that demand continuity of those fields, we once again get reflection and transmission coefficients that involve the mismatch in impedances across the interface.

The reflection coefficient for the electric field is, once again,

$$\Gamma = (Z_2 - Z_1)/(Z_2 + Z_1), \tag{7.179}$$

but now the mismatch is between a resistive impedance $Z_1 = R_1$ (which is $R_1 = \eta_1\sec\theta_1$ for TE or $R_1 = \eta_1\cos\theta_1$ for TM) and a reactive impedance $Z_2 = jX_2$

(which is $X_2 = \omega\mu_2/\alpha$ for TE or $X_2 = -\alpha/\omega\varepsilon_2$ for TM). The reflection coefficient is therefore a complex number, of the form

$$\Gamma = (jX_2 - R_1)/(jX_2 + R_1) \tag{7.180}$$

for either partial wave. Consequently, the complex numerator and denominator have the same magnitude, so that

$$|\Gamma|^2 = 1. \tag{7.181}$$

This states that the magnitude of the reflection coefficient is unity (i.e., the reflection coefficient is merely a phase factor). But that the reflection coefficient has magnitude unity indicates that the electric field reflected back into the denser of the two media has the same strength as that of the incident wave, not merely some fraction of it; only the phase of the wave is altered upon reflection. The process is called *total internal reflection* (internal to the denser medium of the incident wave), abbreviated as TIR.

Under conditions of TIR, the interface between two lossless dielectrics acts as a perfect mirror. However, there do exist fields in the less dense medium that carries the attenuated wave. These fields extend to a depth given by $1/\alpha$, the *e*-folding distance; this depth depends not only on the constitutive parameters of the medium but also on the angle of incidence on the other side, via the quantity p in $\alpha^2 = p^2 - (\omega/c)^2 n_2^2 > 0$ for TIR. By contrast, a perfectly conducting mirror reflects perfectly and has zero field beyond its surface, but is an idealization. Figure 7-25 shows the fields' propagation and attenuation vectors on the two sides of the interface, under conditions of total internal reflection. The two sides are shown separated, for clarity.

Just as the first step toward determining the mismatch for oblique waves involved finding the angle of refraction from Snell's law, that step is also first even when the law fails. It then yields the attenuation rate instead of an angle, and this enters also into the reactive impedances for the two polarizations when total internal reflection is confirmed. Thereafter, the calculations of reflection and transmission coefficients proceed in the usual manner, except that complex numbers enter the calculations and that the transmitted wave is a nonuniform, attenuating wave.

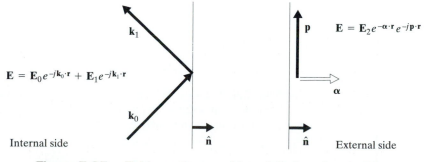

Figure 7-25   Fields on the two sides of the interface in TIR.

Example 7.13

Consider oblique incidence onto the interface between two dissimilar media, with transmission attempted from the denser side, from a nonmagnetic dielectric with refractive index $n_1 = 3.9$ to one with index $n_2 = 2.5$, at incidence angle $\theta_0 = 43°$. What are the reflected and transmitted waves, for both TE and TM polarizations? ◊

Attempting to find the angle of refraction from Snell's law $n_1 \sin \theta_1 = n_2 \sin \theta_2$ yields $n_1 \sin \theta_1 = 2.6598$, which exceeds $n_2 = 2.5$, so that this is indeed a case of total internal reflection.

The resistive impedance of side 1 is

$$Z_1 = \eta_1 \sec \theta_1 = (\eta_0/3.9) \sec 43° = 0.3506 \eta_0 \quad \text{for TE,} \qquad (7.182)$$

or

$$Z_1 = \eta_1 \cos \theta_1 = (\eta_0/3.9) \cos 43° = 0.1875 \eta_0 \quad \text{for TM.} \qquad (7.183)$$

To obtain the reactive impedances of side 2, we need the attenuation coefficient

$$\alpha = \sqrt{p^2 - (\omega/c)^2 n_2^2} = (\omega/c)\sqrt{n_1^2 \sin^2 \theta_1 - n_2^2} = 0.9080(\omega/c). \qquad (7.184)$$

The reactances are

$$Z_2^{\text{TE}} = j\omega\mu/\alpha = j\omega\mu_0/\alpha = j\eta_0/(\alpha c/\omega)$$

$$= j\eta_0/0.9080 = j(1.1013)\eta_0, \qquad (7.185)$$

$$Z_2^{\text{TM}} = \alpha/j\omega\varepsilon = \alpha/j\omega\varepsilon_0 n_2^2 = -j(\alpha c/\omega)\eta_0/n_2^2$$

$$= -j(0.9080/2.5^2)\eta_0 = -j(0.1453)\eta_0.$$

As a check on the calculations, we may verify that

$$Z_2^{\text{TE}} Z_2^{\text{TM}} = \eta_2^2 = \eta_0^2/n_2^2 = 0.1600\eta_0^2, \qquad (7.187)$$

which is indeed the case.

The reflection coefficients are

$$\Gamma_e^{\text{TE}} = (Z_2 - Z_1)/(Z_2 + Z_1) = [j1.1013 - 0.3506]/[j1.1013 + 0.3506]$$

$$= e^{j0.6164} \qquad (7.188)$$

for the electric field of the TE wave and

$$\Gamma_h^{\text{TM}} = (Z_1 - Z_2)/(Z_1 + Z_2) = [0.1875 + j0.1453]/[0.1875 - j0.1453]$$

$$= e^{j1.3183} \qquad (7.189)$$

for the magnetic field of the TM wave. In each case, the reflected field has the same amplitude as the incident one, but is shifted in phase.

The corresponding transmission coefficients are

$$T_e^{\text{TE}} = 1 + \Gamma_e^{\text{TE}} = 1 + e^{j0.6164} = 1.8160 + j0.5781, \qquad (7.190)$$

$$T_h^{\text{TM}} = 1 + \Gamma_h^{\text{TM}} = 1 + e^{j1.3183} = 1.2498 + j0.9683 \qquad (7.191)$$

and these give the transverse fields for each polarization. We cannot simply use the medium's impedance $\eta_2 = \eta_0/2.5$ to get the magnetic field of the transmitted TE wave or the electric field of the transmitted TM wave, because we are dealing not with a uniform wave in medium 2, but an attenuating one. The reactive wave impedances yield only the transverse part of the attenuating wave. To get the complete H^{TE} or E^{TM} from E^{TE} or H^{TM}, we use one or the other of Maxwell's equations, in the forms

$$[\boldsymbol{\alpha} + j\mathbf{p}] \times \mathbf{E} = j\omega\mu\mathbf{H}, \qquad \mathbf{H} \times [\boldsymbol{\alpha} + j\mathbf{p}] = j\omega\varepsilon\mathbf{E}, \qquad (7.192)$$

for the TE and the TM partial waves, respectively. We have already found both $\boldsymbol{\alpha}$ and $\mathbf{p}$; their magnitudes are $\alpha = (\omega/c)0.9080$ and $p = (\omega/c)2.6598$, their directions are normal and tangential to the plane, and both are perpendicular to the transverse fields. Thus, if we let $\hat{\mathbf{n}} = \hat{\mathbf{z}}$ and $\mathbf{p} = p\hat{\mathbf{x}}$, then the transverse fields lie along $\hat{\mathbf{y}}$. The fields that have a longitudinal component (along z) are then

$$\mathbf{H}_2^{TE} = (2.66\hat{\mathbf{z}} + j0.908\hat{\mathbf{x}})E_2^{TE}/\eta_0$$

$$= (2.66\hat{\mathbf{z}} + j0.908\hat{\mathbf{x}})(1.816 + j0.578)E_0^{TE}/\eta_0, \qquad (7.193)$$

$$\mathbf{E}_2^{TM} = -(2.66\hat{\mathbf{z}} + j0.908\hat{\mathbf{x}})\eta_0 H_2^{TM}/6.25$$

$$= -(2.66\hat{\mathbf{z}} + j0.908\hat{\mathbf{x}})(1.250 + j0.968)\eta_0 H_0^{TM}/6.25$$

$$= -(2.66\hat{\mathbf{z}} + j0.908\hat{\mathbf{x}})(0.200 + j0.155)\eta_0 H_0^{TM}, \qquad (7.194)$$

upon substituting the transmission coefficients. These are the fields at $x = 0$, and $z = 0$; elsewhere in medium 2, they propagate along x and decay along z.

SIGNIFICANCE FOR DIELECTRIC WAVEGUIDES

Oblique incidence from an optically dense medium (index n_1) onto a less dense one (index $n_2 < n_1$) results in transmission at an angle θ_2 steeper than the incidence angle θ_1 if θ_1 is less than the critical angle θ_c given by

$$\theta_c = \sin^{-1}(n_2/n_1). \qquad (7.195)$$

For incidence at the critical angle $\theta_1 = \theta_c$, θ_2 attains its maximum value of $\pi/2$ and the refracted wave travels parallel to the interface. Beyond the critical angle, so that the incident wave is more grazing to the interface than θ_c, there results *total internal reflection*. The reflected wave in the denser medium has the same strength as the incident one; there is no propagation across the interface to infinite distance normal to the boundary plane; although there are fields beyond the plane, they propagate parallel to the interface and decay in the direction away from it. Figure 7-26 compares the three cases of incidence, at $\theta < \theta_c$, at $\theta = \theta_c$, and at $\theta > \theta_c$. In the third part of the figure, the fading of the propagation vectors is intended to convey the notion of evanescence of the strength of the transmitted wave, in the direction normal to the interface.

Total internal reflection has applications in optical systems when perfect mirror action (neglecting conductive losses) is needed; prisms in binoculars are one example. TIR is the fundamental process that allows guided propagation of light in

Figure 7-26   Incidence angles less than, equal to, and greater than critical.

dielectric waveguides, which are structures with no metal parts; an outstanding version of these structures is the optical fiber.

If we can arrange for the light to impinge upon the edge of a dielectric waveguide at a sufficiently grazing angle, then the light should be totally reflected into the dielectric and will be prevented from escaping into the medium surrounding the guide. When the reflected wave encounters another wall of the dielectric waveguide, it can again bounce back, perfectly, into the waveguide and remain confined, and guided, by the structure. Light that strikes the boundary at too small an angle of incidence, however, can cross the interface and leak light away from the waveguide.

Despite the perfect reflection $(|\Gamma| = 1)$ under conditions of total internal reflection, fields do appear on the other side but attenuate at right angles to the plane. This fact can be verified by placing another dense medium close to, but not touching the original interface, leaving a gap. The decaying fields (what remains of them at the other side of the gap) can then cross the new interface and may propagate away. Total internal reflection has then been thwarted within the first medium, and power "tunnels" across the otherwise "forbidden" region of the gap. This process can be used to tap some of the light otherwise confined to the dielectric waveguide. The resulting structure has two parallel interfaces and involves three media.

REFLECTION AT GRAZING INCIDENCE

A different sort of total reflection, or nearly total reflection, occurs at grazing incidence, meaning the case of an incidence angle that is close to 90°, so that the incident wave is traveling nearly parallel to the interface. This situation is of special importance for wireless communications, particularly with cellular telephones. It is often the case that the transmitter and receiver are at heights above the ground that may be measured in meters, while their horizontal separation may be kilometers. In such circumstances, the wave emitted by the transmitter may reach the receiver not only along the line of sight (straight line from transmitter to receiver) but also after reflection from the ground, at grazing incidence; see Figure 7-27 but imagine a much

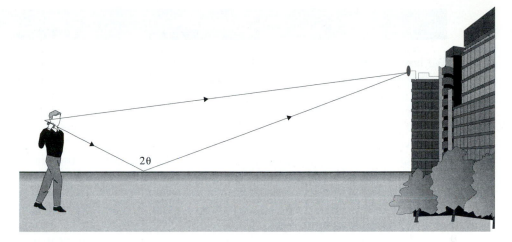

Figure 7-27 Grazing incidence of wave from cellular phone to base station.

wider separation between the cellular phone user and the base station on the building, so that $2\theta \to 180°$.

Grazing incidence is merely a particular case of oblique incidence, and we can use our formulas to analyze this situation directly, by simply going to the limit of $\theta \to \pi/2$. What makes the result special is that the limiting case is independent of the polarization of the wave and of the constitution of the media (the air and the ground, in the example of wireless communications).

For TE polarization, the impedance of the medium of the incident wave is $Z_1 = \eta_1 \sec \theta \to \infty$ in the limit, so that the reflection coefficient becomes

$$\Gamma_e = (Z_2 - Z_1)/(Z_2 + Z_1) \to (Z_2 - \infty)/(Z_2 + \infty) \to -1, \qquad (7.196)$$

regardless of Z_2 (as long as Z_2 is finite) and irrespective of η_1.

For TM polarization, the impedance is $Z_1 = \eta_1 \cos \theta \to 0$ and the reflection coefficient becomes

$$\Gamma_h = (Z_1 - Z_2)/(Z_1 + Z_2) \to (0 - Z_2)/(0 + Z_2) \to -1, \qquad (7.197)$$

again regardless of Z_2 and of η_1. For grazing incidence, no matter how the wave is polarized and no matter what the two dissimilar media are made of, the incident wave's field vectors are reversed by the reflection, suffering a 180° phase shift.

One consequence for mobile telephone communications is that if the wave emitted by the transmitter reaches the receiver along both the line of sight and also after reflection from the ground at grazing incidence, the two are very nearly 180° apart in phase and very nearly cancel each other. Fortunately, the cancellation is not perfect, as the angle of incidence does not quite reach 90°. However, the presence of the ground does cause the signal strength to diminish much more rapidly with distance of separation than would be the case in the absence of the ground.

SUMMARY

We have examined various questions and techniques related to impedance measurement and matching. We looked at the interference of countertraveling waves, which forms partially standing waves, or even pure standing waves if the reflected and incident waves have the same strength. The VSWR is the ratio of the maximum voltage produced by constructive interference to the minimum one at points of destructive interference. It is directly related to the magnitude of the reflection coefficient and it is easily measured. Use of a slotted line identifies the maximum and minimum voltage amplitudes and also pinpoints distances between the load and points of destructive interference. These measurements readily yield both magnitude and phase of the complex reflection coefficient and therefore furnish a measurement of the complex load impedance.

We illustrated the use of the transmission line circuit equations with the design of a tuning stub used to match an otherwise mismatched line. This takes advantage of the impedance transforming properties of a transmission line, allowing us to find a location along the line where the admittance has a conductive part matched to the characteristic admittance $1/Z_0$. Matching then requires adding a parallel susceptance that cancels whatever the susceptance is at that point. The required purely reactive admittance is obtainable by using a shorted length of a similar line attached at the connection point. This parallel element, being purely reactive, cannot absorb any of the power that reaches it, so that all the power emitted by the source will be absorbed by the original mismatched load, without subjecting the source to reflected waves.

We contemplated the use of a Smith chart to exhibit circuit relationships along a transmission line graphically. The chart is the unit circle in the complex plane of reflection coefficient, labeled by the corresponding impedances or admittances. Because $\Gamma(z)$ has a fixed magnitude and a phase linearly related to the distance z, motion along the line appears as movement around a circle centered at the origin of the plane, making it easy to read the values of impedances encountered as the line is traversed. We illustrated the use of the chart with the design of a tuning stub.

We examined waves that encounter planar interfaces between uniform media, at normal incidence, and found that when the media have mismatched impedances, there is reflection back to the source as well as transmission across the boundary. The reflection and transmission coefficients were obtained as

$$\Gamma = (\eta_2 - \eta_1)/(\eta_2 + \eta_1), \qquad T = 2\eta_2/(\eta_2 + \eta_1) \qquad (7.198)$$

at the interface between sourced medium 1 and unsourced medium 2, corresponding to a three-wave combination with only one wave incident onto the interface, the other two being scattered away from it. These expressions are based on continuity of tangential electric and magnetic fields across the planar boundary between media. This continuity condition follows from Maxwell's equations in integral form and allows those equations to be satisfied at points of a surface of discontinuity in constitutive parameters.

We analyzed the action of perfect mirrors and we designed an antireflection coating for mismatched media. For the latter, we noted that when two waves are both incident onto an interface, resulting in a combination of four waves (two incident, two scattered) rather than just three, then the expressions for reflection and transmission involve an effective impedance that depends on the relative strengths of the pairs of waves, not just on the medium's intrinsic impedance.

These results were extended to oblique incidence onto an interface; this case represents the mechanism for the trapping and propagation of light within a dielectric waveguide or an optical fiber. We treated obliquely incident uniform waves as normally incident but nonuniform waves. By this stratagem, we were able to reduce the problem of reflection and transmission to a previously solved case, except that equivalent impedances, dependent on the angle of incidence and on the polarization, replace the impedances of the media.

We found that Snell's laws of reflection and refraction are a direct consequence of the electromagnetic boundary conditions at the interface. They simply allow continuity of the tangential electric and magnetic fields at all points of the interface, once continuity has been achieved at any one point. They require the tangential component of the wave vector, which describes the nonuniformity of the wave, to be identical for the incident, the reflected, and the refracted waves.

By choosing to polarize one or the other of the electric and magnetic fields along the unique direction common to both the plane of the interface and the plane of polarization, we were able to impose continuity of the fields at the interface for each of two partial waves, the transverse electric (TE) and transverse magnetic (TM) polarizations. The continuity requirements led to the Fresnel formulas, which are of the same form as for normal incidence but use the effective impedances of the two polarizations, $\eta \sec \theta$ for TE and $\eta \cos \theta$ for TM. The concept of TE and TM waves will recur later, in our study of waveguiding.

We noted that the freedom provided by the dependence of the impedances on angle of incidence and on polarization allows us to tune to a desired impedance by tilting the interface with respect to the incident wave. In particular, we are able to eliminate reflection at the interface between mismatched media, by matching their *effective* impedances. This is achieved by arranging for incidence at Brewster's angle and for the appropriate polarization—namely, TM for nonmagnetic media. Tilting a beam of light to this angle therefore qualifies as an impedance matching technique.

Refraction of light that is sent from a denser to a less dense medium across a planar interface turns into the phenomenon of total internal reflection when the angle of incidence exceeds the critical angle, the angle beyond which Snell's law fails to yield a real angle of refraction. The formalism that yields the reflection and transmission coefficients in terms of the effective wave impedances on the two sides of the interface is valid even beyond the critical angle. At sufficiently grazing angles, the impedance of the outside medium becomes reactive instead of resistive, inductive for the TE partial wave and capacitive for the TM polarization.

In either case, electromagnetic fields do penetrate to the less dense outer medium, but they do not propagate perpendicularly away from the interface; instead, they attenuate in the direction normal to the plane while traveling parallel to the plane. The attenuation depends on the angle of incidence: At angles just beyond critical, the fields decay only gradually and are loosely coupled to the interface; for more nearly grazing angles, the fields persist over only a short range beyond the boundary plane and are tightly bound to the interface.

This sort of binding of a wave to a surface is an instance of the guiding of electromagnetic waves along a structure, in this case the interface between two dielectrics. Such guiding occurs also at an impenetrable wall formed by a conductor, as we shall see later.

While we have gained dexterity with transmission line circuit calculations, we are not yet well versed in certain fundamental aspects of energy and power transfer. As just one example, we should ask where the energy is during the time interval between emission from the generator and its eventual absorption by the load after a delay. Is the power lost after it is emitted but then reappears upon absorption in the load? Or is it conserved throughout? If it is not lost, where does it reside? Besides exploring these fundamental issues, we need to deal with more practical and more realistic transmission lines, which may be made of imperfect conductors and dielectrics and are therefore lossy.

PROBLEMS

Matching with a capacitor

7.1   A parallel-wire transmission line of characteristic impedance $Z_0 = 300\,\Omega$ is operated at a frequency of 1 GHz and has an unknown but mismatched load. However, it is found that attaching a 1-pF capacitor across the line at a distance of 1/8 wavelength in front of the load causes the reflection to be eliminated in front of the combination. Find the (complex) voltage reflection coefficient Γ_l at the load.

Load impedance from voltage measurements

7.2   A coaxial transmission line of characteristic impedance $Z_0 = 50\,\Omega$ is terminated with an unknown load Z_l. The measured magnitude of the voltage $|V(z)|$ varies monotonically from a minimum of 4 volts, 7 cm from the load, to a maximum of 20 volts, 13 cm from the load and oscillates beyond that point.
(a) What is the wavelength?
(b) What is the VSWR?
(c) What is the magnitude of the reflection coefficient?
(d) What is the phase of the reflection coefficient at the load?
(e) What is the load impedance?

Reactive part of load impedance from voltage measurements

7.3   A 300-Ω parallel-wire transmission line is terminated in a load impedance whose resistive part matches the line but whose reactive part is unknown. The maximum magnitude of the voltage along the line is measured to be 18 volts; the minimum is 6 volts. Find the reactance X of the load. Is it uniquely determined?

Effects of a slight mismatch

7.4   A uniform transmission line (length l, characteristic impedance Z_0, phase constant β) is only slightly mismatched [i.e., it is loaded by a (complex) impedance that differs from the characteristic impedance by a very small difference ΔZ]. To first order in ΔZ (i.e., neglecting $|\Delta Z|^2$ and higher-order terms),
   (a) What is the input impedance of the line?
   (b) What is the VSWR along the line?

Standing wave ratio for voltage and for current

7.5   A 50-ohm transmission line is loaded by $(22 + j96)$ ohms. The interference pattern of the incident and reflected waves can be observed by measuring the magnitude of either the voltage or the current along the line.
   (a) What is the standing wave ratio (SWR) if voltage is measured?
   (b) What is the standing wave ratio (SWR) if current is measured?
   (c) Where (in wavelengths) is the minimum of the interference pattern nearest the load if voltage is measured?
   (d) Where (in wavelengths) is the minimum of the interference pattern nearest the load if current is measured?
   (e) If the interference pattern measured by current were mistakenly interpreted as a voltage pattern, what load impedance would be calculated from the "VSWR" and the position of the nearest minimum?

Locus of impedance along transmission line in complex plane

7.6   It is asserted in the text that the locus of the (normalized) impedance $Z(z)/Z_0 = [1 + \Gamma(z)]/[1 - \Gamma(z)]$ as z varies along the transmission line is a circle in the complex plane. To confirm this, find a (possibly complex) constant C such that the magnitude $r(z) = |Z(z)/Z_0 - C|$ is independent of z. The constant C is then the center of the circle. What is the radius of the circle? Express answers in terms of the constant magnitude of the reflection coefficient, $|\Gamma|$. For this purpose, assume that $0 < |\Gamma| < 1$.

Conditions for physical realizability

7.7   The text asserts that the condition $0 \le |\Gamma| \le 1$ on the reflection coefficient corresponds to that of physical realizability of the load impedance, that Re $Z_l \ge 0$ or, for $Z_l/Z_0 = r + jx$, that $r \ge 0$. Demonstrate this equivalence.

Loci of constant resistance in Γ-plane

7.8   Demonstrate that the constant-r loci of $\Gamma(x)$ in $r + jx = (1 + \Gamma)/(1 - \Gamma)$ are circles, centered at $\gamma = r/(r + 1)$ and with radius $\rho = 1/(r + 1)$ in the Γ-plane, as claimed in the text.

Loci of constant reactance in Γ-plane

7.9   Demonstrate that the constant-x loci of $\Gamma(r)$ in $r + jx = (1 + \Gamma)/(1 - \Gamma)$ are circles, centered at $\gamma = 1 + j/x$ and with radius $\rho = 1/|x|$ in the Γ-plane, as claimed in the text.

Reading the VSWR from the circle of travel on the Smith chart

7.10   Demonstrate that the intersection of the circle of travel with the positive real axis of the Smith chart is at a normalized resistance value equal to the VSWR, as claimed in the text.

Quarter-wave matching section at other than the design frequency

7.11 The quarter-wave matching section designed in the text achieves perfect matching only at certain frequencies, those for which the length of the section is an odd multiple of a quarter wavelength. For the shortest section and the case of $Z_{02} = 2Z_{01}$, obtain the power reflection coefficient $|\Gamma|^2$ when the matching section is operated at
(a) double the design frequency;
(b) half the design frequency.

Bandwidth of quarter-wave matching section

7.12 The quarter-wave matching section designed in the text achieves perfect matching only at certain frequencies, those for which the length of the section is an odd multiple of a quarter wavelength. For the shortest section and the case of $Z_{02} = 2Z_{01}$, obtain the fractional bandwidth $\Delta f / f_0$, around the design frequency f_0, over which the power reflection coefficient $|\Gamma|^2$ is less than 5%.

Faster-than-light speed of a node of a standing wave

7.13 The motion $z(t)$ of a node of a standing wave pattern with reflection coefficient $\Gamma(0) = \Gamma e^{j\varphi}$ was found in the text to be given by

$$\tan\big[\beta z(t)\big] = \frac{\Gamma \cos(\omega t + \varphi) + \cos \omega t}{\Gamma \sin(\omega t + \varphi) - \sin \omega t}$$

and was plotted there. For a transmission line with $\omega/\beta = c$, the plot clearly shows motion with a speed that fluctuates about the speed of light c.
(a) Find the speed dz/dt of the node.
(b) How much faster than the speed of light is the node's maximum speed?
Note: There is no violation of the principles of special relativity here. No object, mass, energy, or information is moving faster than light in this situation. The node is merely a geometrically defined point, without substance.

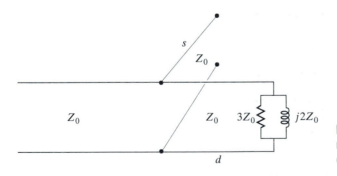

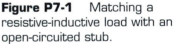

Figure P7-1   Matching a resistive-inductive load with an open-circuited stub.

Matching with open-circuited tuning stub

7.14 A transmission line with characteristic impedance Z_0 and phase constant β is loaded with a parallel combination of a resistance $R = 3Z_0$ and an inductive reactance $X = 2Z_0$, as indicated in Figure P7-1. It is to be matched by use of an open-circuited tuning stub of the same type and of the shortest possible length s, attached at distance d in front of the load. Find βd and βs.

Incidence at critical angle

7.15 The region below the xy-plane $(z < 0)$ is a nonmagnetic dielectric with index of refraction n. The region above $(z > 0)$ is air. A uniform plane wave at frequency ω is incident obliquely from below, at exactly the critical angle for total internal reflection.
(a) The propagation vector of the incident wave can be written as

$$\mathbf{k}_0 = (\omega/c)\big[A(n)\hat{\mathbf{y}} + B(n)\hat{\mathbf{z}}\big].$$

Find $A(n)$, $B(n)$ as functions of n.
(b) If the incident wave is TE, with $\mathbf{E}_0 = E\hat{\mathbf{x}}$, give the propagation vector $\mathbf{k}_2$ and the $\mathbf{E}_2$, $\mathbf{H}_2$ fields of the steady-state uniform plane wave in the air.
(c) If the incident wave is TM, with $\mathbf{H}_0 = H\hat{\mathbf{x}}$, give the propagation vector $\mathbf{k}_2$ and the $\mathbf{E}_2$, $\mathbf{H}_2$ fields of the steady-state uniform plane wave in the air.

Approximately grazing incidence

7.16 Consider incidence from air onto the interface with a nonmagnetic dielectric of refractive index n, at an angle $\theta = \pi/2 - \delta$ that is close to $90°$. Obtain (a) Γ_e^{TE} and (b) Γ_h^{TM} to first order in δ: The results can be expressed as $\Gamma = -1 + A\delta$; find $A = A(n)$ for each polarization.

Scattering by a resistive sheet

7.17 Obtain the reflection and transmission coefficients, Γ_e, T_e, for normal incidence onto a resistive sheet of surface conductivity σ_s, in air.

Oblique incidence onto a resistive sheet

7.18 Obtain the reflection and transmission coefficients, Γ, T, for oblique incidence at angle θ onto a resistive sheet of surface conductivity σ_s, in air, for both TE and TM polarization.

Parallel resistive sheets

7.19 Two resistive sheets each have surface conductivity σ_s and are parallel to each other, in air, a quarter wavelength apart.
(a) For normal incidence of a plane wave, find the reflection and transmission coefficients, Γ_e, T_e, of the pair.
(b) For what value of σ_s are the magnitudes $|\Gamma_e|$, $|T_e|$, equal? What is that magnitude?

Oblique incidence onto perfect absorber

7.20 The perfect absorber of plane waves designed and analyzed in the text operates properly at normal incidence. Consider oblique incidence at angle θ onto the resistive sheet of surface resistivity $r_s = \eta_0$ a quarter wavelength in front of a parallel perfectly conducting sheet. Obtain the reflection coefficients Γ_e and Γ_h for TE and TM polarizations. What are the magnitudes of these reflection coefficients for $\theta = 30°$?

Circularly polarized wave at a mirror

7.21 A perfect, planar mirror coincides with the xy-plane and normally incident as well as reflected plane waves exist in the vacuum region $z < 0$ in front of the mirror, at frequency ω. The magnetic field at the mirror surface is circularly polarized: $\mathbf{H} = H_0(\hat{\mathbf{x}} + j\hat{\mathbf{y}})$. Find the (complex) electric field amplitude $\mathbf{E}(z)$ in the space $z < 0$.

Reflection coefficients for given refraction

7.22 Oblique incidence at a certain incidence angle θ_0, from air onto a lossless nonmagnetic dielectric of refractive index n, is observed to result in refraction at angle $\theta_2 = \pi/2 - \theta_0$. Express, in terms of n alone,
  (a) the reflection coefficient Γ_h for TM polarization of the incident wave;
  (b) the reflection coefficient Γ_e for TE polarization of the incident wave.

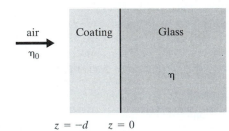

air

η_0

Coating   Glass

η

$z = -d$ $z = 0$

Figure P7-2   Antireflection coating for air–glass interface.

Mistaken design of antireflection coating

7.23 In an attempt to design an antireflection coating for a glass medium (nonmagnetic, permittivity $\varepsilon = 4$), an engineer (now unemployed) used a coating with an impedance that is the geometric mean $\sqrt{\eta_0 \eta}$ of the air and glass impedances, η_0, η, but, mistakenly, with a thickness $\lambda/2$ (i.e., $1/2$ wavelength measured in the coating), instead of $\lambda/4$.
  (a) Find the impedance $Z(z)$ within the coating (between the air–coating interface at $z = -d$ and the coating–glass interface at $z = 0$; see Figure P7-2).
  (b) Find the reflection coefficient Γ_e for normal incidence from the air onto the coated glass.

Wrong polarization for Brewster's angle

7.24 An experimenter wants to transmit a plane wave across the interface between two different nonmagnetic, lossless dielectrics. Because he finds a reflection coefficient $\Gamma_e = -0.5$ at normal incidence, he decides to tilt the interface to get incidence at Brewster's angle, which allows perfect transmission of a TM wave. Mistakenly, however, he shines a TE wave at this angle, instead of TM polarization. What reflection coefficient does he observe?

CHAPTER 8

Poynting Theorems and Lossy Transmission Lines

We have some fundamental questions to explore, regarding conservation of energy in a system that features a delay between emission of the energy by the source and the eventual absorption of that energy by the load. Where does the energy reside after emission but before absorption? Is the energy localized or is it distributed throughout the transmission structure? If the energy delivered to the load differs from the energy produced by the generator, where can the missing energy have gone?

It is especially important to have an appreciation of how the energy is distributed in an electromagnetic system, because if the structure is not the perfectly ideal one in geometry or constitution, or if it needs to be disturbed by inserting probes or drilling holes for access or measurements, then it is usually beneficial to confine the perturbation to low-energy regions, while high-energy locations are kept undisturbed.

To explore these issues, we turn again to Maxwell's equations and combine them in a way that suggests to us an interpretation in terms of energy flow. The result is known as Poynting's theorem, and we will find two distinct versions of it, one in the time domain and a separate one in the frequency domain. We will also examine lossy transmission lines and consider why just about any structure can exhibit transmission line behavior.

POYNTING THEOREM

Since electromagnetic fields are governed by Maxwell's equations, any assertion about the energetics of such fields must be grounded in those equations. Although energy is originally a mechanical concept, we have already concluded, in an earlier chapter, that the mechanical forces exerted on charges by electromagnetic fields lead us to interpret the electrical quantity $\mathbf{E} \cdot \mathbf{J}$ as power density (energy per unit time, per unit volume) at the point and instant where the electric field and the current density co-exist. This expression combines a source, $\mathbf{J}$, and a resultant field, $\mathbf{E}$; what we seek now is a reexpression of this quantity entirely in terms of derivatives of the fields. We can get this through Maxwell's equations, and the result will be an alternative expression for the power density, which we can then equate to the original one to interpret the new result.

The Maxwell equations in differential form are the easiest starting point.

$$\nabla \times \mathbf{E} = -\partial \mathbf{B}/\partial t, \qquad \nabla \times \mathbf{H} = \mathbf{J} + \partial \mathbf{D}/\partial t. \qquad (8.1)$$

We note that dotting the second equation with $\mathbf{E}$ will get us the $\mathbf{E} \cdot \mathbf{J}$ term we are interested in. Let's see where this leads. We get

$$\mathbf{E} \cdot \nabla \times \mathbf{H} = \mathbf{E} \cdot \mathbf{J} + \mathbf{E} \cdot \partial \mathbf{D}/\partial t \qquad (8.2)$$

and we do have the $\mathbf{E} \cdot \mathbf{J}$ term, but we see two other terms, each of the form of one field times the derivative of another. This form should remind us of part of the expression for the derivative of a product.

For the time derivative of the relevant product, we have

$$\partial(\mathbf{E} \cdot \mathbf{D})/\partial t = \partial \mathbf{E}/\partial t \cdot \mathbf{D} + \mathbf{E} \cdot \partial \mathbf{D}/\partial t \qquad (8.3)$$

and we argue that, for a linear medium with $\mathbf{D}$ and $\mathbf{E}$ proportional to each other, the two terms on the right are actually identical, both being proportional to $\mathbf{E} \cdot \partial \mathbf{E}/\partial t$. We will therefore make the replacement

$$\mathbf{E} \cdot \partial \mathbf{D}/\partial t = \partial(\tfrac{1}{2} \mathbf{E} \cdot \mathbf{D})/\partial t \qquad (8.4)$$

in (8.2).

For the spatial derivative of the other product, involving a curl, we turn to a vector identity that entails curls—namely, the divergence of a cross product:

$$\nabla \cdot (\mathbf{E} \times \mathbf{H}) = \mathbf{H} \cdot \nabla \times \mathbf{E} - \mathbf{E} \cdot \nabla \times \mathbf{H}. \qquad (8.5)$$

We are concerned with the last term, which appears in (8.2); we cannot claim that the last two terms here are the same but we can replace the first term on the right by using the other of the two Maxwell equations, after dotting with $\mathbf{H}$:

$$\mathbf{H} \cdot \nabla \times \mathbf{E} = -\mathbf{H} \cdot \partial \mathbf{B}/\partial t. \qquad (8.6)$$

For this new term, we write again the time derivative of a product,

$$\partial(\mathbf{H} \cdot \mathbf{B})/\partial t = \partial \mathbf{H}/\partial t \cdot \mathbf{B} + \mathbf{H} \cdot \partial \mathbf{B}/\partial t \qquad (8.7)$$

and argue again that, for a linear medium with $\mathbf{B}$ and $\mathbf{H}$ proportional to each other, the two terms on the right are actually identical, both being proportional to $\mathbf{H} \cdot \partial \mathbf{H}/\partial t$. We will therefore make the replacement

$$\mathbf{H} \cdot \partial \mathbf{B}/\partial t = \partial(\tfrac{1}{2} \mathbf{H} \cdot \mathbf{B})/\partial t. \qquad (8.8)$$

Now, collecting the terms we have replaced with equivalents, we find

$$
\begin{aligned}
-\nabla \cdot (\mathbf{E} \times \mathbf{H}) &= -\mathbf{H} \cdot \nabla \times \mathbf{E} + \mathbf{E} \cdot \nabla \times \mathbf{H} \\
&= \mathbf{H} \cdot \partial \mathbf{B}/\partial t + \mathbf{E} \cdot \mathbf{J} + \mathbf{E} \cdot \partial \mathbf{D}/\partial t \\
&= \mathbf{E} \cdot \mathbf{J} + \partial(\tfrac{1}{2} \mathbf{E} \cdot \mathbf{D} + \tfrac{1}{2} \mathbf{H} \cdot \mathbf{B})/\partial t,
\end{aligned}
\qquad (8.9)
$$

for a linear medium. This is Poynting's theorem, in differential form. To interpret it, we first integrate it over any fixed volume V. The volume integral of the divergence turns into a surface integral over the closed surface S that bounds the volume V. The volume integral of the (partial) time derivative becomes the (total) time derivative of the integral over the volume (total, because after spatial integration only time variation is left). We get

$$-\oint_S \mathbf{E} \times \mathbf{H} \cdot d\mathbf{S} = \int_V \mathbf{E} \cdot \mathbf{J} \, dV + \frac{d}{dt}\int_V (\tfrac{1}{2}\mathbf{E} \cdot \mathbf{D} + \tfrac{1}{2}\mathbf{H} \cdot \mathbf{B}) \, dV. \qquad (8.10)$$

This is the Poynting theorem, in integral form. What does it mean?

INTERPRETATION

We can begin the interpretation of this vital theorem with the first term on the right, because we already know what it means. It is the integral of $\mathbf{E} \cdot \mathbf{J}$, which we know to represent power per unit volume. The integration over the volume V therefore gives us the total power expended within the volume; this is measured in watts.

The other two quantities in the equation must also represent power. Since power is the rate at which energy is expended, or delivered, or transferred, or converted, or otherwise caused to vary in time, it can be expressed as the time rate of change of energy. In the theorem, the last term is also a time rate of change; it is the rate of increase of the integral of the quantity $(\tfrac{1}{2}\mathbf{E} \cdot \mathbf{D} + \tfrac{1}{2}\mathbf{H} \cdot \mathbf{B})$ over the volume V. Consequently, the last term of the equation must represent the rate of increase of the amount of energy within the volume V. We must therefore interpret the quantity

$$W = W(t) = \int_V (\tfrac{1}{2}\mathbf{E} \cdot \mathbf{D} + \tfrac{1}{2}\mathbf{H} \cdot \mathbf{B}) \, dV \qquad (8.11)$$

as the energy that has been accumulated within the volume V until time t.

The two terms on the right of the theorem account for two processes that can affect the energy in volume V. It can be *expended* within the volume, at the rate $P = \int \mathbf{E} \cdot \mathbf{J} \, dV$, measured in watts and expressing *resistive loss*, or else it can be *accumulated,* or stored, in the volume. The accumulation is at the rate dW/dt, measured in joules/second or watts, so that the quantity W must be the energy stored within volume V, measured in joules.

Before interpreting the left-hand term of the theorem, we note that the stored energy term $W = \int (\tfrac{1}{2}\mathbf{E} \cdot \mathbf{D} + \tfrac{1}{2}\mathbf{H} \cdot \mathbf{B}) \, dV$ comprises two separate contributions, one purely electric (*capacitive*) and the other purely magnetic (*inductive*). Since the interpretation of the energy term is valid for any volume V, the integrand $\tfrac{1}{2}\mathbf{E} \cdot \mathbf{D}$ must represent the density of electric energy stored at a point in space and the second term, $\tfrac{1}{2}\mathbf{H} \cdot \mathbf{B}$, must be the density of magnetic energy stored at the point; both are measured in joules/m³.

The implication of these interpretations is that wherever there is an electric field $\mathbf{E}$, in a medium that makes it proportional to $\mathbf{D}$ at the point, there is stored $\tfrac{1}{2}\mathbf{E} \cdot \mathbf{D}$ joules of electric energy per unit volume at that point. Similarly, wherever there is a magnetic field $\mathbf{H}$, proportional to $\mathbf{B}$, there is $\tfrac{1}{2}\mathbf{H} \cdot \mathbf{B}$ joules of magnetic energy per unit volume stored at the point. Where both types of field coexist, both energies are stored at the point.

Turning back to the Poynting theorem, we now see that it declares that the rate at which the energy in the volume V is being expended plus the rate at which it is being stored, $P + dW/dt$, must equal the surface integral on the left side of the equation; the surface S is the one that encloses the volume. That is one fact. Another fact to consider is that the principle of conservation of energy asserts that the energy expended and stored within a volume V must be supplied, instant by instant, from the exterior to the interior of the volume, which means across its bounding surface S. Both facts refer to the rate at which energy crosses the bounding surface of the volume as it flows from whatever sources of energy may exist in the exterior, across the surface, and then toward the interior, where it may be either stored or used up. Combining the two facts, we conclude that what is measured by the term $- \oint_S \mathbf{E} \times \mathbf{H} \cdot d\mathbf{S}$ in the equation is the rate at which energy crosses the surface S, in watts.

This surface integral requires us to sum up over all area elements of the surface the contributions of the dot product of a vector $\mathbf{E} \times \mathbf{H}$ with the negative of the area element $d\mathbf{S}$. Recalling that, for any closed surface, the direction of the surface element $d\mathbf{S}$ is taken outward, so that $-d\mathbf{S}$ points inward, we realize that we are measuring the alignment of the vector $\mathbf{E} \times \mathbf{H}$ with the inward area element and then summing all these infinitesimal contributions to arrive at the total rate of power transfer inward across the bounding surface. We conclude that the vector $\mathbf{E} \times \mathbf{H}$ measures the flow of electromagnetic power. The magnitude of this "Poynting vector" gives the flow rate, in watts/m², and the direction of this vector specifies the direction of the power flow. The Poynting vector is a field, being a function of position in space, and of time as well.

Example 8.1

Obtain the Poynting vector everywhere in the presence of an infinitely long pair of parallel wires (radius a, separation s) forming a transmission line charged to voltage $V(z, t)$ and carrying current $I(z, t)$. ◇

In a previous chapter, we derived the electric and magnetic fields generated by a pair of parallel wires. We superimposed the contributions from each wire, finding that the individual wires contribute electric and magnetic fields that each vary inversely with the distance from that wire. The electric field is directed radially away from or toward the wire; the magnetic field is at right angles to that direction, as depicted in Figure 8-1, and oriented by the right-hand rule with respect to the current in the wire.

In terms of the voltage and current at position z along the transmission line, the two fields are, as found earlier,

$$\mathbf{E} = E_1 \hat{\mathbf{e}}_1 + E_2 \hat{\mathbf{e}}_2 = \frac{V(z, t)}{2 \ln(s/a)} \left(\frac{\hat{\mathbf{e}}_1}{\rho_1} - \frac{\hat{\mathbf{e}}_2}{\rho_2} \right), \tag{8.12}$$

$$\mathbf{H} = H_1 \hat{\mathbf{h}}_1 + H_2 \hat{\mathbf{h}}_2 = \frac{I(z, t)}{2\pi} \left(\frac{\hat{\mathbf{h}}_1}{\rho_1} - \frac{\hat{\mathbf{h}}_2}{\rho_2} \right). \tag{8.13}$$

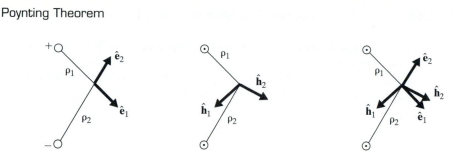

Figure 8-1   Contributions to the Poynting vector from parallel wires.

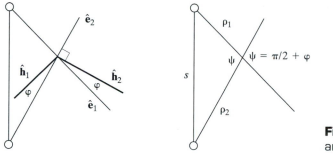

Figure 8-2   Relation between angles in parallel-wire geometry.

There remains only to evaluate the cross product of these two fields. We note first that the unit vectors $\hat{\mathbf{e}}_1$ and $\hat{\mathbf{h}}_1$ are perpendicular, as are $\hat{\mathbf{e}}_2$ and $\hat{\mathbf{h}}_2$; their cross products are both equal to the unit vector in the z-direction, $\hat{\mathbf{z}}$. The mixed cross products, $\hat{\mathbf{e}}_1 \times \hat{\mathbf{h}}_2$ and $\hat{\mathbf{e}}_2 \times \hat{\mathbf{h}}_1$ are, however, in the $-\hat{\mathbf{z}}$ direction and their magnitudes depend on the angle between the pairs of vectors. That angle depends on the observation point and is easily related to the coordinates of that point, as in Figure 8-2. We readily see that the angles are the same for both pairs and that $\sin \varphi = -\cos \psi$, as in the figure. The Poynting vector is therefore

$$\mathbf{E} \times \mathbf{H} = \frac{V(z,t)I(z,t)}{4\pi \ln(s/a)} \left(\frac{\hat{\mathbf{z}}}{\rho_1^2} + \frac{\hat{\mathbf{z}}}{\rho_2^2} - \frac{2\hat{\mathbf{z}} \cos \psi}{\rho_1 \rho_2} \right), \tag{8.14}$$

which simplifies to

$$\mathbf{E} \times \mathbf{H} = \frac{V(z,t)I(z,t)\hat{\mathbf{z}}}{4\pi \ln(s/a)\rho_1^2 \rho_2^2} [\rho_1^2 + \rho_2^2 - 2\rho_1 \rho_2 \cos \psi]. \tag{8.15}$$

We recognize the law of cosines for the triangle in the figure and therefore replace the quantity in the brackets with s^2, leaving

$$\mathbf{E} \times \mathbf{H} = \frac{V(z,t)I(z,t)s^2}{4\pi \ln(s/a)\rho_1^2 \rho_2^2} \hat{\mathbf{z}}. \tag{8.16}$$

This is the Poynting vector, at any point in the space surrounding the wires. We note that it is everywhere along $\hat{\mathbf{z}}$ (or opposite to it if the VI product is negative), indicating that the power flow is along the direction of the wires. We also see that it varies from point to point in space, inversely as the squared distance from each of the two wires. Recall that there are no fields inside the perfectly conducting wires, hence no Poynting vector in them; all the power flow takes place in the space around the wires. It is strongest just outside either wire, where one of ρ_1 and ρ_2 attains its smallest value, a, while the other is s (we continue to use the simplifying assumption that a is negligible compared to s). As our observation point recedes from one of the wires, the strength of the power flow declines rapidly. For example, midway between the wires, the Poynting vector has a magnitude proportional to $1/(s/2)^4$, compared to $1/(a^2 s^2)$ just outside one wire. The strength midway is therefore weaker by the factor $16a^2/s^2$ than at either wire; since $a \ll s$, we sense that the power flow is concentrated near the two wires.

A simple, but instructive, example of power flow arises in considering the fields around a segment of current-carrying wire.

Example 8.2

Find the Poynting vector and the total power flow at the surface of an imperfectly conducting wire (conductivity σ, radius a, segment length l) that carries uniformly distributed current I. ◈

The electric field is evidently $E = J/\sigma = I/\sigma\pi a^2$ inside the wire and, by continuity of tangential electric fields, the electric field just outside the wire is the same. Ampère's law gives us the magnetic field outside the wire ($\rho > a$) as $H = I/2\pi\rho$. Since the electric field is directed axially and the magnetic field azimuthally, the cross product is directed radially inward:

$$\mathbf{E} \times \mathbf{H} = -\hat{\boldsymbol{\rho}}(I/\sigma\pi a^2)(I/2\pi\rho) = -\hat{\boldsymbol{\rho}}(I^2/2\pi^2\sigma a^2\rho), \tag{8.17}$$

which becomes $-\hat{\boldsymbol{\rho}}(I^2/2\pi^2\sigma a^3)$ at the surface of the wire.

The total power crossing the surface of the segment of the wire is given by integrating this quantity, whose magnitude is just a constant at the wire surface, over the area $2\pi al$ of the segment. Since the area element is $d\mathbf{S} = \hat{\boldsymbol{\rho}}a\, d\varphi\, dz$, we get for the surface integral

$$-\oint_S \mathbf{E} \times \mathbf{H} \cdot d\mathbf{S} = -\int_0^l \int_0^{2\pi} \left[-\hat{\boldsymbol{\rho}}(I^2/2\pi^2\sigma a^3)\right] \cdot \hat{\boldsymbol{\rho}}a\, d\varphi\, dz$$

$$= I^2 l/\sigma\pi a^2 = I^2 R. \tag{8.18}$$

This is the power that crosses the surface of the wire, laterally, inward, from the outside to the inside of the wire. We recognize $l/\sigma\pi a^2$ as the resistance R of the segment of the wire, so that this power is exactly what gets dissipated within the wire. The power is carried by the fields surrounding the wire and is converted to heat inside the wire.

Whenever electric and magnetic fields coexist (and are not parallel), we will be interested in calculating the Poynting vector $\mathbf{E} \times \mathbf{H}$ to determine the pattern of power flow. Similarly, we will be able to learn about electric and magnetic energy storage by calculating the distribution of $\frac{1}{2}\mathbf{E} \cdot \mathbf{D}$ and $\frac{1}{2}\mathbf{H} \cdot \mathbf{B}$. Regions of high power flow and of high energy accumulations are particularly important to the operation of the electromagnetic structure or device.

Example 8.3

A uniform, linearly polarized, transverse electromagnetic plane wave, of arbitrary wave-form $F(t)$, in a vacuum, is described by

$$\mathbf{E}(z,t) = \hat{\mathbf{e}}F(t - z/c), \qquad \mathbf{H}(z,t) = \hat{\mathbf{h}}F(t - z/c)/\eta_0, \qquad \hat{\mathbf{e}} \times \hat{\mathbf{h}} = \hat{\mathbf{z}}. \qquad (8.19)$$

Obtain and describe the Poynting vector, electric energy storage density, and magnetic energy storage density for this plane wave. ◈

The Poynting vector is $\mathbf{E}(z,t) \times \mathbf{H}(z,t) = \hat{\mathbf{z}}F^2(t - z/c)/\eta_0$. This is the power flow density carried by the plane wave; being a function of $t - z/c$, it is itself a wave, prop-agating along z at the speed of light. The direction of the power flow is along $\hat{\mathbf{z}}$, which is also the direction of propagation of the plane wave.

The electric energy density is $\frac{1}{2}\mathbf{E} \cdot \mathbf{D} = \frac{1}{2}\varepsilon_0 F^2(t - z/c)$, which also propagates as a wave. The magnetic energy density is, since $\eta_0 = \sqrt{\mu_0/\varepsilon_0}$,

$$\tfrac{1}{2}\mathbf{H} \cdot \mathbf{B} = \tfrac{1}{2}\mu_0 F^2(t - z/c)/\eta_0^2 = \tfrac{1}{2}\varepsilon_0 F^2(t - z/c),$$

which is exactly the same as the electric energy density, at every point and instant.

Note also that the total energy density, $\varepsilon_0 F^2(t - z/c)$, multiplied by the velocity $c\hat{\mathbf{z}}$ of propagation, is just $\hat{\mathbf{z}}F^2(t - z/c)/\eta_0$, the power flow density. That relationship, that energy density times velocity equals power flow, is reminiscent of the properties of fluid flow and invokes a mental image of the plane wave's acting as if it were a stream of electromagnetic energy flowing along its direction of propagation.

COAXIAL LINE

As another important example of a transmission line and of power and energy trans-mission, we consider the coaxial line, this time in detail. This is comprised of a cylin-drical wire, of radius a, surrounded by a coaxial cylindrical conducting shell of inner radius b. The perfect conductors are separated by a perfectly insulating dielectric of permittivity ε and permeability μ; see Figure 8-3.

Both the parallel-wire line and the coaxial line allow easy connection of a volt-age or current source at one end and of a load at the other. An advantage of the coax-ial line over the other is that the outer conductor acts as a shield, confining and protecting the fields in the dielectric from disturbing influences.

Figure 8-3 Coaxial line with generator and load attached.

Physically, what happens when the generator is attached to one end of the coaxial line is that the line gets charged up and a current flows along it. We expect the generator to push charges onto the center conductor and draw them from the outer one; we also expect a current to be injected into the center wire and an equal current to be drawn from the outer conductor. We should therefore have equal and opposite charges spread around the surfaces of the two conductors, meaning the outer surface of the inner conductor and the inner surface of the outer one. We should also have equal and opposite currents flowing on these two surfaces of the perfect conductors.

Corresponding to the equal and opposite charge distributions, we expect an electric field to appear in the dielectric space between the conductors, emanating from the center conductor and ending on the outer one. Because of the axial current in the inner conductor, we also expect a magnetic field to be generated in the dielectric, surrounding the inner wire. There are no fields at all in the perfect conductors and, as long as no special action is taken to disturb the equality of the opposite currents and charges on the two conductors, there should be no fields outside the outer shell. This is how the outer conductor acts as a shield; the fields are entirely confined to the dielectric between the inner and outer conductors.

The time-varying magnetic flux that accompanies the time-varying current will cause the voltage from the inner to the outer conductor to vary from point to point along the coaxial line. The charge and current distributions and the electric and magnetic fields will all vary along the length of the line. We can get the distributions of voltage, current, and charge per unit length, $V(z, t)$, $I(z, t)$, and $q(z, t)$, by finding the electric and magnetic fields that accompany these and then applying the laws of electromagnetics to the electric and magnetic fluxes they generate.

The symmetry of the structure guarantees, provided we also feed it and load it symmetrically, that the electric and magnetic fields will be azimuthally symmetric; their strengths can depend only on the radial distance ρ from the axis of the structure (and on z and t) but not on azimuth φ. The electric field will be radial and the magnetic field will be directed azimuthally; see Figure 8-4.

To find the strength of the electric field as a function of radius ρ, we apply Gauss's law to a cylindrical surface of radius ρ and extended infinitesimally along the line, from z to $z + dz$; see Figure 8-5. The electric flux over this surface is

$$d\Psi(\rho, z, t) = \varepsilon E(\rho, z, t) 2\pi\rho \, dz \tag{8.20}$$

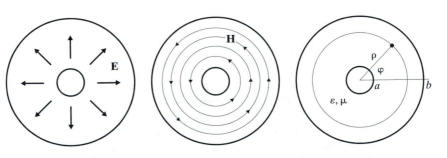

Figure 8-4   Electric and magnetic field patterns in a coaxial line.

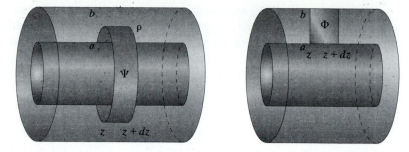

Figure 8-5   Electric and magnetic flux surfaces of infinitesimal width.

and this must equal the enclosed charge $q(z, t)\, dz$. We conclude that the electric field strength is

$$E(\rho, z, t) = q(z, t)/2\pi\varepsilon\rho \tag{8.21}$$

in the space $a < \rho < b$.

We can integrate this electric field to get the voltage or emf from the inner conductor to the outer one:

$$V(z, t) = \int_a^b \frac{q(z, t)}{2\pi\varepsilon\rho}\, d\rho = \frac{q(z, t)}{2\pi\varepsilon} \ln(b/a). \tag{8.22}$$

We see that the distributions of charge per unit length and of voltage are related by a constant of proportionality and we conclude that the capacitance per unit length $C = q/V$ of the coaxial line is

$$C = 2\pi\varepsilon/\ln(b/a). \tag{8.23}$$

This also allows us to rewrite the electric field distribution in terms of that of the voltage.

$$\mathbf{E}(\rho, \varphi, z, t) = \frac{V(z, t)}{\ln(b/a)\rho}\, \hat{\boldsymbol{\rho}} \qquad (a < \rho < b). \tag{8.24}$$

To find the strength of the magnetic field as a function of radius ρ, we apply Ampère's law to a circular disk of radius ρ transverse to the line, at z. The mmf around the edge is

$$U(\rho, z, t) = H(\rho, z, t)2\pi\rho \tag{8.25}$$

and this must equal the current that crosses the disk, which is just $I(z, t)$ since there is no axial displacement current. The magnetic field is therefore

$$\mathbf{H}(\rho, \varphi, z, t) = \frac{I(z, t)}{2\pi\rho}\, \hat{\boldsymbol{\varphi}} \qquad (a < \rho < b). \tag{8.26}$$

We can integrate the magnetic field to get the magnetic flux on a rectangular area from the inner to the outer conductor and extending infinitesimally along the line, from z to $z + dz$, as in Figure 8-5. The infinitesimal magnetic flux over this surface is

$$d\Phi(\rho, z, t) = dz \int_a^b \mu H(\rho, z, t)\, d\rho = dz\, \mu\, \frac{I(z, t)}{2\pi} \int_a^b \frac{d\rho}{\rho}$$

$$= dz\big[\mu \ln(b/a)/2\pi\big] I(z, t). \tag{8.27}$$

We see that the flux per unit length $d\Phi/dz$ and the current are proportional and we conclude that the inductance per unit length $L = [d\Phi/dz]/I$ of the coaxial line is

$$L = \mu\, \ln(b/a)/2\pi. \tag{8.28}$$

Before demonstrating the transmission line properties of the coaxial line, let us use our new-found knowledge of the fields in the dielectric space between the coaxial conductors to inquire into the flow of power in this structure.

Example 8.4

Find the Poynting vector and the total power flow in a coaxial line, in terms of the voltage and current distributions along the line. ◈

The electric and magnetic fields are confined to the dielectric space between the conductors, in $a < \rho < b$. The electric field is radial and the magnetic field is azimuthal, so that the cross product is directed axially. The Poynting vector is

$$\mathbf{E} \times \mathbf{H} = \frac{V(z, t)}{\ln(b/a)\rho}\, \hat{\boldsymbol{\rho}} \times \hat{\boldsymbol{\varphi}}\, \frac{I(z, t)}{2\pi\rho} = \frac{V(z, t)I(z, t)}{2\pi\, \ln(b/a)\rho^2}\, \hat{\mathbf{z}}. \tag{8.29}$$

This power flow density is directed along the axial direction; it is azimuthally symmetric (no dependence on φ); it varies radially as the inverse of the square of the distance from the axis, but exists only in the range $a < \rho < b$. The power flow at some transverse plane is therefore strongest just outside the inner conductor and weaker at the outer one, by a factor of $(b/a)^2$. Note again that the power does not flow within the perfect conductors, where the current resides; it flows where the fields are, in the insulator.

We can integrate the Poynting vector over a transverse plane at z to get the total power that flows across that plane. The integration region reduces to the annular area A defined by $a < \rho < b$, for all φ (i.e., $0 < \varphi < 2\pi$), at the fixed location z. The surface element is $d\mathbf{S} = \hat{\mathbf{z}}\rho\, d\rho\, d\varphi$ and we get

$$\int_A \mathbf{E} \times \mathbf{H} \cdot d\mathbf{S} = \int_0^{2\pi} \int_a^b \frac{V(z, t)I(z, t)}{2\pi\, \ln(b/a)\rho^2}\, \hat{\mathbf{z}} \cdot \hat{\mathbf{z}}\rho\, d\rho\, d\varphi$$

$$= V(z, t)I(z, t). \tag{8.30}$$

This gives the comfortably consistent result that the total power carried across a transverse plane at z by the electric and magnetic fields within the annular space between the conductors of the coaxial line equals the electrical power VI that enters the rest of the line beyond point z.

We have found the power flow inside a coaxial line to be proportional to the product $V(z, t)I(z, t)$, but how do these two distributions vary along the line?

We can apply the Faraday-Maxwell law to the magnetic flux surface in the last figure to find how $V(z + dz, t)$ differs from $V(z, t)$. The emf around the edge of the rectangle is

$$\text{emf} = V(z + dz, t) - V(z, t) = -\partial[d\Phi]/\partial t \qquad (8.31)$$

and we have determined that the magnetic flux $d\Phi$ is proportional to the current in the inner conductor, via the inductance per unit length L, as $d\Phi = L\, dz\, I(z, t)$. Dividing by dz, we get

$$\partial V(z, t)/\partial z = -L\, \partial I(z, t)/\partial t. \qquad (8.32)$$

We can also invoke continuity of total current, as applied to the electric flux surface in the figure but closed by two transverse disks of radius ρ at z and at $z + dz$, to assert that the current into the enclosed region, $I(z, t)$, must equal the current out, $I(z + dz, t)$, plus the displacement current across the cylindrical surface, $\partial[d\Psi]/\partial t$. Since we have determined that the electric flux $d\Psi$ is proportional to the voltage from the inner to the outer conductor, via the capacitance per unit length C, as $d\Psi = C\, dz\, V(z, t)$, we get

$$\partial I(z, t)/\partial z = -C\, \partial V(z, t)/\partial t. \qquad (8.33)$$

These equations, $\partial V/\partial z = -L\, \partial I/\partial t$ and $\partial I/\partial z = -C\, \partial V/\partial t$ are exactly the transmission line equations we studied and solved in connection with the parallel-wire line. The only difference is that the detailed expressions for the transmission line parameters L and C are now

$$L = \mu\, \ln(b/a)/2\pi, \qquad C = 2\pi\varepsilon/\ln(b/a), \qquad (8.34)$$

instead of the corresponding ones for the parallel-wire case.

The solutions are of identical form. We first construct one combination of these parameters

$$u_0 = 1/\sqrt{LC} = 1/\sqrt{\mu\varepsilon} \qquad (8.35)$$

to get the speed of wave propagation; this is exactly as it was for the parallel-wire case and would become the speed of light c if the dielectric in the coaxial line were a vacuum. We also form the alternative combination of L and C

$$Z_0 = \sqrt{L/C} = \sqrt{\mu/\varepsilon}\, \ln(b/a)/2\pi \qquad (8.36)$$

to get the characteristic impedance of the coaxial line.

The solution for the voltage and current distributions is the same—namely, a superposition of incident and reflected waves. The voltage is the sum of the two waves; the characteristic impedance times the current is the difference of the same two waves:

$$V(z, t) = V_1(t - z/u_0) + V_2(t + z/u_0), \qquad (8.37)$$
$$Z_0 I(z, t) = V_1(t - z/u_0) - V_2(t + z/u_0). \qquad (8.38)$$

The two waveforms $V_1(t)$ and $V_2(t)$ are to be found by applying the boundary conditions at the two ends of the coaxial line, imposed by the source and load attached to the line. We conclude that the ideal coaxial line does act as a transmission line; the speed of propagation is the speed of light in the dielectric between the coaxial conductors; the characteristic impedance depends on the constitution of that dielectric and on the geometry of the conductors.

Example 8.5

A coaxial cable is to have polyethylene $(\varepsilon = 2.25\varepsilon_0)$ as its dielectric spacer between conductors. Can we design it to have a characteristic impedance of 300 Ω (a typical impedance for two-wire lines)? How about 50 Ω? ◇

The characteristic impedance will be $Z_0 = (\eta/2\pi)\ln(b/a)$, where $\eta = \sqrt{\mu/\varepsilon}$ $= \eta_0/\sqrt{2.25} = 251\ \Omega$, so that $\eta/2\pi = 40\ \Omega$. For Z_0 to attain 300 Ω, we need the factor $\ln(b/a)$ to be $300/40 = 7.5$, which would require $b/a \approx 1800$, a rather impractical ratio of the diameters of the outer and inner conductors! To get a 50-Ω coaxial line, however, would require $\ln(b/a) = 1.25$ or $b/a = 3.49$, a readily attainable ratio. $Z_0 = 50\ \Omega$ is a standard characteristic impedance for coaxial lines.

COMPLEX POYNTING THEOREM

In the sinusoidal steady state, the Poynting vector will oscillate in time at double the operating frequency but will have a time-averaged value that is of primary interest. That average power flow value can be obtained directly in the frequency domain by taking half the real part of the cross product of the complex electric field amplitude with that of the magnetic field, one of them complex conjugated: $\frac{1}{2}\mathrm{Re}(\mathbf{E} \times \mathbf{H}^*)$. The complex vector $\frac{1}{2}\mathbf{E} \times \mathbf{H}^*$ is called the complex Poynting vector. Its real part gives us the time-averaged power flow; it turns out that the imaginary part of this vector is also highly informative.

To see this, we can rederive the Poynting theorem directly in the frequency domain, starting from Maxwell's equations in that regime:

$$\nabla \times \mathbf{E} = -j\omega\mathbf{B}, \qquad \nabla \times \mathbf{H} = \mathbf{J} + j\omega\mathbf{D}. \tag{8.39}$$

We first complex conjugate the second of these equations

$$\nabla \times \mathbf{H}^* = \mathbf{J}^* - j\omega\mathbf{D}^* \tag{8.40}$$

and substitute the two relations into the divergence of the complex Poynting vector:

$$\nabla \cdot \left(\tfrac{1}{2}\mathbf{E} \times \mathbf{H}^*\right) = \tfrac{1}{2}(\nabla \times \mathbf{E}) \cdot \mathbf{H}^* - \tfrac{1}{2}(\nabla \times \mathbf{H}^*) \cdot \mathbf{E}$$
$$= -j\omega\tfrac{1}{2}\mathbf{B} \cdot \mathbf{H}^* - \tfrac{1}{2}\mathbf{J}^* \cdot \mathbf{E} + j\omega\tfrac{1}{2}\mathbf{D}^* \cdot \mathbf{E}. \tag{8.41}$$

We integrate this equation over a volume V bounded by closed surface S; we also change the factor $1/2$ to $2(1/4)$ to help interpret the result:

$$-\oint_S \tfrac{1}{2}\mathbf{E} \times \mathbf{H}^* \cdot d\mathbf{S} = \int_V \tfrac{1}{2}\mathbf{E} \cdot \mathbf{J}^*\, dV + j2\omega \int_V \left[\tfrac{1}{4}\mathbf{H}^* \cdot \mathbf{B} - \tfrac{1}{4}\mathbf{E} \cdot \mathbf{D}^*\right] dV. \tag{8.42}$$

This is the complex Poynting theorem.

INTERPRETATION

We consider a medium with constitutive parameters $\varepsilon(\omega), \mu(\omega)$, and $\sigma(\omega)$ that are real at the operating frequency ω. In that case, the integrands on the right side are all real:

$$\mathbf{E} \cdot \mathbf{J}^* = \sigma|E|^2, \qquad \mathbf{H}^* \cdot \mathbf{B} = \mu|H|^2, \qquad \mathbf{E} \cdot \mathbf{D}^* = \varepsilon|E|^2. \qquad (8.43)$$

This makes it particularly easy to separate the real and imaginary parts of the equation. For the real part, we have

$$\mathrm{Re}\left\{-\oint_S \tfrac{1}{2}\mathbf{E} \times \mathbf{H}^* \cdot d\mathbf{S}\right\} = \int_V \tfrac{1}{2}\mathrm{Re}[\mathbf{E} \cdot \mathbf{J}^*]\,dV = \langle P \rangle, \qquad (8.44)$$

where we have recognized the integral on the right as the expression for the time-averaged power dissipation $\langle P \rangle$ within the volume V. For the imaginary part we get

$$\mathrm{Im}\left\{-\oint_S \tfrac{1}{2}\mathbf{E} \times \mathbf{H}^* \cdot d\mathbf{S}\right\} = 2\omega \int_V \tfrac{1}{2}\mathrm{Re}\left[\tfrac{1}{2}\mathbf{H}^* \cdot \mathbf{B} - \tfrac{1}{2}\mathbf{E} \cdot \mathbf{D}^*\right]dV$$

$$= 2\omega\left[\langle W_m \rangle - \langle W_e \rangle\right], \qquad (8.45)$$

where we have similarly acknowledged the appearance of the time-averaged densities of magnetic and electric energy in the integrands on the right. The terms we have identified are exactly analogous to the real and reactive power terms we encounter in circuits with steady-state excitation. We conclude that the integral of the complex Poynting vector over a closed surface is a complex number whose real and imaginary parts tell us about time-averaged power dissipation and energy storage within the enclosed volume:

$$-\oint_S \tfrac{1}{2}\mathbf{E} \times \mathbf{H}^* \cdot d\mathbf{S} = \langle P \rangle + j2\omega\left[\langle W_m \rangle - \langle W_e \rangle\right]. \qquad (8.46)$$

Note, however, that what appears on the right is the imbalance, the difference, between the averaged magnetic and electric energies stored in the volume, not the sum. Comparing with the Poynting theorem in the time domain,

$$-\oint_S \boldsymbol{\mathcal{E}} \times \boldsymbol{\mathcal{H}} \cdot d\mathbf{S} = P + d\left[W_m + W_e\right]/dt, \qquad (8.47)$$

we see that the complex version is not obtainable by merely averaging the time-domain version; the sum of the magnetic and electric energies appears in the original Poynting theorem while the difference shows up in the complex version. The complex Poynting theorem is a result related to, but separate from, the original theorem.

Example 8.6

Obtain the complex power distribution along a coaxial line with voltage source V_0 at $z = 0$ and shorted by a transverse conducting plate at $z = l$. ◈

The voltage and current distributions for this case of a shorted line (with $\Gamma_l = -1$) are given by

$$V(z) = V_1\left[e^{j\beta(l-z)} - e^{-j\beta(l-z)}\right] = V_0 \sin\beta(l-z)/\sin\beta l, \qquad (8.48)$$

$$Z_0 I(z) = V_1\left[e^{j\beta(l-z)} + e^{-j\beta(l-z)}\right] = -jV_0 \cos\beta(l-z)/\sin\beta l, \qquad (8.49)$$

where the unknown incident wave amplitude V_1 has been evaluated in terms of the applied voltage V_0. The phase constant β is ω/u_0. The fields in the dielectric and the complex Poynting vector are, in $a < \rho < b$,

$$\mathbf{E}(\rho, \varphi, z) = V(z)\hat{\boldsymbol{\rho}}/[\ln(b/a)\rho], \qquad \mathbf{H}(\rho, \varphi, z) = I(z)\hat{\boldsymbol{\varphi}}/[2\pi\rho],$$

$$\tfrac{1}{2}\mathbf{E} \times \mathbf{H}^* = \hat{\mathbf{z}}\tfrac{1}{2}V(z)I^*(z)/[2\pi \ln(b/a)\rho^2]. \tag{8.50}$$

If we integrate this power flow density across a transverse plane A at z, what we get is the same as if we performed a closed surface integral of the complex Poynting vector over the annular cylindrical region between the inner and outer conductors and from location z to the short at $z = l$. This is because the complex Poynting vector's normal component is zero on the surfaces of the two conductors and of the shorting plate. As a result, we have

$$-\oint_S \tfrac{1}{2}\mathbf{E} \times \mathbf{H}^* \cdot d\mathbf{S} = \int\int_A \tfrac{1}{2}\mathbf{E} \times \mathbf{H}^* \cdot \hat{\mathbf{z}}\, \rho\, d\rho\, d\varphi$$

$$= \tfrac{1}{2}V(z)I^*(z) \int\int_A \left[2\pi \ln(b/a)\rho^2\right]^{-1}\hat{\mathbf{z}} \cdot \hat{\mathbf{z}}\, \rho\, d\rho\, d\varphi$$

$$= \tfrac{1}{2}V(z)I^*(z) \tag{8.51}$$

for the complex power flow into the segment of coaxial line from z to l. In detail, this is

$$\tfrac{1}{2}V(z)I^*(z) = j\big[|V_0|^2/4Z_0\big] \sin 2\beta\,(l - z)/\sin^2 \beta l. \tag{8.52}$$

This is how the total complex power is distributed along the line.

We note that the complex power is purely imaginary, indicating that there is no time-averaged power dissipation in the segment of coaxial line; this is consistent with our assumption of perfect conductors and a perfect dielectric. We note next that the difference between the total, time-averaged magnetic and electric energies stored within the segment from z to l varies as $\sin 2\beta(l - z)$. Near the short, the imbalance favors the magnetic energy, but from a quarter wavelength to a half wavelength from the short, it is the electric energy that dominates the magnetic one. This alternation persists all the way to the source.

We can use the complex Poynting theorem to trace the flow of power when a plane wave encounters a mismatched medium, as in the next example.

Example 8.7

A plane wave $\mathbf{E}_0 e^{-j\beta_1 z}$ is normally incident onto the interface plane $z = 0$ between a medium of impedance η_1 in front and one with impedance η_2 behind the plane.

(a) *Obtain the power transmission coefficient T_p that relates the power density that is transmitted to infinity in medium 2 to the incident power density.*

(b) *Confirm that energy is conserved in the process of reflection and transmission of waves, by calculating the complex power that crosses a closed surface that straddles the interface plane, namely, a cylinder of cross sectional area A from $z = z_1 < 0$ to $z = z_2 > 0$; see Figure 8-6.* ◈

(a) Since $E_2 = T_e E_0$ and $H_2 = T_h H_0$, we get at once

$$\tfrac{1}{2}\mathbf{E}_2 \times \mathbf{H}_2^* = T_e T_h^* \tfrac{1}{2}\mathbf{E}_0 \times \mathbf{H}_0^*, \tag{8.53}$$

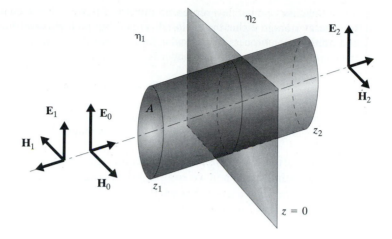

Figure 8-6   Power conservation in plane waves scattered by interface.

which exhibits the power transmission coefficient $T_p = T_e T_h^*$. If the impedance on the far side is real, this reduces to

$$T_p = T_e T_h^* = 4\eta_1\eta_2/(\eta_1 + \eta_2)^2. \qquad (8.54)$$

(b) The quantity to be integrated over the cylinder is $\frac{1}{2}\mathbf{E} \times \mathbf{H}^*$. Along the side of the cylinder, there is no component of this vector along $d\mathbf{S}$. At z_2, the integrand is just

$$\tfrac{1}{2}\mathbf{E}_2 e^{-j\beta_2 z_2} \times \mathbf{H}_2^* e^{j\beta_2 z_2} = T_e T_h^* \tfrac{1}{2}\mathbf{E}_0 \times \mathbf{H}_0^* \qquad (8.55)$$

and $d\mathbf{S}$ is along $\hat{\mathbf{z}}$. At z_1, it is

$$\tfrac{1}{2}\mathbf{E}_0\big[e^{-j\beta_1 z_1} + \Gamma e^{j\beta_1 z_1}\big] \times \mathbf{H}_0^*\big[e^{j\beta_1 z_1} - \Gamma^* e^{-j\beta_1 z_1}\big]$$

$$= \tfrac{1}{2}\mathbf{E}_0 \times \mathbf{H}_0^*\big[1 + \Gamma e^{j2\beta_1 z_1} - \Gamma^* e^{-j2\beta_1 z_1} - |\Gamma|^2\big] \qquad (8.56)$$

and $d\mathbf{S}$ is along $-\hat{\mathbf{z}}$; here, Γ refers to Γ_e. The total complex power across the cylindrical surface is

$$\oint \tfrac{1}{2}\mathbf{E} \times \mathbf{H}^* \cdot d\mathbf{S} = \tfrac{1}{2}\mathbf{E}_0 \times \mathbf{H}_0^* \cdot \hat{\mathbf{z}} AF, \qquad (8.57)$$

where the factor F is, upon expressing T in terms of Γ,

$$F = T_e T_h^* - \big[1 + \Gamma e^{j2\beta_1 z_1} - \Gamma^* e^{-j2\beta_1 z_1} - |\Gamma|^2\big]$$

$$= (1 + \Gamma)(1 - \Gamma^*) - \big[1 + \Gamma e^{j2\beta_1 z_1} - \Gamma^* e^{-j2\beta_1 z_1} - |\Gamma|^2\big]$$

$$= \big(1 - e^{j2\beta_1 z_1}\big)\Gamma - \big(1 - e^{-j2\beta_1 z_1}\big)\Gamma^*. \qquad (8.58)$$

Being the difference between one quantity and its own complex conjugate, this factor is clearly purely imaginary. Hence, assuming the incident wave carries real power, the entire complex power outflow out of the cylinder is only imaginary. There is no net real power across this surface, so that all the power entering at z_1 emerges at z_2, which expresses conservation of power.

Imaginary power flow represents only the difference between magnetic and electric energy storage within the enclosed volume, but no real power loss in the reflection and transmission process. In particular, for real impedances η_1 and η_2, Γ is also real and the complex power across the cylinder is

$$\oint \tfrac{1}{2}\mathbf{E} \times \mathbf{H}^* \cdot d\mathbf{S} = \tfrac{1}{2}\mathbf{E}_0 \times \mathbf{H}_0^* \cdot \hat{z} A\left[-2j \sin 2\beta_1 z_1\right]\Gamma. \tag{8.59}$$

This is independent of z_2, at which there is only the propagating transmitted wave, but does depend on z_1, where there is a standing wave that stores electromagnetic energy.

A subtlety arises when we examine and interpret conservation of power in the case of oblique incidence. When we compare the powers in the scattered waves to the incident wave power, we expect that whatever power is not transmitted must be reflected, the two adding up to the incident power, for lossless media. This is correct, but we must be aware that it is not the power density, or magnitude of the Poynting vector, that satisfies the conservation principle; rather, it is the integral of the Poynting vector over a closed surface that maintains the balance between the power that enters the enclosed volume and the power that leaves it. This integral involves only the component of the Poynting vector normal to the surface of integration. When we examine incident and scattered plane waves at an interface plane, we find that the relevant component of the Poynting vector is only the one normal to the interface. This is borne out in the next example.

Example 8.8

Oblique incidence from air onto a lossless, nonmagnetic dielectric at a certain incidence angle is observed to result in 80% power transmission when TE polarization is used and in 100% power transmission when TM polarization is tried. What are the index of refraction n of the dielectric and the angle of incidence θ? ◇

Since there is perfect transmission when TM polarization is used, the incidence must be at Brewster's angle, so that $\tan \theta_1 = n = \cot \theta_2$, recalling that θ_1 and θ_2 are then complementary angles.

For the TE wave, we can avoid the slight complication of dealing with the normal component of the power density if we use the reflected power fraction, instead of the given transmission factor, and invoke conservation of power flow. The simplification comes from the equality of the angles of reflection and of incidence, so that the $\cos \theta_1$ factor is the same for the incident and the reflected waves and cancels from the power ratio. Thus, we use 20% as the reflected power fraction, instead of 80% transmission, since power must be conserved.

Accordingly, we have $|\Gamma_e|^2 = 20\% = 0.2$. But $\Gamma_e = (Z_2 - Z_1)/(Z_2 + Z_1)$ and $Z = \eta \sec \theta$, with $\eta_2/\eta_1 = 1/n$. We also have that $\sec \theta_2/\sec \theta_1 = 1/n$ (from $\tan \theta_1 = n = \cot \theta_2$), so that $Z_2/Z_1 = 1/n^2$ and $|\Gamma_e| = (n^2 - 1)/(n^2 + 1)$. This inverts to

$$n^2 = (1 + |\Gamma_e|)/(1 - |\Gamma_e|) = (1 + \sqrt{0.2})/(1 - \sqrt{0.2})$$

or $n = 1.618$ and $\theta_1 = 58.3°$, to answer the question.

If we prefer to use the 80% power transmission fraction directly, then we must recognize the $\cos \theta_1$ and $\cos \theta_2$ factors in the incident and refracted wave power flow expressions, in order to utilize only the normal component of the Poynting vector. Then $(|E_2|^2/\eta_2)/(|E_0|^2/\eta_1) = (\eta_1/\eta_2)|T_e|^2$ is *not* the ratio that equals $80\% = 0.8$; rather, it is

$$(|E_2|^2 \cos\theta_2/\eta_2)/(|E_0|^2 \cos\theta_1/\eta_1) = (\eta_1 \sec\theta_1/\eta_2 \sec\theta_2)|T_e|^2$$

or $(Z_1/Z_2)|T_e|^2$ that is the 80% fraction. But $T_e = 2Z_2/(Z_2 + Z_1)$ gives $4Z_1 Z_2/(Z_2 + Z_1)^2 = 0.8$. Using $Z_1/Z_2 = n^2$, we get $4n^2/(n^2 + 1)^2 = 0.8$ or $2n/(n^2 + 1) = \sqrt{0.8}$; this quadratic equation is solved by $n = 1.618$ (as well as 0.618, but that would be less than unity), giving the same answers as before.

GENERAL TRANSMISSION LINES

Many structures behave electrically as transmission lines. We can develop transmission line equations that describe the operation of a wide variety of configurations; we will attempt to be as general as possible.

For a given structure, we need first to identify the following items.

a. A direction of propagation, say, z, as against the transverse directions.
b. A medium for the **E** and **H** fields; we characterize the medium by its constitutive parameters ε, μ, σ.

The equations will relate the following variables.

c. A transverse voltage: $v(z, t)$, obtainable from **E** by integration.
d. A longitudinal current: $i(z, t)$, obtainable from **H** by integration.

We consider a segment of the structure along the longitudinal direction, with one end at a variable position z, say, from z to l. Longitudinal current $i(z, t)$ enters the segment at location z; current $i(l, t)$ exits the segment at $z = l$. The difference between these two values can be accounted for in two ways. Current may flow transversely across the structure all along the segment, in the amount of I_{shunt}, and charge may accumulate along the segment, amounting to a total charge on the segment Q. We then have, as in the schematic Figure 8-7:

e. The continuity equation:

$$i(z, t) - i(l, t) = I_{shunt} + \partial Q/\partial t. \tag{8.60}$$

The figure is deliberately uninformative as to the nature of the structure; by remaining schematic in form, the figure is intended to suggest that it applies to any structure at all.

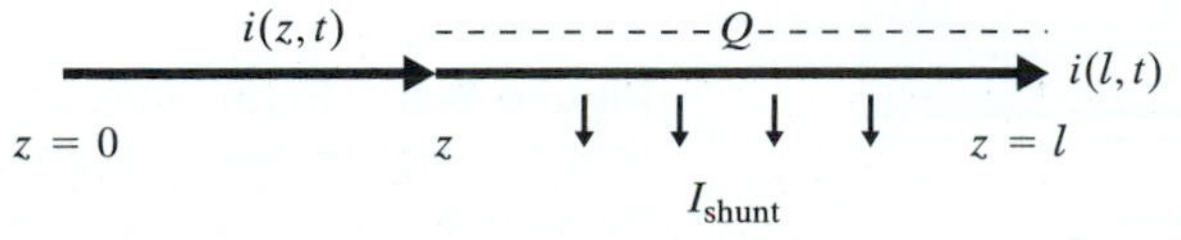

Figure 8-7 Schematic view of current imbalance on segment of structure.

The transverse voltage is $v(z, t)$ at the beginning of the segment and it is $v(l, t)$ at its end. The difference between these two values can be accounted for in two ways. There may be emf in the longitudinal direction, amounting to V_{series}, which complements the two transverse voltages at the ends to form an emf around a complete closed circuit, and there may be an accumulation of magnetic flux across a surface bounded by that closed curve, amounting to a total flux Φ. We then have, as in the schematic Figure 8-8,

f. the Faraday-Maxwell law, observing the right-hand rule,

$$v(z, t) - v(l, t) = V_{series} + \partial\Phi/\partial t. \tag{8.61}$$

Again, the figure is not meant to suggest any particular structure; the concept applies to any structure.

There remains to relate I_{shunt} and Q to the transverse voltage distribution $v(z, t)$, and also V_{series} and Φ to the longitudinal current distribution within the segment. These relationship are obtained through the field patterns, which need to be assessed in order to complete the calculation.

g. The total charge on the segment from z to l is related to the field pattern by Gauss's law:

$$Q = \int_z^l \int \mathbf{D} \cdot d\mathbf{S}. \tag{8.62}$$

We cannot be specific about the integration, because we have no specific structure in mind; nevertheless, since the segment extends from z to l, part of the surface of integration will surely involve our longitudinal coordinate z, over that range. The surface element will be directed transverse to the segment, for the same reason.

But $\mathbf{D} = \varepsilon\mathbf{E}$, where the $\mathbf{E}$ field needed is the transverse one. Since this $\mathbf{E}$ field integrates to the transverse voltage,

$$v(z, t) = \int \mathbf{E} \cdot d\mathbf{l}, \tag{8.63}$$

we should be able to relate Q and v to each other, although the result for Q will still entail the integration from z to l, which we cannot complete at this stage, as it still awaits information about how the voltage varies along the segment:

$$Q = \int_z^l C v(z', t) \, dz'. \tag{8.64}$$

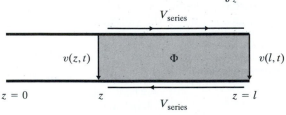

Figure 8-8   Schematic view of voltage imbalance on segment of structure.

This is the form we should expect for any structure capable of accumulating charge Q along a longitudinal segment of the structure, in response to a longitudinal distribution of transverse voltage $v(z, t)$. Note that we have designated the integration variable z' rather than z, to distinguish it from the variable lower limit of integration z. The quantity C is the capacitance per unit length of the structure. We have been careful to keep this factor inside the integral, as it need not be constant; it may well be a function of the longitudinal variable z' and perhaps of time too.

h. The total shunt current flowing transversely along the segment from z to l is related to the field pattern by Ohm's law. We have

$$I_{shunt} = \int_z^l \int \mathbf{J} \cdot d\mathbf{S}. \tag{8.65}$$

Again, we cannot be specific about the integration, because we have not been specific about the structure. However, part of the surface integration along the segment will surely involve our longitudinal coordinate, from z to l.

But $\mathbf{J} = \sigma\mathbf{E}$, where the $\mathbf{E}$ field needed is the transverse one. Since this $\mathbf{E}$ field integrates to the transverse voltage,

$$v(z, t) = \int \mathbf{E} \cdot d\mathbf{l}, \tag{8.66}$$

we should be able to relate I_{shunt} and v to each other. Again, the result for the shunt current will entail the integration from z to l, which still awaits information about how the voltage varies along the segment:

$$I_{shunt} = \int_z^l Gv(z', t)\, dz'. \tag{8.67}$$

This is the form we should expect for any structure capable of leaking a shunt current in a longitudinal segment of the structure, in response to a longitudinal distribution of transverse voltage $v(z, t)$. The quantity G is the conductance per unit length of the structure. In case G depends on z, it needs to be kept inside the integral.

i. The total magnetic flux across the area encompassed by the schematic closed curve illustrated in Figure 8-8 and extending along the segment from z to l is the integral of the $\mathbf{B}$ field:

$$\Phi = \int_z^l \int \mathbf{B} \cdot d\mathbf{S}. \tag{8.68}$$

The element of surface $d\mathbf{S}$ will be directed transversely, so the transverse components of $\mathbf{B}$ are the ones that enter this integration.

But $\mathbf{B} = \mu\mathbf{H}$, where the $\mathbf{H}$ field needed is the transverse one. Since this $\mathbf{H}$ field integrates along some closed curve to give the longitudinal current,

$$i(z, t) = \int \mathbf{H} \cdot d\mathbf{l}, \tag{8.69}$$

we should be able to relate Φ and i to each other. The result for the magnetic flux will entail the integration from z to l, which we cannot complete as yet, as it awaits information about how the longitudinal current varies along the segment:

$$\Phi = \int_z^l Li(z', t) \, dz' \tag{8.70}$$

This is the form we should expect for any structure that allows generation of a magnetic flux around a longitudinal segment of the structure, in response to a longitudinal distribution of longitudinal current $i(z, t)$. The quantity L is the inductance per unit length of the structure. If it depends on z, it must be kept inside the integral.

   j. The longitudinal emf along the axial portion of the schematic closed curve in Figure 8-8 is the line integral of the longitudinal **E** field:

$$V_{\text{series}} = \int_z^l E_z \hat{\mathbf{z}} \cdot d\mathbf{l}. \tag{8.71}$$

There may be both a forward and a return portion of the path for this line integral, but it will surely extend from z to l.

But Ohm's law will relate the longitudinal component of **E** to the longitudinal component of the current density as $J_z = \sigma_m E_z$, where σ_m refers to the conductivity of the part of the structure that carries longitudinal current, typically some metal part of the structure. This σ_m is to be distinguished from the conductivity σ we invoked to relate the transverse leakage current to the transverse electric field, typically in an imperfect dielectric. Since this J_z field integrates across some transverse area to give the longitudinal current,

$$i(z, t) = \int J_z \, dA, \tag{8.72}$$

we should be able to relate V_{series} and i to each other. The result for the longitudinal emf will entail the integration from z to l, which still awaits information about how the longitudinal current varies along the segment:

$$V_{\text{series}} = \int_z^l Ri(z', t) \, dz'. \tag{8.73}$$

This is the form to be expected for any structure that carries a longitudinal current along an imperfectly conducting segment of the structure. The quantity R is the resistance per unit length of the structure. If it depends on z, it must be kept inside the integral.

Now that we have related the quantities $Q, I_{\text{shunt}}, \Phi$, and V_{series} to the primary unknown distributions $v(z, t)$ and $i(z, t)$, we can substitute the integrals along the segment that typically represent these quantities into

$$v(z, t) - v(l, t) = V_{\text{series}} + \partial \Phi / \partial t, \tag{8.74}$$

$$i(z, t) - i(l, t) = I_{\text{shunt}} + \partial Q / \partial t \tag{8.75}$$

to arrive at the following pair of transmission line equations in integral form, applicable with great generality by using appropriate parameters R, L, C, G.

$$v(z, t) = v(l, t) + \int_z^l Ri(z', t)\, dz' + \partial/\partial t \int_z^l Li(z', t)\, dz', \tag{8.76}$$

$$i(z, t) = i(l, t) + \int_z^l Gv(z', t)\, dz' + \partial/\partial t \int_z^l Cv(z', t)\, dz'. \tag{8.77}$$

The time derivatives are partial because the integrals to which they are applied are functions not only of time t but also of longitudinal position z in the lower limit. The coefficients R, L, G, C may be functions of t, if the structure changes in time, and of z, if the structure is not uniform along the longitudinal direction; in the latter case, the parameters must remain inside the integrals.

If the structure is time independent, the time derivatives may be taken inside the integrals and past the coefficients R, L, G, C. We can then differentiate the equations with respect to the variable lower limit z, thereby obtaining the corresponding pair of transmission line equations in differential form,

$$\partial v(z, t)/\partial z = -Ri(z, t) - L\, \partial i(z, t)/\partial t, \tag{8.78}$$

$$\partial i(z, t)/\partial z = -Gv(z, t) - C\, \partial v(z, t)/\partial t. \tag{8.79}$$

These can still apply to a nonuniform structure, for which the parameters R, L, G, C will be functions of z. The negative signs arise from differentiation of the integrals with respect to their lower, not upper, limit.

In the harmonic steady state, with $v(z, t)$ written as $\mathrm{Re}\left[V(z)e^{j\omega t}\right]$ and similarly for all the other quantities involved, and provided the structure is independent of time, the transmission line integral equations become relations among the complex amplitudes of the voltage and current distributions:

$$V(z) = V(l) + \int_z^l (R + j\omega L)I(z')\, dz', \tag{8.80}$$

$$I(z) = I(l) + \int_z^l (G + j\omega C)V(z')\, dz', \tag{8.81}$$

or, more simply,

$$V(z) = V(l) + \int_z^l ZI(z')\, dz', \tag{8.82}$$

$$I(z) = I(l) + \int_z^l YV(z')\, dz', \tag{8.83}$$

where Z is the series impedance per unit length and Y is the shunt admittance per unit length. Typically,

$$Z = R + j\omega L, \qquad Y = G + j\omega C, \tag{8.84}$$

although these could get more complicated, or else simpler, in certain cases. The differential form of these steady-state integral equations is obtainable by differentiating with respect to z, which is the lower limit of the integrals:

$$dV(z)/dz = -ZI(z), \qquad dI(z)/dz = -YV(z). \tag{8.85}$$

Several warnings are in order here, before we look at examples. First, the complex coefficients Z and Y in these equations are separate, unrelated parameters; they are not reciprocals of each other. Second, Z and Y may be functions of z, whereupon they describe a *nonuniform* transmission line; when they are constants, the transmission line is *uniform*. Third, the parameter Z, the series impedance per unit length, is not to be confused with the characteristic impedance of the transmission line, which we denote Z_0.

Within this framework for describing wave-transmitting structures, one transmission line differs from another in having different parameters R, L, G, C or Z, Y. Let us examine some examples of how we can obtain these coefficients; afterward, we will explore solutions to the general transmission line equations.

Example 8.9

Obtain the transmission line parameters for a coaxial line constructed of imperfect conductors and an imperfect insulator, in the steady state. ◇◇

The direction of propagation is clearly that of the common axis, z. We must identify a field pattern that we can properly expect to achieve in the structure. With charge distributed on the surfaces of the conductors, we expect a radial electric field E_ρ. With current flowing along the center conductor, we expect an azimuthal magnetic field H_φ. If the structure were ideal, these would be the only fields present; for the imperfect version of the coaxial line, these are merely the predominant fields. Since the center wire is imperfectly conducting, the current density in it will be accompanied by a small but nonzero axial electric field E_z, so that the field pattern is only approximately transverse. For a reasonably designed coaxial line, despite the imperfect conductivity of the metal parts, the fields will be nearly as in the ideal case and we will use the ideal fields until forced to deal with the small axial electric field.

For the segment of coaxial line from z to l, we apply Gauss's law to the surface of a cylinder coaxial with the two conductors but of some radius ρ between them. We should get the total charge on the inner conductor's segment from z to l no matter what radius ρ we choose, between a and b. Refer to Figure 8-9; the charge on the segment is

$$Q(z) = \int_z^l \int_0^{2\pi} \varepsilon E_\rho(\rho, z') \rho \, d\varphi \, dz' = 2\pi\varepsilon\rho \int_z^l E_\rho(\rho, z') \, dz'. \tag{8.86}$$

However, the transverse voltage $V(z)$ is

$$V(z) = \int_a^b E_\rho(\rho, z) \, d\rho \tag{8.87}$$

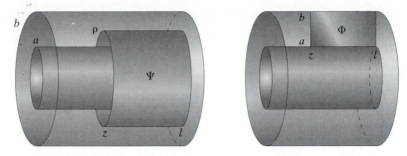

Figure 8-9 Electric and magnetic flux surfaces for coaxial line.

without the factor ρ that appears on the right of the previous equation. We therefore divide $Q(z)$ by ρ and then integrate from $\rho = a$ to $\rho = b$, in order to force $V(z)$ to make its appearance in the integral:

$$\int_a^b Q(z)\, d\rho / \rho = Q(z) \ln(b/a)$$

$$= 2\pi\varepsilon \int_z^l \int_a^b E_\rho(\rho, z')\, d\rho\, dz' = 2\pi\varepsilon \int_z^l V(z')\, dz' \qquad (8.88)$$

and we conclude that the charge on the segment is related to the transverse voltage by

$$Q(z) = \int_z^l [2\pi\varepsilon/\ln(b/a)] V(z')\, dz'. \qquad (8.89)$$

We therefore simply read off the capacitance per unit length

$$C = 2\pi\varepsilon/\ln(b/a). \qquad (8.90)$$

This is a constant, independent of z, and can be placed either inside or outside the integral.

Similarly, the radial electric field in the imperfect dielectric is accompanied by a small radial current density $J_\rho = \sigma E_\rho$ and the resultant flow is a shunt current from the inner to the outer conductor, through the dielectric. Integrating this radial current density over the same coaxial surface at radius ρ that we just used for the electric flux calculation, we get

$$I_{\text{shunt}}(z) = \int_z^l \int_0^{2\pi} \sigma E_\rho(\rho, z')\, \rho\, d\varphi\, dz'$$

$$= 2\pi\sigma\rho \int_z^l E_\rho(\rho, z')\, dz' \qquad (8.91)$$

and the remainder of the calculation is identical to the previous one, giving

$$I_{\text{shunt}} = \int_z^l [2\pi\sigma/\ln(b/a)] V(z')\, dz'. \qquad (8.92)$$

We read the conductance per unit length as the constant

$$G = 2\pi\sigma/\ln(b/a). \qquad (8.93)$$

Referring again to Figure 8-9, we calculate the magnetic flux over the rectangle from z to l as

$$\Phi(z) = \int_z^l \int_a^b \mu H_\varphi(\rho, z') \, d\rho \, dz', \tag{8.94}$$

but the magnetic field is related to the current in the inner conductor as

$$I(z) = \int_0^{2\pi} H_\varphi(\rho, z) \, \rho \, d\varphi = 2\pi\rho H_\varphi(\rho, z). \tag{8.95}$$

We therefore have

$$\Phi(z) = \int_z^l \int_a^b \mu \, \frac{I(z')}{2\pi\rho} \, d\rho \, dz'$$

$$= \int_z^l [(\mu/2\pi) \ln(b/a)] I(z') \, dz' \tag{8.96}$$

and we read the inductance per unit length to be

$$L = (\mu/2\pi) \ln(b/a). \tag{8.97}$$

Finally, we want to relate the series voltage to the current distribution. The emf involved is the integral of the axial electric field along the surface of the two conductors, from z to l on the inner conductor and from l to z for the return current on the outer one. We have, in terms of the axial field $E_z(\rho, z)$,

$$V_{\text{series}} = \int_z^l E_z(a, z') \, dz' - \int_z^l E_z(b, z') \, dz'. \tag{8.98}$$

We know that

$$E_z(a, z) = (1/\sigma_m) J_z(a, z), \qquad E_z(b, z) = (1/\sigma_m) J_z(b, z), \tag{8.99}$$

where σ_m is the conductivity of the two conductors, but this alone does not help us to relate the axial electric field to the longitudinal current

$$I(z) = \int_{A_1} J_z(\rho, z) \, dA = -\int_{A_2} J_z(\rho, z) \, dA. \tag{8.100}$$

Here, A_1 and A_2 are the cross-sectional areas of the inner and outer conductors, respectively. But now we are stuck.

The difficulty is that we do not know the transverse distribution of the current in the two conductors. We do not yet have any ready means for determining this distribution across the conductors; we will look into this in the next example. Meanwhile, we are in a good position to make a sensible approximation, to advance the process of evaluating the transmission line parameters. We declare that the axial electric field is expected to be weak, if the wire and shell are reasonable conductors. A rough approximation to the distribution of the current across the wire and shell should therefore suffice to give us a good answer regarding the operation of the coaxial line.

We therefore adopt a *uniform* distribution of the current in the wire and in the coaxial shell, which is reasonable at low frequencies, at least. Then

$$J_z(r, z) \approx J_z(z) = I(z)/A_1 \qquad \text{(in the wire)}, \tag{8.101}$$

$$J_z(r, z) \approx J_z(z) = -I(z)/A_2 \qquad \text{(in the shell)}. \tag{8.102}$$

With this approximation, the series emf appears as

$$V_{\text{series}} = \int_z^l \left[I(z')/\sigma_m A_1 \right] dz' - \int_z^l \left[-I(z')/\sigma_m A_2 \right] dz'$$

$$= \int_z^l \left[1/\sigma_m A_1 + 1/\sigma_m A_2 \right] I(z') \, dz'. \tag{8.103}$$

We can now read off the resistance per unit length of the line, as

$$R = 1/\sigma_m A_1 + 1/\sigma_m A_2 = 1/\sigma_m A_{\text{eq}}, \tag{8.104}$$

where we have absorbed our ignorance about the details of the cross-sectional distribution of the current by combining the reciprocals of the areas of the inner and outer conductors to define an equivalent area A_{eq}.

We now have all the relevant parameters describing, at least approximately, the coaxial line with imperfect conductors and dielectric. We collect these into the pair of coefficients

$$Z = R + j\omega L = 1/\sigma_m A_{\text{eq}} + j\omega(\mu/2\pi)\ln(b/a), \tag{8.105}$$

$$Y = G + j\omega C = (\sigma + j\omega\varepsilon)\left[2\pi/\ln(b/a) \right]. \tag{8.106}$$

In this case, the impedance and admittance coefficients are constants; the coaxial line is a uniform transmission line.

As a side light, we note that the quantity $\sigma + j\omega\varepsilon$, which shows up frequently in connection with imperfect dielectrics, is often referred to as the complex conductivity of the material, at the operating frequency ω. Alternatively, the quantity $\varepsilon - j\sigma/\omega$ is called the complex permittivity. The ratio $\sigma/\omega\varepsilon$ is known as the *loss tangent* of the material (it is the tangent of the polar angle of the complex permittivity, as plotted in the complex plane).

We have previously analyzed the current distribution inside a cylindrical wire quasistatically. We have also been somewhat vague in approximating this distribution in order to arrive expeditiously at the resistance parameter R associated with the inner wire and outer shell of the imperfect coaxial line. For a more rigorous approach to the radial distribution of the fields inside the wire, we must treat it as a wave-propagating structure, a radial transmission line.

Example 8.10

Obtain transmission line equations describing the radial variation of the electric and magnetic fields and the current density inside a cylindrical conducting wire that carries total current I_0. ◈

Let the wire have radius a and constitutive parameters ε, μ, and σ_m. The current density distribution $J_z(\rho)$ is accompanied by axial electric field $E_z(\rho)$ and generates an azimuthal magnetic field $H_\varphi(\rho)$. We are interested at this stage in only the fields internal to the wire, and we are not considering any axial variation. The wire is azimuthally symmetric, and we assume the source maintains this symmetry. We will consider a finite length l of the wire.

The given quantity is the total current I_0 carried by the wire, and we want to describe how this current is distributed radially across the wire. We know that the radial distribution will not be uniform in the time-varying case, because the associated time-varying magnetic flux will affect the radial distribution of the electric field and hence of the current density.

We choose the radial direction as the direction of propagation, with coordinate ρ. A pair of convenient voltage and current variables is the emf along the length l of the wire, $V(\rho) = l\,E_z(\rho)$, and the current $I(\rho)$ carried within the wire from the axis out to radius ρ. Accordingly, the coordinate ρ will play the role of z in our previous analysis of transmission lines and the radius a will be the "end of the line" as was l in the previous cases. Refer to Figure 8-10.

We can write the current up to radius ρ as

$$I(\rho) = \int_0^\rho \int_0^{2\pi} J_z(\rho')\rho'\,d\varphi\,d\rho' = \int_0^\rho 2\pi\sigma_m E_z(\rho')\rho'\,d\rho'$$

$$= \int_0^\rho [2\pi\sigma_m/l]V(\rho')\rho'\,d\rho' = \int_0^\rho G(\rho')V(\rho')\,d\rho'. \tag{8.107}$$

Note that we are careful to distinguish the integration variable ρ' from the integration limit ρ. The equation gives us the conductance per unit length

$$G(\rho) = 2\pi\sigma_m\rho/l. \tag{8.108}$$

The magnetic field at radius ρ is given in terms of the current enclosed up to radius ρ as

$$H_\varphi(\rho) = I(\rho)/2\pi\rho, \tag{8.109}$$

by Ampère's law. We apply the Faraday-Maxwell law to the rectangular area of length l and of width extending from ρ to a to get the radial variation of the axial voltage $V(\rho)$:

$$V(\rho) - V(a) = -j\omega \int_0^l \int_\rho^a \mu H_\varphi(\rho')\,d\rho'\,dz$$

$$= -j\omega \int_\rho^a [\mu l/2\pi\rho']I(\rho')\,d\rho' \tag{8.110}$$

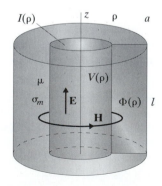

Figure 8-10 Conducting wire as a radial transmission line.

This allows us to read off the inductance per unit length

$$L(\rho) = \mu l / 2\pi\rho \qquad (8.111)$$

or the impedance per unit length

$$Z(\rho) = j\omega\mu l / 2\pi\rho. \qquad (8.112)$$

Note that both the conductance and the inductance parameters are functions of the co-ordinate along the direction of propagation. The admittance and impedance parameters are functions of ρ; the radial transmission line is nonuniform.

The nonuniform transmission line equations can be rewritten in the more standard form by expressing the current $I(\rho)$ in terms of the condition at the "load" at $\rho = a$, as

$$I(\rho) = I(a) - \int_{\rho}^{a} Y(\rho')V(\rho') \, d\rho', \qquad (8.113)$$

$$V(\rho) = V(a) - \int_{\rho}^{a} Z(\rho')I(\rho') \, d\rho', \qquad (8.114)$$

with

$$Y(\rho) = 2\pi\sigma_m\rho/l \qquad \text{and} \qquad Z(\rho) = j\omega\mu l / 2\pi\rho. \qquad (8.115)$$

The boundary conditions are that $I(a) = I_0$, the full current in the wire, and that there can be no current up to zero radius: $I(0) = 0$. Note that what remains less than standard in regard to the form of the transmission line equations in this case are the minus signs in front of the integrals. These are of no great consequence, although we must be sure to retain them. They reflect our choice of the $I(\rho)$ variable; had we chosen as our current variable the current between radii ρ and a instead—that is, $I(a) - I(\rho)$ in our present notation—or even just $-I(\rho)$, we would have had the usual positive signs.

The results take the form of nonuniform transmission line equations, in which the impedance and admittance coefficients are functions of the coordinate in the direction of propagation.

Once we solve these nonuniform transmission line equations, we can use the resulting distributions to determine the proper relation between series voltage and longitudinal current in the center conductor of the coaxial line, for example.

SOLUTION TO GENERAL TRANSMISSION LINE EQUATIONS

The general transmission line equations may be solved by iteration of the integral equations

$$V(z) = V(l) + \int_{z}^{l} Z(z')I(z') \, dz', \qquad I(z) = I(l) + \int_{z}^{l} Y(z')V(z') \, dz', \quad (8.116)$$

just as we did for our quasistatic analysis, or by attacking the corresponding differential equations

$$dV(z)/dz = -Z(z)I(z), \qquad dI(z)/dz = -Y(z)V(z), \qquad (8.117)$$

perhaps by substituting power series as trial functions.

If the transmission line is uniform—that is, if it has the same structure for all distances z—then Z and Y are independent of z and the equations have constant coefficients. These are easily solved by differentiation:

$$dV(z)/dz = -ZI(z), \qquad dI(z)/dz = -YV(z) \tag{8.118}$$

become, on differentiating one of them,

$$d^2V(z)/dz^2 = -Z \, dI(z)/dz = ZYV(z). \tag{8.119}$$

This has the form of

$$d^2V(z)/dz^2 = \gamma^2V(z), \tag{8.120}$$

where the constant in this constant-coefficient, second-order ordinary differential equation is

$$\gamma^2 = ZY. \tag{8.121}$$

Typically, this constant will be

$$\gamma^2 = (R + j\omega L)(G + j\omega C), \tag{8.122}$$

which is a complex number.

The second-order equation is solved by exponentials:

$$V(z) = V_1 e^{-\gamma z} + V_2 e^{\gamma z}, \tag{8.123}$$

where V_1 and V_2 are constants to be evaluated from the boundary conditions. From $dV/dz = -ZI$ we also get, after dividing by $-\gamma$,

$$(Z/\gamma)I(z) = V_1 e^{-\gamma z} - V_2 e^{\gamma z}. \tag{8.124}$$

We provide a symbol for the factor on the left, $Z/\gamma = Z/\sqrt{ZY}$, or

$$Z_0 = \sqrt{Z/Y}, \tag{8.125}$$

and name it the characteristic impedance of the transmission line; we then have

$$V(z) = V_1 e^{-\gamma z} + V_2 e^{\gamma z}, \tag{8.126}$$

$$Z_0 I(z) = V_1 e^{-\gamma z} - V_2 e^{\gamma z}. \tag{8.127}$$

Evidently, the voltage is the sum of two exponentials and the characteristic impedance times the current is the difference of the same two exponentials. Note that both γ and Z_0 are complex constants.

INTERPRETATION

The complex constant

$$\gamma = \sqrt{ZY}, \tag{8.128}$$

which typically appears as

$$\gamma = \sqrt{(R + j\omega L)(G + j\omega C)}, \tag{8.129}$$

is called the *propagation constant* of the line. It can be separated into real and imaginary parts, as

$$\gamma = \alpha + j\beta. \tag{8.130}$$

The exponential $e^{-\gamma z}$ factors as

$$e^{-\gamma z} = e^{-(\alpha + j\beta)z} = e^{-\alpha z}e^{-j\beta z} \tag{8.131}$$

and the term $V_1 e^{-\gamma z} = |V_1|e^{j\varphi_1}e^{-\gamma z}$ can be interpreted in the time domain by evaluating

$$\text{Re}\left[V_1 e^{-\gamma z}e^{j\omega t}\right] = \text{Re}\left[|V_1|e^{j\varphi_1}e^{-\alpha z}e^{j(\omega t - \beta z)}\right]$$

$$= |V_1|e^{-\alpha z}\cos(\omega t - \beta z + \varphi_1). \tag{8.132}$$

This is a sinusoidal traveling wave, $\cos(\omega t - \beta z + \varphi_1)$, that propagates along the z-axis at speed ω/β, but with an amplitude that depends on z, diminishing (attenuating) exponentially from $|V_1|$ at $z = 0$ down to $|V_1|e^{-\alpha z}$ at distance z. Thus, the exponential $e^{-\gamma z}$ is to be interpreted as a sinusoidal wave with exponentially attenuating amplitude.

The real part of γ is α and is called the *attenuation constant*; the imaginary part β is the *phase constant*. The complex combination $\gamma = \alpha + j\beta$ is the *propagation constant*.

By the same process of conversion to the time domain, we find that the term $V_2 e^{\gamma z} = |V_2|e^{j\varphi_2}e^{\gamma z}$ corresponds to

$$\text{Re}\left[V_2 e^{\gamma z}e^{j\omega t}\right] = |V_2|e^{\alpha z}\cos(\omega t + \beta z + \varphi_2). \tag{8.133}$$

This is a sinusoidal wave with an amplitude that grows exponentially along z but note that the wave $\cos(\omega t + \beta z + \varphi_2)$ travels along the $-z$-direction, at speed ω/β, and that in that direction, its amplitude decays. That is, this term is also a wave that travels and attenuates, both processes occurring in the negative z-direction. See Figure 8-11 for a depiction of both types of wave, as each might appear in a snapshot.

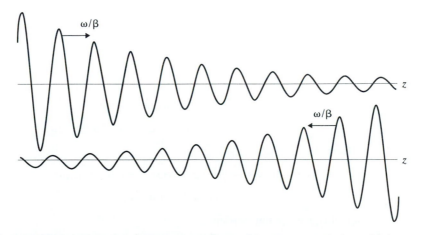

Figure 8-11 Forward- and backward-traveling attenuated sinusoidal waves.

The result is that the voltage along the transmission line is a superposition of an attenuating sinusoidal wave along the positive z-direction and another attenuating sinusoidal wave in the opposite direction; both waves exhibit the same attenuation rate $\alpha = \text{Re}\,\gamma$ and the same wave speed $\omega/\beta = \omega/[\text{Im}\,\gamma]$. The voltage is the sum of the two waves; the characteristic impedance times the current is the difference of the same two attenuating waves.

The characteristic impedance

$$Z_0 = \sqrt{Z/Y} \qquad (8.134)$$

typically appears as

$$Z_0 = \sqrt{(R + j\omega L)/(G + j\omega C)}, \qquad (8.135)$$

and is complex. The sign of the square root is unambiguous because, for a physically realizable transmission line, the resistive (real) part of the characteristic impedance must be nonnegative.

All the previously obtained formulas for reflection coefficient, impedance, and the like remain valid in this general case of a uniform transmission line with complex characteristic impedance and propagation constant, but with β replaced by $\beta - j\alpha$ or $-j\gamma$. However, the speed of wave propagation is determined by the phase $\omega t - \beta z$, not by the amplitude, so that $u_0 = \omega/\beta = \omega/[\text{Im}\,\gamma]$ only, not $\omega/(-j\gamma)$. The real part of γ represents attenuation, due to the loss of energy in heating the imperfectly conducting and imperfectly insulating portions of the line. The complex propagation constant is indicative of a lossy transmission line.

Example 8.11

Consider a coaxial line with perfectly conducting inner wire and outer shell but with a lossy dielectric between them, say plexiglass ($\varepsilon = 2.60\varepsilon_0$, loss tangent 0.0057). Obtain the propagation constant and the characteristic impedance at an operating frequency of 3 GHz. ◈

The parameters of this line are

$$L = (\mu/2\pi)\ln(b/a), \qquad C = 2\pi\varepsilon/\ln(b/a),$$

$$R = 0, \qquad G = 2\pi\sigma/\ln(b/a), \qquad (8.136)$$

so that

$$Z = j\omega L = (j\omega\mu/2\pi)\ln(b/a), \qquad (8.137)$$

$$Y = G + j\omega C = (\sigma + j\omega\varepsilon)2\pi/\ln(b/a). \qquad (8.138)$$

We form the product

$$ZY = j\omega\mu(\sigma + j\omega\varepsilon) = -\omega^2\mu\varepsilon + j\omega\mu\sigma$$

$$= -\omega^2\mu\varepsilon[1 - j(\sigma/\omega\varepsilon)] \qquad (8.139)$$

to get the propagation constant

$$\gamma = \sqrt{ZY} = j\omega\sqrt{\mu\varepsilon}\,\sqrt{1 - j(\sigma/\omega\varepsilon)}. \qquad (8.140)$$

Rather than proceeding at once with a calculation of the square root of the complex number $1 - j(\sigma/\omega\varepsilon)$, we recognize the loss tangent $\sigma/\omega\varepsilon$ and note that it is a small number, so that it suffices to approximate the square root with

$$\sqrt{1 - j(\sigma/\omega\varepsilon)} \approx 1 - j\tfrac{1}{2}(\sigma/\omega\varepsilon), \qquad (8.141)$$

by using its binomial expansion. This is accurate as long as the square of the loss tangent is negligible compared to one. To this precision, we have

$$\gamma = j\omega\sqrt{\mu\varepsilon}\left[1 - j\tfrac{1}{2}(\sigma/\omega\varepsilon)\right]$$
$$= \tfrac{1}{2}\omega\sqrt{\mu\varepsilon}\,(\sigma/\omega\varepsilon) + j\omega\sqrt{\mu\varepsilon}$$
$$= \tfrac{1}{2}\sqrt{\mu/\varepsilon}\,\sigma + j\omega\sqrt{\mu\varepsilon} \qquad (8.142)$$

and for this case we get, from the version $\omega\sqrt{\mu\varepsilon}\left[\tfrac{1}{2}(\sigma/\omega\varepsilon) + j\right]$,

$$\gamma = (\omega/c)\sqrt{2.60}\left[\tfrac{1}{2}(0.0057) + j\right]$$
$$= \left[2\pi(3\cdot10^9)/(2.998\cdot10^8)\right]\sqrt{2.60}\left[0.00285 + j\right]$$
$$= (0.2889 + j101.38) \text{ m}^{-1}. \qquad (8.143)$$

We read off the attenuation and phase constants as

$$\alpha = 0.2889 \text{ m}^{-1}, \qquad \beta = 101.38 \text{ m}^{-1}. \qquad (8.144)$$

This tells us that the amplitude of a unidirectional wave falls by a factor of e over a distance $1/\alpha = 3.46$ m. Since the wavelength is here $2\pi/\beta = 6.20$ cm, this distance corresponds to 55.8 wavelengths, meaning that there are this many spatial oscillations in one e-folding length.

The characteristic impedance is found as

$$Z_0 = \sqrt{Z/Y} = \sqrt{j\omega L/(G + j\omega C)}$$
$$= \sqrt{L/C}/\sqrt{1 + G/j\omega C}$$
$$= \sqrt{L/C}/\sqrt{1 - j(\sigma/\omega\varepsilon)}$$
$$\approx \sqrt{L/C}\left[1 + j\tfrac{1}{2}(\sigma/\omega\varepsilon)\right] \qquad (8.145)$$

to the same approximation. The factor $\sqrt{L/C}$ is $(\eta/2\pi)\ln(b/a)$. We were not given the ratio b/a of the line; for the sake of illustration, assume that were it not for the nonzero loss tangent, the characteristic impedance $\sqrt{L/C}$ would have been the standard 50 Ω. If so, then

$$Z_0 = (50\ \Omega)[1 + j0.00285] = (50 + j0.143) \text{ ohms}. \qquad (8.146)$$

The line is primarily resistive but slightly inductive as well.

To illustrate the concepts behind the operation of a lossy transmission line, let us consider an unlikely line, so lossy as to serve only an educational purpose.

Example 8.12

A lossy transmission line has characteristic impedance $Z_0 = (50 + j5)\ \Omega$ and is loaded by an impedance $Z_l = (300 - j100)\ \Omega$. Its propagation constant γ is such that, in terms of the line's length l, $\gamma l = 1 + j12$. What is its input impedance? ◈

The mismatch between load and characteristic impedance, both of which are complex, gives the reflection coefficient at the load as

$$\Gamma_l = \frac{Z_l - Z_0}{Z_l + Z_0} = \frac{(300 - j100) - (50 + j5)}{(300 - j100) + (50 + j5)} = \frac{250 - j105}{350 - j95}$$

$$= 0.74768e^{-j0.13259}, \tag{8.147}$$

The complex reflection coefficient elsewhere is given by the ratio of the attenuated reflected wave to the attenuated incident one:

$$\Gamma(z) = \Gamma_l e^{-2\gamma(l-z)}. \tag{8.148}$$

At the input, this is

$$\Gamma(0) = \Gamma_l e^{-2\gamma l} = 0.74768e^{-j0.13259}e^{-2(1+j12)}$$

$$= 0.74768e^{-2}e^{-j24.13259}$$

$$= 0.10119e^{-j24.13259} = 0.05466 + j0.08515. \tag{8.149}$$

Consequently, the input impedance is

$$Z_{in} = Z_0 \frac{1 + \Gamma(0)}{1 - \Gamma(0)} = Z_0 \frac{1.05466 + j0.08515}{0.94534 - j0.08515}$$

$$= (50 + j5)\left(1.11475\ e^{j0.17040}\right)$$

$$= (50 + j5)(1.09861 + j0.18904)$$

$$= (53.985 + j14.945)\ \Omega. \tag{8.150}$$

Note that the input impedance of this high-loss line is considerably closer to the characteristic impedance than is the mismatched load. This is a consequence of the attenuation of the incident and reflected waves to and from the load.

Before leaving this example, let us see how the axial distribution of the reflection coefficient appears on the Smith chart. Instead of a circle of travel, we get a spiral, as the exponential decay factor in $\Gamma(z) = \Gamma_l e^{-2(\alpha+j\beta)(l-z)}$ attenuates the amplitude of Γ while it reduces the phase as we approach the generator; see Figure 8-12.

Finally, let us look at a plot of the voltage amplitude interference pattern, as in Figure 8-13. The maxima and minima are now no longer at constant levels, because of the attenuation of both the incident and reflected waves. The VSWR therefore also varies; note that it is greater near the mismatched load, where the reflected wave is a significant fraction of the incident one, and that it weakens near the generator, where the incident wave is strong but the reflected one has been severely attenuated.

The variation of the VSWR along the line, as just mentioned, allows easy measurement of the attenuation constant of the line, as follows.

Example 8.13

A slotted-line measurement made near the load of a lossy transmission line gives a VSWR of 6.5 and another such measurement 500 meters farther from the load gives VSWR = 1.5. What is the attenuation constant α of the line? ◇

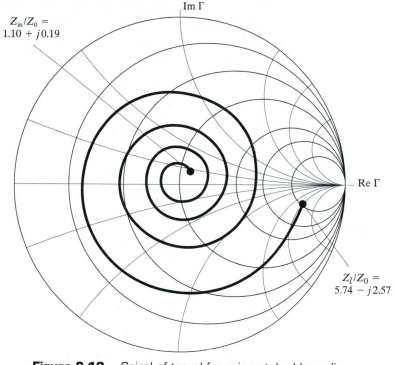

$$Z_{in}/Z_0 = 1.10 + j0.19$$

Im Γ

Re Γ

$$Z_l/Z_0 = 5.74 - j2.57$$

Figure 8-12   Spiral of travel for mismatched lossy line.

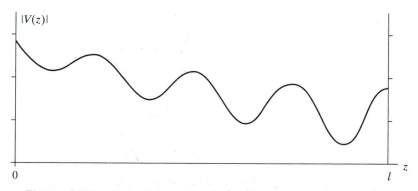

$|V(z)|$

0 l

z

Figure 8-13   Standing-wave pattern for lossy transmission line.

The VSWR measured near the distance $l - z$ from the load is given by

$$[1 + |\Gamma_l|e^{-2\alpha(l-z)}]/[1 - |\Gamma_l|e^{-2\alpha(l-z)}],$$

so that $|\Gamma_l|e^{-2\alpha(l-z)} = (\text{VSWR} - 1)/(\text{VSWR} + 1)$. Near the load, this ratio is $5.5/7.5$, while far from the load it is $0.5/2.5$. The ratio of these two ratios, which is $\exp[2\alpha(z_1 - z_2)]$, is therefore $11/3$, so that

$$2\alpha(500 \text{ m}) = \ln(11/3) = 1.2993 \quad \text{gives} \quad \alpha = 1.30 \,(\text{km})^{-1}.$$

Now let's go to the extreme case of a lossy line, that of attempting to propagate a wave through a good, but imperfect, conductor. To facilitate the comparison with the previous cases, we retain the coaxial-line geometry but consider the "dielectric" between the perfectly conducting inner wire and outer shell to be a good conductor, instead of a good insulator. A material is a good conductor if the conduction current it carries is overwhelmingly stronger than the displacement current that accompanies it. The criterion for a good conductor reduces to $\sigma/\omega\varepsilon \gg 1$. Of course, this is not a realistic use for a device of coaxial-line geometry but it will give us an idea of how waves can propagate inside a conductor.

Example 8.14

Determine the nature of the wave solutions to the transmission line equations for a coaxial line with perfectly conducting inner wire and outer shell, separated by a material that is a good conductor. ◈

We have again that $Z = j\omega L$ but in $Y = G + j\omega C$, we can neglect the capacitive term associated with the shunt displacement current, leaving only the conductive part G that describes the shunt conduction current:

$$Z = j\omega L = (j\omega\mu/2\pi)\ln(b/a), \qquad Y = G = 2\pi\sigma/\ln(b/a). \qquad (8.151)$$

The propagation constant becomes

$$\gamma = \sqrt{ZY} = \sqrt{j\omega LG} = \sqrt{j\omega\mu\sigma}. \qquad (8.152)$$

We need only separate the real and imaginary parts of this expression. Since the complex number j has magnitude unity and phase $\pi/2$, its square root (the one with a positive real part) has magnitude unity and phase $\pi/4$, which gives it equal real and imaginary parts:

$$\sqrt{j} = (1 + j)/\sqrt{2}, \qquad \gamma = \sqrt{\omega\mu\sigma/2}\,[1 + j] = \alpha + j\beta. \qquad (8.153)$$

We see that the attenuation and phase constants are equal in this extreme case:

$$\alpha = \sqrt{\omega\mu\sigma/2}, \qquad \beta = \sqrt{\omega\mu\sigma/2}. \qquad (8.154)$$

The waves propagate, at speed ω/β, but are also attenuated, at a rate α that attains equality with β. This means that, within a single wavelength $\lambda = 2\pi/\beta$ from the source, the wave amplitude is attenuated from its original strength by a factor of $e^{-\alpha\lambda} = e^{-2\pi} = 0.00187$. For practical purposes, the wave does not survive as far as one wavelength's penetration into the good conductor.

The actual length of that wavelength depends on the frequency of the source and on the parameters μ, σ of the medium. For a metal, the attenuation is so considerable as to leave substantial fields only within a thin layer past the source plane, a layer whose thickness $\delta = 1/\alpha = \sqrt{2/\omega\mu\sigma}$ is called a *skin depth*. For copper at 1 MHz, the fields are reduced to 22% of the entrance amplitude within just 0.1 millimeter. Figure 8-14 shows a voltage distribution along the conducting structure we are considering. The figure is drawn for zero initial phase of the voltage and covers just the first wavelength. Clearly, the voltage becomes negligible before it completes even one spatial oscillation.

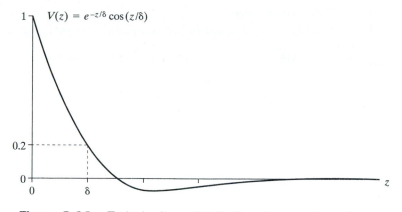

Figure 8-14 Typical voltage distribution along good conductor.

The characteristic impedance in this extreme case of a good conductor is

$$Z_0 = \sqrt{j\omega L/G} = (1 + j)\sqrt{\omega\mu/2\sigma}[\ln(b/a)/2\pi].\qquad (8.155)$$

This is complex, with resistive and inductive parts of equal amounts.

Example 8.15

A lossy transmission line has the following parameters.

$$R = 0.01\ \Omega/\text{m} \qquad G = 10^{-6}\ \text{S/m} \qquad L = 1\ \mu\text{H/m} \qquad C = 10^{-9}\ \text{F/m}$$

At a frequency $f = 1.592\ kHz$,

  (a) *find the characteristic impedance Z_0 of the line;*

  (b) *compare the wave speed with the vacuum speed of light, c;*

  (c) *find the attenuation rate of a traveling wave, in dB/km.* 《》

We need Z and Y first, in order to get Z_0 and γ.

  (a) Using $Z = R + j\omega L = R + j2\pi f L = (0.01 + j0.01)\ \Omega/\text{m} = (1 + j)10^{-2}\ \Omega/\text{m}$
and $Y = G + j\omega C = G + j2\pi f C = (10^{-6} + j10^{-5})\ \text{S/m} = (1 + j10)10^{-6}\ \text{S/m}$,
we get

$$Z_0 = \sqrt{Z/Y} = \sqrt{(1 + j)/(1 + j10)} \cdot 10^2\ \Omega$$

$$= 100\sqrt{[\sqrt{2}e^{j\pi/4}]/[\sqrt{101}e^{j\tan^{-1}10}]} = 100(2/101)^{1/4}e^{j1/2(\pi/4-\tan^{-1}10)}$$

$$= 37.513e^{-j0.342865} = (35.33 - j12.61)\ \Omega.$$

  (b) $\gamma = \alpha + j\beta = \sqrt{ZY} = \sqrt{(1 + j)(1 + j10)} \cdot 10^{-4}$

$$= 10^{-4}(2 \cdot 101)^{1/4}e^{j1/2(\pi/4+\tan^{-1}10)}$$

$$= 10^{-4} \cdot 3.7700e^{j1.12826} = (1.614 + j3.407)10^{-4}\ \text{m}^{-1}.$$

The wave speed is

$$\omega/\beta = 2\pi f/\beta = 10^4/(3.407 \cdot 10^{-4}) = 0.2935 \cdot 10^8 \text{ m/sec} = 0.0979 \, c.$$

(c) $\alpha = 1.614 \cdot 10^{-4} \text{ m}^{-1} = 0.1614 \, (\text{km})^{-1}$. For $l = 1$ km, we have

$$e^{-\alpha l} = e^{-0.1614} = 10^{-(1.402)/20},$$

so that $\alpha = 1.402$ dB/km.

Example 8.16

A transmission line is one quarter wavelength long $(l = \lambda/4)$, is open-circuited at its load, and has a voltage source $V_0 e^{j\omega t}$ applied at $z = 0$.

(a) *If the line is lossless, the steady-state voltage at the open circuit is not defined. Why?*

(b) *If the line is lossy, with $\gamma = \alpha + j\beta$ known, what is the steady-state voltage at the open circuit? Express the answer in terms of only V_0, α, β.* «»

(a) For $l = \lambda/4$, the input impedance is $Z_0^2/Z_l = 0$, a short circuit across the voltage source, so that there is no steady state.

(b) Using $V(z) = V_1 e^{-\gamma z} + V_2 e^{\gamma z}$ and $Z_0 I(z) = V_1 e^{-\gamma z} - V_2 e^{\gamma z}$, we have

$$V_1 + V_2 = V_0, \qquad V_1 e^{-\gamma l} - V_2 e^{\gamma l} = 0.$$

The solution is

$$V_1 = V_0/[1 + e^{-2\gamma l}], \qquad V_2 = V_0 e^{-2\gamma l}/[1 + e^{-2\gamma l}]$$

and the voltage at the open circuit is

$$V(l) = V_1 e^{-\gamma l} + V_2 e^{\gamma l} = 2V_0 e^{-\gamma l}/[1 + e^{-2\gamma l}].$$

But $\gamma = \alpha + j\beta$ and $\beta l = \beta \lambda/4 = \pi/2$, so that $e^{-\gamma l} = e^{-\alpha l} e^{-j\pi/2} = -je^{-\alpha l}$ and we get
$V(l) = 2V_0(-je^{-\alpha l})/[1 - e^{-2\alpha l}] = -2jV_0/[e^{\alpha l} - e^{-\alpha l}] = V_0/[j \sinh(\alpha l)]$
$= V_0/[j \sinh(\pi\alpha/2\beta)]$.

PLANE WAVES IN CONDUCTORS

The plane-wave counterpart of a lossy transmission line is a medium with some nonzero conductivity. All dielectrics have some conductivity; the perfect insulator is an idealization, the limit of zero conductivity. Conductors convert a portion of the field energy into thermal form, through Joule heating, so that waves that travel in imperfect insulators lose energy as they propagate in the medium; their fields eventually diminish to zero. Good conductors, such as metals, are another extreme; their mobile charge carriers almost instantaneously rearrange themselves when electric fields appear, tending to exclude fields from their interior. The perfect conductor is the idealized limit of a good conductor; it reduces any fields to zero immediately.

A perfect dielectric's fields satisfy $\nabla \times \mathbf{E} = -j\omega\mu\mathbf{H}$ and $\nabla \times \mathbf{H} = j\omega\varepsilon\mathbf{E}$ and exhibit transmission line behavior with parameters $Z = j\omega\mu$ and $Y = j\omega\varepsilon$, which give the purely imaginary propagation constant $\gamma = \sqrt{ZY} = j\omega\sqrt{\mu\varepsilon} = j\beta$ and the purely real wave impedance $Z_0 = \sqrt{Z/Y} = \sqrt{\mu/\varepsilon} = \eta$. A medium with nonzero conductivity allows fields that satisfy $\nabla \times \mathbf{E} = -j\omega\mu\mathbf{H}$ and $\nabla \times \mathbf{H} = (\sigma + j\omega\varepsilon)\mathbf{E}$ and therefore have the parameters $Z = j\omega\mu$ and $Y = \sigma + j\omega\varepsilon$. These combine into a complex propagation constant and a complex characteristic impedance:

$$\gamma = \sqrt{ZY} = \sqrt{(-\omega^2\mu\varepsilon) + j(\omega\mu\sigma)} = \alpha + j\beta, \qquad (8.156)$$

$$Z_0 = \sqrt{Z/Y} = j\omega\mu/\gamma = j\omega\mu/(\alpha + j\beta). \qquad (8.157)$$

To extract the attenuation constant α and the phase constant β, we separate the real and imaginary parts in the square of the quantity $\alpha + j\beta$:

$$(\alpha + j\beta)^2 = (\alpha^2 - \beta^2) + j(2\alpha\beta) = \gamma^2 = (-\omega^2\mu\varepsilon) + j(\omega\mu\sigma) \qquad (8.158)$$

yields two equations for the two unknowns α and β.

$$\beta^2 - \alpha^2 = \omega^2\mu\varepsilon, \qquad 2\alpha\beta = \omega\mu\sigma. \qquad (8.159)$$

We have already looked at the solution in the limit of $\omega\varepsilon$ negligible compared to σ but here we want the exact solution, valid for all conductivities.

If we square and add the two equations, we can eliminate the cross term $2\alpha\beta$ and obtain simultaneous equations for α^2 and β^2. We add

$$\beta^4 - 2\beta^2\alpha^2 + \alpha^4 = \omega^4\mu^2\varepsilon^2 \qquad \text{and} \qquad 4\beta^2\alpha^2 = \omega^2\mu^2\sigma^2 \qquad (8.160)$$

and get $(\beta^2 + \alpha^2)^2 = \omega^2\mu^2(\sigma^2 + \omega^2\varepsilon^2)$ or

$$\beta^2 + \alpha^2 = \omega^2\mu\varepsilon\sqrt{1 + (\sigma/\omega\varepsilon)^2}. \qquad (8.161)$$

This can be combined with one of the original equations, $\beta^2 - \alpha^2 = \omega^2\mu\varepsilon$, giving finally

$$\beta = \omega\sqrt{\mu\varepsilon}\sqrt{(\sec\theta + 1)/2}, \qquad (8.162)$$

$$\alpha = \omega\sqrt{\mu\varepsilon}\sqrt{(\sec\theta - 1)/2}, \qquad (8.163)$$

where we have previously defined the *loss tangent*

$$\tan\theta = \sigma/\omega\varepsilon. \qquad (8.164)$$

The loss tangent designation makes sense when the loss properties of the medium are expressed through a complex permittivity ε^0 as $j\omega\varepsilon^0 = \sigma + j\omega\varepsilon$ or $\varepsilon^0 = \varepsilon - j(\sigma/\omega)$, whose polar form has phase angle $-\theta$, with $\tan\theta = \sigma/\omega\varepsilon$.

> **Example 8.17**
>
> *If a certain medium's constitutive parameters are such that $1/\sqrt{\mu\varepsilon} = c/3$ and $\sigma/\omega\varepsilon = 1$ at an operating frequency for which the free-space wavelength is $\lambda_0 = 2$ m, what are the phase and attenuation constants, the wavelength, and the wave speed in the medium?* ◈

We have $\tan\theta = 1$ or $\theta = \pi/4$, so that $\sec\theta = \sqrt{2}$ and we get

$$\beta = (\omega/c)3\sqrt{(\sqrt{2}+1)/2} = (2\pi/\lambda_0)3\sqrt{(\sqrt{2}+1)/2} = 10.355 \text{ m}^{-1},$$

$$\alpha = (\omega/c)3\sqrt{(\sqrt{2}-1)/2} = (2\pi/\lambda_0)3\sqrt{(\sqrt{2}-1)/2} = 4.289 \text{ m}^{-1}.$$

The wavelength in the medium is $2\pi/\beta = 0.6068$ m and the wave speed is $\omega/\beta = c/\{3\sqrt{(\sqrt{2}+1)/2}\} = 0.3034c = (0.9096)10^8$ m/s.

If the constitutive parameters μ, ε, σ can be considered independent of frequency, then another convenient way to express the loss tangent $\sigma/\omega\varepsilon$ is f_0/f, where f_0, called the *transition frequency*, is the operating frequency f at which the loss tangent is unity; for much higher frequencies, the medium is a good insulator (small $\sigma/\omega\varepsilon$); for much lower frequencies, it is a good conductor (high $\sigma/\omega\varepsilon$). For example, ocean water has a transition frequency at about 900 MHz, while earth has it at a few MHz.

Figure 8-15 presents a log-log plot of both $\alpha/\sigma\eta$ (the lower curve) and $\beta/\sigma\eta$ (the upper curve), where $\eta = \sqrt{\mu/\varepsilon}$, against the normalized frequency f/f_0, assuming the parameters are themselves not functions of frequency. The straight-line asymptotes are also indicated.

These curves will be distorted by any frequency dependence of the constitutive parameters. Note that α is always less than β, although they approach equality for good conductors. The fact that the propagation constant is complex for a conducting medium indicates that waves attenuate as they propagate in it. That the char-

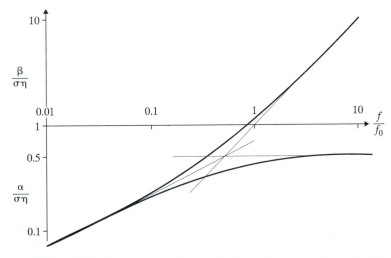

Figure 8-15   Frequency dependence of attenuation and phase constants.

acteristic impedance is complex implies that there is a phase difference between the oscillations of the electric and magnetic fields; they do not oscillate in perfect unison, as they do in a perfect dielectric.

SUMMARY

We have focused on energy in electromagnetic wave-propagating structures, where it resides, how it flows, and how it can get dissipated. The key result is the Poynting theorem, which demonstrates that the cross product $\mathbf{E} \times \mathbf{H}$ of the electric and the magnetic field represents power flow, in watts/m^2, at any point and instant where the two fields coexist.

Integration of this Poynting vector over a closed surface bounding some volume tells us the total power crossing that surface. It therefore accounts for both power expenditure and the rate of energy accumulation within the enclosed volume. That is, we can determine key aspects of the energetics of the electromagnetic processes in a region by calculations limited to the bounding surface of that region. We illustrated this sort of calculation with the fields of a parallel-wire transmission line and also within a current-carrying wire. We discovered that the transfer of energy from source to load, as well as to the portions of the structure that dissipate some of the available energy, takes place through the dielectric spaces surrounding the conducting parts that carry the currents. That is where the predominant fields are and that is where the Poynting vector conveys the power. For the parallel-wire line, the power flows along, but outside, the wires, mostly very near to them. For the current-carrying wire, the power flow is from the exterior of the wire to its interior, laterally across the wire's surface. That power is dissipated inside the wire, converted into heat.

The coaxial line is an example of a shielded cable structure, wherein the fields are confined by a conducting enclosure. It is also particularly easy to analyze; its electric and magnetic fields and the Poynting vector field are mathematically simple, and we demonstrated that their variation, in time and axially, conforms to the wave-propagating action of a transmission line.

For the sinusoidal steady state, the relevant quantity related to energy flow, storage, and dissipation is the time-averaged energy level; the remainder consists of a rapid oscillation to which most circuitry is insensitive or unresponsive. We can arrive at the time-averaged power flow directly from the complex-amplitude description of the fields, by forming the complex Poynting vector $\frac{1}{2}\mathbf{E} \times \mathbf{H}^*$ and taking its real part. Its imaginary part also has significance, being proportional to the difference between the time-averaged magnetic and electric energy storage densities. This is how we interpret the complex Poynting theorem, which is independent of its time-domain counterpart.

The flux of the complex Poynting vector is connected directly to the concept of reactive power in circuit theory. One application of this result is that we can equate the complex power crossing surface S—namely, $\int_S \frac{1}{2} \mathbf{E} \times \mathbf{H}^* \cdot d\mathbf{S}$—to the complex circuit power $\frac{1}{2}|I|^2 Z$ that enters the region beyond that surface when current I is responsible for the existence of the $\mathbf{E}$ and $\mathbf{H}$ fields. Since $\mathbf{E}$ will be proportional to I and $\mathbf{H}^*$ will be proportional to I^*, the ratio of $\int_S \mathbf{E} \times \mathbf{H}^* \cdot d\mathbf{S}$ to $|I|^2$ involves only the geometry and constitution of the medium, not the field strength, and allows us to calculate the complex impedance $Z = R + jX$ presented to the source of the current that sets up the electromagnetic fields.

We argue that just about any structure can behave electromagnetically as a transmission line. We outlined a sequence of steps we can take, starting from a constitutive and geometric description of the structure, to arrive at the electrical parameters of the line, typically L, C, G, R, which combine into an impedance per unit length $Z = R + j\omega L$ and an admittance per unit length $Y = G + j\omega C$. If the structure is uniform along the presumed direction of propagation, z, so that it looks the same geometrically and constitutively for all z, then these two parameters will be independent of z. Otherwise, Z or Y or both will be functions of z and we are describing a nonuniform transmission line.

The transverse voltage $V(z)$ and the longitudinal current $I(z)$ we associate with whatever structure we analyze will satisfy transmission line equations in integral form

$$V(z) = V(l) + \int_z^l Z(z')I(z')\,dz', \tag{8.165}$$

$$I(z) = I(l) + \int_z^l Y(z')I(z')\,dz'. \tag{8.165}$$

Alternatively, they can appear in differential form as

$$dV(z)/dz = -Z(z)I(z), \qquad dI(z)/dz = -Y(z)V(z). \tag{8.167}$$

We illustrated the calculations that yield Z and Y for the cases of a coaxial line with imperfect conductors and dielectric and for a radial transmission line that can describe a conducting wire. The former is a uniform transmission line, with constant Z and Y; the latter is inherently nonuniform.

If the line is nonuniform, the most expedient approach to solving the equations is likely to be iteration of the two integral equations, as was done for quasistatic analysis. This leads to the voltage and current in the guise of a pair of infinite series of successive corrections to their initial approximations as uniform or otherwise simple distributions.

For the case of a uniform line, however, we obtained the full solution to the general transmission line equations. This entails merely forming the complex propagation constant γ and the complex characteristic impedance Z_0 from the transmission line impedance and admittance per unit length, Z and Y:

$$\gamma = \sqrt{ZY}, \qquad Z_0 = \sqrt{Z/Y}. \tag{8.168}$$

The solution is then that the complex voltage amplitude $V(z)$ is the sum of two exponentials $V_1 e^{-\gamma z}$ and $V_2 e^{\gamma z}$ and that the characteristic impedance Z_0 times the current amplitude $I(z)$ is the difference of the same two exponentials. Each exponential is interpreted as a wave propagating along $\pm z$ with an amplitude that decays exponentially in the direction of propagation. The real and imaginary parts of $\gamma = \alpha + j\beta$ are the attenuation constant and the phase constant, respectively. The wavelength is $\lambda = 2\pi/\beta$ and the speed of propagation of the attenuated wave is ω/β.

We illustrated wave propagation along a lossy transmission line with an exaggeratedly dissipative line; practical lines all have some attenuation but are useful as long-distance signal transmission structures only if $\alpha \ll \beta$, for which there are a great many spatial oscillations before the wave amplitude falls to insignificant levels.

We examined the penalties involved in attempting to use a good conductor as a wave-propagating medium. We found that such media exhibit attenuation and phase constants that attain equality. A consequence is that the wave amplitude dies away within less than a wavelength. Penetration of a good conductor is limited to approximately a skin depth. On the one hand, this furnishes an important shielding property of metals, even when they are thin sheets, to keep unwanted fields out of certain regions. On the other hand, the skin effect severely limits communication in conducting environments, such as in the oceans.

Finally, we examined plane wave propagation in an unbounded medium with an arbitrary conductivity. We found conducting media to exhibit a propagation constant and a characteristic impedance that are both complex. The complex propagation constant makes waves attenuate as they propagate; the complex characteristic impedance indicates a phase difference between the oscillations of the electric and magnetic fields. Despite the complex nature of the wave impedance for a medium with conductivity, we can use the same formulas as for perfect dielectrics to assess reflection and transmission at an interface between a dielectric and a conductor, or two conductors.

We will next be examining other waveguiding structures and will find that we can make use of even empty space, with no structure at all, as a transmission line. The question of how we can attach a source or a load to empty space is answered by the use of antennas. We will therefore finish by looking at the operation of elementary radiating structures.

PROBLEMS

Properties of lossy transmission line

8.1   A lossy transmission line has the following characteristics at frequency 31.83 kHz.

$$Z_0 = (50 - j\,10)\ \Omega \quad \gamma = (1.5 + j\,7.5)\ \text{km}^{-1}$$

Find the parameters R, G, L, C of the line.

Open-circuit voltage of lossy half-wave transmission line

8.2   A transmission line is one half wavelength long ($l = \lambda/2$), is open-circuited at its load, and has a voltage source $V_0 e^{j\omega t}$ applied at $z = 0$.
(a) If the line is lossless, what is the steady-state voltage at the open circuit?
(b) If the line is lossy, with $\gamma = \alpha + j\beta$ known, what is the steady-state voltage at the open circuit? Express the answer in terms of only V_0, α, β.

Short-circuit current of lossy quarter-wave transmission line

8.3   A transmission line (Z_0) is one quarter wavelength long ($l = \lambda/4$), is short-circuited at its load at $z = l$, and has a voltage source $V_0 e^{j\omega t}$ applied at $z = 0$. The line is lossy, with $\gamma = \alpha + j\beta$ known. What is the steady-state current at the short circuit? Express the answer in terms of only V_0, α, β, Z_0.

Elaborate transmission line equations (capacitive)

8.4   A special type of transmission line supports a field pattern for which the equations for the voltage and current variations are

$$\frac{dV}{dz} = -j\omega L I(z) - \frac{I(z)}{j\omega C A}, \qquad \frac{dI}{dz} = -j\omega C V(z).$$

The parameters L, C, A are known constants and ω is the frequency.
(a) Find the complex propagation constant γ and the characteristic impedance Z_0 of this line.
(b) Can the field pattern propagate along the line? If so, what is the phase velocity? If not, how do the fields vary along the z-direction?

Elaborate transmission line equations (inductive)

8.5   A special type of transmission line supports a field pattern for which the equations for the voltage and current variations are

$$\frac{dV}{dz} = -j\omega L I(z), \qquad \frac{dI}{dz} = -j\omega C V(z) - \frac{V(z)}{j\omega L A}.$$

The parameters L, C, A are known constants and ω is the frequency.
(a) Find the complex propagation constant γ and the characteristic impedance Z_0 of this line.
(b) Can the field pattern propagate along the line? If so, what is the phase velocity? If not, how do the fields vary along the z-direction?

Passbands and stopbands from elaborate transmission line equations

8.6   A special type of transmission line supports a field pattern for which the equations for the voltage and current variations are

$$\frac{dV}{dz} = -j\omega L I(z) - \frac{I(z)}{j\omega C A_1}, \qquad \frac{dI}{dz} = -j\omega C V(z) - \frac{V(z)}{j\omega L A_2}.$$

The parameters L, C, A_1, A_2 (with $A_1 > A_2$) are known positive constants.
   Over what range of frequencies ω can waves propagate along this line?

Attenuation and operating frequency from transmission line parameters

8.7   Given the parameters L, C, R, G of a uniform, lossy transmission line and given also the wave speed u_0 under certain operating conditions,
   (a) obtain the attenuation constant α, in terms of these given quantities;
   (b) obtain the operating frequency ω, in terms of α and the given quantities.

Wave speed and wavelength from transmission line parameters

8.8   Given the parameters L, C, R, G of a uniform, lossy transmission line and given also the attenuation constant α under certain operating conditions at some undetermined frequency,
   (a) obtain the wave speed u_0, in terms of these given quantities;
   (b) obtain the free-space wavelength λ_0, in terms of u_0 and the given quantities.

Transmission line parameters from measured properties

8.9   The following properties of a certain lossy transmission line are known or measured at some operating frequency.
   (i)   Its characteristic impedance is $Z_0 = 50$ ohms (real).
   (ii) Its wave speed u_0 is 90% of the vacuum speed of light, c.
   (iii) The free-space wavelength at the operating frequency is $\lambda_0 = 2$ m.
   (iv) The amplitude of a unidirectionally propagating wave is reduced to 80% of its original amplitude after traversing $l = 100$ m.
   From this given information, obtain the parameters L, C, R, G of the lossy line.

Effect of attenuation on SWR measurements

8.10   A transmission line has losses of 0.1 dB/m. A slotted-line measurement made near the load of the line gives the SWR as 8.0 . At what distance from the load will the SWR be reduced to 1.25 ?

Poynting vector inside current-carrying wire

8.11   An infinitely long cylindrical wire of radius a and conductivity σ carries a constant axial current I_0, uniformly distributed over the wire's cross section (no skin effect).
   (a) Obtain the Poynting vector (magnitude and direction) at all points inside the wire.
   (b) How much power crosses an internal cylindrical surface of radius ρ and length l, coaxial with the wire?

Power flow around sphere with injected current

8.12   A conducting sphere of radius a (center at $z = 0$) has steady current I_0 injected through a straight wire at its "south pole," so that charge accumulates on the sphere. Assume that this charge spreads uniformly over the sphere's surface. The sphere was neutral at time $t = 0$. In the air space outside the sphere and wire (i.e., for $r > a$ and $0 \le \theta < \pi$, with $\mu = \mu_0, \varepsilon = \varepsilon_0, \sigma = 0$), obtain

(a) $\mathbf{E}(r, \theta, \varphi, t)$
(b) $\mathbf{H}(r, \theta, \varphi, t)$
(c) the Poynting vector $\mathbf{P}(r, \theta, \varphi, t)$
(d) the total power that crosses the "equatorial" plane $z = 0$.

Power flow from second-order fields in a capacitor

8.13 The fields in a cylindrical capacitor (radius a, plate separation h, lossless dielectric with ε, μ) with a voltage $V(t)$ applied symmetrically and externally were found in an earlier chapter to be, up to second order,

$$\mathbf{E} = \hat{z}\{V/h - \mu\varepsilon[d^2V/dt^2](a^2 - \rho^2)/4h\}, \qquad \mathbf{H} = \hat{\varphi}[\varepsilon/2h][dV/dt]\rho.$$

(a) Give the direction and magnitude of the Poynting vector everywhere inside the capacitor, with only these second-order fields included.

(b) Find the total power crossing the outer cylindrical surface $\rho = a$ of the capacitor. In what direction does it flow?

Power flow seen near a falling charge

8.14 A perfectly conducting ball (mass M, radius a) carries charge Q. Starting from rest at time $t = 0$ at the origin, the ball falls (acceleration of gravity $= g$) along the negative z-axis, in air $(\mu = \mu_0, \varepsilon = \varepsilon_0, \sigma = 0)$ (See Figure P8-1.).

(a) The azimuthal magnetic field $H(t)$ at a point along the circumference of a circle of radius ρ_0 in the xy-plane, centered at the origin and with $\rho_0 > a$, can be expressed as a function of time for $t > 0$ as a magnitude

$$H(t) = Kt/(t_0^m + t^m)^n.$$

Obtain the constants K, t_0, m, n.

(b) Obtain the magnitude $E(t)$ of the (lowest-order) electric field $\mathbf{E}(t)$ at the same point. [Write $E(t)$ as a function of time, in a form similar to that of $H(t)$. "Lowest-order" means that no correction terms are sought.]

(c) Obtain the magnitude of the Poynting vector $\mathbf{E}(t) \times \mathbf{H}(t)$ at the same point.

(d) How far has the charge fallen when the Poynting vector seen at ρ_0 in the xy-plane is strongest? Express the distance in terms of ρ_0.

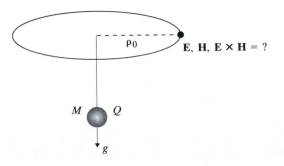

ρ_0

$\mathbf{E}, \mathbf{H}, \mathbf{E} \times \mathbf{H} = ?$

M Q

g

Figure P8-1   Fields and Poynting vector of a falling charge.

Power flow in a solenoid

8.15 Surface current $K_0(t)$ flows azimuthally on the surface of an infinitely long cylinder of radius a, as in Figure P8-2. The medium inside and outside is air. Statically, the generated magnetic field exists only inside the cylinder: $H_0 = K_0$.

(a) Inside and outside, find the electric field generated, to first order.

(b) Inside and outside, find the Poynting vector, using fields up to first order.

(c) What power crosses radius ρ, over a length l? Where does this power go?

Figure P8-2   Transverse surface current flow pattern on ideal solenoid.

Power flow around semi-infinite wire

8.16 A steady current I_0 in a semi-infinite wire (along the positive z-axis) accumulates charge at its tip (at the origin). The magnetic field generated by this current in a vacuum was found in the text to be $H_\varphi = -(I_0/4\pi r)\cot(\theta/2)$. Assume there is zero charge at the tip at time $t = 0$.

(a) Find the Poynting vector (magnitude and direction) everywhere outside the wire.

(b) Calculate the power flow across an infinitely long cylinder of radius ρ, coaxial with the semi-infinite wire. Does this power flow outward or inward?

Power flow around semi-infinite solenoid

8.17 Total magnetic flux Φ_0 emerges at the origin from an infinitely long and infinitesimally thin solenoid that lies along the positive z-axis. The tip acts as a point source and the flux spreads outward in all directions into the infinite medium of permeability μ_0. Assume the thin solenoid interferes negligibly with this outflux. If the total flux Φ_0 supplied by the solenoid varies (slowly) with time, the induced electric field $\mathbf{E}(\mathbf{r}, t)$ was found in a problem of an earlier chapter to be $\hat{\boldsymbol{\varphi}}[d\Phi_0/dt][1 + \cos\theta]/[4\pi r\sin\theta]$.

(a) Find the Poynting vector (magnitude and direction) everywhere outside the solenoid.

(b) Compare the power flow across the upper half $(z > 0)$ of an infinitely long cylinder of radius ρ, coaxial with the semi-infinite solenoid, with the flow across the lower half $(z < 0)$, in terms of flow directions and ratio of flows.

Single wire over parallel ground plane; image method

8.18 Consider the transmission line formed by a single, horizontal, perfectly conducting, thin wire (radius a) in air, at a height $h \gg a$ above a parallel, perfectly conducting plane of infinite extent, as in Figure P8-3.

When the wire is charged, the electric field that emanates from the wire is modified by the presence of the conducting plane so as to make the field perpendicular to that plane. We know the same condition applies, because of symmetry, when there are two, parallel, equally and oppositely charged wires and no conducting plane at all, if the separation between the two wires is $2h$. We can therefore get the field pattern in the single-wire case by adopting that of the two-wire configuration, unchanged, but only in the

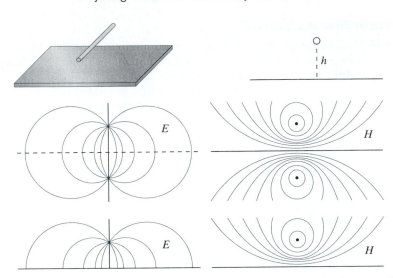

Figure P8-3   Single wire over parallel ground plane; image fields; actual fields.

upper half of the space and replacing the field below the midplane with zero. The "phantom" oppositely charged wire below the midplane is called the *image* of the real wire.

When the single wire carries current, the magnetic field that surrounds the wire is modified by the presence of the conducting plane so as to make the field parallel to that plane, to ensure that there be no discontinuity in normal B field, from zero in the conducting plane to any nonzero value above it. We know the same condition applies, because of symmetry, when there are two parallel wires carrying equal and opposite currents and no conducting plane at all, if the separation between the two wires is $2h$. We can therefore get the field pattern in the single-wire case by adopting that of the two-wire configuration, unchanged, but only in the upper half of the space and replacing the field below the midplane with zero.

(a) From the electric field pattern of the single wire above the plane, find the capacitance per unit length of the transmission line.

(b) From the magnetic field pattern of the single wire above the plane, find the inductance per unit length of the transmission line.

(c) Obtain the characteristic impedance Z_0 of the line.

(d) What is the numerical value of Z_0 for a wire of radius 0.5 mm at height 1 cm above the conducting plane?

(e) What is the numerical value of Z_0 for a power line of radius 0.5 cm at height 5 m above the ground? Assume the wire is perfectly horizontal and the ground is perfectly conducting.

Conducting bracket as transmission line

8.19 A perfectly conducting U-shaped bracket, in air, is fed a total current $I_0 \exp[j\omega t]$ at one leg (with an equal current extracted from the other leg). Neglecting fringing, this results in a magnetic field $H(z)$ in the air space within the bracket, parallel to the width w of the bracket, and an electric field $E(z)$ in the same space, parallel to the height h of the bracket. Obtain

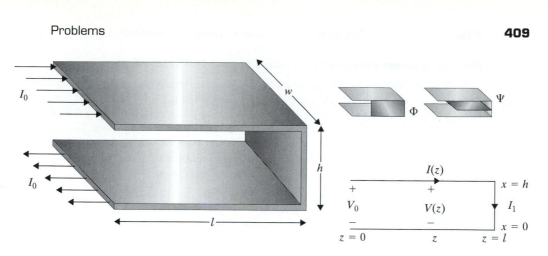

Figure P8-4   A conducting bracket acts as a transmission line.

(a) the inductance coefficient L in the magnetic flux relation $\partial\Phi/\partial z = -LI(z)$ [for $\Phi(z)$ on the surface indicated in Figure P8-4, from z to l];

(b) the capacitance coefficient C in the electric flux relation $\partial\Psi/\partial z = -CV(z)$ [for $\Psi(z)$ on the surface indicated in the figure, from z to l];

(c) the characteristic impedance Z_0 and wave velocity u_0 for the structure;

(d) the voltage $V(z)$ and current $I(z)$ along the structure;

(e) the current I_1 along the end of the bracket, at $z = l$;

(f) the voltage V_0 at the open end of the bracket, at $z = 0$.

Parallel-plate capacitor as radial transmission line

8.20 The fields inside a circular parallel-plate capacitor (area $A = \pi a^2$, plate separation h, perfect dielectric ε, μ) can be related to a voltage distribution $V(\rho)$ from one plate to the other and to a "current" $I(\rho)$ defined as the displacement current from the axis up to radius ρ: $I(\rho) = j\omega\Psi(\rho)$; see Figure P8-5. If the equations relating $V(\rho)$ and $I(\rho)$ are written

$$V(\rho) = V(a) - \int_{\rho}^{a} Z(\rho')I(\rho')\,d\rho', \qquad I(\rho) = \int_{0}^{\rho} Y(\rho')V(\rho')\,d\rho',$$

obtain the coefficients $Z(\rho)$, $Y(\rho)$ of this radial transmission line.

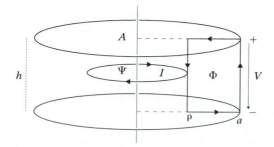

Figure P8-5   A capacitor acts as a nonuniform, radial transmission line.

Air gap between pole pieces as radial transmission line

8.21 The "north" and "south" pole pieces of a cylindrical electromagnet are brought close to each other, leaving an air gap of height h, as in Figure P8-6. The parallel circular faces of the two pole pieces are of radius a. The electromagnet delivers magnetic flux to the air gap; neglect fringing of the fields beyond the edge of the gap space. The fields in the air gap can be related to a voltage distribution $V(\rho) = -j\omega\Phi(\rho)$ around a horizontal circle of radius ρ and to a "current" $I(\rho)$ defined as the mmf from one pole face to the other, at radius ρ: $I(\rho) = hH(\rho)$. If the equations relating $V(\rho)$ and $I(\rho)$ are written

$$V(\rho) = -\int_0^\rho Z(\rho')I(\rho')\,d\rho', \qquad I(\rho) = I(a) + \int_\rho^a Y(\rho')V(\rho')\,d\rho',$$

obtain the coefficients $Z(\rho)$, $Y(\rho)$ of this radial transmission line.

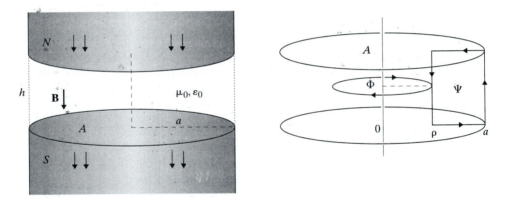

Figure P8-6   Cylindrical air gap between pole pieces as radial transmission line.

Vertical wire ending on horizontal plane as radial transmission line

8.22 Current I_0 flows along the surface of a perfectly conducting, vertical, semi-infinite cylindrical wire (radius $= a$) until it reaches a perfectly conducting, infinitely extended, horizontal plane, whereupon the current spreads out along the surface of the plane, equally in all directions; see Figure P8-7. The space outside the wire and above the plane is air.

We claim that, as far as the radial field dependence is concerned, this structure behaves as a radial transmission line. If $V(\rho)$ is the emf around the edge of the rectangular area labeled Φ and if $I(\rho)$ is the mmf around the outer edge of the annular area labeled Ψ, these satisfy transmission line equations. Obtain the impedance and admittance coefficients Z, Y.

Significance of loss tangent

8.23 For many structures that involve a lossy dielectric, the impedance and admittance coefficients appear as

$$Z = j\omega\mu g \qquad \text{and} \qquad Y = (\sigma + j\omega\varepsilon)/g = j\omega\varepsilon[1 - j(\sigma/\omega\varepsilon)]/g,$$

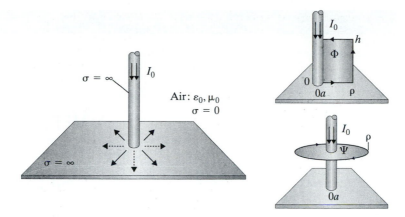

Figure P8-7 Vertical wire ending on horizontal plane as radial transmission line.

where g is a geometric factor. The ratio $\sigma/\omega\varepsilon$ is called the loss tangent of the lossy dielectric, because when the complex number $1 - j(\sigma/\omega\varepsilon)$ is plotted in the complex plane, the ratio represents the tangent of the (negative) polar angle, θ, of the complex number. That is, $1 - j(\sigma/\omega\varepsilon)$ can be written as a magnitude and phase in the form $\sec\theta \, e^{-j\theta}$, where $\tan\theta = \sigma/\omega\varepsilon$ is the loss tangent.

Obtain the complex propagation constant $\gamma = \alpha + j\beta$ for a structure with parameters $Z = j\omega\mu g$ and $Y = j\omega\varepsilon[1 - j\tan\theta]/g$.

Low-frequency equivalent skin-effect resistance of wire

8.24 The electric field in a wire (conductivity σ, area $A = \pi a^2$) carrying current $I(t)$ was obtained in an earlier chapter, to first order at radius ρ, as

$$E(\rho) = I/\sigma A + (\mu/8\pi)(dI/dt)\left[-1 + 2(\rho/a)^2\right].$$

To this order,

(a) obtain the power $p(t)$ dissipated in a length l of the wire;

(b) specialize the result to the case of a sinusoidal current at frequency ω and deduce a low-frequency equivalent skin-effect resistance R_{eq} from the time-averaged power $\langle p \rangle = R_{eq}\langle I^2 \rangle$. Express the result as $R_{eq} = (l/\sigma A)\left[1 + \xi(a/\delta)^m\right]$, where $\delta = \sqrt{2/\omega\mu\sigma}$ is the skin depth. What are the constants ξ and m?

Poynting vector of circularly polarized wave at a mirror

8.25 A perfect, planar mirror coincides with the xy-plane and normally incident as well as reflected plane waves exist in the vacuum region $z < 0$ in front of the mirror, at frequency $\omega = kc$. The magnetic field at the mirror surface is circularly polarized: $\mathbf{H} = H_0(\hat{\mathbf{x}} + j\hat{\mathbf{y}})$. In an earlier problem, the electric field amplitude in the space $z < 0$ was found to be $\mathbf{E}(z) = \eta_0 H_0(\hat{\mathbf{x}} + j\hat{\mathbf{y}})\sin kz$. Find the complex Poynting vector in the space $z < 0$.

CHAPTER 9

Waveguiding and Radiating Structures

We have developed transmission line equations and their solutions for parallel-wire and coaxial lines. The recipe we outlined for arriving at the parameters of a wave-propagating system was designed to be as general as possible and suggests that just about any structure of conductors and dielectrics can serve as a channel for the transmission of signals and energy from one point to another. Of course, some structures are more effective than others, and we must be concerned about how to attach a source and a load to the transmission line we design.

We want to explore some additional structures, some similar to but some significantly different from the ones we have studied. We will find that we can utilize even nothing at all, empty space, to transmit electromagnetic waves. The design problem in that case is to connect the source and load to no structure at all. This is done with antennas, and we will conclude our survey of concepts in electromagnetics by seeing how a rudimentary antenna can radiate a signal and energy into space.

PARALLEL-PLATE TRANSMISSION LINE

Yet another two-conductor transmission line is built of a dielectric slab sandwiched between two parallel conducting plates, like an extended capacitor. This is a rudimentary form of the *strip line*, which has acquired great importance in connection with integrated electronic circuitry and in integrated optics. The strip line has an extended planar conducting substrate, a dielectric layer on this substrate, and a narrow conducting line parallel to the substrate. The strip line has significant fringing fields that complicate the analysis, and we will restrict ourselves to the simple case of an infinitely extended dielectric slab separating two infinitely extended conducting plates.

As in Figure 9-1, the plates are separated by height b and we deal with the infinite lateral extent of the structure by restricting our attention to a width a, so that we may avoid the complications of infinite width and also of fringing. The structure looks like and behaves as a capacitor, with equal and opposite charge distributions

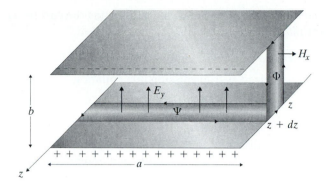

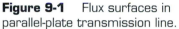

Figure 9-1   Flux surfaces in parallel-plate transmission line.

on the two plates and an electric field directed from one plate to the other. But the resultant time-varying electric flux will generate a magnetic field, whose own time-varying flux may in turn generate and sustain the electric field we postulated. The dielectric has parameters μ and ε and no conductivity. We take the intended direction of propagation along the z-axis.

The electric flux Ψ across the infinitesimally wide strip from z to $z + dz$ is

$$\Psi = \varepsilon E_y a\, dz \tag{9.1}$$

and the mmf around the edge of that strip is

$$U = \oint \mathbf{H} \cdot d\mathbf{l} = aH_x(z + dz) - aH_x(z). \tag{9.2}$$

Equating U and $j\omega\Psi$ and dividing by $a\, dz$ gives us

$$\frac{dH_x(z)}{dz} = j\omega\varepsilon E_y. \tag{9.3}$$

The horizontal magnetic field has flux Φ across the vertical strip of width dz, with

$$\Phi = \mu H_x b\, dz \tag{9.4}$$

and the time variation of this magnetic flux induces the emf

$$V = \oint \mathbf{E} \cdot d\mathbf{l} = bE_y(z) - bE_y(z + dz). \tag{9.5}$$

Equating V and $-j\omega\Phi$ and dividing by $b\, dz$ gives us

$$\frac{dE_y(z)}{dz} = j\omega\mu H_x. \tag{9.6}$$

We can reexpress these in terms of the voltage from the lower plate to the upper one and of the current along z in the lower plate. These are

$$V(z) = bE_y(z) \qquad \text{and} \qquad I(z) = -aH_x(z), \tag{9.7}$$

where the negative sign is dictated by the right-hand rule in equating the current $I(z)$ to the mmf around the lower plate. The results are

$$dV(z)/dz = -j\omega\mu(b/a)I(z), \qquad dI(z)/dz = -j\omega\varepsilon(a/b)V(z). \qquad (9.8)$$

These should be immediately recognized as transmission line equations, and we read off the impedance and admittance parameters as

$$Z = j\omega\mu(b/a), \qquad Y = j\omega\varepsilon(a/b). \qquad (9.9)$$

These are constants, the line being uniform.

The parallel-plate line will therefore behave as a transmission line, with propagation constant

$$\gamma = \sqrt{ZY} = \sqrt{j\omega\mu(b/a)j\omega\varepsilon(a/b)} = j\omega\sqrt{\mu\varepsilon}, \qquad (9.10)$$

so that $\gamma = \alpha + j\beta$ makes $\alpha = 0$, which restates that the line with perfectly conducting plates and a perfectly insulating dielectric is lossless, and $\beta = \omega\sqrt{\mu\varepsilon}$, just as for the parallel-wire line. However, the characteristic impedance is

$$Z_0 = \sqrt{Z/Y} = \sqrt{j\omega\mu(b/a)/j\omega\varepsilon(a/b)}$$
$$= \sqrt{\mu/\varepsilon}(b/a) = \eta(b/a). \qquad (9.11)$$

Without further ado, we can declare that the parallel-plate line will carry incident and reflected sinusoidal waves, as

$$V(z) = V_1 e^{-j\beta z} + V_2 e^{j\beta z}, \qquad (9.12)$$

$$Z_0 I(z) = V_1 e^{-j\beta z} - V_2 e^{j\beta z}. \qquad (9.13)$$

The complex amplitudes of the two waves are to be determined from the conditions imposed by the load and the generator. We conclude that signals do propagate in the dielectric sandwich and will get reflected if the load is mismatched.

We note that the parallel-plate line is a two-conductor transmission line, a close relative of the parallel-wire line. It is also open on the two sides when it is not infinitely wide, unlike the coaxial line, which is shielded and confines the fields within a dielectric inside a conducting container. The open geometry of the parallel-plate structure exposes the fields between the plates to disturbing external influences. It raises the question of whether we could close the two sides with conducting side plates and still retain the wave-transmitting property of the original structure. This would turn the parallel-plate geometry into that of a hollow pipe, which is a single-conductor structure. Can this guide a wave?

WAVEGUIDES

We start with the parallel-plate geometry, which has two separate conductors that can carry equal and opposite charges and currents but is open at the sides, where fringing can occur and the fields may be affected by surrounding objects. We propose

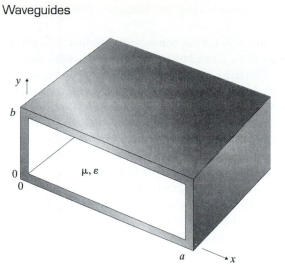

Figure 9-2   Rectangular
waveguide geometry.

to remedy the last complaints by closing off both sides with shorting plates perpendicular to the original ones. The geometry then turns into a hollow conducting pipe
filled with the dielectric; see Figure 9-2. The cross-sectional geometry is rectangular;
this structure may succeed in transmitting waves within and along the pipe, under
proper conditions, and it is referred to as a rectangular waveguide.

The original two plates lie at $y = 0$ and $y = b$ and we now introduce additional plates at $x = 0$ and $x = a$. Before we added the shorting plates at the sides, we
had a vertical E_y field and a horizontal H_x field; the electric field began and ended on
equal and opposite charges on the two horizontal plates and the magnetic field was
generated by the equal and opposite currents on those two plates. With the shorting
plates in place, there is only a single conductor. The question is, Can the fields survive
within the pipe without a second conductor for the return currrent and for the opposite charge?

Comparing the electrical provisions without and with the shorting plates, we
see in the first case equal and opposite charges attracted to each other but kept apart
by the dielectric but then in the second instance an opportunity for the capacitor to
discharge through the shorting plates; refer to Figure 9-3.

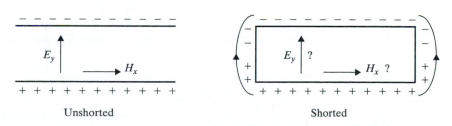

Unshorted   Shorted

Figure 9-3   Charges on parallel plates before and after shorting.

We expect the charges that would have remained separated to rush to their opposite kind along the new side walls, canceling each other when they meet and thereby discharging the capacitor. The field E_y should collapse as the capacitor discharges, and the H_x field that depended on it should collapse along with it. It seems that the side walls should make the structure inoperative as a transmission line.

But wait. The charges that rush around the sides of the structure to discharge the capacitor constitute a current. That shunt discharge current will generate a magnetic field, in the axial direction, H_z, and its magnetic flux, being time varying, will induce an emf. That emf will include a vertical electric field, E_y, as in Figure 9-4, and the question now becomes, Can this vertical electric field created by the time-varying discharge currents on the two sides sustain the otherwise collapsing vertical field?

We know this cannot succeed at dc or at low frequencies, as we need time variation of the axial magnetic flux rapid enough to sustain the vertical field that accompanies the disappearing opposite charges. How fast must the fields vary for the fields not to collapse? To quantify the requirement, let us evaluate the emf around the rectangular area from variable position x to a, the location of the end wall. Since there cannot be a vertical electric field at the perfectly conducting vertical wall itself, and since there is no horizontal electric field at all, the emf around the rectangle reduces to just the contribution at x:

$$\oint \mathbf{E} \cdot d\mathbf{l} = E_y(x)b - E_y(a)b = E_y(x)b - 0$$

$$= +j\omega\Phi = j\omega\mu \int_x^a H_z(x')b\,dx' \tag{9.14}$$

or

$$E_y(x) = j\omega\mu \int_x^a H_z(x')\,dx'. \tag{9.15}$$

This asserts that, while the end wall at $x = a$ imposes zero vertical electric field there, it need not demand zero vertical field elsewhere; E_y need not collapse, if the frequency ω is high enough to furnish enough emf to sustain it. Note also that E_y must also vanish at $x = 0$ where there is another vertical conducting wall. This requires $\int_0^a H_z\,dx = 0$, so that H_z must have zero area under its own curve. This too is consistent with our picture of how H_z is generated; Figure 9-4 shows that the shunt currents at the two walls generate axial magnetic fields in opposite directions near those walls.

But E_y must now become a function of x, being zero at $x = a$ and at $x = 0$ but not elsewhere, as well as a function of z for the pipe to act as a transmission line. The

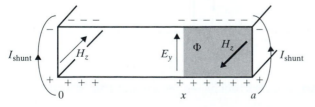

Figure 9-4 Shunt discharge currents create an axial magnetic field.

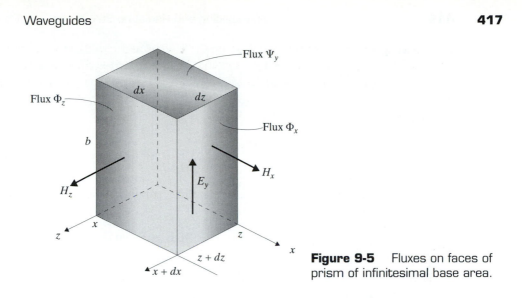

Figure 9-5   Fluxes on faces of prism of infinitesimal base area.

same will be the case for H_x and the newly created H_z. How can we deal with spatial dependence on both x and z?

Consider an infinitesimal area, dx by dz, located at (x, z) and extended in the y-direction the full height b of the structure, as suggested by Figure 9-5. If our design succeeds, there should be created the field components E_y, H_x, H_z within this region. Correspondingly, there will be fluxes labeled Ψ_y, Φ_x, Φ_z in the figure (the notation is not to imply that these fluxes are components of some vector; the subscripts are merely labels). The three fluxes are simply given by

$$\Psi_y = \varepsilon E_y \, dx \, dz, \qquad \Phi_x = \mu H_x b \, dz, \qquad \Phi_z = \mu H_z b \, dx. \tag{9.16}$$

We can now apply Maxwell's equations to these fluxes, by evaluating the emf and mmf around the appropriate faces of the prism in the figure. We get first

$$j\omega \Psi_y = j\omega\varepsilon E_y \, dx \, dz = \oint \mathbf{H} \cdot d\mathbf{l}$$

$$= H_x(z + dz)\, dx - H_z(x + dx)\, dz - H_x(z)\, dx + H_z(x)\, dz. \tag{9.17}$$

Dividing by $(dx\, dz)$, we find

$$\frac{\partial H_x}{\partial z} - \frac{\partial H_z}{\partial x} = j\omega\varepsilon E_y. \tag{9.18}$$

Next, we get

$$-j\omega\Phi_x = -j\omega\mu H_x b \, dz = \oint \mathbf{E} \cdot d\mathbf{l} = bE_y(z) - bE_y(z + dz). \tag{9.19}$$

Dividing by $(b\, dz)$, we find

$$\frac{\partial E_y}{\partial z} = j\omega\mu H_x. \tag{9.20}$$

Finally, we get

$$-j\omega\Phi_z = -j\omega\mu H_z b\, dx = \oint \mathbf{E} \cdot d\mathbf{l} = bE_y(x + dx) - bE_y(x). \quad (9.21)$$

Dividing by $(b\, dx)$, we find

$$\frac{\partial E_y}{\partial x} = -j\omega\mu H_z. \quad (9.22)$$

We should note that these three partial differential equations could have been obtained directly from

$$\nabla \times \mathbf{H} = j\omega\varepsilon\mathbf{E}, \qquad \nabla \times \mathbf{E} = -j\omega\mu\mathbf{H} \quad (9.23)$$

by applying these operators to $E_y(x, z)$, $H_x(x, z)$, $H_z(x, z)$. What we have just done is merely to rederive the expressions for the curl operator in rectangular coordinates, for this case of fields that lack any y-variation.

The three partial differential equations are coupled (the equation for E_y involves H_x or H_z); we can avoid juggling all three by first uncoupling them. We simply substitute the last two into the first of the three, getting

$$(1/j\omega\mu)\partial^2 E_y/\partial z^2 - (-1/j\omega\mu)\partial^2 E_y/\partial x^2 = j\omega\varepsilon E_y \quad (9.24)$$

or

$$\partial^2 E_y/\partial x^2 + \partial^2 E_y/\partial z^2 = -\omega^2\mu\varepsilon E_y. \quad (9.25)$$

We now have uncoupled the equations, at the cost of dealing with a second-order partial differential equation, instead of coupled first-order equations. Once we solve for E_y, we can get H_x and H_z at once from (9.20) and (9.22).

We note in passing that we could have obtained this second-order equation for E_y directly from Maxwell's equations, by taking the curl of one of them and applying a vector identity: From $\nabla \times \mathbf{E} = -j\omega\mu\mathbf{H}$ and $\nabla \times \mathbf{H} = j\omega\varepsilon\mathbf{E}$ we get

$$\nabla \times \nabla \times \mathbf{E} = -j\omega\mu\nabla \times \mathbf{H} = -j\omega\mu j\omega\varepsilon\mathbf{E} = \omega^2\mu\varepsilon\mathbf{E} \quad (9.26)$$

and the identity

$$\nabla \times \nabla \times \mathbf{E} = \nabla\nabla \cdot \mathbf{E} - \nabla^2\mathbf{E}, \quad (9.27)$$

together with $\nabla \cdot \mathbf{E} = 0$ (from $j\omega\varepsilon\mathbf{E} = \nabla \times \mathbf{H}$ and the identity $\nabla \cdot \nabla \times \mathbf{H} = 0$), gives us

$$\nabla^2\mathbf{E} = -\omega^2\mu\varepsilon\mathbf{E}. \quad (9.28)$$

This is called the *wave equation in the steady state* or also the Helmholtz equation. The differential operator ∇^2 is just the sum of the second partial derivatives in the three rectangular coordinates; in our case, there is no y-dependence and we get just (9.25). We now undertake to solve this equation for the vertical electric field in the rectangular waveguide.

SEPARATION OF VARIABLES

The simplest approach to solving the partial differential equation

$$\partial^2 E_y/\partial x^2 + \partial^2 E_y/\partial z^2 + \omega^2\mu\varepsilon E_y = 0 \qquad (9.29)$$

is by *separation of variables*. We are searching for a function $E_y(x, z)$ of two variables, x and z, but the task would be considerably simpler if we could restrict ourselves to finding a product form

$$E_y(x, z) = F(x)G(z) \qquad (9.30)$$

that is a function of x alone multiplied by a function of z alone. This seems highly restrictive, but if we can decompose the unknown E_y in this way, it may turn out that there are many such solutions and, as for any linear equation, we can add or superimpose them to form other solutions of greater generality. Note that we are now after two unknown functions, F and G, rather than just one, E_y, but the dependence of each on only one variable is a simplification that far outweighs this inconvenience.

Accordingly, we substitute the assumed product form into the partial differential equation and obtain

$$(d^2F/dx^2)G + F(d^2G/dz^2) + \omega^2\mu\varepsilon FG = 0 \qquad (9.31)$$

and the technique entails an attempt to separate the variables x and z by making them appear in separate terms. If we divide the last equation by $F(x)G(z)$, then the first term will depend on only x, the second one on only z, and the third one on neither. This represents success of our attempt to separate the variables, and we find

$$(d^2F/dx^2)/F + (d^2G/dz^2)/G + \omega^2\mu\varepsilon = 0. \qquad (9.32)$$

We have here some function of x, plus some other function of z, plus a constant, all adding up to zero. We argue that the only way this can happen is if the first two of these terms are also constants (independent of either x or z). If this is not obvious, we need only differentiate this entire equation with respect to one of the two variables, say x. Since the second and third terms then give zero, we see that the x-derivative of the first term is zero, which means that it is actually independent of x (as well as of z); call this constant K. With the first and third terms of the equation acknowledged to be constants, we see that the second term must also be a constant. We are left with

$$d^2F/dx^2 = KF(x), \qquad d^2G/dz^2 = -[\omega^2\mu\varepsilon + K]G(z), \qquad (9.33)$$

which are two separate ordinary differential equations, with constant coefficients, for the two unknown functions $F(x)$ and $G(z)$ of a single variable each, instead of the original single partial differential equation in two variables.

These two equations are so simple that we could just proceed to solve them and complete our search for product solutions of the original partial differential equation. From an educational standpoint, however, it is desirable to slow our headlong plunge into the intricacies of the solutions and first try to learn more about the constant K in each of the two ordinary differential equations. We can do this by interpreting these equations.

Recall that the second derivative of a function measures the curvature of that function, the rate of change of its slope. The first of the two ordinary differential equations, $d^2F/dx^2 = KF$, asserts that the curvature of the unknown $F(x)$ function is proportional to the value of the function itself, with K as the proportionality constant. Recall that the graph of a function with positive curvature (positive rate of change of slope) curves upward, while one with negative curvature curves downward. If K is a positive constant, then $F(x)$ is a function that curves upward when its value is positive and curves downward when it lies below the x-axis. It is possible for the graph of such a function to cross the x-axis but, to either side of that crossing point, the graph will turn away from the axis. Such behavior is illustrated in Figure 9-6. On the other hand, if the constant K is negative, then $F(x)$ is a function that curves downward when it lies above the x-axis and curves upward when below it; that is, it always curves toward the x-axis. That behavior is also depicted in the figure. The key point is that if the constant K is positive, the graph of the $F(x)$ function can cross the x-axis no more than once but if K is negative, the graph can cross the axis again and again.

We are therefore in a position to declare that the constant K must be negative. This is because the electric field $E_y = F(x)G(z)$ is required to vanish at both walls, at $x = 0$ and at $x = a$, for all z. The graph of the x-dependence of this field, $F(x)$, must therefore cross the x-axis at least twice: $F(0) = 0$ and $F(a) = 0$. This precludes a positive value of the constant K, the ratio of the curvature of $F(x)$ to its value. Note also that a zero value of the constant K would call for no curvature at all, a straight-line graph, which can also not cross the axis at two points. We conclude that the con-

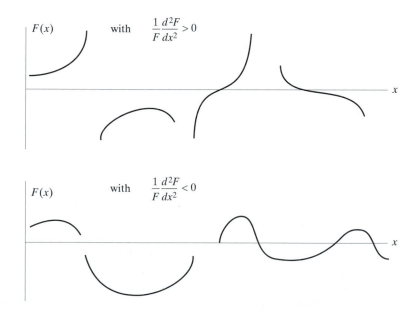

Figure 9-6   Typical graphs with positive and negative curvature/value ratios.

stant K must be negative; we will emphasize this by renaming it $K = -p^2$, with p another unknown but real constant.

We are left with the two equations

$$d^2F/dx^2 + p^2F(x) = 0, \qquad d^2G/dz^2 = [p^2 - \omega^2\mu\varepsilon]G(z) \qquad (9.34)$$

and the first of these is subject to the pair of boundary conditions

$$F(0) = 0 \quad \text{and} \quad F(a) = 0. \qquad (9.35)$$

We of course recognize the equation for $F(x)$ as solvable by trigonometric functions, $\sin px$ and $\cos px$, which indeed oscillate and cross the axis repeatedly. Because of the condition $F(0) = 0$, we retain only the $F(x) = A \sin px$ solution; we reject the $\cos px$ possibility as being nonzero at $x = 0$. To satisfy the second boundary condition, $F(a) = 0$, we must impose $A \sin pa = 0$, but we cannot allow the amplitude constant A to be zero, since that would leave no electric field at all. Instead we declare that we must have $\sin pa = 0$, or

$$pa = m\pi \quad (m = \text{integer}) \quad \text{or} \quad p = m\pi/a. \qquad (9.36)$$

This gives us the value of the previously undetermined *separation constant* p. The solution for the $F(x)$ factor of the electric field is therefore

$$F(x) = A \sin(m\pi x/a), \qquad (9.37)$$

with any amplitude A of this sinusoidal distribution. We see that we have infinitely many such solutions, one for each (nonzero, positive) integer m; we could superimpose any number of such solutions, each with its own amplitude A_m; note that negative integers give us no additional solutions, as $-m$ offers the same solution as m but with a negative amplitude, and that $m = 0$ leaves no field at all.

Turning our attention to the z-dependence $G(z)$ of the electric field will answer the question of whether this field can or cannot propagate along the hollow pipe. The equation for $G(z)$ is

$$d^2G/dz^2 = [p^2 - \omega^2\mu\varepsilon]G(z), \qquad (9.38)$$

in which we now know that p is one of the infinitely many permissible values $m\pi/a$; evidently, there will be a separate $G(z)$ function for each choice of the integer m and the corresponding $F(x)$ factor. The quantity in the brackets is just a known constant, so that the equation is solved by exponentials:

$$G(z) = G_1 e^{-\gamma z} + G_2 e^{\gamma z}, \qquad (9.39)$$

where G_1 and G_2 are constants and

$$\gamma^2 = p^2 - \omega^2\mu\varepsilon = (m\pi/a)^2 - \omega^2\mu\varepsilon. \qquad (9.40)$$

The product solution for the electric field, corresponding to the integer m, is therefore

$$E_y(x, z) = F(x)G(z) = A \sin(m\pi x/a)[G_1 e^{-\gamma z} + G_2 e^{\gamma z}] \qquad (9.41)$$

or, combining constants,

$$E_y(x, z) = \sin(m\pi x/a)[E_1 e^{-\gamma z} + E_2 e^{\gamma z}]. \qquad (9.42)$$

Does this field propagate or not?

The answer depends on whether γ is real, imaginary, or complex. We know how to interpret a complex exponent $\gamma = \alpha + j\beta$; the real part α represents the spatial rate of attenuation of the field and the imaginary part β is the phase constant for a propagating wave, as in $e^{-\alpha z}e^{j(\omega t - \beta z)}$. We need only find the real and imaginary parts of γ for our case of the hollow rectangular pipe to determine whether, and how well, the field can be transmitted along the pipe.

From

$$\gamma = \sqrt{(m\pi/a)^2 - \omega^2\mu\varepsilon}, \tag{9.43}$$

we see that γ is, for the case of a hollow pipe comprised of a perfect conductor surrounding a perfect dielectric, either purely real or purely imaginary. Which one prevails depends on the frequency ω. If the frequency is too low, the exponent γ is purely real, $\gamma = \alpha$, and there is only attenuation of the field, without propagation. If the frequency is high enough, the exponent γ is purely imaginary, $\gamma = j\beta$, and there is propagation, without attenuation. The frequency below which there is only attenuation and above which there is propagation is called the *cutoff* frequency, ω_c. For our case, this is

$$\omega_c = m\pi/\left[a\sqrt{\mu\varepsilon}\right] \tag{9.44}$$

and depends on the integer m that distiguishes one x-dependence of the electric field, $\sin(m\pi x/a)$, from another permissible one.

The analysis has confirmed precisely the behavior we anticipated for the hollow rectangular pipe, that we would need a sufficiently high frequency of operation in order to get enough induced emf from the rate of change of magnetic flux to sustain the electric field that would otherwise collapse. We now have a specific minimum frequency $\omega = \omega_c$ for the pipe to act as a waveguide. We conclude that signals do propagate in the hollow rectangular pipe, provided we operate above the cutoff frequency of the desired permissible field pattern.

Example 9.1

An air-filled rectangular waveguide in common use (designated WR-90) has inner dimensions $a = 0.900$ in by $b = 0.400$ in. What is the minimum frequency f_{min} at which it can be operated as a waveguide? What is the wavelength of the propagating wave if we operate at $f = (3/2)f_{min}$? How far along the pipe will the field strength be at least 1% of its value at the emitter if we operate instead at $f = (2/3)f_{min}$ and there is no reflected field? ◈

Taking air as electrically equivalent to a vacuum, we have for the lowest cutoff frequency ($m = 1$)

$$\omega_c = \pi c/a = 2\pi f_{min} \tag{9.45}$$

or

$$f_{min} = c/2a$$

$$= \left[2.998 \cdot 10^8 \text{ m/sec}\right]/\left[(2 \cdot 0.900 \text{ in})(2.54 \text{ cm/in})(10^{-2} \text{ m/cm})\right]$$

$$= 6.557 \text{ GHz.}$$

Next, from

$$\gamma^2 = (\pi/a)^2 - \omega^2/c^2 = \omega_c^2/c^2 - \omega^2/c^2 = -\beta^2, \qquad (9.46)$$

we get, with $\omega = (3/2)\omega_c$,

$$\beta^2 = (3\omega_c/2c)^2 - (\omega_c/c)^2 = (5/4)(\omega_c/c)^2 = (5/4)(\pi/a)^2$$

or $2\pi/\lambda = [\sqrt{5}/2]\pi/a$, or $\lambda = 4a/\sqrt{5} = 1.610$ in as the wavelength at frequency $f = (3/2)f_{\min}$.

Last, from

$$\gamma^2 = (\pi/a)^2 - \omega^2/c^2 = \omega_c^2/c^2 - \omega^2/c^2 = \alpha^2, \qquad (9.47)$$

we get, with $\omega = (2/3)\omega_c$,

$$\alpha^2 = (\omega_c/c)^2 - (2\omega_c/3c)^2 = (5/9)(\omega_c/c)^2 = (5/9)(\pi/a)^2$$

or $\alpha = [\sqrt{5}/3]\pi/a$ and we want the distance z such that $e^{-\alpha z} = 0.01$, so that we calculate $z = \ln(100)/\alpha = a \cdot 3 \cdot \ln(100)/[\sqrt{5}\pi] = 1.9667a = 1.770$ in as the distance into the waveguide over which the field drops to 1% of its strength at the entrance, at frequency $f = (2/3)f_{\min}$.

TE AND TM MODES

We have found infinitely many product solutions to the partial differential equation that describes the transverse and axial dependence of the electric field in the rectangular pipe, one for each positive integer m in the preceding equations. Each such solution has its own minimum frequency for propagation; if we fall below that cutoff frequency, the field attenuates axially, instead of propagating along the pipe.

For an empty pipe of width a, the cutoff frequency is $\omega_c = m\pi c/a$ and the lowest frequency at which we can begin to achieve propagation along the waveguide is for the case $m = 1$. The electric field's transverse variation is then proportional to $\sin(\pi x/a)$; this is illustrated in Figure 9-7, along with that of the $m = 5$ version for comparison. Each such solution, and the field patterns and properties associated with it, is called a *mode* of wave propagation.

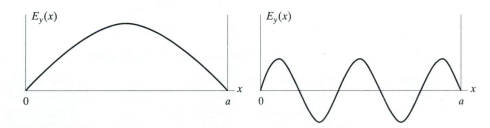

Figure 9-7   Transverse field variation in rectangular waveguide, for $m = 1$ and $m = 5$.

We can superimpose different modes to arrive at solutions that are not restricted to be product solutions. For example, at the emitter at $z = 0$, we may have a field described by

$$E_y(x, 0) = E_1 \sin(\pi x/a) + E_2 \sin(2\pi x/a) + E_3 \sin(3\pi x/a) + \cdots \quad (9.48)$$

and we recognize this to be simply a Fourier series, which can represent any reasonable field distribution that is zero at both $x = 0$ and $x = a$. What we have discovered is that the z-dependence of this emitted field is obtained by appending the appropriate $e^{\pm\gamma z}$ factor to each of the sinusoids. To be more specific, suppose we have precluded reflections and that we are operating at a frequency ω that is above $2\pi c/a$ but below $3\pi c/a$. Then γ will be imaginary for only $m = 1$ and $m = 2$ and will be real for all higher modes. The field at some location z in the pipe will then be

$$E_y(x, y) = E_1 e^{-j\beta_1 z} \sin(\pi x/a) + E_2 e^{-j\beta_2 z} \sin(2\pi x/a)$$
$$+ E_3 e^{-\alpha_3 z} \sin(3\pi x/a) + E_4 e^{-\alpha_4 z} \sin(4\pi x/a) + \cdots, \quad (9.49)$$

where

$$\beta_m = \sqrt{\omega^2/c^2 - (m\pi/a)^2}, \qquad \alpha_m = \sqrt{(m\pi/a)^2 - \omega^2/c^2}. \quad (9.50)$$

Evidently, the first two modes can propagate indefinitely far along the waveguide while all higher-order modes attenuate, or *evanesce*, along it. At a sufficient distance from the emitter, we are left with a field described by only the two harmonics, $m = 1$ and $m = 2$, in the Fourier series.

It is often desirable to allow only one mode to propagate, while all others are below their respective cutoff frequencies. That mode, the one with the lowest cutoff frequency ($m = 1$), is called the *dominant* mode and we simply operate at a frequency that is above its cutoff but below the cutoff of all other modes. Knowing the field distribution of the lone surviving mode of propagation along the waveguide eases the task of designing a suitable load to receive the transmitted signal or energy. Let us examine the fields of the dominant mode.

Assuming the operating frequency is above cutoff and that reflections have been averted by using a matched load, the electric field is

$$E_y(x, z) = E_1 e^{-j\beta z} \sin(\pi x/a). \quad (9.51)$$

From $H_x = (1/j\omega\mu)\partial E_y/\partial z$ and $H_z = (-1/j\omega\mu)\partial E_y/\partial x$, we get the components of the magnetic field that accompanies this electric field as

$$H_x(x, z) = (-\beta/\omega\mu)E_1 e^{-j\beta z} \sin(\pi x/a), \quad (9.52)$$

$$H_z(x, z) = (j\pi/\omega\mu a)E_1 e^{-j\beta z} \cos(\pi x/a). \quad (9.53)$$

The phase constant β is given by $\beta^2 = (\omega/c)^2 - (\pi/a)^2$. A notable difference between the electric and magnetic field patterns of this mode is that the electric field is only transverse, while the magnetic field has both transverse and axial components. To emphasize this feature, the mode is called *transverse electric*, or TE, and to identify the mode more specifically, it is labeled TE_{10}. The subscripts tell us that the transverse electric field displays one sinusoidal peak along the width of the waveguide but none along its height. The alternate mode illustrated in the last figure is labeled TE_{50}.

The sinusoidal variation of the field across the pipe can be considered a standing wave pattern along x, one that propagates along z as a wave if the mode is above cutoff. That standing wave can be thought of as a wave and its reflection bouncing back and forth from the side walls at $x = 0$ and $x = a$. In the same way, we could also have a standing wave bouncing vertically between the floor and ceiling of the pipe, at $y = 0$ and $y = b$. In our calculation, we explicitly excluded any y-variation of the fields, but there is no necessity for this.

As discussed earlier, the two Maxwell equations in the steady state—namely, $\nabla \times \mathbf{E} = -j\omega\mu\mathbf{H}$ and $\nabla \times \mathbf{H} = j\omega\varepsilon\mathbf{E}$—can be combined into the Helmholtz equation for either the electric or the magnetic field, and hence for any of their rectangular components, as

$$\nabla^2\mathbf{E} = -\omega^2\mu\varepsilon\mathbf{E}, \qquad \nabla^2\mathbf{H} = -\omega^2\mu\varepsilon\mathbf{H}. \qquad (9.54)$$

Separation of variables as, say, $H_z = X(x)Y(y)Z(z)$, leads to sinusoids in both x and y and exponentials in z. To impose zero tangential electric field at $x = 0$ and $x = a$ and at $y = 0$ and $y = b$, the separation constants in the sinusoids have to be multiples of π/a and π/b and the Helmholtz equations are satisfied if the variation in z—namely, $e^{\pm\gamma z}$—is given by

$$\gamma^2 = (m\pi/a)^2 + (n\pi/b)^2 - \omega^2\mu\varepsilon, \qquad (9.55)$$

where m and n are integers. Once again, γ is imaginary and there is propagation only if the operating frequency ω is higher than the cutoff frequency ω_c given by

$$\omega_c^2 = (m\pi c/a)^2 + (n\pi c/b)^2 \qquad (9.56)$$

for an empty pipe of width a and height b. Transverse electric modes with such variation in x and y are designated TE_{mn}.

> **Example 9.2**
>
> *We want to operate a rectangular waveguide at a frequency that is to be at least 30% higher than the cutoff frequency of the TE_{10} mode and at least 30% below the cutoff frequency of the TE_{01} mode. What is the smallest aspect ratio a/b of the dimensions of the waveguide that we can choose?* ◈
>
> From $\omega_c^2 = (m\pi c/a)^2 + (n\pi c/b)^2$ for a rectangular waveguide, we find that $f_c = c/2a$ for the TE_{10} mode and that $f_c = c/2b$ for the TE_{01} mode. We want $f \geq 1.30(c/2a)$ and $f \leq 0.70(c/2b)$, so the smallest ratio a/b uses the smallest a and largest b, giving $(a/b)_{\min} = 13/7 = 1.857$.

To enrich further the possibilities for wave propagation along a rectangular waveguide, we can solve the Helmholtz equation as it applies to the axial electric field E_z instead of H_z. It is the same equation with the same separation of variables, but we still impose the vanishing of the tangential electric field at the walls, floor, and ceiling of the pipe, not of the magnetic field. What we get then is a set of *transverse magnetic* modes, designated TM_{mn} when the axial field displays m sinusoidal peaks across the width and n across the height of the pipe. It is the magnetic field that is then only transverse; there is an axial component to the electric field. This axial electric field furnishes axial displacement current, and if the time variation of the fields, their frequency, is sufficiently high, this displacement current in the empty space inside the pipe can substitute for the conduction current carried by the center wire of a coaxial line, for example. The general solution for fields inside a waveguide is a superposition of all possible TE and all possible TM modes.

Example 9.3

The preceding discussion of transverse magnetic (TM) modes of a rectangular waveguide suggests an axial electric field to carry the longitudinal displacement current, of the form $E_z = E_0 X(x)Y(y)e^{-\gamma z}$, with $X(x) = \sin(m\pi x/a)$ and $Y(y) = \sin(n\pi y/b)$ to ensure that the axial electric field vanish at the four walls of the waveguide. The magnetic field that surrounds the axial displacement current includes $H_x = \xi E_0 X(x)Y'(y)e^{-\gamma z}$ and $H_y = -\xi E_0 X'(x)Y(y)e^{-\gamma z}$, for some appropriate constant ξ; there is no axial magnetic field in a TM mode.

- **(a)** *On the basis of the Helmholtz equation $\nabla^2 E_z + \omega^2 \mu \varepsilon E_z = 0$, determine the relation of the propagation constant γ to the frequency ω.*
- **(b)** *Find the constant ξ that makes the magnetic field consistent with E_z.*
- **(c)** *Obtain the E_x and E_y field components.*
- **(d)** *Why is there not a TM_{10} mode, a counterpart of the dominant TE_{10}?* ◈

- **(a)** We will need $\nabla^2 E_z = E_0[X''(x)Y(y) + X(x)Y''(y) + \gamma^2 X(x)Y(y)]e^{-\gamma z}$ and, since $X'' = -(m\pi/a)^2 X$ and $Y'' = -(n\pi/b)^2 Y$, we have

$$\nabla^2 E_z = E_0[-(m\pi/a)^2 - (n\pi/b)^2 + \gamma^2]X(x)Y(y)e^{-\gamma z}$$
$$= -\omega^2 \mu \varepsilon E_z = E_0[-\omega^2 \mu \varepsilon]X(x)Y(y)e^{-\gamma z}.$$

   This is satisfied if $\gamma^2 + \omega^2 \mu \varepsilon = (m\pi/a)^2 + (n\pi/b)^2 \equiv p^2$.

- **(b)** The z-component of $\nabla \times \mathbf{H}$ must agree with $j\omega\varepsilon E_z$. This is $\partial H_y/\partial x - \partial H_x/\partial y$ and equals $-\xi E_0 X''(x)Y(y)e^{-\gamma z} - \xi E_0 X(x)Y''(y)e^{-\gamma z} = \xi p^2 E_0 X(x)Y(y)e^{-\gamma z}$. This will be $j\omega\varepsilon E_z = j\omega\varepsilon E_0 X(x)Y(y)e^{-\gamma z}$ if $\xi p^2 = j\omega\varepsilon$ or $\xi = j\omega\varepsilon/(\gamma^2 + \omega^2\mu\varepsilon)$.

- **(c)** We have $j\omega\varepsilon E_x = (\nabla \times \mathbf{H})_x = -\partial H_y/\partial z = \gamma H_y$ and $j\omega\varepsilon E_y = (\nabla \times \mathbf{H})_y = \partial H_x/\partial z = -\gamma H_x$. Since $\gamma\xi/j\omega\varepsilon = \gamma/(\gamma^2 + \omega^2\mu\varepsilon)$, we find

$$E_x = -[\gamma/(\gamma^2 + \omega^2\mu\varepsilon)]E_0 X'(x)Y(y)e^{-\gamma z},$$
$$E_y = -[\gamma/(\gamma^2 + \omega^2\mu\varepsilon)]E_0 X(x)Y'(y)e^{-\gamma z}.$$

- **(d)** For $n = 0$, we have $Y(y) = \sin(n\pi y/b) = 0$ and $Y'(y) = 0$, which makes all components of $\mathbf{E}$ and $\mathbf{H}$ equal to zero. There is no field, and therefore no such mode.

   For TM modes, we must exclude the possibility of $m = 0$ or $n = 0$ in the expression for the cutoff frequencies in terms of the integers m and n. For TE modes, one of the two integers can be zero.

For both the TE and TM cases, there is a cutoff frequency for each mode; above this frequency the wave can propagate indefinitely, below it the mode is evanescent, decaying exponentially along the pipe instead of propagating. In that sense, the wave-

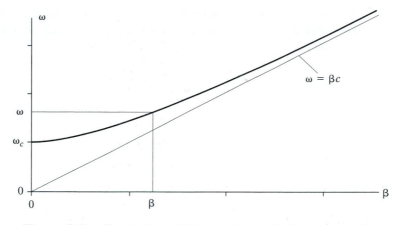

Figure 9-8   Dispersion relation, ω versus β, for a waveguide.

guide behaves as a high-pass filter for each mode. To get propagation, with phase constant β, the frequency must exceed the cutoff frequency ω_c and the phase constant is then given by the *dispersion relation*

$$\omega^2 = \omega_c^2 + \beta^2 c^2. \tag{9.57}$$

As a consequence of the symmetry of the rectangular shape of the waveguide, the expression for the cutoff frequency is the same for both the TE and TM modes, as given in (9.56). The dispersion relation is illustrated in Figure 9-8; it is a hyperbola, asymptotic to a straight line of slope c and with an apex at the cutoff frequency. The curve clearly shows the need to exceed the cutoff frequency to get a real phase constant β and it furnishes the wavelength $2\pi/\beta$ and the phase velocity ω/β for any frequency above cutoff.

Example 9.4

At frequency $f_1 = 10.0\ GHz$, the guide wavelength along a certain lossless dielectric-filled rectangular waveguide is measured as $\lambda_1 = 4.00$ cm, for a particular mode. At $f_2 = 11.0\ GHz$, the guide wavelength is $\lambda_2 = 3.00$ cm, for the same mode.

(a) *What is the cutoff frequency for that mode?*

(b) *What is the relative permittivity $\varepsilon/\varepsilon_0$ of the dielectric?* ◈

When the waveguide is filled with a dielectric of permittivity ε, the dispersion relation uses u_0 instead of a vacuum's c and is altered to

$$\omega^2 = \omega_c^2 + \beta^2 u_0^2 \qquad \text{with} \qquad u_0^2 = 1/\mu\varepsilon. \tag{9.58}$$

(a) From the dispersion relation $\omega^2 = \omega_c^2 + \beta^2/\mu_0\varepsilon = \omega_c^2 + \beta^2 c^2/(\varepsilon/\varepsilon_0)$ and from $\lambda = 2\pi/\beta$ for the guide wavelength, we get $f^2\lambda^2 = f_c^2\lambda^2 + c^2/(\varepsilon/\varepsilon_0)$. Subtracting the equations for the two operating frequencies to eliminate the unknown ε, we find $f_1^2\lambda_1^2 - f_2^2\lambda_2^2 = f_c^2(\lambda_1^2 - \lambda_2^2)$, which immediately gives $f_c = 8.544$ GHz.

(b) From $f^2 = f_c^2 + c^2/[\lambda^2(\varepsilon/\varepsilon_0)]$, subtraction of the two results to eliminate the cutoff frequency gives $[(c/\lambda_2)^2 - (c/\lambda_1)^2]/(f_2^2 - f_1^2) = \varepsilon/\varepsilon_0 = 2.08$.

As a matter of interest, we note that the dispersion curve lies above the asymptote, so that the phase velocity ω/β in an empty waveguide, at any frequency that allows propagation, is greater than the speed of light. This may give us pause. According to the theory of relativity, no bit of matter, or energy, or information can move that fast. But all remains well: The result implies only that no material influence can move fast enough along the waveguide to keep up with a fixed phase of the wave. This does not imply that the wave in the pipe violates the dictates of relativity; no material object moves at the phase velocity of the wave, only a geometrically defined entity, the phase front, does so, and this is unrestricted by relativistic requirements. The waveguide does carry energy and information, but these travel at a speed slower than that of light, as we will now verify.

Example 9.5

> *Obtain the time-averaged power flow across a transverse plane of a rectangular waveguide under dominant mode operation.*
> *Find the total (electric and magnetic) energy stored per unit length in the dominant mode of the waveguide.*
> *Obtain the ratio of the power flow to the energy per unit length. This can be interpreted as the speed of energy flow in the waveguide.* ◈

The time-averaged power flow density is given by $\mathrm{Re}\left[\frac{1}{2}\mathbf{E} \times \mathbf{H}^*\right]$. For the propagating TE_{10} mode, this is

$$\mathrm{Re}\left[\tfrac{1}{2}\mathbf{E} \times \mathbf{H}^*\right] = \mathrm{Re}\left[\tfrac{1}{2}\hat{\mathbf{z}}(-E_y H_x^*) + \tfrac{1}{2}\hat{\mathbf{x}}(E_y H_z^*)\right]$$

$$= \mathrm{Re}\left[\tfrac{1}{2}\hat{\mathbf{z}}(\beta/\omega\mu)|E_1|^2 \sin^2(\pi x/a) \right.$$

$$\left. + \tfrac{1}{2}\hat{\mathbf{x}}(-j\pi/\omega\mu a)|E_1|^2 \sin(\pi x/a)\cos(\pi x/a)\right]$$

$$= \tfrac{1}{2}\hat{\mathbf{z}}(\beta/\omega\mu)|E_1|^2 \sin^2(\pi x/a) \tag{9.59}$$

and this integrates across the area of the waveguide to the power flow

$$P = \int \mathrm{Re}\left[\tfrac{1}{2}\mathbf{E} \times \mathbf{H}^*\right] \cdot d\mathbf{S} = \int_0^b \int_0^a \tfrac{1}{2}\hat{\mathbf{z}}(\beta/\omega\mu)|E_1|^2 \sin^2(\pi x/a) \cdot \hat{\mathbf{z}}\, dx\, dy$$

$$= (ab/4)(\beta/\omega\mu)|E_1|^2. \tag{9.60}$$

Next, the time-averaged electric energy per unit volume is

$$w_e = \tfrac{1}{4}\varepsilon|E|^2 = \tfrac{1}{4}\varepsilon|E_1|^2 \sin^2(\pi x/a) \tag{9.61}$$

and this integrates across the waveguide to the electric energy per unit length

$$W_e = (ab/8)\varepsilon|E_1|^2. \tag{9.62}$$

For the time-averaged magnetic energy per unit volume, we have

$$w_m = \tfrac{1}{4}\mu|H|^2 = \tfrac{1}{4}\mu\left[|H_x|^2 + |H_z|^2\right]$$

$$= \tfrac{1}{4}\mu|E_1|^2\left[(\beta/\omega\mu)^2 \sin^2(\pi x/a) + (\pi/\omega\mu a)^2 \cos^2(\pi x/a)\right] \tag{9.63}$$

and this integrates across the waveguide to the magnetic energy per unit length

$$W_m = (ab/8)\mu|E_1|^2[(\beta/\omega\mu)^2 + (\pi/\omega\mu a)^2]. \tag{9.64}$$

But since $\beta^2 + (\pi/a)^2 = \omega^2\mu\varepsilon$ according to the dispersion relation, this is also

$$W_m = (ab/8)\mu|E_1|^2[\omega^2\mu\varepsilon/\omega^2\mu^2] = (ab/8)\varepsilon|E_1|^2. \tag{9.65}$$

The total time-averaged energy per unit length stored in the guide is

$$W = W_e + W_m = (ab/4)\varepsilon|E_1|^2. \tag{9.66}$$

Finally, the ratio of the power flow to the energy per unit length is

$$v = P/W = (ab/4)(\beta/\omega\mu)|E_1|^2/(ab/4)\varepsilon|E_1|^2$$
$$= \beta/\omega\mu\varepsilon = c^2/(\omega/\beta) \tag{9.67}$$

for an empty waveguide. This is the square of the speed of light, divided by the phase velocity of the propagating wave. Since power flow is the rate of transfer of energy, the ratio P/W can be interpreted as the speed of this transfer of energy across the transverse plane.

We saw that the phase velocity ω/β exceeds the speed of light; hence, v is less than c and we conclude that the energy velocity is less than the speed of light, in conformity with relativity.

A *cavity* is a hollow space completely enclosed by a conducting wall. Harmonic fields can exist within a cavity if they oscillate at certain frequencies. A simple cavity is formed by a rectangular waveguide closed at both ends by transverse conducting walls, forming a box. A cavity, or cavity resonator, is characterized by a set of natural frequencies of oscillation of its internal fields.

Example 9.6

An air-filled rectangular waveguide has dimensions $a = 3.0$ cm by $b = 1.5$ cm. It is closed by transverse walls $l = 5.0$ cm apart. What is the lowest frequency at which the cavity can sustain oscillations? ◈

The TE_{10} mode of the rectangular waveguide becomes a standing wave, with transverse electric field $E(x, z) = E_0 \sin(\pi x/a) \sin\beta z$, with $\beta = \pi/l$ in order to satisfy the boundary condition of zero transverse electric field at both $z = 0$ and $z = l$ with the smallest possible β. But, for the TE_{10} waveguide mode, we have $\omega^2 = (\pi c/a)^2 + \beta^2 c^2$ or $f = \sqrt{(c/2a)^2 + (c/2l)^2}$. Here, for $a = 3.0$ cm and $l = 5.0$ cm, we get $f = 5.827$ GHz as the smallest natural frequency.

INTERPRETATION OF WAVEGUIDE MODES

We noted earlier that the sinusoidal variation of the field across the rectangular waveguide can be considered a standing wave pattern along x, one that propagates along z. We can be more explicit about this, and thereby gain some insight into the operation of the waveguide.

The simplest of the rectangular waveguide modes we have examined is the TE_{10} mode. Its fields are

$$\mathbf{H} = C[\hat{\mathbf{z}}\cos px + \hat{\mathbf{x}}j(\beta/p)\sin px], \quad \mathbf{E} = (-j\omega\mu/p)C\hat{\mathbf{y}}\sin px, \quad (9.68)$$

where $p = \pi/a$ and C is the amplitude of the axial magnetic field. For purposes of interpretation, we can rewrite the sinusoidal pattern as a sum of two exponentials, as in

$$\mathbf{E} = (-j\omega\mu/p)C\hat{\mathbf{y}}\sin px = C(\omega\mu/2p)\hat{\mathbf{y}}[e^{-jpx} - e^{jpx}] \qquad (9.69)$$

$$\mathbf{H} = C[\hat{\mathbf{z}}\cos px + \hat{\mathbf{x}}j(\beta/p)\sin px]$$

$$= \tfrac{1}{2}C\hat{\mathbf{z}}[e^{-jpx} + e^{jpx}] - \tfrac{1}{2}C(\beta/p)\hat{\mathbf{x}}[e^{-jpx} - e^{jpx}], \qquad (9.70)$$

so that, with the axial propagation factor $e^{-j\beta z}$, the entire field is seen to be a superposition of two obliquely traveling plane waves:

$$\mathbf{E} = C(\omega\mu/2p)\hat{\mathbf{y}}[e^{-j(px+\beta z)} - e^{j(px-\beta z)}], \qquad (9.71)$$

$$\mathbf{H} = (C/2p)[\hat{\mathbf{z}}p - \hat{\mathbf{x}}\beta]e^{-j(px+\beta z)} + (C/2p)[\hat{\mathbf{z}}p + \hat{\mathbf{x}}\beta]e^{j(px-\beta z)}. \qquad (9.72)$$

The first of these oblique waves travels along $\mathbf{k}_1 = p\hat{\mathbf{x}} + \beta\hat{\mathbf{z}}$ and the other along $\mathbf{k}_2 = -p\hat{\mathbf{x}} + \beta\hat{\mathbf{z}}$. The fields can be written as

$$\mathbf{E} = C(\omega\mu/2p)\hat{\mathbf{y}}[e^{-j\mathbf{k}_1\cdot\mathbf{r}} - e^{-j\mathbf{k}_2\cdot\mathbf{r}}], \qquad (9.73)$$

$$\mathbf{H} = (C/2p)[\mathbf{k}_1 \times \hat{\mathbf{y}}e^{-j\mathbf{k}_1\cdot\mathbf{r}} - \mathbf{k}_2 \times \hat{\mathbf{y}}e^{-j\mathbf{k}_2\cdot\mathbf{r}}]. \qquad (9.74)$$

These should be recognized as superpositions of two plane waves. Because the $\mathbf{k}$'s differ only in the sign of their transverse component, the two plane waves are directed at equal and opposite angles to the waveguide's axis. They reflect and bounce back and forth between the parallel walls at $x = 0$ and $x = a$, each becoming the other upon reflection at either wall, while the combination of two oblique waves progresses along the z-axis in this zig-zag fashion. Figure 9-9 shows how the two oblique waves that comprise the TE_{10} mode are related to the waveguide geometry.

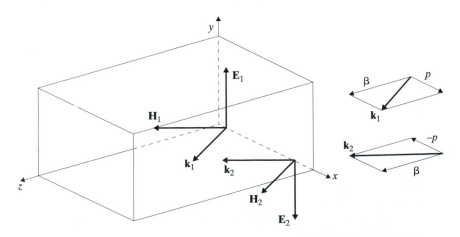

Figure 9-9 TE_{10} mode as a pair of oblique plane waves.

WAVE DISPERSION AND GROUP VELOCITY

Having learned how electromagnetic fields travel along a waveguide, by bouncing back and forth between the side walls while progressing forward along the guide's axis, and having seen that energy is conveyed by the wave, we can inquire now about the transmission of information along a waveguide. First, we must review basic features of narrow-band signals.

Of itself, a single-frequency sinusoid, forever oscillating with fixed amplitude, phase, and frequency, conveys almost no information at all. It carries no message other than its own existence and the duration of its period. An information-bearing signal is comprised of an entire spectrum of frequencies. How the spectrum is affected when the signal is conveyed from a transmitter to a receiver is an important engineering concern.

In a vacuum, the single-frequency exponential $\exp[j\omega t]$ is transmitted as $\exp[j(\omega t - kz)]$ and the phase constant k is proportional to the frequency ω: $k = k(\omega) = \omega/c$, so that the phase velocity $\omega/k = c$ is the same for all frequency components. All the frequencies in the spectrum are transmitted at the same speed and the entire signal arrives undistorted.

For a medium in which the constitutive parameters μ, ε are dependent on frequency, the phase constant is a more complicated function of frequency: $\beta = \beta(\omega) = \omega\sqrt{\mu(\omega)\varepsilon(\omega)}$, so that the phase velocity $\omega/\beta = 1/\sqrt{\mu(\omega)\varepsilon(\omega)} = v_p(\omega)$ is no longer the same for all frequencies. Different frequencies in the spectrum of the signal are transmitted at different speeds and arrive at different times. The reconstructed signal at the receiver therefore appears distorted. We say the medium is *dispersive* when the phase velocity depends on the frequency. A waveguide is also a dispersive transmission system.

For an important class of information-bearing signals, the distortion caused by dispersion is benign. The full signal does get distorted in a dispersive medium, but the information in the signal can remain undistorted, being merely transmitted at a speed different from that of any single frequency component. This class of signals comprises narrow-band signals, meaning that the spectrum is confined to a relatively narrow range of frequencies.

To interpret this class of narrow-band signals, we can verify that they represent slow modulations of a high-frequency sinusoidal carrier oscillation. If the spectrum $F(\omega)$ of the signal is significantly different from zero only in a narrow band of frequencies near $\omega = \omega_0$, as shown in Figure 9-10, then it can be expressed as a shifted

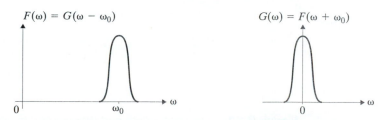

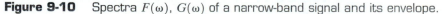

Figure 9-10   Spectra $F(\omega)$, $G(\omega)$ of a narrow-band signal and its envelope.

version $F(\omega) = G(\omega - \omega_0)$ of a spectrum $G(\omega)$ that is confined to the vicinity of zero frequency. The nature of the time-domain signal is revealed by the change of variable $\omega = \omega_0 + \Omega, d\omega = d\Omega$ in the inverse Fourier transform:

$$f(t) = \int F(\omega)e^{j\omega t}\, d\omega/2\pi \qquad (\omega \text{ near } \omega_0)$$

$$= \int F(\omega_0 + \Omega)\exp\left[j(\omega_0 + \Omega)t\right]d\Omega/2\pi \qquad (\Omega \text{ near } 0)$$

$$= e^{j\omega_0 t}\int G(\Omega)e^{j\Omega t}\, d\Omega/2\pi \qquad (\Omega \text{ near } 0)$$

$$= e^{j\omega_0 t}g(t), \qquad (9.75)$$

where $g(t)$ is the inverse Fourier transform of the spectrum $G(\omega)$ that has the same shape as the $F(\omega)$ spectrum, but is centered about zero frequency, not about the central frequency ω_0. The last form of the narrow-band signal $f(t)$ shows it to be a complex exponential $e^{j\omega_0 t}$ multiplied by an amplitude factor $g(t)$ that is a function of time. The product is a sinusoidal oscillation whose amplitude varies in time. Since the spectrum of $g(t)$, which is $G(\omega)$, contains only low frequencies (low compared to ω_0), the time signal $g(t)$ is slowly varying (slow compared to the period $2\pi/\omega_0$). The underlying oscillation of the carrier is therefore still recognizable as a sinusoidal swing over short time intervals, but the amplitude of this oscillation changes slowly, over longer time spans. The slowly varying amplitude is called the envelope of the modulated carrier signal; its nature is suggested schematically in Figure 9-11. Thus, a narrow-band signal is equivalent to a slowly modulated sinusoid. We can now ask how the narrow-band signal $f(t)$ gets transmitted in a given medium.

In a vacuum, the signal $f(t) = g(t)e^{j\omega t}$ emitted by the transmitter is received at distance z as $f(t - z/c) = g(t - z/c)e^{j\omega(t-z/c)}$, where $c = 1/\sqrt{\mu_0\varepsilon_0}$. In a dielectric, however, with μ and ε functions of ω, there is no single value of $1/\sqrt{\mu\varepsilon}$ that will be valid for all frequencies. We do know that each individual frequency will propagate with its own phase constant $\beta = \beta(\omega)$, however. This means that $e^{j\omega t}$ at $z = 0$

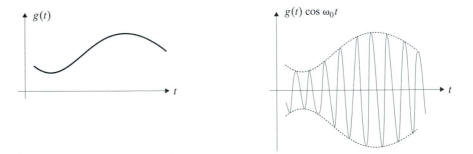

Figure 9-11   Envelope signal and modulated carrier.

becomes $\exp j[\omega t - \beta(\omega)z]$ at $z \neq 0$, where $\beta(\omega) = \omega \sqrt{\mu(\omega)\varepsilon(\omega)}$. Hence, the effect of the linear medium is obtainable by superimposing all the frequency components, each individually affected by its own propagation factor. That is, we know what happens to each component of the spectrum of the emitted signal and we can reconstruct what the receiver sees by reverting to the time domain, through the inverse Fourier transform:

$$f(t) = \int F(\omega)e^{j\omega t}\, d\omega/2\pi \qquad (\text{at } z = 0) \qquad (9.76)$$

becomes

$$f(z, t) = \int F(\omega) \exp j[\omega t - \beta(\omega)z]\, d\omega/2\pi \qquad (\text{at } z \neq 0). \qquad (9.77)$$

This is valid for any signal in a linear medium, even a wideband one, but we can go further for the special case of a narrow-band signal, using the same change of variable $\omega = \omega_0 + \Omega$ applied earlier.

$$f(z, t) = \int F(\omega) \exp j[\omega t - \beta(\omega)z]\, d\omega/2\pi \qquad (\omega \text{ near } \omega_0)$$

$$= \int F(\omega_0 + \Omega) \exp j[(\omega_0 + \Omega)t - \beta(\omega_0 + \Omega)z]\, d\Omega/2\pi$$

$$= e^{j\omega_0 t} \int G(\Omega) \exp j[\Omega t - \beta(\omega_0 + \Omega)z]\, d\Omega/2\pi \qquad (\Omega \text{ near } 0). \qquad (9.78)$$

Since $\omega = \omega_0 + \Omega$ is near ω_0 and assuming that the function $\beta(\omega)$ is well behaved in the vicinity of the carrier frequency ω_0, we can express $\beta(\omega)$ as a Taylor series about ω_0:

$$\beta(\omega) = \beta(\omega_0 + \Omega) = \beta(\omega_0) + [d\beta/d\omega]_0\Omega + \tfrac{1}{2}[d^2\beta/d\omega^2]_0\Omega^2 + \cdots$$

$$\approx \beta_0 + [d\beta/d\omega]_0\Omega. \qquad (9.79)$$

The approximation that retains only the first two terms of the series is valid if terms quadratic, and of higher order, in the small quantity Ω are negligible, compared to the linear terms retained. The quantity β_0 is just the phase constant, $\beta(\omega_0)$, at the central frequency ω_0; the derivatives of $\beta(\omega)$ are all evaluated at this carrier frequency.

Provided the approximation is valid in the narrow band near ω_0, substitution of $\beta(\omega_0 + \Omega) = \beta_0 + [d\beta/d\omega]_0\Omega$ into the last of the inverse transform expressions for $f(z, t)$ yields

$$f(z, t) = e^{j\omega_0 t} \int G(\Omega) \exp j[\Omega t - (\beta_0 + [d\beta/d\omega]_0\Omega)z]\, d\Omega/2\pi$$

$$= e^{j\omega_0 t}e^{-j\beta_0 z} \int G(\Omega) \exp j[\Omega t - [d\beta/d\omega]_0\Omega z]\, d\Omega/2\pi$$

$$= e^{j(\omega_0 t - \beta_0 z)} \int G(\Omega) \exp j[\Omega\{t - [d\beta/d\omega]_0 z\}]\, d\Omega/2\pi. \qquad (9.80)$$

The integral in the last expression is already known, however, since the envelope signal $g(t)$ was defined in terms of its spectrum $G(\Omega)$ by the inverse transform in (9.75):

Since $\int G(\Omega) \exp j\Omega t \, d\Omega/2\pi = g(t)$, we have

$$\int G(\Omega) \exp j[\Omega\{t - [d\beta/d\omega]_0 z\}]d\Omega/2\pi = g(t - [d\beta/d\omega]_0 z), \quad (9.81)$$

so that, finally,

$$\begin{aligned} f(z, t) &= e^{j(\omega_0 t - \beta_0 z)}g(t - [d\beta/d\omega]_0 z) \\ &= \exp j[\omega_0(t - [\beta_0/\omega_0]z)]g(t - [d\beta/d\omega]_0 z) \end{aligned} \quad (9.82)$$

is the signal received at $z \neq 0$ when the narrow-band signal

$$f(0, t) = f(t) = e^{j\omega_0 t}g(t) \quad (9.83)$$

is emitted at $z = 0$.

To interpret this result, we note first that the emitted narrow-band signal is a product of the carrier oscillation $e^{j\omega_0 t}$ and the envelope signal $g(t)$. We note next that the received signal is also a narrow-band signal, the product of a delayed carrier oscillation $\exp j[\omega_0(t - z/u_0)]$ and a delayed envelope signal $g(t - z/v_0)$. But the delays that affect the carrier and the envelope in their transmission from the emitter at $z = 0$ to the receiver at z are not the same! Both delays are proportional to the distance z, but with different coefficients, different speeds of propagation:

$$u_0 = \omega_0/\beta_0 \qquad v_0 = 1/[d\beta/d\omega]_0 = [d\omega/d\beta]_0. \quad (9.84)$$

The former is just the phase velocity, evaluated at the carrier frequency; the latter is called the *group velocity*, also evaluated at the carrier frequency. At any frequency, we define the phase and group velocities

$$v_p = \omega/\beta, \qquad v_g = d\omega/d\beta. \quad (9.85)$$

Mathematically, the distinction is that the phase velocity is the *ratio* of the frequency ω to the phase constant β, while the group velocity is the *derivative* of the frequency with respect to the phase constant; since β is a function of ω, ω is also a function of β and its derivative $d\omega/d\beta$ can be obtained from either the function $\omega = \omega(\beta)$ or the inverse relation $\beta = \beta(\omega)$.

Physically, the distinction is that the carrier oscillation propagates as a wave at the phase velocity $v_p(\omega_0)$, while the information-bearing envelope or modulation propagates as a wave at another speed, the group velocity $v_g(\omega_0)$. There is hence progressive slippage between the timing of the underlying carrier and of its envelope. Each factor of the overall signal, the carrier and the envelope, arrives undistorted, but each with its own delay. The product signal is therefore not precisely identical in shape to the emitted signal, but it does preserve the information that was imposed upon the carrier by the modulation. For a narrow-band signal, then, the distortion caused by dispersion does not affect the information or message carried by the envelope or modulating signal and, in this sense, this distortion is benign.

The term *group velocity* refers to the group of frequencies, in the vicinity of the carrier frequency, that comprise the narrow-band signal. Another term used to designate a narrow-band signal wave, particularly when the modulation takes the form of a pulse, is *wave packet*. If the envelope is a pulse of finite duration (but much longer than a carrier period), the resultant wave is localized in space as it travels, at the group velocity, forming a wave packet with a recognizable identity. Generally, since the information carried by a modulated carrier wave resides in its envelope, the group velocity represents the velocity of transmission of information.

The functional relation between frequency ω and phase constant β in a medium, or in any wave-propagating structure or system, is the *dispersion relation* for that medium or system. If ω and β are proportional, as they are in a vacuum, then $d\omega/d\beta = \omega/\beta$ and the phase and group velocities are actually the same (in a vacuum, both are c). In that special case, the medium or system is nondispersive. For any other variation of ω with β, the medium is dispersive. When the wave-propagating features of a medium are being considered, its dispersion relation is one of the most important properties to establish about that medium.

Example 9.7

There are many waveguiding media or systems for which the vacuum relationship between frequency ω and phase constant k—namely, $\omega^2 = k^2c^2$—is found to be modified by a constant shift, say ω_c^2, so that the dispersion relation turns out to be (with k for a vacuum renamed β for other media)

$$\omega^2 = \omega_c^2 + \beta^2 c^2, \tag{9.86}$$

where ω_c is a fixed frequency characteristic of the system or medium. Note that, in such cases, ω_c is the minimum frequency for which waves that satisfy this dispersion relation can propagate: Below that frequency, no real value of the phase constant β can be found.

Examples of such waveguiding systems include metallic waveguides, in the form of hollow pipes of any shape, and also a cold, tenuous plasma (electrified gas). For waveguides, the fixed quantity ω_c is called the cutoff frequency, because waves are unable to propagate below that frequency; this cutoff frequency depends on the size and geometry of the waveguide. For a plasma, typically a gas of ions and unbound electrons, the quantity ω_c is called the plasma frequency; the square of that frequency measures the number of electrons or ions per unit volume.

Obtain the phase and group velocities at any frequency, for a waveguide with cutoff frequency ω_c, and determine how the two velocities are related. ◈

For a waveguide, the phase velocity at frequency ω is

$$v_p = v_p(\omega) = \omega/\beta = c\omega/\sqrt{\omega^2 - \omega_c^2}. \tag{9.87}$$

Note again that the phase velocity exceeds the vacuum speed of light $(v_p > c)$ for all frequencies $(\omega > \omega_c)$ at which the wave can propagate. In fact, the phase velocity becomes infinitely high as the operating frequency ω drops closer to and approaches the cutoff frequency ω_c. Asymptotically, far above the cutoff frequency, the phase velocity approaches c, from above.

To obtain the group velocity, we need the derivative $d\omega/d\beta$ at any frequency. It is sufficient simply to differentiate the complete dispersion relation, (9.86), with respect to β, recalling that ω_c is constant, and to solve for the required derivative: $\omega^2 = \omega_c^2 + \beta^2 c^2$ yields $2\omega(d\omega/d\beta) = 2\beta c^2$, so that $d\omega/d\beta = \beta c^2/\omega$ or

$$v_g = v_g(\omega) = c\sqrt{\omega^2 - \omega_c^2}/\omega. \tag{9.88}$$

Note that the group velocity is less than the vacuum speed of light $(v_g < c)$ for all frequencies $(\omega > \omega_c)$ at which the wave can propagate. In fact, the group velocity approaches zero as the operating frequency ω approaches the cutoff frequency ω_c. Asymptotically, far above the cutoff frequency, the group velocity approaches c, from below.

Note also that, for this dispersion relation, the phase and group velocities are related by

$$d\omega/d\beta = c^2/(\omega/\beta) \quad \text{or} \quad v_p v_g = c^2, \tag{9.89}$$

so that the product of the phase and group velocities is the same for all frequencies. This is not a general law, however; it is a consequence of the particular form (i.e., hyperbolic) of the dispersion relation in this example. Although it is valid for waveguides of any shape, and also for a plasma, it is by no means universal.

Note that the group velocity represents the speed at which information is conveyed by a narrow-band propagating wave. We know from the theory of relativity that information cannot travel faster than light. We breathe a sigh of relief at having found that the group velocity for the waveguide is less than the speed of light, so that the results are consistent with the requirements of relativity.

It is important to remember that our conclusion that the envelope of a narrow-band signal propagates undistorted in a dispersive medium was dependent on an approximation. In expanding the function $\beta(\omega)$ in a Taylor series prior to using it in the exponential factor of the Fourier transform, we truncated that series after the linear term. Neglecting the higher-order terms in that series implies that, near the carrier frequency, we feel justified in approximating the dispersion curve by a straight line tangent to the curve at the carrier frequency. This is valid only if the neglected terms are sufficiently small, not merely in relation to the terms retained, but small enough to allow their exponential factor in the integrand to be approximated by unity. For the leading neglected term, $\frac{1}{2}[d^2\beta/d\omega^2](\omega - \omega_0)^2 z$, to be small, the transmission distance must not be too large, the curvature of the dispersion curve, $d^2\beta/d\omega^2$, must be sufficiently small, and the range of frequencies for which there is significant spectral strength must be narrow enough. The results therefore apply to sufficiently narrow-band signals. The carrier must be at a frequency at which the dispersion curve is well behaved, exhibiting no resonance or discontinuity nearby. At sufficiently great distances z, the neglected terms no longer remain small enough and the envelope signal does suffer distortion in the dispersive medium. Typically, a wave packet will eventually lose its original shape and stretch out as it travels.

Example 9.8

If a waveguide is operated with a carrier frequency 10% above the cutoff frequency, how fast will a pulse, or other modulating signal, propagate? ◇

From (9.88), the group velocity is

$$v_g = c\left[(\omega_0^2 - \omega_c^2)^{1/2}\right]/\omega_0, \qquad \text{or} \qquad v_g = c\left[1 - (\omega_c/\omega_0)^2\right]^{1/2}$$

with $\omega_0 = 1.1\omega_c$, so that the pulse travels at $(0.4166)c$, or 12.5 cm/nsec.

ANTENNAS AND RADIATION

We concluded from looking at the propagation of plane waves that even "nothing" can behave as a transmission line. To make use of this property requires us to be able to attach a generator and a load to empty space. This is achieved with structures called antennas.

A transmitting antenna is a structure that can generate electromagnetic fields and allow them to propagate into space as a wave. A receiving antenna is a structure that allows an electromagnetic wave that impinges upon it to induce a current that can be fed to a load impedance.

We have seen that time-varying electric and magnetic fluxes generate mmfs and emfs that can sustain the fields that comprise the fluxes. We have also verified that a consequence of the interaction of electric and magnetic fields is that they tend to propagate as waves. We can conclude that a time-varying current will radiate electromagnetic waves. On a more fundamental level, since current is moving charge, time-varying current is accelerated charge and we can declare that a charge that undergoes accelerated motion will radiate an electromagnetic wave.

Figure 9-12 shows one typical antenna structure. The time-varying generator injects a current into the wire and the time-varying current in the wire radiates fields and energy into space. The open, unconfined nature of the structure makes it efficient as a transmitting antenna. To analyze its operation, we decompose the current distribution along the antenna wire into infinitesimal elements and inquire into the radiation from each such element. An element carrying current I and of infinitesimal vector length $d\boldsymbol{l}$ is also shown in Figure 9-12. This affords an opportunity for us to tie up a loose end. In an earlier chapter, we derived the quasistatic fields of a current element but we did not explore what becomes of these fields in the case of a rapidly varying current. What are the exact fields of a current element, in the time-varying case?

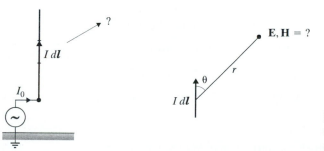

Figure 9-12 Vertical wire antenna above ground plane; current element.

For a conducting medium, we found the quasistatic fields from a current element $I\,dl$ located at the origin and directed along the z-axis to be, in spherical coordinates,

$$H_\varphi = (I\,dl/4\pi r^2)\sin\theta,$$

$$J_r = (I\,dl/4\pi r^3)2\cos\theta, \qquad J_\theta = (I\,dl/4\pi r^3)\sin\theta. \qquad (9.90)$$

Now that we are dealing with a vacuum instead of a conducting medium, we can rewrite these with the displacement current density $\partial\mathbf{D}/\partial t$ or $j\omega\varepsilon_0\mathbf{E}$ replacing the current density $\mathbf{J}$ of the dipole:

$$H_\varphi = (I\,dl/4\pi r^2)\sin\theta,$$

$$j\omega\varepsilon_0 E_r = (I\,dl/4\pi r^3)2\cos\theta, \qquad j\omega\varepsilon_0 E_\theta = (I\,dl/4\pi r^3)\sin\theta. \qquad (9.91)$$

but these are still only quasistatic results, applicable for slow time variation, meaning only low frequencies. Note that the magnetic field falls off from the source as $1/r^2$ and the electric field decays away even faster, as $1/r^3$. Such fields don't get very far from the source element before becoming insignificantly weak. How might the situation change if we allow the frequency of the source to become high?

To answer this, we will need to solve the Maxwell equations for the current element at any frequency, not just low ones. As a trial solution to the equations, let us assume that the fields retain the same variation in θ but that they may have more complicated r-dependences. We will therefore assume the following form for the fields and try to find functions $f(r)$, $g(r)$, and $h(r)$ that will make these trial functions satisfy the full Maxwell equations.

$$H_\varphi = (I\,dl/4\pi r^2)\sin\theta\,h(r), \qquad (9.92)$$

$$j\omega\varepsilon_0 E_r = (I\,dl/4\pi r^3)2\cos\theta\,g(r), \qquad (9.93)$$

$$j\omega\varepsilon_0 E_\theta = (I\,dl/4\pi r^3)\sin\theta\,f(r). \qquad (9.94)$$

We know that, at low frequencies, as $\omega \to 0$, $f(r)$, $g(r)$, and $h(r)$ must all approach unity, to leave the known quasistatic results. Maxwell's equations for the vacuum region are, in terms of the displacement current $j\omega\varepsilon_0\mathbf{E}$ and with $\omega/c = k$,

$$\nabla \times \mathbf{H} = j\omega\varepsilon_0\mathbf{E}, \qquad \nabla \times j\omega\varepsilon_0\mathbf{E} = k^2\mathbf{H}. \qquad (9.95)$$

Expressing the curl operation in spherical coordinates, we need

$$\nabla \times \mathbf{H} = \hat{\mathbf{r}}[1/(r\sin\theta)]\partial[\sin\theta\,H_\varphi]/\partial\theta - \hat{\boldsymbol{\theta}}[1/r]\partial[rH_\varphi]/\partial r$$

$$= \hat{\mathbf{r}}[(I\,dl/4\pi)2\cos\theta\,h/r^3] - \hat{\boldsymbol{\theta}}\sin\theta\,[(I\,dl/4\pi)(1/r)d(h/r)/dr] \qquad (9.96)$$

to be

$$j\omega\varepsilon_0\mathbf{E} = \hat{\mathbf{r}}[(I\,dl/4\pi)2\cos\theta\,g/r^3] + \hat{\boldsymbol{\theta}}\sin\theta\,[(I\,dl/4\pi)f/r^3]. \qquad (9.97)$$

This will be so if

$$h(r) = g(r) \qquad \text{and} \qquad r^2 d[h(r)/r]/dr = -f(r). \qquad (9.98)$$

We also need, noting that $\nabla \times \mathbf{E}$ has only a φ-component because there is no φ-dependence of the trial functions,

$$\nabla \times j\omega\varepsilon_0\mathbf{E} = \hat{\boldsymbol{\varphi}}\left[(1/r)\partial(rj\omega\varepsilon_0 E_\theta)/\partial r - (1/r)\partial(j\omega\varepsilon_0 E_r)/\partial\theta\right]$$
$$= \hat{\boldsymbol{\varphi}}\sin\theta\,(I\,dl/4\pi)\left[(1/r)d(f/r^2)/dr + (2/r^4)g\right] \quad (9.99)$$

to be

$$k^2\mathbf{H} = \hat{\boldsymbol{\varphi}}k^2\sin\theta\,(I\,dl/4\pi)\left[h/r^2\right]. \quad (9.100)$$

This will be so if

$$k^2h(r) = rd\left[f(r)/r^2\right]/dr + (2/r^2)g(r). \quad (9.101)$$

We now have three equations for the three unknown functions $f(r)$, $g(r)$, and $h(r)$. Since $f(r)$ and $g(r)$ are already expressed in terms of $h(r)$ in (9.98), we can eliminate them from (9.101) to get a single equation for $h(r)$ as

$$k^2h(r) - (2/r^2)h(r) = rd\{-d[h(r)/r]/dr\}/dr, \quad (9.102)$$

which reduces to

$$d^2h/dr^2 - (2/r)dh/dr + k^2h = 0. \quad (9.103)$$

This is a linear, nonconstant-coefficient, ordinary differential equation for $h(r)$; once we solve it, we can get the other two unknown functions at once as

$$g(r) = h(r) \quad \text{and} \quad f(r) = -r^2d[h/r]/dr = h - r\,dh/dr. \quad (9.104)$$

The boundary condition on the equation comes from the need for $f(r)$, $g(r)$, and $h(r)$ all to become unity as $k \to 0$, so as to revert to the quasistatic result.

The solution to (9.103), the differential equation for $h(r)$, under these boundary conditions, is readily verified to be

$$h(r) = (1 + jkr)e^{-jkr}. \quad (9.105)$$

This is accompanied by $g(r) = h(r)$ and by

$$f(r) = \left(1 + jkr - k^2r^2\right)e^{-jkr}. \quad (9.106)$$

Since $j\omega\varepsilon_0$ can be rewritten as jk/η_0, the exact fields of a current element are, when we reinsert the $e^{j\omega t}$ time dependence,

$$\mathbf{E}(r,\theta,\varphi,t) = \eta_0\left[I\,dl/4\pi r^2\right]e^{j(\omega t - kr)}\{\hat{\mathbf{r}}\,2\cos\theta\,[1/jkr + 1]$$
$$+\,\hat{\boldsymbol{\theta}}\sin\theta\,[1/jkr + 1 + jkr]\}, \quad (9.107)$$

$$\mathbf{H}(r,\theta,\varphi,t) = \left[I\,dl/4\pi r^2\right]e^{j(\omega t - kr)}\hat{\boldsymbol{\varphi}}\sin\theta[1 + jkr]. \quad (9.108)$$

This electromagnetic field of a single current element $I\,dl\,e^{j\omega t}$ directed along the z-axis can be superimposed to get the fields generated by any current distribution. Let us examine and interpret these fields.

Because of the complexity of the field expressions, it is useful to examine a couple of limiting cases. The *near fields* refer to the fields near the origin, in the sense that

$kr \ll 1$. Since $kr = \omega r/c$, making $r \to 0$ is the same as making $\omega \to 0$ and we recover the quasistatic results. In this limit, the magnetic field decays as $1/r^2$ and the electric field as $1/r^3$; both die away rather rapidly. In the opposite extreme, however, something magical happens.

The *far fields* refer to the fields at great distances from the origin, in the sense that $kr \gg 1$, which calls for either large distances or high frequencies. In this limit, the fields approach

$$\mathbf{E}(r, \theta, \varphi, t) = \eta_0[jkI\ dl]\hat{\boldsymbol{\theta}} \sin \theta\ e^{j(\omega t - kr)}/4\pi r, \tag{9.109}$$

$$\mathbf{H}(r, \theta, \varphi, t) = [jkI\ dl]\hat{\boldsymbol{\varphi}} \sin \theta\ e^{j(\omega t - kr)}/4\pi r. \tag{9.110}$$

These are waves; they travel in the direction of increasing distance r from the source element, that is, in the radial direction. We see that the current element radiates waves out to large distances from itself. The radiated field amplitude does diminish with distance, but only as $1/r$. The field strength also varies with elevation angle θ, being strongest at right angles to the direction of the current element. The electric and magnetic fields are perpendicular to each other and to the direction of propagation; these field orientation properties are shared with the plane waves we examined earlier.

Before continuing with the interpretation of the fields we found for the current element, we should note that we are using one solution to the differential equation for $h(r)$—namely, $h(r) = (1 + jkr)e^{-jkr}$. Since (9.103) is a second-order differential equation, it should have two independent solutions; why have we used only one? In fact, an independent other solution, also satisfying the boundary conditions, is the complex conjugate $h^*(r) = (1 - jkr)e^{+jkr}$, but we reject this for our problem, because it gives us waves that travel along the direction of decreasing distances r (that is, radially inward from infinity toward the source). This is mathematically admissible but not physically acceptable; it would violate the radiation condition, discussed in Chapter 6, which allows a source to emit energy but not to suck it in from infinity toward itself.

Returning to our outbound wave, the fact that the field strength falls off with distance as only $1/r$, rather than $1/r^2$ and $1/r^3$ for the near fields, has the momentous consequence that power is radiated out to infinity. We can verify this by calculating the Poynting vector.

The time-averaged power flow density in the far-field region $(kr \gg 1)$ is

$$\text{Re}\left[\tfrac{1}{2}\mathbf{E} \times \mathbf{H}^*\right] = \hat{\mathbf{r}}(\eta_0/2)[kI\ dl]^2 \sin^2 \theta/(4\pi r)^2. \tag{9.111}$$

The power flow is in the radial direction, away from the source point, and varies inversely as the square of the distance from the current element. It also has a squared-sinusoid dependence on the elevation angle, so that there is no power flow in the direction of the element and maximum power at right angles to it. The variation of the power flow with direction, here as $\sin^2 \theta$, is the far-field *antenna pattern* for the structure, here a current element, that is the cause of the radiation.

To verify that power is radiated out to infinity, let us calculate the total time-averaged power that crosses a (large) sphere of radius r_0. This is

$$P = \oint_S \text{Re}\left[\tfrac{1}{2}\mathbf{E} \times \mathbf{H}^*\right] \cdot d\mathbf{S}$$

$$= \int_0^{2\pi} \int_0^{\pi} \left\{(\eta_0/2)[kI\, dl]^2 \sin^2\theta / (4\pi r_0)^2\right\}\hat{\mathbf{r}} \cdot \hat{\mathbf{r}} r_0^2 \sin\theta\, d\theta\, d\varphi$$

$$= \left\{(\eta_0/2)[kI\, dl]^2 / (4\pi)^2\right\}2\pi \int_0^{\pi} \sin^3\theta\, d\theta. \tag{9.112}$$

Since

$$\int_0^{\pi} \sin^3\theta\, d\theta = \int_{-1}^{1} \sin^2\theta\, d(\cos\theta) = \int_{-1}^{1}(1 - x^2)\, dx = 4/3,$$

we get

$$P = \left\{(\eta_0/2)[kI\, dl]^2 / (4\pi)^2\right\}2\pi\,(4/3) \tag{9.113}$$

and replacing k with $2\pi/\lambda$, where λ is the free-space wavelength at the frequency $\omega = kc$, we get

$$P = (\eta_0/3)\pi|I|^2(dl/\lambda)^2. \tag{9.114}$$

Note that this power, which crosses a sphere of radius r_0, is independent of r_0. This expresses conservation of energy and indicates that this much power is radiated to all radii, even out to infinite distances.

Another way to express this result is that the power radiated to infinity by a current element of length dl with current I is the same as if the current I were fed to an equivalent load or *radiation resistance* R_{rad} that we can define by $P = \tfrac{1}{2}|I|^2 R_{\text{rad}}$, so that for the current element,

$$R_{\text{rad}} = (2\pi/3)\eta_0(dl/\lambda)^2. \tag{9.115}$$

The coefficient has the numerical value $(2\pi/3)\eta_0 = 789$ ohms, but the radiation resistance will be small because the element has been taken to be infinitesimally small. When we know the current distribution on an antenna of finite size, we can superimpose the fields of the current elements and obtain the radiation resistance and radiated power pattern for the antenna, whatever its structure.

Antennas are often deployed as arrays, their constituent elements spaced judiciously and their currents phased appropriately to achieve a desired shaping of the overall radiation pattern. The radiation from the individual antennas in the array interferes constructively and destructively in different directions, so that the resultant radiation pattern can be quite different from that of any one element of the array.

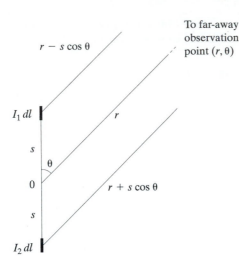

To far-away
observation
point (r, θ)

$r - s \cos \theta$

$I_1\, dl$

r

s

θ

0

$r + s \cos \theta$

s

$I_2\, dl$

Figure 9-13   Pair of aligned current elements, spaced $2s$ apart.

Example 9.9

Two current elements of moments $I_1\, dl$ and $I_2\, dl$ have their axes aligned and are spaced apart by distance $2s$ along the common z-axis, as illustrated in Figure 9-13. Obtain the radiation pattern of this rudimentary array, for the special case of equal complex currents $I_1 = I_2 = I_0$ and for elements spaced one wavelength apart. ◇

We can superimpose the fields (but not the Poynting vectors!) of each element in the far-field region and then calculate the complex Poynting vector to obtain the radiation pattern of the array. Since the two aligned elements share their z-axis, they also share the elevation angle θ and the azimuthal unit vector $\hat{\boldsymbol{\varphi}}$. Their magnetic fields therefore superimpose in the far-field region as

$$\mathbf{H} = jk\, dl\, \hat{\boldsymbol{\varphi}} \sin\theta \left\{ [I_1/4\pi(r - s\cos\theta)]e^{-jk(r-s\cos\theta)} \right.$$
$$\left. + [I_2/4\pi(r + s\cos\theta)]e^{-jk(r+s\cos\theta)} \right\}. \tag{9.116}$$

Note that we use $r \pm s\cos\theta$ as the distances from the two current elements to the far-away observation point. For the far-field region, it is safe to neglect the small distance s compared to the enormous distance r in the two denominators. However, we cannot be so cavalier about it in the exponents, because the exponentials factor as $e^{-jkr}e^{\pm jks\cos\theta}$ and there is no justification for considering $ks = 2\pi s/\lambda$ to be small in the latter exponential; s is small compared to r but not compared to λ. We therefore approximate the superimposed magnetic field for large r as

$$\mathbf{H} = jk\, dl\, \hat{\boldsymbol{\varphi}}[\sin\theta/4\pi r]e^{-jkr}\{I_1 e^{jks\cos\theta} + I_2 e^{-jks\cos\theta}\}. \tag{9.117}$$

In the far-field region, the electric field is perpendicular to both the magnetic field and the radial direction and its magnitude is η_0 times that of the magnetic field. The complex Poynting vector is therefore radial and equals

$$\mathrm{Re}[\tfrac{1}{2}\mathbf{E} \times \mathbf{H}^*] = \hat{\mathbf{r}}(\eta_0/2)[k\, dl/4\pi r]^2 \sin^2\theta \left\{ |I_1|^2 + |I_2|^2 + 2\,\mathrm{Re}[I_1 I_2^* e^{j2ks\cos\theta}] \right\}$$
$$= \hat{\mathbf{r}}(\eta_0/2)[k|I_0|\, dl/4\pi r]^2 \sin^2\theta \left\{ 2 + 2\cos(2ks\cos\theta) \right\}$$
$$= \hat{\mathbf{r}}(4\eta_0/2)[k|I_0|\, dl/4\pi r]^2 \sin^2\theta \, \cos^2(ks\cos\theta). \tag{9.118}$$

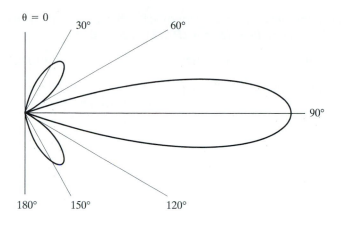

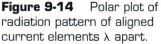

Figure 9-14 Polar plot of radiation pattern of aligned current elements λ apart.

Comparing with the corresponding expression for the radiated power density for a single current element, we see the factor of four from the feeding of the current I_0 to two elements instead of just one and we also see the factor $\cos^2(ks \cos \theta)$ that represents the interference of the radiation from the two individual current elements. This factor displays constructive interference in the direction $\theta = 90°$, at right angles to the array and elements, but destructive interference, to a null, in those directions for which $ks \cos \theta = \pm \pi/2$. For the case at hand, $ks = 2\pi s/\lambda = \pi$ and the nulls occur where $\cos \theta = \pm 0.5$, at $\theta = 60°$ and at $\theta = 120°$. A polar plot of the radiation pattern $\sin^2 \theta \cos^2(\pi \cos \theta)$ is depicted in Figure 9-14. It gives the relative strength of the radiation in different directions in the far-field region. The plot should be imagined spun about the vertical z-axis to illustrate the uniform azimuthal dependence. The strongest portion of the plot is the *main lobe* and the lesser ones are *sidelobes*.

ANTENNA TERMINOLOGY

We collect here a number of terms and symbols used to describe the characteristics and operation of antennas and antenna systems. We have already encountered several such terms; others are new.

Viewed as a system, a transmitting antenna has as its input an amount of power P and as its output the radiation it produces in different regions of space. For a receiving antenna, the input is the power density incident on it from some direction and the output is the power delivered to its load. There are a number of ways to describe the radiation from an antenna. The real part of the complex Poynting vector, $\text{Re}\left[\frac{1}{2}\mathbf{E} \times \mathbf{H}^*\right]$, gives the time-averaged power flow as a function of position in space. Dotting this with an element of area $d\mathbf{S}$ and integrating over some surface furnishes the power that crosses that surface. An antenna radiates power nonuniformly in different directions; conservation of power dictates that the radial variation must be inverse square, so that only the directional properties of the radiation distinguish one antenna from another.

Accordingly, we denote the *power flow density* by $dP/d\mathbf{S}$, which reminds us that the power across an area element $d\mathbf{S}$ is $dP = (dP/d\mathbf{S}) \cdot d\mathbf{S}$, so that

$$dP/d\mathbf{S} = \text{Re}\left[\tfrac{1}{2}\mathbf{E} \times \mathbf{H}^*\right]. \tag{9.119}$$

This is a function of r, θ, φ, but we know its radial dependence is as r^{-2} and it is useful to extract that particular variation from the radial power flow. This leaves the *radiation intensity* $dP/d\Omega$ to express the directional properties of the radiation. This is the power per unit solid angle, a function of elevation θ and azimuth φ. Recall that the solid angle subtended at a point by some surface in space is a measure of its three-dimensional angular opening, defined by the ratio of the area of the surface, projected onto a sphere about the point, to the square of the radius of that sphere. The element of solid angle associated with the surface element is

$$d\Omega = \hat{\mathbf{r}} \cdot d\mathbf{S}/r^2. \tag{9.120}$$

Hence, on the surface of a sphere, $d\Omega = \sin\theta \, d\theta \, d\varphi$. Solid angle is measured in steradians (sr); there are 4π steradians in the full space surrounding a point, since the area of a sphere of radius r is $4\pi r^2$. As a possibly helpful benchmark, the continent of Africa subtends about 0.75 sr at the center of the earth.

Correspondingly, the radiation intensity is found as

$$dP/d\Omega = r^2\hat{\mathbf{r}} \cdot dP/d\mathbf{S} = r^2\hat{\mathbf{r}} \cdot \text{Re}\left[\tfrac{1}{2}\mathbf{E} \times \mathbf{H}^*\right]. \tag{9.121}$$

For example, for the pair of aligned current elements, spaced $2s$ apart, the radiation intensity is, from (9.118),

$$dP/d\Omega = 2\eta_0\left[k|I_0|\,dl/4\pi\right]^2 \sin^2\theta \cos^2(ks\cos\theta). \tag{9.122}$$

One convenient way to isolate the directional properties from geometrical and other factors is to express them through the antenna's *directivity pattern* $D(\theta, \varphi)$. This is defined as the ratio of the radiation intensity to the average intensity, the average being taken over all directions.

$$D(\theta, \varphi) = \frac{dP/d\Omega}{\int (dP/d\Omega)\, d\Omega/4\pi}. \tag{9.123}$$

Since $dP/d\Omega$ integrates to the total radiated power P, this is also

$$D(\theta, \varphi) = (4\pi/P)\, dP/d\Omega. \tag{9.124}$$

The *directivity D* of an antenna is simply the peak of its directivity pattern. It furnishes an indication of how narrowly the radiation is directed by the antenna. The factor $P/4\pi$ would be the radiation intensity $(dP/d\Omega)_0$ of an *isotropic antenna*, one that radiates equally in all directions, if such an antenna existed. The directivity therefore compares the antenna's radiation pattern to the hypothetical one of an isotropic radiator.

Example 9.10

Find the directivity of the pair of aligned current elements one wavelength apart. ◈

From $dP/d\Omega = C\sin^2\theta\cos^2(\pi\cos\theta)$, the peak value occurs at $\theta = \pi/2$ and equals C. We need the average value of the intensity, $\langle dP/d\Omega \rangle$, in order to get the directivity as $D = (dP/d\Omega)_{\text{max}}/(dP/d\Omega)_{\text{avg}} = C/\langle dP/d\Omega \rangle$. The average value is just $\langle dP/d\Omega \rangle = P/4\pi$, which we find as

$$\langle dP/d\Omega \rangle = (1/4\pi)\int\int C\sin^2\theta\cos^2(\pi\cos\theta)\sin\theta\,d\theta\,d\varphi$$

$$= (C/2)\int_0^\pi \sin^2\theta\cos^2(\pi\cos\theta)\sin\theta\,d\theta$$

$$= (C/2)\int_{-1}^1 (1-\cos^2\theta)\cos^2(\pi\cos\theta)\,d(\cos\theta)$$

$$= (C/2\pi^3)\int_{-\pi}^\pi (\pi^2 - u^2)\cos^2 u\,du,$$

upon changing variables to $u = \pi\cos\theta$. The last form is readily found in tables of integrals or else handled by integration by parts. The result is

$$\langle dP/d\Omega \rangle = C[1/3 - 1/(4\pi^2)],$$

so that the directivity is $D = 1/[1/3 - 1/(4\pi^2)] = 3.247$.

The *gain* of an antenna is defined similarly to the directivity, except that the input power is the normalization factor, instead of the radiated power:

$$G(\theta,\varphi) = (4\pi/P_{\text{in}})dP/d\Omega; \qquad G = G(\theta,\varphi)_{\text{max}}. \tag{9.125}$$

The relation between the input power and the radiated power defines the antenna's *efficiency*, and this also relates the gain to the directivity. That the efficiency is less than 100% may be due to losses in the imperfectly conducting or insulating structure or to reflections if the antenna is mismatched to the transmission line that feeds it.

For a receiving antenna, the key parameter is the *aperture*, A, defined as the ratio of the power received to the power density incident onto the antenna from a particular direction. If no direction is specified, the peak such ratio is meant. The actual power delivered to the antenna's load is affected by the antenna's efficiency. When that factor is included, we speak of the antenna's *effective area*.

To tease the student to continue studying electromagnetics, let's take just the first step toward proving the reciprocity theorem. This result implies that, under optimized conditions, the transmitting and receiving patterns of an antenna are identical. It also implies that, when optimized, the power flow from one antenna to another is the same, no matter which one is the transmitter and which is the receiver. A consequence of reciprocity is that, if antenna 1, with directivity D_1, radiates power P_1, then the power flow at distance r is, optimally,

$$(dP/dS)_1 = (P_1/4\pi r^2)D_1. \tag{9.126}$$

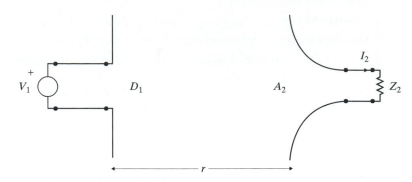

Figure 9-15   Transmitting and receiving antennas.

If receiving antenna 2, with aperture A_2, is located there, the power it receives will be

$$P_2 = (dP/dS)_1 A_2 = P_1 D_1 A_2/4\pi r^2. \tag{9.127}$$

Figure 9-15 shows the two antennas, of arbitrary structure.

Reciprocity proclaims that, under optimum conditions and provided the sources are at the same frequency and the intervening medium is linear, passive, and isotropic, the ratio of power received to power transmitted is the same, regardless of which antenna is sending to the other: $P_2/P_1 = D_1 A_2/4\pi r^2$ is the same as $D_2 A_1/4\pi r^2$. As a result, we have $D_1 A_2 = D_2 A_1$ or

$$A_1/D_1 = A_2/D_2 \tag{9.128}$$

for any pair of antennas. This says that the ratio of aperture to directivity is the same for any two antennas, so that this ratio must be a constant independent of the nature of the antenna; it depends only on the operating frequency.

By calculating the aperture and directivity of just one simple antenna, we can find the value of the ratio for all antennas. It is

$$A/D = \lambda^2/4\pi. \tag{9.129}$$

As a hint as to what's behind this remarkable result, called the *reciprocity theorem*, we take just the first step toward deriving it. It is based on Maxwell's equations, of course. If current density $\mathbf{J}_1$ generates fields $\mathbf{E}_1, \mathbf{H}_1$ and source $\mathbf{J}_2$ at the same frequency gives rise to fields $\mathbf{E}_2, \mathbf{H}_2$ then we get for the divergence of the mixed expression $\mathbf{E}_1 \times \mathbf{H}_2$

$$\nabla \cdot (\mathbf{E}_1 \times \mathbf{H}_2) = \mathbf{H}_2 \cdot \nabla \times \mathbf{E}_1 - \mathbf{E}_1 \cdot \nabla \times \mathbf{H}_2$$

$$= -j\omega\mu\mathbf{H}_2 \cdot \mathbf{H}_1 - j\omega\varepsilon\mathbf{E}_1 \cdot \mathbf{E}_2 - \mathbf{E}_1 \cdot \mathbf{J}_2. \tag{9.130}$$

This tells us that, despite appearances, the quantity $\mathbf{E}_1 \cdot \mathbf{J}_2 + \nabla \cdot (\mathbf{E}_1 \times \mathbf{H}_2)$ is symmetric under interchange of the subscripts 1 and 2:

$$-\nabla \cdot (\mathbf{E}_1 \times \mathbf{H}_2 - \mathbf{E}_2 \times \mathbf{H}_1) = \mathbf{E}_1 \cdot \mathbf{J}_2 - \mathbf{E}_2 \cdot \mathbf{J}_1, \tag{9.131}$$

the differential form of the *Lorentz reciprocity theorem*. In integral form, it is

$$-\oint_S (\mathbf{E}_1 \times \mathbf{H}_2 - \mathbf{E}_2 \times \mathbf{H}_1) \cdot d\mathbf{S} = \int_V (\mathbf{E}_1 \cdot \mathbf{J}_2 - \mathbf{E}_2 \cdot \mathbf{J}_1) \, dV. \qquad (9.132)$$

This is the starting point for obtaining the reciprocity properties cited for antennas, as well as other important results, but we will not pursue these here.

SUMMARY

In this chapter we have been interested in structures, other than the transmission lines we investigated earlier, that can also convey signals and energy from a source to a load. As a crude model for a strip line, we examined a parallel-plate line and compared it to the parallel-wire transmission line. We found that both had the same phase constant but that the characteristic impedances differ, reflecting the different geometry.

We noted the unshielded feature of the parallel-plate line and conceptually modified it, closing it on the two open sides and thereby forming a rectangular waveguide structure. This is fundamentally different from two-conductor transmission lines, and we argued that the hollow-pipe geometry could serve to convey waves only if there is rapid enough time variation of magnetic or electric fluxes to sustain the fields that tend to collapse for lack of separated charges and currents. The analysis we carried out confirmed this expectation, for a particular kind of mode.

The method of analysis entailed solution of the steady-state wave equation, or Helmholtz equation, by the technique of separation of variables. The separation of the transverse and axial variables gave us a sinusoidal standing wave transversely and either propagation or attenuation axially. As predicted, we get propagation only if we operate at a frequency that exceeds the cutoff frequency, for each possible mode. There are transverse electric (TE) and transverse magnetic (TM) modes; the former have axial magnetic flux while the latter feature axial displacement current.

We also examined the relation between frequency and phase constant, the dispersion relation for the waveguide. This informs us of the wavelength and phase velocity for any frequency above cutoff. We noted that the phase velocity for a waveguide exceeds the speed of light but does not violate relativistic principles, the phase velocity being defined only geometrically. We confirmed that the energy velocity, defined as the ratio of power flow to energy density, is indeed less than the speed of light, as it must be. We undertook a Fourier analysis of narrow-band signals such as are used in communications and of wave packets and found that such signals are also transmitted along a waveguide at a rate less than the speed of light. We decomposed a rectangular waveguide mode into a pair of obliquely traveling plane waves that bounce back and forth between the walls of the waveguide.

Waves in empty space led us to examine the question of how we might attach a source and a load to a vacuum. This is done with antennas, which are any structures that can carry time-varying currents. The more efficient such structures are open, so as to release the radiated waves rather than confine them. As an elementary version of a radiating structure, we analyzed the fields generated by a current element that carries a time-varying current. We started with the quasistatic fields and modified them to satisfy the Maxwell equations at any frequency. We were able to get the exact fields of a current element, and we noted the natural division between the near fields and the far fields. We found that the current element radiates fields, and therefore signals and energy, all the way to infinity. Antennas can be analyzed as superpositions of current elements, if we know or can discover the currents that flow along them.

As a final example, we looked at the interference of the radiation from just two separate current elements and calculated the radiation pattern for two aligned current elements a wavelength apart, noting that nulls may appear, separating the main radiation lobe from lesser sidelobes, as is typical of most antennas. We reviewed some nomenclature applied to antenna systems and ended with a hint about the underpinnings of the reciprocity theorem, which informs us that the transmitting and receiving antennas can be interchanged, when the system is optimized, without affecting the transfer of power.

PROBLEMS

Nonparallel-plate transmission line

9.1   A transmission line is formed of flat but nonparallel conducting plates, as in Figure P9-1.

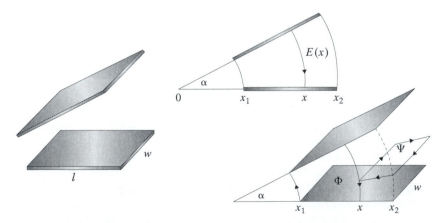

Figure P9-1   Transmission line with nonparallel plates.

The plates' dimensions are l by w, they are in air, and they would meet at angle α if they were extended. We can expect electric field lines that are circular, perpendicular to the plates, and a magnetic field parallel to both plates. Neglect fringing beyond the edges of the plates.

(a) We can write the transmission line equations for this structure as

$$V(x) = V(x_1) - \int_{x_1}^{x} Z(x')I(x')\, dx' \quad \text{and} \quad I(x) = \int_{x}^{x_2} Y(x')V(x')\, dx',$$

where $V(x)$ is the voltage from the top to the bottom plate and $I(x)$ is the current into the segment of the upper plate from x to x_2, which must equal the displacement current $j\omega\Psi$ and fall to zero at $x = x_2$. Obtain $Z(x)$ and $Y(x)$.

(b) Verify that the parameters $Z(x)$ and $Y(x)$ become those of the parallel-plate line (width w, height h) in the limit $\alpha \to 0$.

Parallel-plate line closed by parallel cylinders

9.2   While in the rectangular waveguide, a parallel-plate line closed by side walls, the longitudinal magnetic field shares the space with the transverse electric and magnetic fields, the configuration in Figure P9-2 largely separates the transverse and longitudinal fields. The parallel-plate line (width a, height b) is closed at each side by large, parallel cylinders (each of area A). As a result, the shunt currents that tend to discharge the plates flow around the periphery of the cylinders and generate the longitudinal magnetic flux almost entirely within these cylinders, thereby maintaining the transverse fields between the plates (the capacitive part) by use of physically separated axial magnetic fields (in the inductive cylindrical portions).

(a) Obtain Z and Y in the transmission line equations $dV/dz = -ZI$, $dI/dz = -YV$.

(b) What is the cutoff frequency for the resultant waveguide mode?

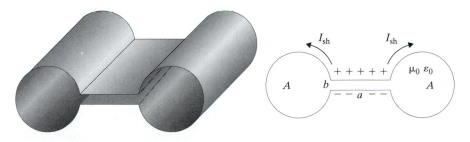

Figure P9-2   Parallel-plate line closed by large cylinders.

Concentric spherical waveguide

9.3   Two concentric, perfectly conducting spheres of radii a, b are separated by air, as in Figure P9-3. The structure is to serve as a waveguide for waves propagating in the direction of the elevation angle θ. We seek the mode that has the radial variation of the static case, with an inwardly radial electric field. The suggested flux surfaces shown in the

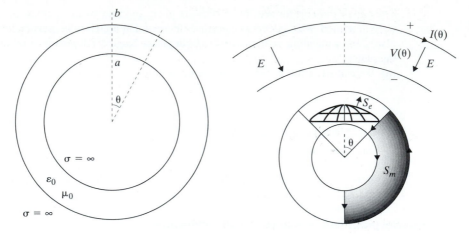

Figure P9-3   Concentric spherical waveguide.

figure are the spherical cap S_e at radius r, extending from angle 0 to θ, and the planar area S_m between radii a and b and extending from angle θ to π. Consider the voltage $V(\theta)$ across the gap between the spheres and the current $I(\theta)$ into the spherical cap beyond angle θ.

(a) Express the radial electric field $E(r, \theta)$ in terms of $V(\theta)$.

(b) Express the azimuthal magnetic field $H(r, \theta)$ in terms of $I(\theta)$.

(c) Obtain the impedance and admittance coefficients $Z(\theta)$ and $Y(\theta)$.

Properties of a waveguide at a given operating frequency

9.4   An air-filled rectangular waveguide has a cutoff frequency $f_c = 7.50$ GHz for its dominant mode and is operated at $f = 9.00$ GHz.

(a) What is the phase velocity of the wave?

(b) What is the energy velocity of the wave?

(c) What is the wavelength of the wave?

Cutoff frequency of air-filled waveguide mode

9.5   At frequency $f = 10.0$ GHz, the guide wavelength along a certain air-filled rectangular waveguide is measured as $\lambda_g = 4.00$ cm, for a particular mode. What is the cutoff frequency for that mode?

Cutoff frequency of below-cutoff air-filled waveguide mode

9.6   At frequency $f = 10.0$ GHz, the attenuation constant along a certain air-filled rectangular waveguide is measured as $\alpha = 2.5$ cm^{-1}, for a particular mode. What is the cutoff frequency for that mode?

Properties of dielectric-filled waveguide

9.7 At frequency $f_1 = 10.0$ GHz, the group velocity along a certain lossless dielectric-filled rectangular waveguide is measured as 0.25c, for a particular mode. At $f_2 = 11.0$ GHz, the group velocity is 0.35c, for the same mode.
(a) What is the cutoff frequency for that mode?
(b) What is the relative permittivity $\varepsilon/\varepsilon_0$ of the dielectric?

Square waveguide of given cutoff frequency

9.8 The dominant mode of an air-filled square waveguide has a cutoff frequency of $f_0 = 2$ GHz.
(a) What is the width of the square?
(b) What is the next higher cutoff frequency?

Coincident cutoffs in a rectangular waveguide

9.9 Give the dimensions a, b (in cm) of an air-filled rectangular waveguide for which the cutoff frequencies of the TE_{30} and TM_{21} modes both equal 9 GHz.

Design of aspect ratio of rectangular waveguide

9.10 We want to operate an air-filled rectangular waveguide at a frequency such that the phase velocity for the TE_{10} mode is four times the group velocity and the operating frequency is still below the cutoff frequency of the TE_{01} mode.
(a) What is the operating frequency, in terms of the cutoff frequency?
(b) What is the smallest ratio a/b of the dimensions of the waveguide that we can choose?

Natural frequency of a rectangular cavity with movable wall

9.11 An empty rectangular waveguide has dimensions a by b, with $a > b$. One end is closed by a transverse conducting wall, the other by a movable conducting plunger. As the plunger is withdrawn, the distance l between the end walls increases, from zero to infinity. Give the lowest frequency at which the resulting cavity can sustain oscillations, as a function of l.
Hint: The lowest-frequency (dominant) cavity mode has its electric field oriented (polarized) along the shortest dimension of the rectangular enclosure. (Why?)

Power flow and energy storage in a TM mode of a rectangular waveguide

9.12 (a) Obtain the time-averaged power flow across a transverse plane of a rectangular waveguide in which the TM_{11} mode propagates.
(b) Find the total (electric and magnetic) energy stored per unit length in this mode of the waveguide.
(c) Obtain the ratio of the power flow to the energy per unit length (this is the speed of energy flow in the waveguide).

Wavelength or attenuation, far from cutoff

9.13 For an empty rectangular waveguide of width a, height b, length l,

    (a) what is the guide wavelength λ_g of the TE_{10} mode if the operating frequency f is far above the cutoff frequency?

    (b) What is the attenuation of the TE_{10} mode, from one end of the guide to the other, if the operating frequency f is far below the cutoff frequency? Express the answer in dB.

Estimate of number of modes in overmoded waveguide

9.14 We want to estimate the number of modes of a rectangular waveguide (width a, height b, air filled) that can propagate, at high frequencies. There is a TE and a TM mode for every pair of integers m, n (excluding TE_{00} and TM_{mn} with $m = 0$ or $n = 0$) and the ones that can propagate are those for which m and n are such that the dispersion relation $\gamma^2 + \omega^2/c^2 = (m\pi/a)^2 + (n\pi/b)^2$ yields an imaginary γ. In Figure P9-4, each dot represents a pair of integers (m, n) and corresponds to two modes (except for the relatively few along the axes). The number of propagating modes is therefore approximately twice the number of dots up to the curve marked $\gamma = 0$.

    (a) What is the equation for the boundary curve that separates the propagating modes from the evanescent ones, in the figure? What geometric curve is it?

    (b) Why is the number of dots below the boundary curve approximately the area under the curve?

    (c) Estimate the number of propagating modes at a high operating frequency ω.

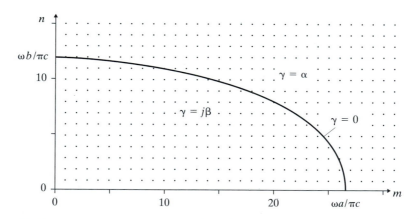

Figure P9-4   For estimating the number of propagating modes in a waveguide.

Air-filled waveguide joined to dielectric-filled guide

9.15 A rectangular waveguide of width a and height b is empty for all $z < 0$ but has a dielectric filler with permittivity $\varepsilon = 2\varepsilon_0$ for all $z > 0$. A TE_{10} mode incident from

$z = -\infty$ will be reflected from the dielectric at $z = 0$, so that the electric field in the empty guide will be

$$E(x, z) = E_0 \sin(\pi x/a)\left[e^{-j\beta z} + \Gamma e^{j\beta z}\right].$$

Find the reflection coefficient Γ.

Plane wave in arbitrary direction

9.16 Recognizing that a plane wave in free space has electric and magnetic fields perpendicular to each other and to the direction of propagation, write the x, y, z-dependence of the electric field of a plane wave at frequency ω that travels along the direction of the vector $\mathbf{b} = 4\hat{\mathbf{x}} - 3\hat{\mathbf{y}} + 12\hat{\mathbf{z}}$ (instead of along $\hat{\mathbf{z}}$ as in the text) and has an electric field directed along the unit vector $\hat{\mathbf{e}}$ in the xy-plane (with positive x- and y-components), with complex amplitude E_0. Find $\hat{\mathbf{e}}$ and the direction $\hat{\mathbf{h}}$ of the magnetic field.

Strengths of natural electromagnetic fields

9.17 The sun irradiates the earth at the rate of 2.00 calories/cm² per minute, a rate known as the solar constant. Assume that sunlight arrives as a plane wave.
  (a) How is the electric field amplitude $|\mathbf{E}|$ related to the time-averaged power density (in watts/m²) of a plane wave?
  (b) What is the magnitude of the Poynting vector (in W/m²) in sunlight?
  (c) What is the amplitude of the electric field (in V/m) in sunlight?
  (d) What is the amplitude of the magnetic field (in A/m) in sunlight?
  (e) What is the power output of the sun, in the form of sunlight alone, in watts?
  Note: 1 cal = 4.186 J; distance to sun = $(1.50)10^{11}$ m.

Plane wave generator

9.18 An idealized antenna that generates plane waves consists of an infinitely thin and infinitely extended current sheet, coincident with the xy-plane, that carries a uniform surface current $\mathbf{K} = -\hat{\mathbf{x}}K_0 e^{j\omega t}$. Describe the electric and magnetic fields of the plane waves generated by this current sheet, in the air on both sides of the xy-plane.

Plane wave incident onto a mirror

9.19 A plane wave $\hat{\mathbf{x}}E_0 e^{-j\omega z/c}$ is incident from $z = -\infty$ in air and encounters a perfectly conducting plane at $z = 0$, which acts as a mirror. This results in a reflected wave and in an induced surface current on the xy-plane.
  (a) Describe the electric and magnetic fields of the resultant superposition of plane waves in the space $z < 0$.
  (b) Find the magnitude and direction of the induced surface current on the mirror.

Plane wave incident onto a dielectric

9.20 A plane wave $\hat{\mathbf{x}}E_0 e^{-j\omega z/c}$ is incident from $z = -\infty$ in air and encounters a dielectric of permittivity ε that fills the half-space $z > 0$. This results in reflection of a wave back into the air and in transmission of a wave beyond the xy-plane into the dielectric. Describe the resultant electric and magnetic fields in both the air and the dielectric.

Phased array of aligned current elements

9.21 The radiation pattern of a pair of current elements can be modified not only by varying their spacing but also by phasing their currents. Consider two current elements of moments $I_1\,dl$ and $I_2\,dl$ with their axes aligned, spaced apart by distance $2s$ along the common z-axis, and with currents of the same magnitude but phased so that $I_2 = I_1 e^{j\psi}$. Their radiation pattern was found in the text to be given by

$$f(\theta) = \sin^2\theta \left\{ |I_1|^2 + |I_2|^2 + 2\,\text{Re}\left[I_1 I_2^* e^{j2ks\cos\theta} \right] \right\}.$$

(a) Specialize this function to the case $I_2 = I_1 e^{j\psi}$ and confirm that the function $f(\theta)$ is proportional to $\sin^2\theta\,\cos^2\left[g(\theta, ks, \psi) \right]$. What is $g(\theta, ks, \psi)$?

From this, we can argue that the radiation pattern should peak at angles such that $g(\theta, ks, \psi) = 0$ or multiples of π, provided these angles are not too close to $\theta = 0$ or $\theta = \pi$, where the $\sin^2\theta$ factor suppresses the peak.

(b) Exactly where would the peaks of the radiation pattern in Figure P9-5 be if they were not affected by the $\sin^2\theta$ factor of the individual current elements? The figure is drawn for the case $ks = 3\pi/4$, $\psi = \pi/2$.

(c) Find the precise angle of the null of the radiation pattern (other than along the z-axis) in the figure.

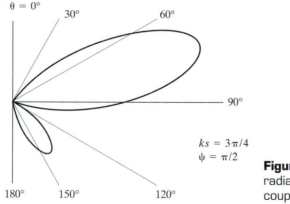

$ks = 3\pi/4$
$\psi = \pi/2$

Figure P9-5 Polar plot of radiation pattern of phased couplet.

Distortionless transmission line

9.22 A transmission line is *distortionless* if the phase velocity and the attenuation constant are both independent of frequency. This condition implies that all frequency components of a signal emitted by the source arrive at the load with the same delay and equally attenuated, so that the signal shape is not distorted. Verify that a lossy transmission line with parameters R, L, C, G is distortionless if $RC = LG$, by finding the phase velocity v_p and the attenuation constant α and confirming that both are independent of frequency.

Dispersion of gravity waves

9.23 Gravity waves are not electromagnetic but induce periodic variations in atmospheric density and temperature; they can be observed in wavy or striated cloud patterns. They are characterized by the dispersion relation $\omega^2 = gk$, where $g = 9.8$ m/s^2 is the acceleration of gravity.

For gravity waves at frequency $f = 0.08$ Hz, find

(a) the wavelength λ;

(b) the phase velocity v_p;

(c) the group velocity v_g.

CHAPTER 10

Review, Epilogue, and Appendix

We have come to the conclusion of our examination of basic concepts in electro-magnetics. A concise review may be timely, and a brief assessment of the many areas one may explore further is in order.

REVIEW OF CONCEPTS

We deal with fields when we need to consider quantities at all points in space and at all instants in time. Mathematically, this need is satisfied when we describe functions of position and time, $f(\mathbf{r}, t)$. The systems to which such descriptions apply are termed *distributed*, as opposed to the *lumped* systems of circuit theory, for example.

Electromagnetics is a discipline marked by action at a distance and after a delay. The former is obvious even in the interaction of separated charges at rest; the latter is seen in the pure delay that is inherent in the operation of an ideal transmission line. The two phenomena are intertwined in virtually all electromagnetic devices and systems, other than purely static ones.

We found that the propagation of action takes place as a wave motion. This is expressed in mathematical terms as any function of position $\mathbf{r}$ and time t of the form $f(\mathbf{r} - \mathbf{v}t)$. The factor $\mathbf{v}$ is the velocity of propagation of this wave. As time elapses, the location $\mathbf{r}$ at which the function f has some specific argument $\mathbf{s}$ moves, as $\mathbf{r} = \mathbf{r}(t) = \mathbf{s} + \mathbf{v}t$, uniformly at velocity $\mathbf{v}$.

We found that electromagnetic fields propagate as waves, that they natural-ly "want" to do so, in the sense that nothing is needed to induce them to move as waves, but that we can guide, transmit, or radiate such fields by introducing suit-able structures of dielectrics and conductors that affect the propagation of the fields.

The dynamics of electromagnetic fields are most readily expressed in terms of integrals of the fields over surfaces S and curves C in space. Specifically, the integrals include emf $V\{C\}$, mmf $U\{C\}$, electric flux $\Psi\{S\}$, magnetic flux $\Phi\{S\}$, and current

456

$I\{S\}$. Maxwell's equations relate these integrals when they apply to an open surface S whose boundary is the closed curve C, mutually oriented by the right-hand rule:

$$V = -d\Phi/dt, \qquad U = d\Psi/dt + I. \tag{10.1}$$

By relating the mechanical effects of electric and magnetic fields to mechanical concepts of force, energy, and power, we found that the Poynting vector $\mathbf{E} \times \mathbf{H}$ represents power flow, in watts/m^2. The disposition of the power that may cross a closed surface S to enter the enclosed region of space V is expressed by the Poynting theorem,

$$-\oint_S \mathbf{E} \times \mathbf{H} \cdot d\mathbf{S} = \int_V \mathbf{E} \cdot \mathbf{J} \, dV + d/dt \int_V \left(\tfrac{1}{2}\mathbf{H} \cdot \mathbf{B} + \tfrac{1}{2}\mathbf{E} \cdot \mathbf{D}\right) dV, \tag{10.2}$$

which accounts for the influx of power across S in terms of the expenditure of power and the accumulation of stored energy within V.

We examined waves both in the transient case, such as $f(z - ct)$, which has shape $f(z)$ and travels at speed c, and also in the harmonic steady state, expressed as complex exponentials, so that a sinusoidal wave showed up as $e^{j(\omega t - \beta z)}$, with phase speed ω/β. We found that transmission lines carry fields, and voltages, and currents, and charges, and fluxes that propagate as waves. In the transient case, the transmission line equations are of the form

$$V(z, t) = V(l, t) - \partial\Phi(z, t)/\partial t, \qquad \Phi = \int LI \, dz, \tag{10.3}$$

$$I(z, t) = I(l, t) - \partial Q(z, t)/\partial t, \qquad Q = \int CV \, dz. \tag{10.4}$$

The solution was expressible as forward- and backward-traveling waves, in terms of a characteristic speed of propagation c and a characteristic impedance Z_0, in the form

$$V(z, t) = f(z - ct) + g(z + ct), \tag{10.5}$$

$$Z_0 I(z, t) = f(z - ct) - g(z + ct). \tag{10.6}$$

For the steady state, the equations appear as

$$V(z) = V(l) + \int_z^l ZI(z') \, dz', \tag{10.7}$$

$$I(z) = I(l) + \int_z^l YV(z') \, dz' \tag{10.8}$$

and the superposition of waves for the solution as

$$V(z) = V_1 e^{-\gamma z} + V_2 e^{\gamma z}, \tag{10.9}$$

$$Z_0 I(z) = V_1 e^{-\gamma z} - V_2 e^{\gamma z}, \tag{10.10}$$

where the propagation constant γ and the characteristic impedance Z_0 are

$$\gamma = \sqrt{ZY}, \qquad Z_0 = \sqrt{Z/Y}, \tag{10.11}$$

provided that Z and Y are independent of z, the defining feature of a uniform transmission line. The two complex constants V_1 and V_2 are determined by the source and load at the two ends of the line. The key quantities needed to specialize the solution to the particular problem at hand are the reflection coefficient at the load, given in terms of the mismatch between the load and the characteristic impedance as

$$\Gamma_l = (Z_l - Z_0)/(Z_l + Z_0) \qquad (10.12)$$

and the impedance at any point along the line, given as

$$Z(z) = Z_0\{1 + \Gamma_l e^{-j2\beta(l-z)}\}/\{1 - \Gamma_l e^{-j2\beta(l-z)}\}. \qquad (10.13)$$

These equations allow the solution of virtually any transmission line problem. For a lossy line, the real phase constant becomes merely the imaginary part of a complex propagation constant $\gamma = \alpha + j\beta$, whose real part gives the attenuation per unit length that accompanies the propagation, as a consequence of energy loss. We studied a number of impedance matching techniques in transmission systems, aimed at eliminating undesired reflections.

We examined uniform and nonuniform plane waves in unbounded media. They exhibit electric and magnetic fields that are perpendicular to each other and to the direction of propagation. We studied the behavior of the field vectors in the sinusoidal steady state. That behavior is called the polarization of the wave, and we found the general polarization to be elliptical, meaning that the field vectors trace out an ellipse in the plane perpendicular to the direction of propagation. The limiting cases of linear and of circular polarization are particularly useful. We found that plane waves that encounter an interface between distinct media are reflected and transmitted in accordance with the same phenomena observable on transmission lines when a discontinuity or load is encountered. The key factor is the mismatch in impedances on the two sides of the interface. We compared impedance matching techniques for plane waves and for transmission lines, such as antireflection coatings and quarter-wave matching sections.

Oblique incidence of plane waves onto an interface provides opportunities to tune the mismatched impedances by merely tilting the interface plane. Snell's laws describe the directions of reflected and refracted waves at an interface. Decomposing an arbitrary incident polarization into its TE and TM components allows us to calculate the strengths of the reflected and refracted waves. Brewster's angle represents matching for a TM wave encountering a mismatched medium at oblique incidence. We looked at total internal reflection as a means of confining light within a dielectric medium, such as an optical fiber or waveguide.

We analyzed propagation along a rectangular waveguide, a hollow pipe that acts as a transmission line at high frequencies but as an attenuator at low frequencies. We found that each possible mode of propagation has a cutoff frequency, ω_c, determined by the geometry, the dielectric in the pipe, and the particular mode, and that propagation occurs only when the operating frequency exceeds this cutoff. Above cutoff, the phase constant and the frequency are related by the dispersion relation

$$\omega^2 = \omega_c^2 + \beta^2 c^2. \qquad (10.14)$$

The phase constant $\beta = 2\pi/\lambda_g = \omega/v_p$ tells us the guide wavelength λ_g and the phase velocity v_p for the given frequency.

We confirmed that fields can propagate in empty space, as a plane wave or a superposition of plane waves. These travel, in any direction, at any frequency, at the speed of light. The frequency f and wavelength λ of a plane wave are related by $f\lambda = c$; when an operating frequency is specified by giving a wavelength, the two specifications are related in this canonical way.

We examined a rudimentary antenna, in the form of a time-varying current element, which we showed radiates power to infinity. We obtained the exact fields of this elementary source. We also looked at interfering radiation from two current elements, representing a rudimentary array of antennas, and saw the shaping of the radiation pattern that could be achieved. We glanced at the vocabulary of antenna systems and hinted at the provenance and utility of the reciprocity theorem.

VISTAS AHEAD

We begin to become knowledgeable only when we realize how little we know. Concepts in electromagnetics are of such wide applicability and utility as to present us with a vast panoply of further areas of study.

In our pursuit of the rules of the game, we have illustrated the influence of uniform, constant electric and magnetic fields on the motion of an electron, but the trajectories we can achieve and utilize in more complicated field configurations remain to be explored. This may be particulary relevant in analyzing space charge effects and in the design of oscillators and amplifiers that use electron beams, especially when the motion of the electrons is relativistic. The motion of electrons in the presence of fields is also crucial to understanding the action of semiconductors. We have also touched on the properties of plasmas, but their behavior exhibits a bewildering variety of fluid, thermal, elastic, mechanical, and electromagnetic phenomena, resulting in an interplay of many types of waves that rewards further study. We have also hinted at the characteristics of superconductors but have not delved into their properties.

For the mathematical analysis of Maxwell's equations, we have pursued the technique of successive approximations, primarily because it requires no advanced mathematics. There is a vast array of techniques for dealing with partial differential equations and with integral equations that are directly relevant to the solution of boundary value problems in electromagnetics. We have also avoided the introduction of scalar and vector potentials and such often helpful representations as multipole expansions and Green's functions.

While we have explored transmission lines and waveguides as examples of wave-propagating systems, we have not delved further into the properties of a variety of structures. Even the coaxial line can support more complex modes of wave propagation than the elementary transverse electromagnetic ones we have considered. Waveguides of various cross-sectional shapes share many important properties; periodic structures and nonuniform ones are even more exotic. We have discussed resonant cavities only briefly, and we have not done nearly enough about the problems of launching and detecting signals along transmission lines and waveguides, of matching, and of coupling between lines. The analysis and design of microwave circuitry is a highly developed area of inquiry. Measurements of the properties of materials are often based on electromagnetic principles.

We have hinted at the capabilities of antenna structures and arrays but have not explored their properties and parameters in much detail. Antennas with known current distributions are readily analyzed by applying the theory of Fourier transforms. The more challenging task is that of determining what currents flow on an antenna in response to excitation at its terminals.

The entire field of optics is based on applications of the principles of electromagnetics, particularly in the limit of extremely small wavelengths. Reflection and refraction at interfaces between different media are crucial to the operation of optical instruments. We have not touched upon scattering and diffraction effects, nor on quantum effects in optics, nor on the properties of lasers and other light sources. Dielectric waveguides and optical fibers constitute another direct extension of the principles we have reviewed.

If this atlas of unexplored territory is daunting, it should be taken as an invitation to further study, for the sake of all the philosophical beauty and the practical utility of electromagnetics.

APPENDIX

We collect in this appendix various mathematical formulas and relations for convenient reference. They include especially the mathematical identities and vector manipulations that we used in the body of the text. Proving them can serve as valuable review exercises. This compilation is intended to remind the reader of matters that were once familiar and to establish the notation we use. It is not intended to teach utterly new material.

Complex Numbers. Our notation for complex numbers uses j as the imaginary unit. The rectangular and polar versions are related by

$$z = x + jy = re^{j\theta}, \tag{10.15}$$

where

$$r^2 = x^2 + y^2; \qquad \tan\theta = y/x. \tag{10.16}$$

Multiplication and division of complex numbers become

$$z_1 z_2 = (x_1 x_2 - y_1 y_2) + j(x_1 y_2 + x_2 y_1) = r_1 r_2 e^{j(\theta_1 + \theta_2)}, \tag{10.17}$$

$$z_1/z_2 = \{(x_1 x_2 + y_1 y_2) + j(x_2 y_1 - x_1 y_2)\}/(x_2^2 + y_2^2) = (r_1/r_2)e^{j(\theta_1 - \theta_2)}. \tag{10.18}$$

The real and imaginary parts are denoted

$$x = r\cos\theta = \operatorname{Re} z; \qquad y = r\sin\theta = \operatorname{Im} z. \tag{10.19}$$

Euler's formula gives the exponential function of a complex argument:

$$e^z = e^{x + jy} = e^x(\cos y + j \sin y). \tag{10.20}$$

Special cases of this include

$$e^{\pm j\pi} = -1; \qquad e^{\pm j(\pi/2)} = \pm j. \tag{10.21}$$

Combinations of exponentials with equal and opposite imaginary exponents form trigonometric functions:

$$e^{j\theta} + e^{-j\theta} = 2\cos\theta; \qquad e^{j\theta} - e^{-j\theta} = 2j\sin\theta. \tag{10.22}$$

Coordinate Systems. Points in space are identified by giving their coordinates, as many as the dimensionality of the space. In two dimensions, the main coordinate systems are the rectangular (x, y) and the polar (r, θ). These are related by

$$x = r\cos\theta, \quad y = r\sin\theta; \qquad r^2 = x^2 + y^2, \quad \tan\theta = y/x. \tag{10.23}$$

In three dimensions, the primary coordinate systems are the rectangular (x, y, z), the cylindrical (ρ, φ, z), and the spherical (r, θ, φ) systems. They are related by the following transformations.

Cylindrical

$$x = \rho\cos\varphi, \quad y = \rho\sin\varphi; \qquad \rho^2 = x^2 + y^2, \quad \tan\varphi = y/x; \tag{10.24}$$

Spherical

$$x = r\sin\theta\cos\varphi, \qquad y = r\sin\theta\sin\varphi, \qquad z = r\cos\theta; \tag{10.25}$$

$$r^2 = x^2 + y^2 + z^2, \qquad \cos\theta = z/r, \qquad \tan\varphi = y/x. \tag{10.26}$$

Figure 10-1 aids in interpreting ρ as the distance between the point and the z-axis, r as the distance between the point and the origin, θ as the *elevation* angle between the position vector and the z-axis, and φ as the *azimuthal* angle between the xz-plane and the plane that contains the position vector and the z-axis.

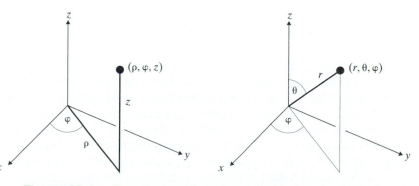

Figure 10-1   Coordinates of a point, cylindrical and spherical.

The element of area in two dimensions is

$$dA = dx\, dy = r\, dr\, d\theta \tag{10.27}$$

and the volume element in three dimensions is given by

$$dV = dx\, dy\, dz = \rho\, d\rho\, d\varphi\, dz = r^2 \sin\theta\, dr\, d\theta\, d\varphi. \tag{10.28}$$

Vector Algebra.   A vector has both a magnitude and a direction in space. The zero vector has zero magnitude (and hence no defined direction); a unit vector has magnitude unity and is fully specified by its direction. When we have wanted to emphasize that a vector is a unit vector, we have indicated this with a circumflex accent, as in $\hat{n}$. Three mutually perpendicular unit vectors, such as those along the directions of the rectangular coordinate axes, $\hat{x}, \hat{y}, \hat{z}$, form a *basis* and any vector **A** can then be expressed as a sum of multiples of these: $\mathbf{A} = a\hat{x} + b\hat{y} + c\hat{z}$ has *components* a, b, c with respect to that basis.

Addition of vectors is performed by adding their individual components to get those of the sum and multiplication of a vector by a *scalar* (i.e., a number) gives a vector each of whose components is that scalar times the components of the original vector.

The *dot product* (or scalar product) of two vectors **A**, **B** is a *scalar*, given by the sum of the products of corresponding components of the two vectors, as in

$$\mathbf{A} \cdot \mathbf{B} = (a_1\hat{x} + a_2\hat{y} + a_3\hat{z}) \cdot (b_1\hat{x} + b_2\hat{y} + b_3\hat{z}) = a_1 b_1 + a_2 b_2 + a_3 b_3. \tag{10.29}$$

The geometrical interpretation (when the vectors are real) is that this scalar is the product of the two magnitudes and of the cosine of the angle between the two vectors. Two vectors are perpendicular (or orthogonal) if their dot product is zero. The dot product of a vector with itself is the square of the magnitude of the vector: $\mathbf{A} \cdot \mathbf{A} = |\mathbf{A}|^2$.

The *cross product* (or vector product) of two vectors **A**, **B** (in three dimensions) is a *vector* whose components are given by the differences of the products of the non-corresponding components of the two vectors, as in

$$\mathbf{A} \times \mathbf{B} = (a_1\hat{x} + a_2\hat{y} + a_3\hat{z}) \times (b_1\hat{x} + b_2\hat{y} + b_3\hat{z})$$

$$= (a_2 b_3 - a_3 b_2)\hat{x} + (a_3 b_1 - a_1 b_3)\hat{y} + (a_1 b_2 - a_2 b_1)\hat{z}. \tag{10.30}$$

The geometrical interpretation is that this vector has a magnitude that is the product of the two magnitudes and of the sine of the angle between the two vectors, and a direction that is perpendicular to the plane formed by the two vectors; the right-hand rule orients the cross-product vector with respect to the original vectors: If the fingers of the right hand indicate the rotation of vector **A** toward vector **B**, then the thumb points along the vector **A** × **B**. It follows that the cross product is not commutative: **A** × **B** = −**B** × **A**. Two vectors are parallel (or collinear) if their cross product is zero. The cross product of a vector with itself is the zero vector.

The scalar triple product of three vectors is the dot product of one with the cross product of the other two; if the cyclic order of the three vectors is preserved, the order of the dot and cross is immaterial:

$$\mathbf{A} \cdot \mathbf{B} \times \mathbf{C} = \mathbf{A} \times \mathbf{B} \cdot \mathbf{C}. \tag{10.31}$$

The geometric interpretation is that this is the volume of the parallelepiped formed by the three vectors (or the negative of that volume).

The vector triple product is the cross product of one with the cross product of the other two; the order does matter. The vector triple product can be simplified by use of the identity

$$\mathbf{A} \times (\mathbf{B} \times \mathbf{C}) = \mathbf{B}(\mathbf{A} \cdot \mathbf{C}) - \mathbf{C}(\mathbf{A} \cdot \mathbf{B}). \tag{10.32}$$

An important special case of this allows us to express any vector $\mathbf{A}$ as the sum of two vectors, one perpendicular to a specified unit vector $\hat{\mathbf{n}}$ and the other along that unit vector:

$$\mathbf{A} = \hat{\mathbf{n}} \times \mathbf{A} \times \hat{\mathbf{n}} + \hat{\mathbf{n}}\mathbf{A} \cdot \hat{\mathbf{n}}. \tag{10.33}$$

Vectors with components that are complex numbers are treated in the same way as real vectors, but their geometric interpretation requires more care. While a real vector has a magnitude and a direction, a complex one generally cannot be assigned a single direction, since the real and imaginary parts may not share one direction. There is also the peril of confusing the notion of the magnitude of a (real) vector and that of the magnitude of a complex number. Finally, we must abandon concepts like the angle between two complex vectors.

The dot product of two complex vectors is a complex number. If that complex number is zero, we say the two complex vectors are orthogonal. The dot product of a complex vector with itself need not be a positive number, but the dot product of the complex conjugate of a vector with the original vector is positive and can define the magnitude (squared) of the complex vector: $|\mathbf{A}|^2 = \mathbf{A}^* \cdot \mathbf{A}$. This is the sum of the squared magnitudes of the complex components of the vector.

Dyadics. Besides a dot product, which is a scalar, and a cross product, which is a vector, we often find it convenient to form yet another product of two vectors, called an *outer product*, or *dyad*, or *dyadic*, or simply a matrix. It is denoted by a juxtaposition of the two vectors, with no dot or cross or other symbol between them, such as $\mathbf{AB}$, and is calculated as the ordinary matrix product of the column vector $\mathbf{A}$ with the row vector $\mathbf{B}$, forming a square matrix. As an example, we may consider the vectors $\mathbf{A}$ and $\mathbf{B}$ and the dyad $\mathbf{AB}$ in

$$\mathbf{A} = \begin{bmatrix} 2 \\ -1 \\ 3 \end{bmatrix} \quad \mathbf{B} = \begin{bmatrix} 5 & 2 & -4 \end{bmatrix} \quad \mathbf{AB} = \begin{bmatrix} 2 \\ -1 \\ 3 \end{bmatrix} \begin{bmatrix} 5 & 2 & -4 \end{bmatrix} = \begin{bmatrix} 10 & 4 & -8 \\ -5 & -2 & 4 \\ 15 & 6 & -12 \end{bmatrix}$$

obtained by the usual row-by-column rule of matrix multiplication.

This *outer product*, which comprises all products of elements of both vectors in every combination, should be compared with the *inner product* or scalar or dot product $\mathbf{B} \cdot \mathbf{A} = 10 - 2 - 12 = -4$, which is a sum of products of corresponding elements, again by row-by-column matrix multiplication. Technically, a single outer product like $\mathbf{AB}$ is a *dyad* and a sum of such products is a dyadic. We often do not distinguish between a row vector and a column vector; in a dot product, the first is automatically a row vector and the second a column vector, to conform to the rules of matrix multiplication. In a dyad, the first is automatically a column vector and the second factor is a row vector, for the same reason.

If the resultant matrix operates on another vector (a column vector) on its right side in the usual matrix product (dot product) fashion, as in $\mathbf{AB} \cdot \mathbf{C}$, the outcome is a vector along the direction of vector $\mathbf{A}$; it is just $\mathbf{A}(\mathbf{B} \cdot \mathbf{C})$. If, however, the same dyad is (dot) multiplied on its left by row vector $\mathbf{D}$, as in $\mathbf{D} \cdot \mathbf{AB}$, the result is $(\mathbf{D} \cdot \mathbf{A})\mathbf{B}$, a row vector along the direction of $\mathbf{B}$. Clearly, the order of the two factors in the dyad matters: generally, $\mathbf{AB} \neq \mathbf{BA}$ even if we don't distinguish row and column vectors.

Vector Differential Calculus. The gradient operator ∇ is defined by its effect in forming the differential of any scalar function $f(\mathbf{r})$ of the position vector $\mathbf{r} = (x, y, z)$ subjected to an infinitesimal change of position $d\mathbf{r}$:

$$f(\mathbf{r} + d\mathbf{r}) - f(\mathbf{r}) = d\mathbf{r} \cdot \nabla f(\mathbf{r}). \tag{10.34}$$

Consequently, in rectangular coordinates, the ∇ operator is given by

$$\nabla = \hat{\mathbf{x}}(\partial/\partial x) + \hat{\mathbf{y}}(\partial/\partial y) + \hat{\mathbf{z}}(\partial/\partial z). \tag{10.35}$$

The operator can be applied to a vector function of position, as a *divergence*, or as a *curl* operation. For the vector function of position

$$\mathbf{F}(\mathbf{r}) = F_1(\mathbf{r})\hat{\mathbf{x}} + F_2(\mathbf{r})\hat{\mathbf{y}} + F_3(\mathbf{r})\hat{\mathbf{z}},$$

the divergence of $\mathbf{F}(\mathbf{r})$ is the scalar

$$\nabla \cdot \mathbf{F} = \partial F_1/\partial x + \partial F_2/\partial y + \partial F_3/\partial z \tag{10.36}$$

and the curl of $\mathbf{F}(\mathbf{r})$ is the vector

$$\nabla \times \mathbf{F} = \hat{\mathbf{x}}(\partial F_3/\partial y - \partial F_2/\partial z) + \hat{\mathbf{y}}(\partial F_1/\partial z - \partial F_3/\partial x) + \hat{\mathbf{z}}(\partial F_2/\partial x - \partial F_1/\partial y). \tag{10.37}$$

Second-order operators include the *Laplacian*, which is the divergence of the gradient when applied to a scalar $f(\mathbf{r})$,

$$\nabla^2 f = \nabla \cdot \nabla f = \partial^2 f/\partial x^2 + \partial^2 f/\partial y^2 + \partial^2 f/\partial z^2, \tag{10.38}$$

but is defined by the difference between the gradient of the divergence and the curl of the curl when applied to a vector:

$$\nabla^2 \mathbf{F} = \nabla\nabla \cdot \mathbf{F} - \nabla \times \nabla \times \mathbf{F}. \tag{10.39}$$

The reason for the use of the same notation for what are fundamentally different operations is that, in rectangular coordinates (and only in rectangular coordinates), the operation has the same appearance when applied to scalars and to vectors:

$$\nabla^2 f = \partial^2 f/\partial x^2 + \partial^2 f/\partial y^2 + \partial^2 f/\partial z^2,$$
$$\nabla^2 \mathbf{F} = \partial^2 \mathbf{F}/\partial x^2 + \partial^2 \mathbf{F}/\partial y^2 + \partial^2 \mathbf{F}/\partial z^2, \tag{10.40}$$

where the latter means the vector

$$\nabla^2 \mathbf{F} = \hat{\mathbf{x}}\nabla^2 F_1 + \hat{\mathbf{y}}\nabla^2 F_2 + \hat{\mathbf{z}}\nabla^2 F_3. \tag{10.41}$$

Some second-order derivatives vanish identically: The divergence of a curl is identically zero and the curl of a gradient is identically the zero vector. Thus,

$$\nabla \cdot \nabla \times \mathbf{F} = 0, \qquad \nabla \times \nabla f = 0. \tag{10.42}$$

We have existence conditions as well: If a vector field $\mathbf{B(r)}$ has no divergence, then there exists another field $\mathbf{A(r)}$ such that $\mathbf{B(r)} = \nabla \times \mathbf{A(r)}$. If a vector field $\mathbf{F(r)}$ has no curl, then there exists a scalar field $U(\mathbf{r})$ such that $\mathbf{F(r)} = \nabla U(\mathbf{r})$.

The gradient operator applied to a product yields results that combine the properties of vectors and of derivatives. If the factors are scalars, we have the usual property of a derivative:

$$\nabla(\varphi \psi) = \varphi \nabla\psi + \psi \nabla\varphi. \tag{10.43}$$

If one factor is a scalar and the other a vector, we get

$$\nabla \cdot (\varphi \mathbf{F}) = \varphi \nabla \cdot \mathbf{F} + \nabla\varphi \cdot \mathbf{F}, \tag{10.44}$$
$$\nabla \times (\varphi \mathbf{F}) = \varphi \nabla \times \mathbf{F} + \nabla\varphi \times \mathbf{F}. \tag{10.45}$$

When both factors are vectors, we must be careful of their order. The *divergence of a cross product* simplifies to

$$\nabla \cdot (\mathbf{F} \times \mathbf{G}) = \mathbf{G} \cdot \nabla \times \mathbf{F} - \mathbf{F} \cdot \nabla \times \mathbf{G}. \tag{10.46}$$

The *curl of a cross product* expands as

$$\nabla \times (\mathbf{F} \times \mathbf{G}) = \mathbf{F}\nabla \cdot \mathbf{G} - \mathbf{G}\nabla \cdot \mathbf{F} + \mathbf{G} \cdot \nabla\mathbf{F} - \mathbf{F} \cdot \nabla\mathbf{G}, \tag{10.47}$$

in which the expression $\mathbf{F}\nabla \cdot \mathbf{G}$ means $\mathbf{F}(\nabla \cdot \mathbf{G})$, which is the product of the scalar $\nabla \cdot \mathbf{G}$ by the vector $\mathbf{F}$, and the quantity $\mathbf{F} \cdot \nabla\mathbf{G}$ can be interpreted as $(\mathbf{F} \cdot \nabla)\mathbf{G}$, which means the operator $(F_1\partial/\partial x + F_2\partial/\partial y + F_3\partial/\partial z)$ applied to each of the components of $\mathbf{G}$. The *gradient of a dot product* has the expansion

$$\nabla(\mathbf{F} \cdot \mathbf{G}) = \mathbf{F} \cdot \nabla\mathbf{G} + \mathbf{G} \cdot \nabla\mathbf{F} + \mathbf{F} \times \nabla \times \mathbf{G} + \mathbf{G} \times \nabla \times \mathbf{F}. \tag{10.48}$$

Differential Operations in Various Coordinate Systems. An important unifying concept in dealing with different coordinate systems is that if a scalar function of position represents a distance along some direction, its gradient is a unit vector oriented toward increasing values of the distance along that direction. Thus, if $s = s(\mathbf{r})$ indicates the distance along a line in the direction of unit vector $\hat{\mathbf{n}}$, as observed at point $\mathbf{r}$, and if we change position to a nearby other point $\mathbf{r} + d\mathbf{r}$, then the distance has changed by $ds = \hat{\mathbf{n}} \cdot d\mathbf{r}$, the component of the position increment along the direction of $\hat{\mathbf{n}}$. But, as for any function of position, $ds = d\mathbf{r} \cdot \boldsymbol{\nabla} s$. These two expressions for the distance increment ds are compatible for all position increments $d\mathbf{r}$ only if

$$\boldsymbol{\nabla} s = \hat{\mathbf{n}} \qquad (s = \text{distance along direction of } \hat{\mathbf{n}}). \tag{10.49}$$

The important vector differential operations in the main curvilinear coordinate systems appear different from the corresponding ones in rectangular coordinates because the basis vectors, which are unit vectors in the direction of increasing values of the coordinates, are then themselves functions of position. They therefore have derivatives, as follows.

The radial unit vector in polar or cylindrical coordinates is

$$\hat{\boldsymbol{\rho}} = \hat{\boldsymbol{\rho}}(\varphi) = \hat{\mathbf{x}}\cos\varphi + \hat{\mathbf{y}}\sin\varphi. \tag{10.50}$$

The azimuthal unit vector in polar, cylindrical, and spherical coordinates is

$$\hat{\boldsymbol{\varphi}} = \hat{\boldsymbol{\varphi}}(\varphi) = -\hat{\mathbf{x}}\sin\varphi + \hat{\mathbf{y}}\cos\varphi. \tag{10.51}$$

The unit vectors in the directions of increasing radius and elevation angle in spherical coordinates are

$$\hat{\mathbf{r}} = \hat{\mathbf{r}}(\theta, \varphi) = \hat{\mathbf{x}}\sin\theta\cos\varphi + \hat{\mathbf{y}}\sin\theta\sin\varphi + \hat{\mathbf{z}}\cos\theta, \tag{10.52}$$

$$\hat{\boldsymbol{\theta}} = \hat{\boldsymbol{\theta}}(\theta, \varphi) = \hat{\mathbf{x}}\cos\theta\cos\varphi + \hat{\mathbf{y}}\cos\theta\sin\varphi - \hat{\mathbf{z}}\sin\theta. \tag{10.53}$$

Consequently, we have the derivatives

$$d\hat{\boldsymbol{\rho}}/d\varphi = \hat{\boldsymbol{\varphi}}, \qquad d\hat{\boldsymbol{\varphi}}/d\varphi = -\hat{\boldsymbol{\rho}} \tag{10.54}$$

and

$$\partial\hat{\mathbf{r}}/\partial\theta = \hat{\boldsymbol{\theta}}, \qquad \partial\hat{\mathbf{r}}/\partial\varphi = \sin\theta\,\hat{\boldsymbol{\varphi}}, \tag{10.55}$$

$$\partial\hat{\boldsymbol{\theta}}/\partial\theta = -\hat{\mathbf{r}}, \qquad \partial\hat{\boldsymbol{\theta}}/\partial\varphi = \cos\theta\,\hat{\boldsymbol{\varphi}}. \tag{10.56}$$

These are useful for the extraction of the following vector differential operators.

In cylindrical coordinates,

gradient: $\boldsymbol{\nabla} f(\rho, \varphi, z) = \hat{\boldsymbol{\rho}}\,\partial f/\partial\rho + \hat{\boldsymbol{\varphi}}(1/\rho)\partial f/\partial\varphi + \hat{\mathbf{z}}\,\partial f/\partial z;$ (10.57)

divergence: $\boldsymbol{\nabla} \cdot \mathbf{F} = (1/\rho)\partial(\rho F_\rho)/\partial\rho + (1/\rho)\partial F_\varphi/\partial\varphi + \partial F_z/\partial z;$ (10.58)

curl: $(\boldsymbol{\nabla} \times \mathbf{F})_\rho = (1/\rho)\partial F_z/\partial\varphi - \partial F_\varphi/\partial z$

$(\boldsymbol{\nabla} \times \mathbf{F})_\varphi = \partial F_\rho/\partial z - \partial F_z/\partial\rho$ (10.59)

$(\boldsymbol{\nabla} \times \mathbf{F})_z = (1/\rho)\partial(\rho F_\varphi)/\partial\rho - (1/\rho)\partial F_\rho/\partial\varphi$

Laplacian: $\nabla^2 f(\rho, \varphi, z) = (1/\rho)\partial[\rho\,\partial f/\partial\rho]/\partial\rho + (1/\rho^2)\partial^2 f/\partial\varphi^2 + \partial^2 f/\partial z^2.$ (10.60)

In spherical coordinates,

gradient: $\boldsymbol{\nabla} f(r, \theta, \varphi) = \hat{\mathbf{r}}\,\partial f/\partial r + \hat{\boldsymbol{\theta}}(1/r)\partial f/\partial\theta + \hat{\boldsymbol{\varphi}}(\partial f/\partial\varphi)/(r\sin\theta);$ (10.61)

divergence: $\boldsymbol{\nabla} \cdot \mathbf{F} = (1/r^2)\partial(r^2 F_r)/\partial r + (1/[r\sin\theta])\{\partial[\sin\theta\,F_\theta]/\partial\theta + \partial F_\varphi/\partial\varphi\};$ (10.62)

curl:
$$(\nabla \times \mathbf{F})_r = (1/[r \sin \theta])\{\partial[\sin \theta \, F_\varphi]/\partial\theta - \partial F_\theta/\partial\varphi\}$$

$$(\nabla \times \mathbf{F})_\theta = (1/[r \sin \theta])\partial F_r/\partial\varphi - (1/r)\partial[r F_\varphi]/\partial r \qquad (10.63)$$

$$(\nabla \times \mathbf{F})_\varphi = (1/r)\partial(r F_\theta)/\partial r - (1/r)\partial F_r/\partial\theta$$

Laplacian: $\nabla^2 f(r, \theta, \varphi) = (1/r^2)\partial[r^2 \partial f/\partial r]/\partial r$

$$+ (1/[r^2 \sin \theta])\partial[\sin \theta \, \partial f/\partial\theta]/\partial\theta + (1/[r^2 \sin^2 \theta])\partial^2 f/\partial\varphi^2. \qquad (10.64)$$

Vector Integral Calculus. Integrations of vectors are performed over lines, surfaces, and volumes in space. Both the vector field $\mathbf{F}(\mathbf{r})$ and the domain of integration C or S or V must be given. The integral is evaluated by converting it into an ordinary single, double, or triple integral over the parameters that describe the curve, surface, or volume of integration.

To specify a curve C, we give the position vector $\mathbf{r} = \mathbf{r}(u)$ as a function of a single parameter u that ranges from u_1 to u_2 as the curve is traversed. The vector field then becomes a function of u, as $\mathbf{F}(\mathbf{r}) = \mathbf{F}(\mathbf{r}(u))$, and the element of length along the curve, $d\mathbf{l}$, is expressed in terms of the differential du by $d\mathbf{l} = (d\mathbf{r}/du)du$, so that the line integral is evaluated as

$$\int \mathbf{F} \cdot d\mathbf{l} = \int \mathbf{F}(\mathbf{r}(u)) \cdot [d\mathbf{r}/du] \, du, \qquad (10.65)$$

between the limits u_1 and u_2.

To specify a surface S, we give the position vector $\mathbf{r} = \mathbf{r}(u, v)$ as a function of two parameters u, v that range between appropriate limits as the surface is traversed. The vector field then becomes a function of u, v as $\mathbf{F}(\mathbf{r}) = \mathbf{F}(\mathbf{r}(u, v))$, and the element of area over the surface, $d\mathbf{S}$, is expressed in terms of the differentials du, dv by

$$d\mathbf{S} = (\partial\mathbf{r}/\partial u) \times (\partial\mathbf{r}/\partial v) \, du \, dv, \qquad (10.66)$$

so that the surface integral is evaluated as the double integral

$$\int \mathbf{F} \cdot d\mathbf{S} = \int\int \mathbf{F}(\mathbf{r}(u, v)) \cdot [(\partial\mathbf{r}/\partial u) \times (\partial\mathbf{r}/\partial v)] \, du \, dv. \qquad (10.67)$$

To specify a volume V, we give the position vector $\mathbf{r} = \mathbf{r}(u, v, w)$ as a function of three parameters u, v, w that range between appropriate limits as the volume is traversed. The scalar (or vector) field then becomes a function of u, v, w as $f(\mathbf{r}) = f(\mathbf{r}(u, v, w))$, and the element of volume, dV, is expressed in terms of the differentials du, dv, dw by

$$dV = (\partial\mathbf{r}/\partial u) \times (\partial\mathbf{r}/\partial v) \cdot (\partial\mathbf{r}/\partial w) \, du \, dv \, dw, \qquad (10.68)$$

so that the volume integral is evaluated as the triple integral

$$\int f \, dV = \int\int\int f(\mathbf{r}(u, v, w))[(\partial\mathbf{r}/\partial u) \times (\partial\mathbf{r}/\partial v) \cdot (\partial\mathbf{r}/\partial w)] \, du \, dv \, dw. \qquad (10.69)$$

In each case, the sign of the element of integration must be made positive, by choosing the direction of traversal of the curve C along its orientation, or by ensuring that the parameters u, v give an element of area $d\mathbf{S}$ that conforms to the orientation of the surface S (or else we reverse the parameters' order), or by choosing the order of the three parameters u, v, w to make the volume element dV positive (or else we interchange one pair of them).

If the integrand, and therefore the integral, is a vector, then it may be dotted with each of the basis vectors of the rectangular coordinate system to yield the three components of the integral. Basis vectors of curvilinear coordinate systems cannot be moved outside of the integral, because they are not constant vectors.

Two theorems relate integrals of derivatives of fields over a region to integrals of the fields themselves over the boundary of the region. The *divergence theorem* is

$$\int_V \boldsymbol{\nabla} \cdot \mathbf{F} \, dV = \oint_S \mathbf{F} \cdot d\mathbf{S}, \tag{10.70}$$

where S is the closed surface that bounds the volume V. For a closed surface, the convention is that the element of area $d\mathbf{S}$ is directed outward. *Stokes's theorem* is

$$\int_S \boldsymbol{\nabla} \times \mathbf{F} \cdot d\mathbf{S} = \oint_C \mathbf{F} \cdot d\boldsymbol{l}, \tag{10.71}$$

where C is the closed curve that is the edge of the open surface S, with the curve and surface mutually oriented to conform with the right-hand rule: If the fingers of the right hand curl along the direction of the oriented curve C, then the thumb points along the orientation of the surface S.

The *solid angle* subtended by a surface S at a point is defined by

$$\Omega = \int_S \mathbf{r} \cdot d\mathbf{S}/r^3, \tag{10.72}$$

where $\mathbf{r}$ is the position vector from the point in question to the surface element $d\mathbf{S}$; equivalently, $d\Omega = \hat{\mathbf{r}} \cdot d\mathbf{S}/r^2$ is the element of solid angle (in steradians).

Fourier Transforms. The Fourier transform pair converts the description of a signal from and to the time and frequency domains. If $g(t)$ is a signal, its Fourier transform or spectrum is given by $G(\omega)$, as in

$$G(\omega) = \int_{-\infty}^{\infty} g(t)e^{-j\omega t} \, dt, \qquad g(t) = \int_{-\infty}^{\infty} G(\omega)e^{j\omega t} \, d\omega/2\pi. \tag{10.73}$$

If t is time, in seconds, then ω is frequency, in radians per second. We often use cycles per second, or hertz, to measure frequency f, so that $\omega = 2\pi f$, because there are 2π radians in a cycle. When expressed directly in terms of frequency in hertz, the Fourier transform appears as the more symmetric pair

$$G(f) = \int_{-\infty}^{\infty} g(t)e^{-j2\pi ft} \, dt, \qquad g(t) = \int_{-\infty}^{\infty} G(f)e^{j2\pi ft} \, df. \tag{10.74}$$

Fourier series are superpositions of discrete exponentials or sinusoids that can represent a periodic signal:

$$f(t) = \sum_{-\infty}^{\infty} F_n e^{jn\omega_0 t}, \qquad F_n = (1/T) \int_{-T/2}^{T/2} f(t) e^{-jn\omega_0 t}\, dt, \qquad (10.75)$$

where the sum is over all positive and negative integers n, the integral is over one period, of duration T, of the periodic signal $f(t)$, and the fundamental frequency is $\omega_0 = 2\pi/T$.

Convolution. The *convolution* of two functions $u(t)$ and $v(t)$ is another function $w(t)$ defined by the integral

$$w(t) = \int_{-\infty}^{\infty} u(\lambda) v(t - \lambda)\, d\lambda. \qquad (10.76)$$

This is a (continuous) sum of products of values of the two functions, at all arguments that add up to t. By an obvious change of variable of integration, we find that, despite appearances, the convolution of $u(t)$ and $v(t)$ is the same as the convolution of $v(t)$ with $u(t)$; that is, the operation is commutative:

$$w(t) = \int_{-\infty}^{\infty} u(t - \xi) v(\xi)\, d\xi. \qquad (10.77)$$

We also define convolutions for functions of a discrete argument, as in

$$w_n = \sum_{-\infty}^{\infty} u_m v_{n-m}. \qquad (10.78)$$

Again, this discrete convolution is commutative.

The primary significance of a convolution is that the response of a linear, time-invariant system to an input signal $x(t)$ is an output signal $y(t)$ given as the convolution of $h(t)$ and $x(t)$, where $h(t)$ is the *impulse response* of the system. Convolution is often abbreviated symbolically as $w(t) = u(t) * v(t)$.

Convolutions are often seen with limits of integration 0 and t instead of $-\infty$ and ∞; this is simply a special case that applies when both functions are zero for negative arguments; in that case, the output function has the same property.

A result of major importance is that convolution in the time domain corresponds to simple, algebraic multiplication in the frequency domain. That is, if $w(t)$ is the convolution of $u(t)$ and $v(t)$ in the time domain, $w(t) = u(t) * v(t)$, and if $U(\omega)$, $V(\omega)$, and $W(\omega)$ are the Fourier transforms of $u(t)$, $v(t)$, and $w(t)$, then $W(\omega) = U(\omega)V(\omega)$ in the frequency domain. One consequence is that a relation like $D(\omega) = \varepsilon(\omega)E(\omega)$ in the frequency domain implies that $D(t)$ and $E(t)$ are related by a convolution, not a multiplication, in the time domain.

Decibel Scale. The decibel scale expresses factors or ratios logarithmically, with the additional quirk that ratios of power or energy and their corresponding ratios of voltage or current are expressed by the same numerical value in dB. For the original unit, the bel (not in common use), the base of the logarithms is 10 but for the decibel, it is $10^{0.1}$. Thus a (power) ratio of 4 is $10^{0.6} = (10^{0.1})^6$ and is expressed as 6 dB. [More precisely, $4 = 10^{0.602} = (10^{0.1})^{6.02}$ corresponds to 6.02 dB.] A voltage ratio of 3 gives a power ratio of 9 and is therefore expressed as 9.54 dB, since $9 = 10^{0.954} = (10^{0.1})^{9.54}$. Thus, for quantities like voltage or current or field amplitudes, to which power is related quadratically, the base of logarithms is effectively $10^{0.05}$. Hence, a voltage ratio of $3 = 10^{0.477} = (10^{0.05})^{9.54}$ is 9.54 dB. It follows that a given ratio of voltage, R, is converted to the decibel scale by giving $20 \log R$ but a ratio of power is expressed in dB by giving $10 \log R$.

When the factor being expressed in dB is not dimensionless, an indication of the unit is commonly appended to the dB designation. For example, a power level of 50 mW may be expressed as 17 dBm (more precisely, 16.99 dBm); the m in dBm refers to 1 milliwatt. We may also say this power level is 17 dB above a milliwatt, or 13 dB below a watt. If the factor is merely a pure number, the base of logarithms is taken to be $10^{0.1}$; for example, a factor of 4π may be expressed as 11 dB (more precisely, 10.99 dB), because $10 \log 4\pi = 10.9921$.

We also use dB to express exponents in powers or exponential functions. For the scale based on natural logarithms, the exponent is in nepers (Np); for the bases $10^{0.1}$ (for power) and $10^{0.05}$ (for amplitudes), the exponent is in decibels (dB). For example, a factor $e^{-\alpha z}$ may be expressed as αz nepers of attenuation, and the attenuation coefficient α may be reported as α Np/m (if z is in meters). This same factor is also $10^{-\alpha z \log e} = (10^{0.05})^{-(20 \log e)\alpha z} = (10^{0.05})^{-8.686\,\alpha z}$, so that the exponent may be expressed as $8.686\alpha z$ dB and the attenuation coefficient as (8.686α) dB/m. That is, 1 Np = 8.686 dB for amplitudes, or 1 Np = 4.343 dB for power.

PROBLEMS

Maxwell integral equations in the time domain

10.1 Write the two Maxwell's equations in integral form in the time domain, relating the **E** and **H** fields in a medium with constitutive parameters μ, ε, σ.

Maxwell differential equations in the time domain

10.2 Write the two Maxwell's equations in differential form in the time domain, relating the **E** and **H** fields in a medium with constant, uniform constitutive parameters μ, ε, σ.

Maxwell integral equations in the frequency domain

10.3 Write the two Maxwell's equations in integral form in the frequency domain, relating the **E** and **H** fields in a medium with constitutive parameters μ, ε, σ.

Maxwell differential equations in the frequency domain

10.4 Write the two Maxwell's equations in differential form in the frequency domain, relating the **E** and **H** fields in a medium with constitutive parameters μ, ε, σ.

Poynting's theorem in the time domain

10.5 Write the Poynting theorem in integral form in the time domain, involving the **E** and **H** fields in a medium with constitutive parameters μ, ε, σ.

Poynting's theorem in the frequency domain

10.6 Write the Poynting theorem in integral form in the frequency domain, involving the **E** and **H** fields in a medium with real constitutive parameters μ, ε, σ.

Recognizing a wave

10.7 What are the speed v and the direction of propagation $\hat{\mathbf{n}}$ of the wave
$$f(\mathbf{r}, t) = \text{sech}^2(\alpha x + \beta y + \gamma z + \delta t), \text{ where } \alpha, \beta, \gamma, \delta \text{ are constants?}$$

When is a wave not a wave?

10.8 Consider the function $f(\mathbf{r}, t) = \exp\{-\frac{1}{2}[\mathbf{k} \cdot (\mathbf{r} - \mathbf{u}t)]^2\}$, where $\mathbf{k}$ and $\mathbf{u}$ are nonzero constant vectors, $\mathbf{r}$ is the position vector, and t is time. Under what conditions is $f(\mathbf{r}, t)$ not a wave?

Operations on complex numbers

10.9 Let $u = 3 + j2$ and $v = 5e^{j0.5}$. Find, in both rectangular and polar form,
(a) $u + v$ (b) $u - v$ (c) uv (d) u/v (e) $\sqrt{u}$ (f) u^*v

Cylindrical and spherical coordinates of a point

10.10 The rectangular coordinates of a point are $(1, 2, 3)$.
(a) What are its cylindrical coordinates?
(b) What are its spherical coordinates?

Operations on unit vectors

10.11 Let the radius of the earth be one unit; let the unit vector $\hat{\mathbf{n}}_1$ point from the center of the earth to New York ($40°45'06''$N, $73°59'39''$W) and $\hat{\mathbf{n}}_2$ to Paris ($48°50'14''$N, $2°20'14''$E).
(a) What is the angle between $\hat{\mathbf{n}}_1$ and $\hat{\mathbf{n}}_2$?
(b) Where on earth is the unit vector along $\hat{\mathbf{n}}_1 \times \hat{\mathbf{n}}_2$?

Vector differential calculus

10.12 (a) Find the gradient of the azimuthal coordinate φ, in cylindrical coordinates.
(b) Find the gradient of the azimuthal coordinate φ, in spherical coordinates.
(c) Find the divergence of the azimuthal unit vector $\hat{\varphi}$, in cylindrical coordinates.
(d) Find the curl of the azimuthal unit vector $\hat{\varphi}$, in spherical coordinates.

Line integral

10.13 By direct integration, find the closed line integral of the vector field $\mathbf{F}(\mathbf{r}) = \mathbf{r} \cdot \hat{\mathbf{y}} \, \hat{\mathbf{x}}$ around the semicircle of radius a in the half-plane $y \geq 0$, centered at the origin of the xy-plane and closed by the diameter on the x-axis, oriented counterclockwise.

Line integral by Stokes's theorem

10.14 By using Stokes's theorem, find the closed line integral of the vector field $\mathbf{F}(\mathbf{r}) = \mathbf{r} \cdot \hat{\mathbf{y}}\,\hat{\mathbf{x}}$ around the semicircle of radius a in the half-plane $y \geq 0$, centered at the origin of the xy-plane and closed by the diameter on the x-axis, oriented counterclockwise.

Surface integral

10.15 By direct integration, find the closed surface integral of the vector field $\mathbf{F}(\mathbf{r}) = \hat{\mathbf{z}} \cdot \mathbf{r}\,\mathbf{r}$ over the hemisphere $r = a, z > 0$ (closed by the disk $r < a, z = 0$).

Surface integral by divergence theorem

10.16 By using the divergence theorem, find the closed surface integral of the vector field $\mathbf{F}(\mathbf{r}) = \hat{\mathbf{z}} \cdot \mathbf{r}\,\mathbf{r}$ over the hemisphere $r = a, z > 0$ (closed by the disk $r < a, z = 0$).

Fourier transform

10.17 Find the Fourier transform $F(\omega)$ of the signal $f(t) = e^{-a|t|}$.

Convolution

10.18 Find the convolution $w(t) = u(t) * u(t)$ of the signal $u(t)$ with itself, where $u(t) = e^{-at}$ for $t > 0$ but $u(t) = 0$ for $t < 0$.

Dyadics

10.19 (a) Without making a distinction between row and column vectors, let $\mathbf{A} = \begin{bmatrix} 2 & -1 & 3 \end{bmatrix}$ and $\mathbf{B} = \begin{bmatrix} 5 & 2 & -4 \end{bmatrix}$. The dyad $\mathbf{AB}$ was calculated in the text; compare the dyad $\mathbf{BA}$ with $\mathbf{AB}$.

(b) The dyadic $\mathbf{Z} = \hat{\mathbf{y}}\hat{\mathbf{x}} - \hat{\mathbf{x}}\hat{\mathbf{y}}$ operates on any 3-vector $\mathbf{u}$ to form another vector $\mathbf{v} = \mathbf{Z} \cdot \mathbf{u}$. This operation can be expressed in a more familiar form, as a cross product of a certain vector $\mathbf{w}$ with the vector $\mathbf{u}$. What is the vector $\mathbf{w}$?

Vector and scalar potentials

10.20 The fact that the magnetic flux density $\mathbf{B}$ has no divergence guarantees that there exists a vector field $\mathbf{A}(\mathbf{r}, t)$, called the vector potential, such that $\mathbf{B} = \nabla \times \mathbf{A}$. Then the Faraday-Maxwell law can be rewritten

$$\nabla \times \mathbf{E} = -\partial \mathbf{B}/\partial t = -\partial(\nabla \times \mathbf{A})/\partial t \qquad \text{or} \qquad \nabla \times (\mathbf{E} + \partial \mathbf{A}/\partial t) = 0.$$

The fact that the field $\mathbf{E} + \partial \mathbf{A}/\partial t$ has no curl guarantees that there exists a scalar field $\varphi(\mathbf{r}, t)$, called the scalar potential, such that $\mathbf{E} + \partial \mathbf{A}/\partial t = -\nabla\varphi$. Then Gauss's law can be rewritten, for source charge density $\rho(\mathbf{r}, t)$ in a vacuum,

$$\nabla \cdot (-\nabla\varphi - \partial \mathbf{A}/\partial t) = \rho/\varepsilon_0 \qquad \text{or} \qquad \nabla^2\varphi = -\rho/\varepsilon_0 - \partial(\nabla \cdot \mathbf{A})/\partial t$$

and the Ampère-Maxwell law, for source current density $\mathbf{J}(\mathbf{r}, t)$ in a vacuum, can be written

$$\nabla \times (\nabla \times \mathbf{A}/\mu_0) = \mathbf{J} + \partial[\varepsilon_0(-\nabla\varphi - \partial \mathbf{A}/\partial t)]/\partial t$$

or

$$\nabla \nabla \cdot \mathbf{A} - \nabla^2\mathbf{A} = \mu_0\mathbf{J} - (1/c^2)\partial^2\mathbf{A}/\partial t^2 - (1/c^2)\nabla(\partial\varphi/\partial t)$$

or

$$\nabla^2\mathbf{A} - (1/c^2)\partial^2\mathbf{A}/\partial t^2 = -\mu_0\mathbf{J} + \nabla[\nabla \cdot \mathbf{A} + (1/c^2)\partial\varphi/\partial t].$$

Because $\mathbf{A}$ is not unique (we can add $\nabla\xi$, for any field $\xi(\mathbf{r}, t)$, to $\mathbf{A}$ and not change $\mathbf{B}$ and then also add $-\partial\xi/\partial t$ to φ and not change $\mathbf{E}$ either), we have the freedom to choose the quantity $\nabla \cdot \mathbf{A} + (1/c^2)\partial\varphi/\partial t$ to be any scalar field we wish. The choice $\nabla \cdot \mathbf{A} + (1/c^2)\partial\varphi/\partial t = 0$ is known as *Lorentz gauge*; it uncouples the two equations for $\mathbf{A}$ and φ.

Write the two differential equations for $\mathbf{A}$ and for φ, for given sources $\mathbf{J}$ and ρ, in Lorentz gauge.

Hertz potential

10.21 The vector and scalar potentials of the previous problem, $\mathbf{A}$ and φ, are related to each other through the Lorentz gauge, $\nabla \cdot \mathbf{A} + (1/c^2)\partial\varphi/\partial t = 0$. Their sources, $\mathbf{J}$ and ρ, are also related to each other through charge conservation, $\nabla \cdot \mathbf{J} + \partial\rho/\partial t = 0$. We can therefore create a new vector field, $\mathbf{\Pi}(\mathbf{r}, t)$, called the Hertz potential, such that $\mathbf{A} = \mu_0 \partial\mathbf{\Pi}/\partial t$ and $\varphi = (-1/\varepsilon_0)\nabla \cdot \mathbf{\Pi}$ and thereby satisfy Lorentz gauge automatically. We can also adopt a new source field $\mathbf{p}(\mathbf{r}, t)$ such that $\mathbf{J} = \partial\mathbf{p}/\partial t$ and $\rho = -\nabla \cdot \mathbf{p}$ and thereby satisfy charge conservation automatically. Assume the sources and fields are in a vacuum.

(a) Write the differential equation relating the Hertz potential $\mathbf{\Pi}$ to the source $\mathbf{p}$.

(b) What are the $\mathbf{D}$ and $\mathbf{H}$ fields, in terms of the Hertz potential $\mathbf{\Pi}$?

Note: Fields obtained as in (b) from a Hertz potential that satisfies the wave equation in (a) automatically satisfy Maxwell's equations and all the associated constraints.

Hertz potential for plane wave

10.22 Among the simplest solutions to the wave equation for the Hertz potential, $\nabla^2\mathbf{\Pi} - (1/c^2)\partial^2\mathbf{\Pi}/\partial t^2 = 0$ in a vacuum and in the absence of sources (compare the previous problem), is $\mathbf{\Pi}(\mathbf{r}, t) = \mathbf{Q}f(t - \mathbf{\Lambda} \cdot \mathbf{r})$, where $\mathbf{Q}$ and $\mathbf{\Lambda}$ are constant vectors and $f(t)$ is any function of time.

(a) What restriction on $\mathbf{\Lambda}$ makes this field a solution to the wave equation?

(b) What $\mathbf{D}$ and $\mathbf{H}$ fields correspond to this Hertz potential?

Answers To Problems

1.1   **(a)** $f(x, y, z) = \sqrt{x^2 + y^2 + (z - s/2)^2} + \sqrt{x^2 + y^2 + (z + s/2)^2}$.

      (b) $f(r, \theta, \varphi) = \sqrt{r^2 - rs \cos\theta + s^2/4} + \sqrt{r^2 + rs \cos\theta + s^2/4}$.

      (c) $f(\mathbf{r}) = |\mathbf{r} - \mathbf{s}/2| + |\mathbf{r} + \mathbf{s}/2|$.

1.2 $\mathbf{r}(t) = (u_0/2\Omega)\left[\hat{\mathbf{x}}(\sinh\Omega t + \sin\Omega t) + \hat{\mathbf{z}}(\sinh\Omega t - \sin\Omega t)\right]$.

1.3 $\Psi\{S_1\} = \rho_0 a^2 b$.

1.4 $\Psi\{S_2\} = (2/3)\rho_0 a^3$.

1.5 $\Psi\{S\} = \frac{1}{2}ab^2 h$.

1.6 $\Psi\{S\} = (2/3)\pi a^3$.

1.7   **(a)** $D_0 = (Q_0/4\pi R^2)N^2 e$.

      (b) $Q_n = Q_0 e^{-n}\left[1 + 2n - (e - 1)n^2\right]$.   **(c)** $Q = Q_0 N^2 e^{-(N-1)}$.

1.8   Flat-top pulse of height 0.2 coul, from time 0.400 sec to 0.500 sec.

1.9   **(a)** $M/m = (2.0)10^{21}$.   **(b)** $F_e/F_g = (4.2)10^{42}$.

1.10 $x/l = 1/\left[1 + \sqrt{Q_2/Q_1}\right]$.

1.11   **(a)** $\mathbf{D}(\mathbf{r}) = f(r)\hat{\mathbf{r}}$ for all r.

      (b) $\mathbf{D}(\mathbf{r}) = Q\mathbf{r}/(4\pi a^3)$ if $r \leq a$; $\mathbf{D}(\mathbf{r}) = Q\mathbf{r}/(4\pi r^3)$ if $r \geq a$.

1.12   **(a)** $\mathbf{E}(x) = (Q_s/2\varepsilon_0)\hat{\mathbf{x}}$ for $x > 0$; $\mathbf{E}(x) = -(Q_s/2\varepsilon_0)\hat{\mathbf{x}}$ for $x < 0$.

      (b) The two fields cancel beyond the planes but add between them. Between the planes, $E(x) = Q_s/\varepsilon_0$, directed from the positively charged plane to the other.

1.13   **(a)** Verify by integrating the charge density over the sphere of radius a.

      (b) $\mathbf{D}(\mathbf{r}) = Q\hat{\mathbf{r}}/(4\pi r^2)$ for $r > a$; $\mathbf{D}(\mathbf{r}) = \left[Q\hat{\mathbf{r}}/(4\pi a^2)\right](r/a)^{\lambda+1}$ for $r < a$.

1.14   **(a)** Verify by integrating the charge density over the cylinder of radius a.

      (b) $\mathbf{D}(\mathbf{r}) = Q_l\hat{\boldsymbol{\rho}}/(2\pi\rho)$ for $\rho > a$; $\mathbf{D}(\mathbf{r}) = \left[Q_l\hat{\boldsymbol{\rho}}/(2\pi a)\right](\rho/a)^{\lambda+1}$ for $\rho < a$.

1.15   **(a)** Verify by integrating the charge density over the slab of thickness a.

      (b) $\mathbf{D}(\mathbf{r}) = -\hat{\mathbf{z}}Q_s/2$ for $z < 0$;
             $\mathbf{D}(\mathbf{r}) = \hat{\mathbf{z}}Q_s\left[(z/a)^{\lambda+1} - 1/2\right]$ for $0 < z < a$;
             $\mathbf{D}(\mathbf{r}) = \hat{\mathbf{z}}Q_s/2$ for $z > a$.

1.16 $l = -(m + 1)$.

1.17 See Figure A1-1.

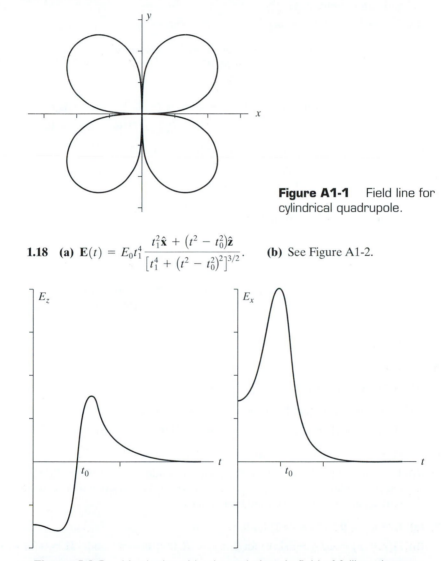

Figure A1-1   Field line for cylindrical quadrupole.

1.18 **(a)** $\mathbf{E}(t) = E_0 t_1^4 \dfrac{t_1^2 \hat{\mathbf{x}} + (t^2 - t_0^2)\hat{\mathbf{z}}}{[t_1^4 + (t^2 - t_0^2)^2]^{3/2}}$.   **(b)** See Figure A1-2.

Figure A1-2   Vertical and horizontal electric field of falling charge.

1.19 $\mathbf{E}(z) = \dfrac{q}{4\pi\varepsilon_0}\dfrac{z\hat{\mathbf{z}}}{(z^2 + a^2)^{3/2}}$.

1.20 $\omega = \{(1/2\pi\varepsilon_0 l^3 m)(q/\sqrt{Q_1 Q_2})\}^{1/2}[Q_1^{1/2} + Q_2^{1/2}]^2$.

1.21 $a/h = \sqrt{2/3} = 0.8165$.

2.1 (a) Both execute circular cyclotron orbits that are horizontal and tangential to the origin but on opposite sides. The electron orbit has relatively high frequency $\omega = (e/m)B$ and small radius v_0/ω; the ion orbit has lower frequency $\Omega = (e/M)B$ and larger radius v_0/Ω.

(b) $d_{max} = (2v_0/eB)(M + m)$.

2.2 (a) Helix: $\mathbf{r}(t) = \mathbf{r}_0 + \mathbf{v}_1 t + \mathbf{v}_0 \sin \omega t/\omega + \mathbf{a}_0[1 - \cos \omega t]/\omega^2$, with $\omega = (e/m)B$ and $\mathbf{a}_0 = (e/m)\mathbf{B} \times \mathbf{v}_0$.

(b) Steady drift: $\langle \mathbf{r}(t) \rangle = \mathbf{r}_0 + \mathbf{a}_0/\omega^2 + \mathbf{v}_1 t$.

2.3 (a) $\mathbf{r}(t) = \mathbf{r}_0 + \sin \omega_0 t (\mathbf{v}_0/\omega_0) - [1 - \cos \omega_0 t]\hat{\mathbf{z}} \times (\mathbf{v}_0/\omega_0) - \frac{1}{2}(e/m)\mathbf{E}_0 t^2$.

(b) $T = 2\pi m/eB_0$. (c) $\mathbf{r}(T/2) = \mathbf{r}_0 - (m/eB_0^2)[(\pi^2/2)\mathbf{E}_0 + 2\mathbf{v}_0 \times \mathbf{B}_0]$.

2.4 (a) $\mathbf{r}(t) = -(\alpha/\omega^2)[1 - \cos \omega t]\hat{\mathbf{x}} + (\alpha/\omega^2)[\omega t - \sin \omega t]\hat{\mathbf{z}}$.

(b) Maximum excursion along $-x$ is $2\alpha/\omega^2$; along y, none; along z, $2\pi\alpha/\omega^2$ per period. The temporal period is $2\pi/\omega$. The drift velocity is $\Delta z/\Delta t = \alpha/\omega = E/B$.

(c) Circle's radius is $a = \alpha/\omega^2 = mE/eB^2$, it rolls at angular velocity $\omega = (e/m)B$ along the direction of $\mathbf{E} \times \mathbf{B}$, and its height is along $-x$.

2.5 $U\{C\} = 2\pi J_0 ab$.

2.6 $U = J_0 \pi b^2$.

2.7 $\nabla \times \mathbf{H} = 2J_0\hat{\mathbf{z}}$, so that only the projection onto the xy-plane of any surface bounded by the semiellipse affects the Stokes surface integral. Therefore, the height h does not affect the mmf.

2.8 (a) $B = 10^{-4}$T. (b) $U = (2B_0 R/\mu_0)\cos \theta_0$.

2.9 (a) $\mathbf{H} = \hat{\boldsymbol{\varphi}}[I\rho/2\pi a^2]$. (b) $\mathbf{H} = \hat{\boldsymbol{\varphi}}[I/2\pi\rho]$.

(c) $\mathbf{H} = \hat{\boldsymbol{\varphi}} I[(c^2 - \rho^2)/(c^2 - b^2)]/2\pi\rho$. (d) $\mathbf{H} = 0$.

2.10 (a) Verify by integrating the current density over the circle of radius a.

(b) $\mathbf{H}(\mathbf{r}) = [I\hat{\boldsymbol{\varphi}}/(2\pi a)](\rho/a)^{\lambda+1}$ for $\rho < a$; $\mathbf{H}(\mathbf{r}) = [I\hat{\boldsymbol{\varphi}}/(2\pi a)](a/\rho)$ for $\rho > a$.

2.11 (a) $\mathbf{H}(x) = (K/2)\hat{\mathbf{y}}$ for $x > 0$, $\mathbf{H}(x) = -(K/2)\hat{\mathbf{y}}$ for $x < 0$.

(b) The two fields cancel beyond the planes but add between them. $\mathbf{H}(x) = K\hat{\mathbf{y}}$ between the planes, for $x_1 < x < x_2$ if the current sheet $K\hat{\mathbf{z}}$ is at x_1 and the sheet $-K\hat{\mathbf{z}}$ is at x_2, and $\mathbf{H}(x) = 0$ beyond both planes.

2.12 (a) $\mathbf{K}(\theta, \varphi) = \hat{\boldsymbol{\theta}} I/(2\pi R \sin \theta)$ for $0 < \theta < \pi$.

(b) $\mathbf{H}(r, \theta, \varphi) = \hat{\boldsymbol{\varphi}} I/(2\pi r \sin \theta)$ for $0 < r < R, 0 < \theta < \pi$ and $\mathbf{H} = 0$ for $r > R$.

2.13 (a) $z(v) = h - (v_1^2/g)\ln[1/(1 - v/v_1)] + (v_1/g)v$.

(b) $z(\rho) = h + (v_1^2/g)\ln[1 + (Ne/Av_1)/\rho] - (v_1 Ne/Ag)/\rho$.

(c) $z(E) = h - (v_1\varepsilon_0 A/Ne)E + (v_1^2/g)\{1 - \exp(-[g\varepsilon_0 A/Nev_1]E)\}$.

2.14 $U(z) = I_0$.

2.15 $U(z) = mI_0$, where $m = m(z)$ is the largest integer no greater than z/h.

2.16 (a) $U\{C_1\} = 0.$ (b) $U\{C_2\} = Kz.$

(c) $\mathbf{H} = K\hat{\mathbf{z}}$ inside; $\mathbf{H} = 0$ outside.

2.17 (a) $\mathbf{D}(\mathbf{r},t) = q(t)\hat{\mathbf{r}}/4\pi r^2, \quad \mathbf{J}(\mathbf{r},t) = \sigma q(t)\hat{\mathbf{r}}/4\pi\varepsilon r^2.$

(b) $dq(t)/dt = -(\sigma/\varepsilon)q(t).$ (c) $q(t) = q_0 e^{-(\sigma/\varepsilon)t}.$ (d) 0.652 nsec.

2.18 (a) $\mathbf{E}(\mathbf{r},t) = q(t)\hat{\mathbf{r}}/4\pi\varepsilon r^2.$

(b) $dq(t)/dt = -(\sigma/\varepsilon)q(t); \quad dQ(t)/dt = -dq(t)/dt.$

(c) $q(t) = q_0 e^{-(\sigma/\varepsilon)t}; \quad Q(t) = Q_0 + q_0[1 - e^{-(\sigma/\varepsilon)t}].$

(d) The original charges ignore each other.

2.19 $\mathbf{H} = -\hat{\boldsymbol{\varphi}}(I_0/4\pi r)[(1 + \cos\theta)/\sin\theta]$ for all three surfaces.

2.20 Separation = 25.5 cm; diameter = 51 cm.

2.21 (a) $\mathbf{H}(z) = (I_0 ab\hat{\mathbf{z}}/\pi)[1/(b^2 + z^2) + 1/(a^2 + z^2)](a^2 + b^2 + z^2)^{-1/2}.$

(b) $\mathbf{H}(0) = (I_0\hat{\mathbf{z}}/\pi ab)(a^2 + b^2)^{1/2}.$ (c) $\mathbf{H}(0) \to I_0\hat{\mathbf{z}}/\pi b;$ yes.

2.22 (a) $I\,dl = 4\pi a^3 J_0.$ (b) $I\,dl = -2\pi a^3 J_0.$

3.1 (a) $V = RF(R)(\pi - \sqrt{2})/2.$ (b) $V = -RF(R)/\sqrt{2}.$

3.2 $C = 4\pi\varepsilon_0 a; \quad$ 11.13 pF;   709 μF.

3.3 $R = [1/a - 1/b]/[4\pi\sigma \sin^2(\alpha/4)].$

3.4 $I_0 = 2\pi a\delta J_0.$ (b) $V_0 = I_0 l/2\pi\sigma a\delta.$

(c) $R = l/2\pi\sigma a\delta; \quad R_0 = l/\sigma\pi a^2; \quad R/R_0 = a/2\delta \gg 1.$

(d) $K = \delta J_0.$ (e) $r_s = 1/\delta\sigma.$

3.5 (a) $P = I_0^2/4\pi\sigma a.$ (b) $R = 1/4\pi\sigma a; \quad G = 4\pi\sigma a.$

3.6 (a) $\mathbf{E}(\mathbf{r},t) = \hat{\mathbf{r}}[I_1 - I_2]t/4\pi\varepsilon_0 r^2.$

(b) $\mathbf{H}(\mathbf{r}) = \hat{\boldsymbol{\varphi}}[(I_1 + I_2) - (I_1 - I_2)\cos\theta]/[4\pi r\sin\theta].$

3.7 (a) $q(t) = q_0 e^{-(\sigma/\varepsilon)t}.$ (b) $p(\rho,t) = [\sigma q_0^2/4\pi^2\varepsilon^2 l^2\rho^2]e^{-2(\sigma/\varepsilon)t}.$

(c) $W = [q_0^2/4\pi\varepsilon l]\ln(b/a).$

3.8 (a) $V(t) = B_0[l^2/8\pi T]e^{-t/T}.$ (b) $W = \sigma Al^3 B_0^2/128\pi^2 T.$

3.9 (a) $I(t) = B_0 v^2 t/[2R_0 + 2r_0 vt].$

(b) $W = [B_0^2 vR_0^2/2r_0^3][\frac{1}{2}\theta^2 - \theta + \ln(1 + \theta)],$ where $\theta = r_0 vT/R_0.$

3.10 (a) $I(t) = [B_0 lv/R_0 T]t/[1 + r_0 vt/R_0].$

(b) $W = [2B_0^2 l^2 v/r_0][\frac{1}{2}\theta^2 - \theta + \ln(1 + \theta)]/\theta^2,$ where $\theta = r_0 vT/R_0.$

3.11 (a) $I(t) = [B_0 v/r_0]/[1 + \csc(\alpha/2)].$

(b) $W = [B_0^2 vl^2/4r_0]/[\tan(\alpha/2) + \sec(\alpha/2)].$

3.12 (a) $I(t) = B_0 lat/[2R_0 + r_0 at^2].$

(b) $W = [B_0^2 l^2 aT/r_0][1 - \tan^{-1}\theta/\theta],$ where $\theta = T\sqrt{r_0 a/2R_0}.$

3.13 **(a)** $I(t) = [7B_0 a^2/12 r_0 LT] t^4/[1 + \csc(\alpha/2)]$.

(b) $W = (49/1584)(B_0^2 a^5/r_0 L^2 T^2) A\tau^{11}$, where $\tau = \sqrt{(l/a)}\cot(\alpha/2)$ and $A = \tan(\alpha/2)/[1 + \csc(\alpha/2)]$.

3.14 **(a)** $q(t) = I_0 t e^{-(\sigma/\varepsilon)t}$. **(b)** $t_{\max} = \varepsilon/\sigma$; $q_{\max} = (\varepsilon/\sigma)I_0 e^{-1}$.

3.15 $\sigma = ne\mu_n + pe\mu_p$.

3.16 **(a)** $U(t) = Q(v_0 - gt)4g^3\rho_0^2/[4g^2\rho_0^2 + (v_0 - gt)^4]^{3/2}$.

(b) $\mathbf{H}(\rho, \varphi, z_0, t) = \hat{\varphi}[2Q/\pi]g^3\rho_0(v_0 - gt)/[4g^2\rho_0^2 + (v_0 - gt)^4]^{3/2}$.

3.17 $\theta_2 = \tan^{-1}[(\sigma_2/\sigma_1)\tan\theta_1]$.

3.18 $\theta_2 = \tan^{-1}[(\sigma_2/\sigma_1)\tan\theta_1]$; $q_s = [(\varepsilon_2/\sigma_2) - (\varepsilon_1/\sigma_1)]J_1\cos\theta_1$.

3.19 **(a)** $\mathbf{B}(\mathbf{r}) = \hat{\mathbf{r}}\Phi_0/4\pi r^2$. **(b)** $\mathbf{E}(\mathbf{r}, t) = \hat{\varphi}[d\Phi_0/dt][1 + \cos\theta]/[4\pi r \sin\theta]$.

3.20 **(a)** $V(t) = (B_0/T)e^{-t/T}\pi R^2$. **(b)** $E_\varphi = (B_0/2T)e^{-t/T}(R^2/\rho)$; $J_\varphi = (\sigma B_0/2T)e^{-t/T}(R^2/\rho)$.

(c) $I_0(t) = (\sigma h R^2 B_0/2T)e^{-t/T}\ln(b/a)$. **(d)** $p = (\sigma B_0^2/4T^2)e^{-2t/T}(R^4/\rho^2)$.

(e) $W = \pi\sigma h R^4 B_0^2 \ln(b/a)/4T$.

3.21 **(a)** $V(t) = (B_0/T)e^{-t/T}\pi\rho_0^2$. **(b)** $E_\varphi = (B_0/2T)e^{-t/T}\rho$; $J_\varphi = (\sigma B_0/2T)e^{-t/T}\rho$.

(c) $I_0(t) = (\sigma h B_0/4T)e^{-t/T}[b^2 - a^2]$. **(d)** $p = (\sigma B_0^2/4T^2)e^{-2t/T}\rho^2$.

(e) $W = \pi\sigma h B_0^2[b^4 - a^4]/16T$.

3.22 **(a)** $V(t) = (\pi B_0/T)e^{-t/T}\min(\rho_0^2, R^2)$.

(b) $E_\varphi = (B_0 R/2T)e^{-t/T}\min(\rho/R, R/\rho)$; $J_\varphi = (\sigma B_0 R/2T)e^{-t/T}\min(\rho/R, R/\rho)$.

(c) $I_0(t) = (\sigma h B_0/4T)e^{-t/T}[R^2 - a^2 + 2R^2\ln(b/R)]$.

(d) $p = (\sigma B_0^2 R^2/4T^2)e^{-2t/T}\min(\rho^2/R^2, R^2/\rho^2)$.

(e) $W = [\pi\sigma h B_0^2/16T][R^4 - a^4 + 4R^4\ln(b/R)]$.

3.23 **(a)** $U(t) = (q\omega/2)[(a/\rho_0)\cos\omega t]/[1 + (a/\rho_0)^2\sin^2\omega t]^{3/2}$.

(b) $H(\rho, t) = \dfrac{q\omega}{4\pi a}\dfrac{(a/\rho)^2\cos\omega t}{[1 + (a/\rho)^2\sin^2\omega t]^{3/2}}$.

3.24 **(a)** $U = (e/2\rho)[v_h\sin^3\theta_h(t) + v_e\sin^3\theta_e(t)]$, or, explicitly,

$$U(t) = (e\rho^2/2)\{v_h[\rho^2 + (s - v_h t)^2]^{-3/2} + v_e[\rho^2 + (s + v_e t)^2]^{-3/2}\}.$$

(b) By the Ampère-Maxwell law. **(c)** $Q = e$.

3.25 **(a)** $I_c = 2\pi a H_c$. **(b)** $R = aI_c/I$.

(c) The model has a discontinuous tangential electric field at the surface $\rho = R$.

3.26 **(a)** $J_r(a, \theta) = 0$. **(b)** $I\,d\mathbf{l} = -2\pi a^3\mathbf{J}_0$.

3.27 **(a)** $J_\theta(a, \theta) = 0$. **(b)** $I\,d\mathbf{l} = 4\pi a^3\mathbf{J}_0$.

4.1 $(\nabla \times \mathbf{F})_\theta = [1/(r\sin\theta)]\partial F_r/\partial\varphi - [1/r]\partial[rF_\varphi]/\partial r$.

4.2 $(\nabla \times \mathbf{F})_\varphi = [1/r]\partial[rF_\theta]/\partial r - [1/r]\partial F_r/\partial\theta$.

4.3 $(\nabla \times \mathbf{F})_\rho = [1/\rho]\partial[F_z]/\partial\varphi - \partial[F_\varphi]/\partial z$.

4.4 $(\nabla \times \mathbf{F})_z = [1/\rho]\partial[\rho F_\varphi]/\partial\rho - [1/\rho]\partial[F_\rho]/\partial\varphi$.

4.5 (a) $E(\rho,t) = V(t)/h$ to order 0; $\qquad$ $E(\rho,t) = V\{C_m\}/h$ to all higher orders.

$\quad$ (b) $\Phi\{S_m\} = \mu\varepsilon[dV/dt][a^2 - \rho^2]/4 - \mu^2\varepsilon^2[d^3V/dt^3][a^2 - \rho^2][3a^2 - \rho^2]/64;$

$\qquad$ $V\{C_m\} = -\mu\varepsilon[d^2V/dt^2][a^2 - \rho^2]/4 + \mu^2\varepsilon^2[d^4V/dt^4][a^2 - \rho^2][3a^2 - \rho^2]/64.$

$\quad$ (c) Verify by comparing tabulated entries for E and H.

4.6 (a) $E(\rho) = E(a),$ $\qquad$ $\Psi(\rho) = \varepsilon E(a)\pi\rho^2,$ $\qquad$ $\Psi(a) = \varepsilon E(a)\pi a^2;$ $\qquad$ $\Psi(a) = Q(t),$

$\qquad$ $E(\rho) = Q(t)/\varepsilon\pi a^2, \Psi(\rho) = Q(t)\rho^2/a^2.$

$\quad$ (b) $U(\rho) = I(t)\rho^2/a^2, H(\rho) = I(t)\rho/2\pi a^2, \Phi(\rho) = \mu h I(t)(a^2 - \rho^2)/4\pi a^2.$

$\quad$ (c) $V\{C_m\} = -[\mu h(dI/dt)/4\pi](1 - \rho^2/a^2),$

$\qquad$ $E(\rho) = E(a) - (\mu/4\pi)(dI/dt)(1 - \rho^2/a^2),$

$\qquad$ $\Psi(\rho) = \varepsilon\pi\rho^2 E(a) - (\mu\varepsilon a^2/8)(dI/dt)(2\rho^2/a^2 - \rho^4/a^4),$

$\qquad$ $\Psi(a) = \varepsilon\pi a^2 E(a) - (\mu\varepsilon a^2/8)(dI/dt) = 0;$

$\qquad$ $E(a) = (\mu/8\pi) dI/dt,$

$\qquad$ $E(\rho) = -(\mu/8\pi)(dI/dt)[1 - 2\rho^2/a^2],$

$\qquad$ $\Psi(\rho) = -(\mu\varepsilon a^2/8)(dI/dt)[\rho^2/a^2 - \rho^4/a^4].$

$\quad$ (d) $U(\rho) = -(\mu\varepsilon a^2/8)(d^2I/dt^2)[\rho^2/a^2 - \rho^4/a^4],$

$\qquad$ $H(\rho) = -(\mu\varepsilon a/16\pi)(d^2I/dt^2)[\rho/a - \rho^3/a^3].$

$\quad$ (e) $E(\rho,t) = Q(t)/\varepsilon\pi a^2 - (\mu/8\pi)(dI/dt)[1 - 2\rho^2/a^2],$

$\qquad$ $H(\rho,t) = [I(t)/2\pi a][\rho/a] - (\mu\varepsilon a/16\pi)(d^2I/dt^2)[\rho/a - \rho^3/a^3].$

$\quad$ (f) $V(a) = Q(t)h/\varepsilon\pi a^2 + (\mu h/8\pi)(dI/dt) = Q(t)/C + L\, dI/dt;$

$\qquad$ $C = \varepsilon\pi a^2/h, L = \mu h/8\pi.$

4.7 (a) $E(\rho,t) = V(t)/h - (\mu\sigma a^2/4h)[dV/dt](1 - \rho^2/a^2),$

$\qquad$ $H(\rho,t) = (\sigma a/2h)V(t)(\rho/a) - (\mu\sigma^2 a^3/16h)[dV/dt](2\rho/a - \rho^3/a^3).$

$\quad$ (b) $I(t) = (\sigma\pi a^2/h)V(t) - (\mu\sigma^2\pi a^4/8h)[dV/dt].$

$\qquad$ $G = \sigma\pi a^2/h, \qquad L = \mu h/8\pi.$

4.8 (a) $E_1(\rho) = -[\mu/8\pi + \varepsilon/\sigma^2 A - (\mu/4\pi)(\rho^2/a^2)]dI/dt.$

$\quad$ (b) $H_1(\rho,t) = -[\mu\sigma a/16\pi][dI/dt][\rho/a - \rho^3/a^3].$

4.9 $E(\rho,t) = -(1/2\pi a)[d\Phi/dt]\rho/a,$

$\quad$ $H(\rho,t) = \Phi(t)/\mu_0\pi a^2 - (\varepsilon_0/8\pi)[d^2\Phi/dt^2][1 - 2\rho^2/a^2].$

4.10 (a) $U_1\{C_1\} = NI(t), \qquad H(t) = NI(t)/h.$

$\quad$ (b) $E(\rho,t) = -\mu_0 N[dI/dt]\rho/2h,$

$\qquad$ $H(\rho,t) = NI(t)/h - \mu_0\varepsilon_0 N[d^2I/dt^2][a^2 - \rho^2]/4h.$

$\quad$ (c) $V\{C_2\} = \mu_0 N^2[dI/dt]\pi a^2/h, \qquad L = \mu_0 N^2\pi a^2/h.$

4.11 (a) $\mathbf{K} = \hat{\boldsymbol{\rho}} I_0/2\pi\rho.$

$\quad$ (b) $\mathbf{E}_1 = -\hat{\mathbf{z}}[\mu_0/2\pi][dI_0/dt] \ln(\rho/a), \qquad \mathbf{H}_1 = 0.$

$\quad$ (c) $\mathbf{E}_2 = 0, \qquad \mathbf{H}_2 = \hat{\boldsymbol{\varphi}}[\mu_0\varepsilon_0 a/8\pi][d^2I_0/dt^2]\{(\rho/a)[1 - 2\ln(\rho/a)] - (a/\rho)\}.$

$\quad$ (d) $\rho_s = 0$ on the wire; $\rho_s = -[\mu_0\varepsilon_0/2\pi][dI_0/dt] \ln(\rho/a)$ on the plane.

4.12 (a) $E(\rho,t) = V(t)/h - (\mu\varepsilon a^2/4h)[d^2V/dt^2][2\ln(\rho/b) - (\rho^2 - b^2)/a^2]$,

$H(\rho,t) = (\varepsilon a/2h)[dV/dt](a/\rho - \rho/a)$.

(b) $I(t) = [\varepsilon\pi(a^2 - b^2)/h]dV/dt$.

(c) $V(a) = V(t) - (\mu\varepsilon a^2/4h)[d^2V/dt^2][2\ln(a/b) - 1 + b^2/a^2]$.

4.13 (a) $E_0 = Q(t)/\varepsilon\pi(a^2 - b^2)$; $\quad \Psi_0(\rho) = -Q(t)(\rho^2 - b^2)/(a^2 - b^2)$;

$V_0 = Q(t)h/\varepsilon\pi(a^2 - b^2)$.

(b) $U_1(\rho) = I(t)(a^2 - \rho^2)/(a^2 - b^2)$;

$H_1(\rho) = [I(t)a/2\pi(a^2 - b^2)][a/\rho - \rho/a]$.

$\Phi_1(\rho) = \int_b^\rho \mu H_1(\rho)h\,d\rho = \dfrac{\mu I(t)a^2 h}{2\pi(a^2 - b^2)}\left(\ln(\rho/b) - \dfrac{\rho^2 - b^2}{2a^2}\right)$.

(c) $E_2(\rho) = V_2(t)/h - (1/h)\,d\Phi_1(\rho)/dt$. $\quad$ Set $\Psi_2(a) = 0$ to extract $V_2(t)$.

(d) $V(a) = Q(t)h/\varepsilon\pi(a^2 - b^2) + V_2(t) - [\mu h/4\pi][dI/dt]\{[2a^2/(a^2 - b^2)]\ln(a/b) - 1\}$.

4.14 (a) $\Phi_0 = (\mu h/2\pi)I_0\ln(\rho/a)$.

(b) $E_1(\rho) = (\mu/2\pi)[dI_0/dt]\ln(\rho/a)$; $\quad H_1(\rho) = 0$.

(c) $\Psi_1 = (\mu\varepsilon a^2/4)[dI_0/dt][2(\rho^2/a^2)\ln(\rho/a) - (\rho^2/a^2 - 1)]$.

(d) $H_m(b) = 0$ for $m > 0$: continuity requires $I + d\Psi/dt = I_0(t)$ for $\rho = b$.

(e) $E_2(\rho) = 0$;

$H_2(\rho) = -(\mu\varepsilon b/8\pi)[d^2I_0/dt^2][2(b/\rho)\ln(b/a) - 2(\rho/b)\ln(\rho/a) - (b/\rho - \rho/b)]$.

4.15 (a) $E = 0$ and $H = 0$ for $z < 0$; E is continuous but H is discontinuous by K_0.

(b) $\Phi_0 = \mu K_0 lz$. $\qquad$ (c) $E_1(z,t) = \mu[dK_0/dt]z$.

(d) $\Psi_1 = \varepsilon\mu[dK_0/dt]wz^2/2$; $\quad I_1 = \sigma\mu[dK_0/dt]wz^2/2$.

(e) $H_1(z,t) = \sigma\mu[dK_0/dt]z^2/2$.

(f) $E_2(z,t) = \sigma\mu^2[d^2K_0/dt^2]z^3/3!$;

$H_2(z,t) = [d^2K_0/dt^2][\sigma^2\mu^2z^4/4! + \varepsilon\mu z^2/2!]$.

4.16 (a)

	0	1	2	3
$U = d\Psi/dt$	0	0	$-\dfrac{\mu_0\varepsilon_0 l}{4}\dfrac{d^2K_0}{dt^2}(a^2 - \rho^2)$	0
$H(\rho) = H(a) + U/l$	K_0	0	$-\dfrac{\mu_0\varepsilon_0}{4}\dfrac{d^2K_0}{dt^2}(a^2 - \rho^2)$	0
$\Phi = \int_0^\rho 2\pi\mu_0 H\rho\,d\rho$	$\mu_0 K_0\pi\rho^2$	0	$-\dfrac{\pi\mu_0^2\varepsilon_0}{8}\dfrac{d^2K_0}{dt^2}(2a^2\rho^2 - \rho^4)$	0
$V = -d\Phi/dt$	0	$-\mu_0\dfrac{dK_0}{dt}\pi\rho^2$	0	$\dfrac{\pi\mu_0^2\varepsilon_0}{8}\dfrac{d^3K_0}{dt^3}(2a^2\rho^2 - \rho^4)$
$E(\rho) = V/2\pi\rho$	0	$-\dfrac{\mu_0}{2}\dfrac{dK_0}{dt}\rho$	0	$\dfrac{\mu_0^2\varepsilon_0}{16}\dfrac{d^3K_0}{dt^3}(2a^2\rho - \rho^3)$
$\Psi = l\int_\rho^a \varepsilon_0 E\,d\rho$	0	$-\dfrac{\mu_0\varepsilon_0 l}{4}\dfrac{dK_0}{dt}(a^2 - \rho^2)$	0	$\dfrac{\mu_0^2\varepsilon_0^2 l}{64}\dfrac{d^3K_0}{dt^3}(3a^4 - 4a^2\rho^2 + \rho^4)$

(b) $H_z(0, t) = K_0 - [\mu_0 \varepsilon_0 a^2/4][d^2 K_0/dt^2].$

(c) $E_\phi(a, t) = -[\mu_0 a/2][dK_0/dt] + [\mu_0^2 \varepsilon_0 a^3/16][d^3 K_0/dt^3].$

4.17 **(a)** $\Phi(0) = \Phi_0(t), \qquad H_0(z) = \Phi_0(t)/\mu_0 hl, \qquad \Phi(z) = \Phi_0(t)[1 - z/l].$

(b) $V_1(z) = V_0(t)(1 - z/l), \qquad E_1(z) = [V_0(t)/h][1 - z/l],$
$\Psi_1(z) = \varepsilon_0 V_0(t)[w/h][z - z^2/2l].$

(c) $H_2(0) = [\varepsilon_0 l/3h][dV_0/dt], \qquad H_2(z) = (\varepsilon_0/6hl)(dV_0/dt)[2l^2 - 6lz + 3z^2].$

(d) $I_0(t) = \Phi_0(t)w/\mu_0 hl + [\varepsilon_0 lw/3h][dV_0/dt]; \qquad L = \mu_0 hl/w, \quad C = \varepsilon_0 lw/3h.$

4.18 **(a)** $E_0 = [ab/(b - a)]V(t)/r^2.$

(b) $H_1 = \varepsilon_0[ab/(b - a)][dV/dt](1 + \cos\theta)/(r\sin\theta).$

(c) $E_2(r, \theta) = \mu_0 \varepsilon_0[a^2 b^2/(b - a)r^2][d^2 V/dt^2] \ln[(1 - \cos\theta_0)/(1 - \cos\theta)].$

(d) $I_1(t) = 2\pi\varepsilon_0[ab/(b - a)][dV/dt][1 + \cos\theta_0];$
$V(\pi, t) = V(t) + \mu_0 \varepsilon_0 ab[d^2 V/dt^2] \ln[(1 - \cos\theta_0)/2].$

4.19 **(a)** $E_0(x) = V(t)/\alpha x.$ $\qquad\qquad$ **(b)** $H_1(x) = [\varepsilon_0/\alpha][dV/dt]\ln(x_2/x).$

(c) $I_1 = [\varepsilon_0 w/\alpha][dV/dt]\ln(x_2/x_1); \qquad C = [\varepsilon_0 w/\alpha]\ln(x_2/x_1).$

4.20 $C_{11} = C_{22} = \varepsilon lw/h + (\varepsilon_0 l/\pi)\ln[1 + 2w/s];$
$C_{12} = C_{21} = -(\varepsilon_0 l/\pi)\ln[1 + 2w/s].$

5.1 **(a)** Both $d\mathbf{r}/d\xi$ and $\hat{\mathbf{e}}_1/\rho_1 - \hat{\mathbf{e}}_2/\rho_2$ are proportional to the vector
$\hat{\mathbf{y}}(1 + \cosh\xi\cos\theta) - \hat{\mathbf{x}}\sinh\xi\sin\theta.$

(b) The vector $\mathbf{r}(\xi) - \mathbf{c}$ has constant magnitude $R = (s/2)|\csc\theta|.$

5.2 **(a)** Both $d\mathbf{r}/d\theta$ and $\hat{\mathbf{h}}_1/\rho_1 - \hat{\mathbf{h}}_2/\rho_2$ are proportional to the vector
$\hat{\mathbf{x}}[1 + \cosh\xi\cos\theta] + \hat{\mathbf{y}}\sin\theta\sinh\xi.$

(b) The vector $\mathbf{r}(\theta) - \mathbf{c}$ has constant magnitude $R = (s/2)|\operatorname{csch}\xi|.$

5.3 **(a)** $\partial\mathbf{r}/\partial\xi \cdot \partial\mathbf{r}/\partial\theta = 0.$ $\qquad\qquad$ **(b)** $C = \pi\varepsilon_0/\cosh^{-1}(s/2a).$

5.4 **(a)** $V'(l, s) = e^{-sT}V'(s).$ $\qquad\qquad$ **(b)** $V(l, t) = V(t - T).$

5.5 **(a)** See Figure A5-1.

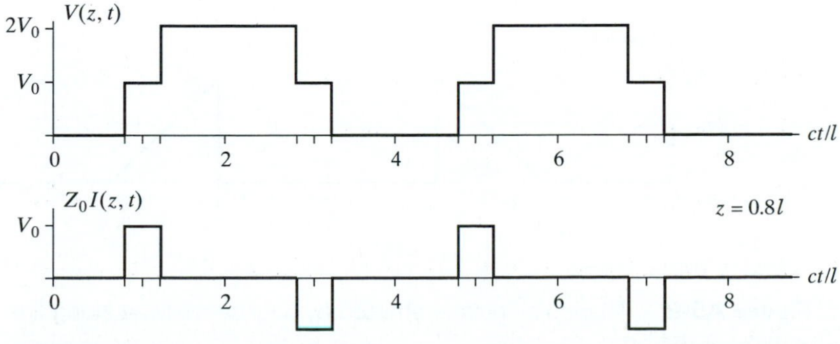

Figure A5-1 Voltage and current waveforms for open-circuited line.

(b) $\langle V \rangle = V_0$; $\langle I \rangle = 0$.

5.6 **(a)** Width $2(l - z)/c$, height $Z_0 I_0$, centered at l/c and width $2(l - z)/c$, height $-Z_0 I_0$, centered at $3l/c$.

(b) Broad pulse: width $(4l - 2z)/c$, height I_0, centered at $2l/c$. Narrower pulse: width $2z/c$, height $2I_0$, centered at $2l/c$.

(c) $\langle I \rangle = I_0$; $\langle V \rangle = 0$.

5.7 $V_{\max}$ is exceeded at $t = ml/c$, where m is the largest integer smaller than $V_{\max}/Z_0 I_0$. If m is even, at the source; if m is odd, at the open circuit.

5.8 **(a)** Use $f(t) = \Gamma_1 g(t)$, $V = -R_1 I$, $V = f(t) + g(t)$, $Z_0 I = f(t) - g(t)$.

(b) Initially $V = Z_0 I$; use $V = V_0 - R_1 I$, $\Delta V = V$.

(c) $V(\infty) = \Delta V (1 + \Gamma_2)/(1 - \Gamma_1 \Gamma_2)$.

(d) $V(\infty) = V_0 R_2/(R_1 + R_2)$, independent of Z_0.

5.9 **(a)** $I(\infty) = (\Delta V/Z_0)(1 - \Gamma_2)/(1 - \Gamma_1 \Gamma_2)$.

(b) $I(\infty) = V_0/(R_1 + R_2)$, independent of Z_0.   **(c)** $t = (8/3)T$.

5.10 **(a)** $I_2(\alpha t - \beta z)$.   **(b)** $\alpha/\beta = c$.

(c) $V(z, t) = Z_0 I_2(\alpha t - \beta z) - Z_0 I_1(\alpha t + \beta z)$.

5.11 **(a)** Along $-z$.   **(b)** $v = \beta/\alpha$.   **(c)** $Z_l = Z_0$.

5.12 **(a)** $V(l/2, nl/c) = V_1 + V_2$ for n odd, $V(l/2, nl/c) = 0$ for n even.

(b) $I(l/2, nl/c) = n(V_1 - V_2)/Z_0$.

5.13 **(a)** For $0 \le z \le l/2$,

$$v(z, t) = f_1(t - z/c) + g_1(t + z/c), \qquad Z_0 i(z, t) = f_1(t - z/c) - g_1(t + z/c).$$

For $l/2 \le z \le l$,

$$v(z, t) = f_2(t - z/c) + g_2(t + z/c), \qquad Z_0 i(z, t) = f_2(t - z/c) - g_2(t + z/c).$$

(b) See Figure A5-2.

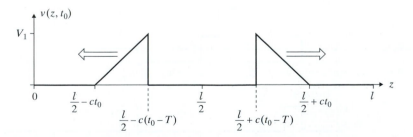

Figure A5-2   Triangular pulses emitted by source midway along line.

5.14 **(a)** and **(b)** See Figure A5-3.

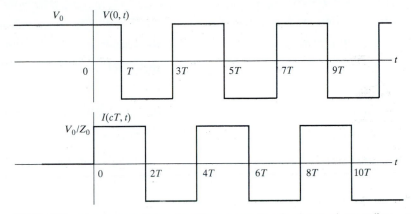

Figure A5-3 Square waves from shorting a previously charged open line.

5.15 **(a)** and **(b)** See Figure A5-4.

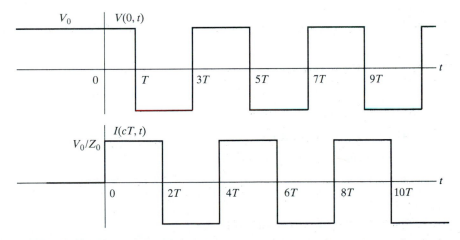

Figure A5-4 Square waves from joining two oppositely charged open lines.

5.16 $Z_0 I_0(0) = 1/2,$ $Z_0 I_0(nl/c) = 1$ for $n \geq 1$.

5.17 $Z_0 I(nT) = \pm 2(n + 1),$ for n even and odd, respectively.

5.18 **(a)** $A = Z_0/(2R_0 + Z_0),$ $B = (2R_0 - Z_0)/(2R_0 + Z_0).$

   (b) $R_0 = Z_0/2;$ $C = 1/2.$

   (c) $V(z, t) = \frac{1}{2} V_0(t - z/c) + \frac{1}{2} V_0(t - [l - z]/c),$
   $Z_0 I(z, t) = \frac{1}{2} V_0(t - z/c) - \frac{1}{2} V_0(t - [l - z]/c).$

5.19 $W_{\text{dissip}}/W_{\text{drain}} = 0.8985.$

5.20 $h(t) = (1/2)\delta(t - l/2c) + (1/6)\delta(t - 3l/2c)$.

6.1 $(360 + j120)\Omega$.

6.2 0.1937 m.

6.3 **(a)** 0.4172 m.   **(b)** 13.66 Ω.

6.4 **(a)** 16.455 MHz.   **(b)** $(27.27 - j41.91)\Omega$.

6.5 **(a)** $\Gamma_0 = -0.53585$ **(b)** $Z_0 = 75 \Omega$.

6.6 **(a)** $Z_0 = \sqrt{Z_1 Z_2}$.   **(b)** Fails if $\beta l = m\pi/2$ (m = integer).

6.7 **(a)** $G_{max} = \max[1, R/Z_0]$.   **(b)** $G_{max} > 1$ if $R > Z_0$ and $\cos \beta l = 0$.

6.8 **(a)** $V(z) = V_0 \sin \beta z/\sin \beta l_0,$ $I(z) = j(V_0/Z_0) \cos \beta z/\sin \beta l_0$ $(0 < z < l_0)$
$V(z) = V_0 e^{-j\beta(z - l_0)},$ $I(z) = (V_0/Z_0)e^{-j\beta(z - l_0)}$ $(l_0 < z < l)$.
      (b) $I_0 = (V_0/Z_0)[1 - j \cot \beta l_0]$.

6.9 **(a)** $V(z) = jZ_0 I_0 e^{-j\beta l_0} \sin \beta z,$ $I(z) = -I_0 e^{-j\beta l_0} \cos \beta z$ $(0 < z < l_0)$
$V(z) = jZ_0 I_0 e^{-j\beta z} \sin \beta l_0,$ $I(z) = jI_0 e^{-j\beta z} \sin \beta l_0$ $(l_0 < z < l)$.
      (b) $V_0 = jZ_0 I_0 e^{-j\beta l_0} \sin \beta l_0$.

6.10 **(a)** $V(z) = V_0 \sin \beta z/\sin \beta l,$ $I(z) = j(V_0/Z_0) \cos \beta z/\sin \beta l$ $(0 < z < l)$
$V(z) = V_0 \cos \beta(2l - z)/\cos \beta l,$ $I(z) = j(V_0/Z_0) \sin \beta(2l - z)/\cos \beta l$ $(l < z < 2l)$
      (b) $I_0 = j(V_0/Z_0)[\tan \beta l - \cot \beta l]$.

6.11 $I = -j(2/\sqrt{3})(V_0/Z_0)$.

6.12 **(a)** $I(0) = j(V_0/Z_0) \tan(\beta l/2)$.   **(b)** $\omega = \pi c/l$.

6.13 **(a)** $I_0 = 2j(V_0/Z_0) \tan(\beta l/2)$.   **(b)** $\omega = \pi c/l$.

6.14 $\omega_0' = \omega_0/\sqrt{1 + l/cZ_0 C}$.

6.15 maximum = 1.576;   minimum = 0.6345.

6.16 $V_{max} V_{min}/2Z_0$.

6.17 1.495 W.

6.18 $p_1 = 0.3287$ W, $p_2 = 0.7705$ W.

6.19 **(a)** $p = \dfrac{|V_0|^2}{2R_0} \dfrac{(2R_0/Z_0)^2 \tan^2(\beta l/2)}{1 + (2R_0/Z_0)^2 \tan^2(\beta l/2)}$.   **(b)** $R_0 = (Z_0/2)|\cot(\beta l/2)|$.

6.20 **(a)** $Z_l = Z_g^*$ maximizes $\text{Re}[Z_l]/|Z_l + Z_g|^2$. **(b)** $Z_l = [Z_0^2/|Z_g|^2]Z_g$.

6.21 **(a)** 5000 km.   **(b)** 300 km.   **(c)** 300 m.
      (d) 3 m.   **(e)** 12.2 cm.   **(f)** 0.62 μm.

6.22 $\mathbf{E}_{11} LP$ along $\hat{\mathbf{x}} + \hat{\mathbf{y}}$. $\mathbf{E}_{12} LHCP$. $\mathbf{E}_{13} LP$ along $\hat{\mathbf{x}} - \hat{\mathbf{y}}$.
       $\mathbf{E}_{21} RHCP$. $\mathbf{E}_{22} LP$ along $\hat{\mathbf{x}} + \hat{\mathbf{y}}$. $\mathbf{E}_{23} LHCP$.
       $\mathbf{E}_{31} LP$ along $\hat{\mathbf{x}} - \hat{\mathbf{y}}$. $\mathbf{E}_{32} RHCP$. $\mathbf{E}_{33} LP$ along $\hat{\mathbf{x}} + \hat{\mathbf{y}}$.

6.23 **(a)** $l = (\lambda_0/4)/|n_1 - n_2|$.   **(b)** circular.

6.24 **(a)** scattered **E**'s are antiparallel;   **(b)** scattered **H**'s are antiparallel.

6.25 **(a)** $E_2/E_1 = 2\eta_2/(\eta_2 - \eta_1)$.   **(b)** $E_3/E_1 = (\eta_2 + \eta_1)/(\eta_2 - \eta_1)$.

6.26 **(a)** $E_2/E_1 = (3\eta_2 + \eta_1)/(\eta_2 - \eta_1)$.   **(b)** $E_3/E_1 = (\eta_2 + 3\eta_1)/(\eta_2 - \eta_1)$.

6.27 **(a)** $\Delta\omega = -\omega[2u/(c + u)]$.   **(b)** $z_n = l_0 + ut - n(\lambda/2)[1 + u/c]$.

6.28 $\mathrm{Re}(\tfrac{1}{2}EE^*) \approx |E_0|^2\{1 - \cos[2kut + 2kl_0]\}$; $\Omega = 2ku$.

6.29 $\mathrm{Re}(\tfrac{1}{2}EE^*) = |E_0|^2\{1 - \cos[2kl_0 + 2ks(t)]\}$.

7.1 $\Gamma_l = (-0.4991 - j0.4704)$.

7.2 **(a)** $\lambda = 24$ cm.   **(b)** VSWR = 5.   **(c)** $|\Gamma| = 2/3$.

  (d) $\varphi = \pi/6$.   **(e)** $Z_l = (95.87 + j115.0)\,\Omega$.

7.3 $X = \pm 346.41\,\Omega$;   no.

7.4 **(a)** $Z(0) \approx Z_0 + \Delta Z e^{-j2\beta l}$.   **(b)** VSWR $\approx 1 + |\Delta Z|/Z_0$.

7.5 **(a)** 11.   **(b)** 11.   **(c)** $(l - z)/\lambda = 0.3238$.

  (d) $(l - z)/\lambda = 0.0738$.   **(e)** "Z_l" = $(5.67 - j24.74)\,\Omega$.

7.6 $C = \dfrac{1 + |\Gamma|^2}{1 - |\Gamma|^2}$; $r = \dfrac{2|\Gamma|}{1 - |\Gamma|^2}$.

7.7 Require $1 - |\Gamma|^2 = 4r/[(r + 1)^2 + x^2] \geq 0$.

7.8 Verify that $\Gamma(x) - \gamma = -\rho f(x)$, where $|f(x)| = 1$.

7.9 Verify that $\Gamma(r) - \gamma = -\rho f(r)$, where $|f(r)| = 1$.

7.10 At the intersection, $Z/Z_0 = (1 + |\Gamma|)/(1 - |\Gamma|) =$ VSWR.

7.11 **(a)** 1/9.   **(b)** 1/17.

7.12 $\Delta f/f_0 = 0.89906$.

7.13 **(a)** $dz/dt = c[1 - \Gamma^2]/[1 + 2\Gamma\cos(2\omega t + \varphi) + \Gamma^2]$.

  (b) $(dz/dt)_{\max} = c[1 + \Gamma]/[1 - \Gamma] = c\,\mathrm{VSWR}$.

7.14 $\beta d = 2.5448$; $\beta s = 0.9649$.

7.15 **(a)** $A(n) = 1$, $B(n) = \sqrt{n^2 - 1}$.   **(b)** $\mathbf{E}_2 = 2E\hat{\mathbf{x}}$, $\mathbf{H}_2 = -(2E/\eta_0)\hat{\mathbf{z}}$, $\mathbf{k}_2 = (\omega/c)\hat{\mathbf{y}}$.

  (c) $\mathbf{H}_2 = 2H\hat{\mathbf{x}}$, $\mathbf{E}_2 = (2\eta_0 H)\hat{\mathbf{z}}$, $\mathbf{k}_2 = (\omega/c)\hat{\mathbf{y}}$.

7.16 **(a)** $A(n) = 2/\sqrt{n^2 - 1}$,   **(b)** $A(n) = 2n^2/\sqrt{n^2 - 1}$.

7.17 $\Gamma_e = -\eta_0\sigma_s/(2 + \eta_0\sigma_s)$, $T_e = 2/(2 + \eta_0\sigma_s)$.

7.18 $\Gamma_e = -\eta_0\sigma_s \sec\theta/(2 + \eta_0\sigma_s \sec\theta)$, $T_e = 2/(2 + \eta_0\sigma_s \sec\theta)$ for TE,
  $\Gamma_h = \eta_0\sigma_s \cos\theta/(2 + \eta_0\sigma_s \cos\theta)$, $T_h = 2/(2 + \eta_0\sigma_s \cos\theta)$ for TM.

7.19 $\Gamma_e = -(\eta_0\sigma_s)^2/[2 + 2\eta_0\sigma_s + (\eta_0\sigma_s)^2]$,
  $T_e = -2j/[2 + 2\eta_0\sigma_s + (\eta_0\sigma_s)^2]$.
  (b) $\sigma_s = \sqrt{2}/\eta_0$; $|T| = 0.2929$.

7.20 $\Gamma_e = -\dfrac{\sec\theta - 1 - j\cot[(\pi/2)\cos\theta]}{\sec\theta + 1 - j\cot[(\pi/2)\cos\theta]}$ for TE polarization.

$\Gamma_h = -\dfrac{1 - \cos\theta + j\cot[(\pi/2)\cos\theta]}{1 + \cos\theta - j\cot[(\pi/2)\cos\theta]}$ for TM polarization.

$|\Gamma_e| = 0.1218$ for TE; $\quad |\Gamma_h| = 0.1342$ for TM.

7.21 $\mathbf{E}(z) = \eta_0 H_0(\hat{\mathbf{x}} + j\hat{\mathbf{y}})\sin(\omega z/c)$.

7.22 **(a)** $\Gamma_h = 0$;   **(b)** $\Gamma_e = -(n^2 - 1)/(n^2 + 1)$.

7.23 **(a)** $Z(z) = \eta_2\big[e^{-j\pi z/d} + \Gamma_{32}e^{j\pi z/d}\big]\big/\big[e^{-j\pi z/d} - \Gamma_{32}e^{j\pi z/d}\big]$, with $\Gamma_{32} = (\eta - \eta_2)/(\eta + \eta_2)$

and $\eta_2 = \sqrt{\eta\eta_0}$.   **(b)** $\Gamma_e = -1/3$.

7.24 $\Gamma_e = -0.8$.

8.1 $R = 0.15\,\Omega/\text{m}, \quad L = 1.8\mu\text{H/m}, \quad G = 0, \quad C = 750\text{pF/m}$.

8.2 **(a)** $-V_0$.   **(b)** $-V_0/\cosh(\pi\alpha/\beta)$.

8.3 $I(l) = V_0/[jZ_0\cosh(\pi\alpha/2\beta)]$.

8.4 **(a)** $\gamma = \sqrt{1/A - \omega^2 LC}, \quad Z_0 = \sqrt{L/C - 1/\omega^2 C^2 A}$.

(b) Propagation if $\omega > 1/\sqrt{LCA}$, at phase velocity $1/\sqrt{LC - 1/\omega^2 A}$; attenuation if $\omega < 1/\sqrt{LCA}$.

8.5 **(a)** $\gamma = \sqrt{1/A - \omega^2 LC}, \quad Z_0 = \sqrt{(L/C)/(1 - 1/\omega^2 LCA)}$.

(b) Propagation if $\omega > 1/\sqrt{LCA}$, at phase velocity $1/\sqrt{LC - 1/\omega^2 A}$; attenuation if $\omega < 1/\sqrt{LCA}$.

8.6 All frequencies except in $1/\sqrt{LCA_1} < \omega < 1/\sqrt{LCA_2}$.

8.7 **(a)** $\alpha = (u_0/2)(RC + LG)$.   **(b)** $\omega = \sqrt{(RG - \alpha^2)/(LC - 1/u_0^2)}$.

8.8 **(a)** $u_0 = 2\alpha/(RC + LG)$.

(b) $\lambda_0 = (2\pi c/u_0)\sqrt{(u_0^2 LC - 1)/(RG - \alpha^2)}$.

8.9 $R = 0.1116\,\Omega/\text{m}; L = 0.1853\,\mu\text{H/m}; G = 44.63\,\mu\text{S/m}; C = 74.13\,\text{pF/m}$.

8.10 84.5 m.

8.11 **(a)** $\mathbf{E} \times \mathbf{H} = -\hat{\boldsymbol{\rho}}\, I_0^2\rho/2\pi^2 a^4\sigma$.   **(b)** $P = [I_0^2 l/\sigma\pi a^2](\rho^2/a^2)$.

8.12 **(a)** $\mathbf{E} = \hat{\mathbf{r}} I_0 t/4\pi\varepsilon_0 r^2$.   **(b)** $\mathbf{H} = \hat{\boldsymbol{\varphi}}[I_0/4\pi r]\tan(\theta/2)$.

(c) $\mathbf{P} = -\hat{\boldsymbol{\theta}}[I_0^2 t/16\pi^2\varepsilon_0][\tan(\theta/2)/r^3]$.   **(d)** $p = I_0^2 t/8\pi\varepsilon_0 a$.

8.13 **(a)** $\mathbf{E} \times \mathbf{H} = -\hat{\boldsymbol{\rho}}\{[\varepsilon/2h^2]V(dV/dt)\rho - [\mu\varepsilon^2/8h^2](dV/dt)(d^2V/dt^2)\rho(a^2 - \rho^2)\}$.

(b) $P = [\varepsilon\pi a^2/h](dV/dt)V$, flowing inward.

8.14 **(a)** $K = 2Q\rho_0/\pi g^2$, $t_0 = \sqrt{2\rho_0/g}$, $m = 4$, $n = 3/2$.

(b) $E = [Q/\pi\varepsilon_0 g^2]/(t_0^4 + t^4)$.

(c) $|\mathbf{E} \times \mathbf{H}| = [2Q^2\rho_0/\pi^2\varepsilon_0 g^4]t/(t_0^4 + t^4)^{5/2}$. **(d)** $\rho_0/3$.

8.15 **(a)** $E_1 = -\frac{1}{2}\mu_0[dK_0/dt]\rho$ for $\rho < a$,

$E_1 = -\frac{1}{2}\mu_0[dK_0/dt]a^2/\rho$ for $\rho > a$.

(b) $\mathbf{E} \times \mathbf{H} = -\hat{\boldsymbol{\rho}}\frac{1}{2}\mu_0 K_0(dK_0/dt)\rho$ inside, $\mathbf{E} \times \mathbf{H} = \mathbf{0}$ outside.

(c) $P = -\pi\rho^2 l\mu_0 K_0(dK_0/dt)$ inside, directed inward, and $P = 0$ outside. The power goes into accumulation of magnetic energy.

8.16 **(a)** $\mathbf{E} \times \mathbf{H} = \hat{\boldsymbol{\theta}}[I_0^2 t/16\pi^2\varepsilon_0]\cot(\theta/2)/r^3$.

(b) $I_0^2 t/16\varepsilon_0\rho$, outward.

8.17 **(a)** $\mathbf{E} \times \mathbf{H} = \hat{\boldsymbol{\theta}}[\Phi_0(d\Phi_0/dt)/16\pi^2\mu_0][1 + \cos\theta]/r^3 \sin\theta$.

(b) Outward above, inward below; ratio $= 8.32$.

8.18 **(a)** $C = 2\pi\varepsilon_0/\ln(2h/a)$. **(b)** $L = (\mu_0/2\pi)\ln(2h/a)$.

(c) $Z_0 = (\eta_0/2\pi)\ln(2h/a)$. **(d)** $221\ \Omega$. **(e)** $318\ \Omega$.

8.19 **(a)** $L = \mu_0 h/w$. **(b)** $C = \varepsilon_0 w/h$. **(c)** $Z_0 = \eta_0(h/w)$; $u_0 = c$.

(d) $V(z) = jZ_0 I_0 \sin[\omega(l - z)/c]/\cos(\omega l/c)$;

$I(z) = I_0 \cos[\omega(l - z)/c]/\cos(\omega l/c)$.

(e) $I_1 = I_0/\cos(\omega l/c)$. **(f)** $V_0 = jZ_0 I_0 \tan(\omega l/c)$.

8.20 $Z(\rho) = j\omega\mu h/2\pi\rho$; $Y(\rho) = j\omega\varepsilon 2\pi\rho/h$.

8.21 $Z(\rho) = j\omega\mu_0 2\pi\rho/h$; $Y(\rho) = j\omega\varepsilon_0 h/2\pi\rho$.

8.22 $Z(\rho) = j\omega\mu_0 h/2\pi\rho$; $Y(\rho) = j\omega\varepsilon_0 2\pi\rho/h$.

8.23 $\alpha = \omega\sqrt{\mu\varepsilon}\ \sqrt{\sec\theta}\ \sin(\theta/2)$, $\beta = \omega\sqrt{\mu\varepsilon}\ \sqrt{\sec\theta}\ \cos(\theta/2)$.

8.24 **(a)** $p(t) = I^2 l/\sigma A + (dI/dt)^2(l\sigma\mu^2 a^2/192\pi)$.

(b) $m = 4$; $\xi = 1/48$.

8.25 $\frac{1}{2}\mathbf{E}(z) \times \mathbf{H}^*(z) = -j\frac{1}{2}\eta_0|H_0|^2\hat{\mathbf{z}}\sin 2kz$.

9.1 **(a)** $Z(x) = j\omega\mu_0\alpha x/w$, $Y(x) = j\omega\varepsilon_0 w/\alpha x$.

(b) $Z(x) \rightarrow j\omega\mu_0 h/w, Y(x) \rightarrow j\omega\varepsilon_0 w/h$.

9.2 **(a)** $Z = j\omega\mu_0 b/a, Y = j\omega\varepsilon_0 a/b + 2/j\omega\mu_0 A$. **(b)** $\omega_c = c\sqrt{2b/aA}$.

9.3 **(a)** $E(r, \theta) = abV(\theta)/[(b - a)r^2]$. **(b)** $H(r, \theta) = I(\theta)/2\pi r \sin\theta$.

(c) $Z(\theta) = j\omega\mu_0(b - a)/2\pi \sin\theta$, $Y(\theta) = j\omega\varepsilon_0 2\pi ab \sin\theta/(b - a)$.

9.4 **(a)** $(5.423)10^8$ m/sec. **(b)** $(1.657)10^8$ m/sec. **(c)** 6.026 cm.

9.5 $f_c = 6.620$ GHz.

9.6 $f_c = 15.566$ GHz.

9.7   **(a)** $f_c = 9.203$ GHz.   **(b)** $\varepsilon/\varepsilon_0 = 2.450$.

9.8   **(a)** $a = 7.495$ cm.   **(b)** $f_c = 2.828$ GHz.

9.9 $a = 4.997$ cm, $b = 2.235$ cm.

9.10   **(a)** $f = 1.1547 f_c$.   **(b)** $(a/b)_{min} = 1.1547$.

9.11 $f_{min} = (c/2)\sqrt{1/a^2 + 1/b^2}$ for $l < b$; $f_{min} = (c/2)\sqrt{1/a^2 + 1/l^2}$ for $l > b$.

9.12   **(a)** $P = (ab/8)(\omega\varepsilon\beta/p^2)|E_0|^2$.   **(b)** $W = (ab/8)(\omega^2\mu\varepsilon/p^2)\varepsilon|E_0|^2$.

        (c) $P/W = \beta/\omega\mu\varepsilon$.

9.13   **(a)** $\lambda_g = c/f$.   **(b)** $27.3(l/a)$ dB.

9.14   **(a)** $\left[m/(\omega a/\pi c)\right]^2 + \left[n/(\omega b/\pi c)\right]^2 = 1$, an ellipse.

        (b) The dots are one unit apart.   **(c)** $(\omega^2 ab/c^2)/2\pi$.

9.15 $\Gamma = -0.1716$.

9.16 $\mathbf{E} = \hat{\mathbf{e}}E_0 e^{-j\omega(4x - 3y + 12z)/13c}$;

        $\hat{\mathbf{e}} = (3\hat{\mathbf{x}} + 4\hat{\mathbf{y}})/5$; $\hat{\mathbf{h}} = (-48\hat{\mathbf{x}} + 36\hat{\mathbf{y}} + 25\hat{\mathbf{z}})/65$.

9.17   **(a)** $|\mathbf{E}| = \sqrt{2\eta_0 P_0}$.   **(b)** 1.40 kW/m^2.

        (c) $|\mathbf{E}| = 1.03$ kV/m.   **(d)** $|\mathbf{H}| = 2.73$ A/m.   **(e)** $(4.0)10^{26}$ W.

9.18 $\mathbf{E}_1 = \hat{\mathbf{x}}(\eta_0 K_0/2)e^{-j\omega z/c}$ for $z > 0$, $\mathbf{E}_2 = \hat{\mathbf{x}}(\eta_0 K_0/2)e^{j\omega z/c}$ for $z < 0$

        and $\mathbf{H}_1 = \hat{\mathbf{y}}(K_0/2)e^{-j\omega z/c}$ for $z > 0$, $\mathbf{H}_2 = -\hat{\mathbf{y}}(K_0/2)e^{j\omega z/c}$ for $z < 0$.

9.19   **(a)** $\mathbf{E} = -2j\hat{\mathbf{x}}E_0 \sin(\omega z/c)$, $\mathbf{H} = 2\hat{\mathbf{y}}(E_0/\eta_0)\cos(\omega z/c)$.

        (b) $\mathbf{K} = 2\hat{\mathbf{x}}(E_0/\eta_0)$.

9.20 $\mathbf{E} = \hat{\mathbf{x}}E_0\left[e^{-j\omega z/c} + \Gamma e^{j\omega z/c}\right]$, $\mathbf{H} = \hat{\mathbf{y}}(E_0/\eta_0)\left[e^{-j\omega z/c} - \Gamma e^{j\omega z/c}\right]$ in the air and

        $\mathbf{E} = \hat{\mathbf{x}}E_0[1 + \Gamma]e^{-j\omega z\sqrt{\mu_0\varepsilon}}$, $\mathbf{H} = \hat{\mathbf{y}}(E_0/\eta_0)[1 - \Gamma]e^{-j\omega z\sqrt{\mu_0\varepsilon}}$ in the dielectric, with

        $\Gamma = \left[1 - \sqrt{\varepsilon/\varepsilon_0}\right]/\left[1 + \sqrt{\varepsilon/\varepsilon_0}\right]$.

9.21   **(a)** $g(\theta, ks, \psi) = ks\cos\theta - \psi/2$.

        (b) $\theta = 70.5°, \theta = 180°$.   **(c)** $\theta = 109.5°$.

9.22 $v_p = \omega/\beta = 1/\sqrt{LC}$ and $\alpha = R/\sqrt{L/C}$ are both independent of ω.

9.23   **(a)** $\lambda = 244$ m.   **(b)** $v_p = 19.5$ m/sec.   **(c)** $v_g = 9.75$ m/sec.

10.1 $\oint_C \mathbf{E} \cdot d\mathbf{l} = -\dfrac{d}{dt}\displaystyle\int_S \mu\mathbf{H} \cdot d\mathbf{S},$ $\qquad \oint_C \mathbf{H} \cdot d\mathbf{l} = \displaystyle\int_S \sigma\mathbf{E} \cdot d\mathbf{S} + \dfrac{d}{dt}\displaystyle\int_S \varepsilon\mathbf{E} \cdot d\mathbf{S}.$

10.2 $\nabla \times \mathbf{E} = -\mu\, \partial\mathbf{H}/\partial t,$ $\qquad\qquad \nabla \times \mathbf{H} = \sigma\mathbf{E} + \varepsilon\, \partial\mathbf{E}/\partial t.$

10.3 $\oint_C \mathbf{E} \cdot d\mathbf{l} = -\displaystyle\int_S j\omega\mu\mathbf{H} \cdot d\mathbf{S},$ $\qquad \oint_C \mathbf{H} \cdot d\mathbf{l} = \displaystyle\int_S (\sigma + j\omega\varepsilon)\mathbf{E} \cdot d\mathbf{S}.$

10.4 $\nabla \times \mathbf{E} = -j\omega\mu\mathbf{H},$ $\qquad\qquad \nabla \times \mathbf{H} = (\sigma + j\omega\varepsilon)\mathbf{E}.$

10.5 $-\oint_S \mathbf{E} \times \mathbf{H} \cdot d\mathbf{S} = \displaystyle\int_V \sigma E^2 \, dV + \dfrac{d}{dt}\displaystyle\int_V \left[\tfrac{1}{2}\varepsilon E^2 + \tfrac{1}{2}\mu H^2\right] dV.$

10.6 $-\oint_S \tfrac{1}{2}\mathbf{E} \times \mathbf{H}^* \cdot d\mathbf{S} = \displaystyle\int_V \tfrac{1}{2}\sigma|E|^2 \, dV + j2\omega \displaystyle\int_V \left[\tfrac{1}{4}\mu|H|^2 - \tfrac{1}{4}\varepsilon|E|^2\right] dV.$

10.7 $v = -\delta/\sqrt{\alpha^2 + \beta^2 + \gamma^2},$ $\qquad \hat{\mathbf{n}} = (\alpha\hat{\mathbf{x}} + \beta\hat{\mathbf{y}} + \gamma\hat{\mathbf{z}})/\sqrt{\alpha^2 + \beta^2 + \gamma^2}.$

10.8 If $\mathbf{k} \cdot \mathbf{u} = 0.$

10.9 (a) $u + v = 7.388 + j4.397 = 8.597e^{j0.5369}.$

(b) $u - v = 1.388 + j0.397 = 1.444e^{j0.2787}.$

(c) $uv = 18.028e^{j1.088} = 8.369 + j15.967.$

(d) $u/v = 0.7211e^{j0.088} = 0.7183 + j0.0634.$

(e) $\sqrt{u} = \pm1.8988e^{j0.294} = \pm(1.8174 + j0.5503).$

(f) $u^*v = 18.028e^{-j0.088} = 17.958 - j1.584.$

10.10 (a) $\rho = 2.236, \varphi = 1.107, z = 3.$

(b) $r = 3.742, \theta = 0.6405, \varphi = 1.107.$

10.11 (a) $\Phi = 52.464°.$

(b) Latitude 37.66°N, longitude 154.32°E, in the Pacific, east of Japan.

10.12 (a) $\nabla\varphi = \hat{\boldsymbol{\varphi}}/\rho.$ $\qquad\qquad\qquad$ (b) $\nabla\varphi = \hat{\boldsymbol{\varphi}}/r\sin\theta.$

(c) $\nabla \cdot \hat{\boldsymbol{\varphi}} = 0.$ $\qquad\qquad\qquad$ (d) $\nabla \times \hat{\boldsymbol{\varphi}} = \hat{\mathbf{r}}\cot\theta/r - \hat{\boldsymbol{\theta}}/r.$

10.13 $-\tfrac{1}{2}\pi a^2.$

10.14 $-\tfrac{1}{2}\pi a^2.$

10.15 $\pi a^4.$

10.16 $\pi a^4.$

10.17 $F(\omega) = 2a/(a^2 + \omega^2).$

10.18 $w(t) = 0$ for $t < 0$ and $w(t) = te^{-at}$ for $t > 0.$

10.19 (a) **BA** is the transpose of **AB**.

(b) $\mathbf{w} = \hat{\mathbf{z}}$

10.20 $\nabla^2\mathbf{A} - (1/c^2)\partial^2\mathbf{A}/\partial t^2 = -\mu_0\mathbf{J}$,

$\nabla^2\varphi - (1/c^2)\partial^2\varphi/\partial t^2 = -\rho/\varepsilon_0$.

10.21 **(a)** $\nabla^2\mathbf{\Pi} - (1/c^2)\partial^2\mathbf{\Pi}/\partial t^2 = -\mathbf{p}$.

(b) $\mathbf{D} = \nabla\nabla\cdot\mathbf{\Pi} - (1/c^2)\partial^2\mathbf{\Pi}/\partial t^2$, $\mathbf{H} = \nabla\times\partial\mathbf{\Pi}/\partial t$.

10.22 **(a)** $|\mathbf{\Lambda}|^2 = 1/c^2$.

(b) $\mathbf{D} = \mathbf{\Lambda}\times(\mathbf{\Lambda}\times\mathbf{Q})f''(t - \mathbf{\Lambda}\cdot\mathbf{r})$, $\mathbf{H} = -(\mathbf{\Lambda}\times\mathbf{Q})f''(t - \mathbf{\Lambda}\cdot\mathbf{r})$.

Index

energy velocity, 429
engineering standard, 280
envelope, 431, 432, 434, 435
equation of motion, 8
equivalent circuit, 162, 163, 195
equivalent impedance, 340
Euler's formula, 249, 461
evanescence, 353, 424
extraordinary boundary condition, 127

far fields, 440
farad, 27, 138
Faraday rotation, 285
Faraday-Maxwell law, 99, 136, 148, 159, 206, 333, 380, 388
ferroelectrics, 10
ferromagnetics, 50
field, 2, 4, 6, 147
field equations, 147
field intensity, 147
field line, 41, 158, 200, 203, 238, 239
field pattern, 155, 291
filament, 59, 70, 71, 108
filamentary current, 67
flow field, 13
fluctuations, 273
fluid flow, 133, 369
flux, 11, 13
flux density, 148
forbidden region, 354
force density, 148
force field, 3, 53
Fourier analysis, 248, 296, 298
Fourier series, 424, 468
Fourier transform, 432, 436, 460, 468, 472
frequency, 252
frequency domain, 9, 50, 238, 248, 250, 264, 276, 278, 296, 374
Fresnel formulas, 341
fringing, 156, 183, 186, 412
functional relation, 213
fundamental frequency, 469

gain, 260, 299, 445
Gauss's law, 20, 88, 149, 153, 380
Gauss's magnetic law, 117, 153
geomagnetic field, 86
good conductor, 396
gradient, 56, 272, 464, 466
gradient of dot product, 465
gravitational constant, 40
gravity waves, 455
grazing incidence, 354, 361
Green's function, 459
group velocity, 434, 435, 436, 455
guiding center, 84

handedness, 280
harmonic factor, 250
harmonic oscillator, 36, 42
harmonic plane wave, 279
harmonic signal, 250, 278
heat, 97, 368, 392, 401
Helmholtz coils, 79, 89
Helmholtz equation, 418, 425
henry, 62
hertz, 253, 275, 468
Hertz potential, 473
high-pass filter, 427
hole, 142
hollow pipe, 415, 421, 422, 435
homogeneous solutions, 172, 187, 272
horizontal polarization, 339
hyperbola, 279, 427
hysteresis, 10, 51

image, 145, 146, 408
imaginary power, 378
imaginary unit, 460
impedance distribution, 255, 257, 297, 458
impedance matching, 328, 357, 458
impedance of free space, 274
improper integral, 201, 204
impulse response, 9, 50, 247, 248, 469
incident wave, 230, 254, 291
indeterminate circuit, 243
index of refraction, 337
inductance, 180, 191, 205, 229, 372, 382, 386, 389
inductive reactance, 259
inductor, 180, 229
initial condition, 240
initial value problems, 187
inner product, 464
input impedance, 255, 263, 267, 297, 393
integral equation, 389, 459
integrated circuit, 412
integrated optics, 412
interface, 118, 333
interference, 266, 307, 443
interference pattern, 308, 310, 394
internal impedance, 267
internal inductance, 180
internal resistance, 241
intrinsic impedance of free space, 217
intrinsic impedance of medium, 328
inverse cube law, 82
inverse Fourier transform, 432, 433
inverse of matrix, 279
inverse square law, 26, 30, 37, 443
inverted impedance, 256
ionosphere, 48, 283
isotropic antenna, 444
isotropic medium, 446